EDITED BY JOHN T. HOUGHTON
Director General, Meteorological Office, Bracknell, UK

CAMBRIDGE UNIVERSITY PRESS
Cambridge
London New York New Rochelle
Melbourne Sydney

Published by the Press Syndicate of the University of Cambridge
The Pitt Building, Trumpington Street, Cambridge CB2 1RP
32 East 57th Street, New York, NY 10022, USA
10 Stamford Road, Oakleigh, Melbourne 3166, Australia

First published 1984
First paperback edition 1985

Printed in Great Britain by the
University Press, Cambridge

Library of Congress catalogue card number: 83-7602

British Library cataloguing in publication data

The global climate.
1. Climatology
I. Houghton, J. T.
551.6 QC981

ISBN 0 521 25138 9 hard covers
ISBN 0 521 31256 6 paperback

UP

Contents

Preface

The World Climate Research Programme (WCRP), which is the research component of the World Climate Programme, is one of the largest coordinated scientific enterprises man has tackled this century. This book has arisen from the desire of scientists involved with the World Climate Research Programme (WCRP) to explain to a wider scientific community the background, aims and main lines of research being pursued under the Programme. It is concerned with a very complex system. The Global Atmospheric Research Programme from which the WCRP developed was largely concerned with the atmosphere alone. Climate research is of necessity concerned with all components of the climate system, namely the atmosphere, oceans, ice and land surfaces and with the exchanges of heat, momentum and material (especially water) which occur between these different components.

Study of the climate system requires, on the one hand, observations with global coverage and, on the other hand, the development of sophisticated numerical models. Observations can be acquired through both remote sounding instruments mounted on satellites and through *in situ* measurements from ships, buoys, balloons, aircraft and land-based stations. Many of the techniques are at an early stage of development; for many of the observations, better coverage and higher accuracy need to be achieved. Regarding numerical models, progress is being made with the description of the physical and dynamical behaviour of various parts of the system, although in some cases serious work has only recently begun, particularly insofar as models with a global emphasis are concerned.

As will be clear from the first chapter, the emphasis of the World Climate Research Programme is on those climate changes which occur over periods of a few weeks to decades of years and on those changes which could be introduced by man's activities. Paramount among the latter is the change which could occur due to the increasing carbon dioxide content arising from the burning of fossil fuels.

The first two chapters in the volume introduce the World Climate Research Programme with its particular problems and major thrusts; after that, status reports are given of research, first in numerical modelling of the atmosphere, then in some of the major physical and dynamical processes which occur in the overall climate system or its different components. Experts from different scientific disciplines and various nations have contributed the articles.

I wish to thank members of the Joint Scientific Committee of the World Climate Research Programme for suggesting the different topics. I also wish to thank the contributors for the trouble they have taken to present their research areas in the general context of the global climate problem. Thanks are also due to the staff of the Cambridge University Press for the care they have taken and the courtesy they have shown in the preparation of the volume.

The scientific challenge of climate research is very large. By its very nature it is international in its scope; all countries of the world would be affected by the impact of major climate change. It is therefore appropriate that climate research should call for international effort. If through these articles which follow, the nature of the scientific problem and the degree of international challenge are more widely understood, the volume will have achieved its aim.

J T Houghton
Chairman
Joint Scientific Committee for the World Climate
Research Programme

John T. Houghton
Meteorological Office
Bracknell, UK
Pierre Morel
World Meterological Organisation
Geneva, Switzerland

1 The World Climate Research Programme

Abstract

The objectives of the World Climate Research Programme (WRCP) are to determine (a) to what extent climate can be predicted and (b) the extent of man's influence on climate. The programme is primarily concerned with changes which occur on timescales from several weeks to several decades. WCRP has developed from the Global Atmospheric Research Programme (GARP) and like GARP is jointly sponsored by the International Council of Scientific Unions (ICSU) and the World Meteorological Organisation (WMO). The WCRP is concerned with the whole climate system, the main components of which are the atmosphere, the oceans, the cryosphere and the land, and with interactions and feedbacks which occur between these components.

Three objectives or streams of research in the WCRP have been identified, namely (1) long-range weather predictions over periods of several weeks, (2) interannual variability of the global atmosphere and the tropical oceans over periods of several years, (3) longer-term variations. Two major experiments, the TOGA (Tropical Oceans and Global Atmosphere) and the WOCE (World Ocean Circulation Experiment) have been identified as the foci of the second and third streams. For these, new modelling efforts (especially coupled atmosphere ocean models) and global observations, especially from satellites of all components (in particular the ocean) of the climate system are required.

1.1 The possibility of climate change

That climate changes on a wide range of timescales is well known from the record of the past. Over the past half million years (Fig. 1.1(*a*)), glacial and interglacial periods have alternated at intervals of about 100000 years. At the time of the last glacial maximum, 20000 years ago, ice sheets covered Canada and large areas of northern Europe and Asia. Sea ice was greatly extended (Untersteiner, 1983, this volume) and sea level was nearly 80 m lower than at present. During the last 1000 years (Fig. 1.1(*b*)) the most pronounced climatic feature has been the 'little ice age' between about 1300 AD and 1800 AD. Since that time, so far as we are able to ascertain, there has been a small general warming, at least in the northern hemisphere (Fig. 1.1(*c*), and Angell & Gruza, 1983, this volume).

Climate variation has a large impact on man's activities and on the economy of human populations. Occurrences of extreme variations in precipitation leading to droughts and floods have always been a cause of concern, increasingly so in recent years as the greater world demand for food resources, especially in developing countries, has created a greater vulnerability.

Even quite small, average hemispheric temperature changes can be reflected in large regional variatons. In the 'little ice age', for instance, winters in Europe were, on average, much more severe than now, glaciers and the cover of sea ice advanced considerably although the hemispheric average temperature only changed by a little over 1 K.

That climate changes can be induced by man's activities is an important current concern. Examples are the possibilities of man's activity influencing the growth of deserts (Rasool, 1983, this volume) and the likely change due to increasing carbon dioxide as a result of burning of fossil fuels. The concentration of atmospheric carbon dioxide will almost certainly nearly double its present value before the year 2100. One model estimate of the increase of

Fig. 1.1. (*a*) Climate of last half million years deduced from measurements of oxygen isotope ratio in plankton shells which relate to global ice volume. (After Hays, Imbrie & Shackleton, 1976.) (*b*) Climate of last 1000 years estimated from evidence relating to east European winters. (After Lamb, 1966.) (*c*) Climate of last 100 years as evidenced by changes in average annual temperature of northern hemisphere. (After Mitchell, 1977.)

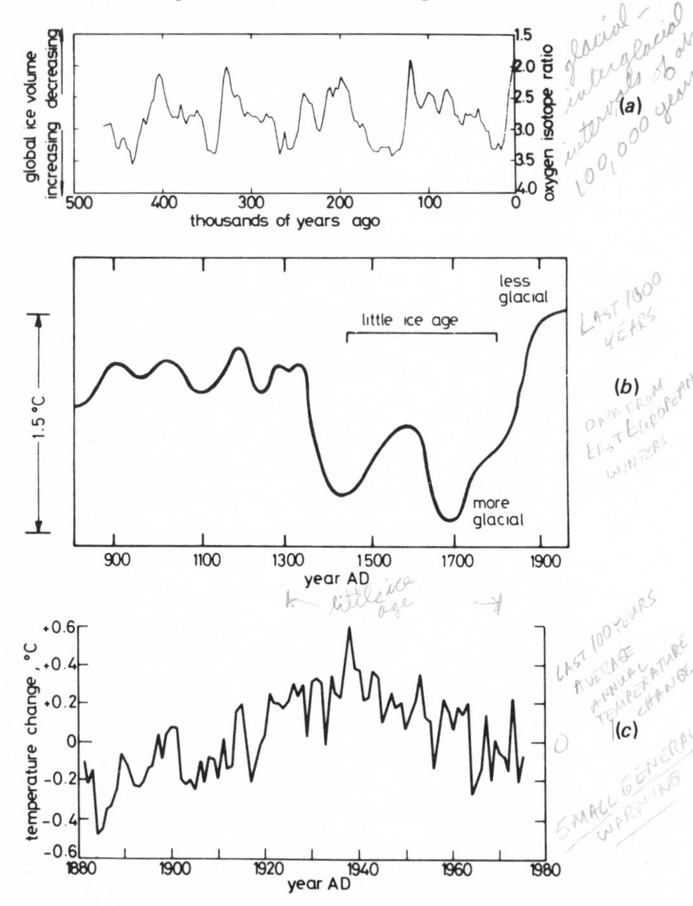

Fig. 1.2. Changes in atmospheric temperature (K) which would be expected to occur if the atmospheric CO_2 content were doubled as predicted from a model. (After Manabe & Wetherald, 1980.)

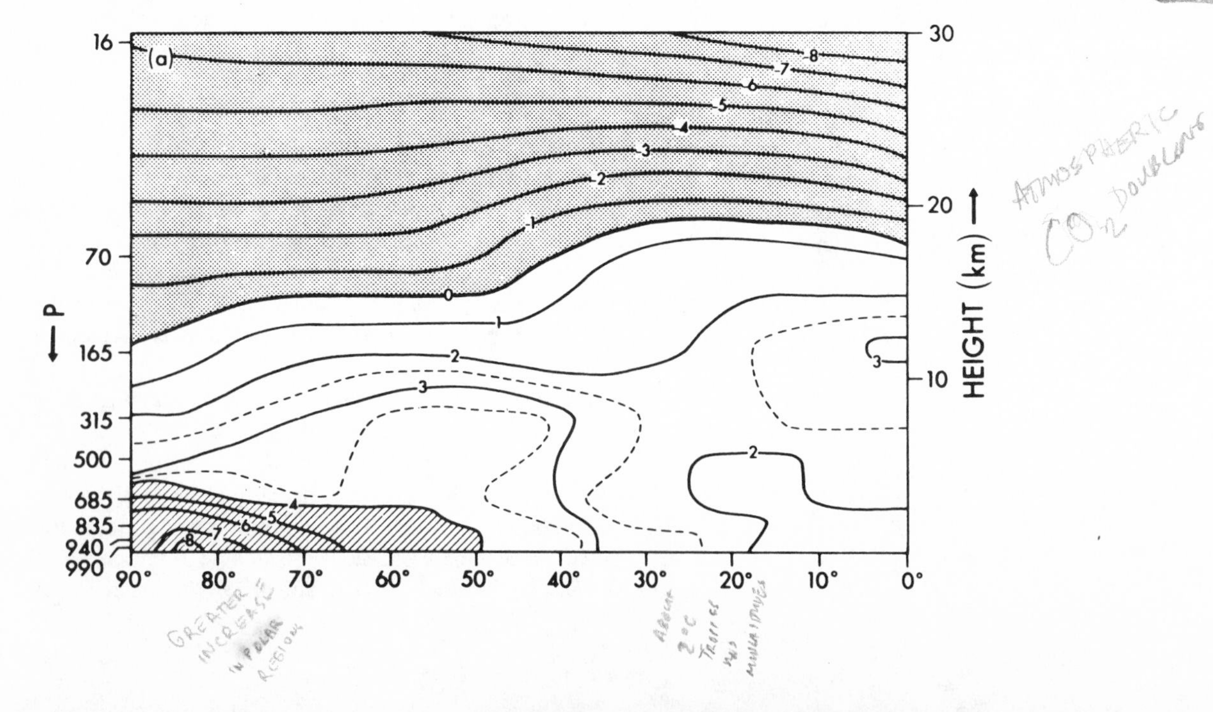

atmospheric temperature resulting from such a doubling is shown in Fig. 1.2. About 2 °C rise in surface temperature may be expected in the tropics and mid-latitudes with a rather greater increase in polar regions. Changes in the distribution of precipitation would also be likely to occur with serious consequences in some parts of the world (Mitchell, 1983). There would also be changes in polar ice with consequent effects on the sea level. Changes in the concentrations of other minor atmospheric constituents, for instance the oxides of nitrogen or aerosol, can also have climatic consequences of a similar amplitude (Kondratyev & Moskalenko, 1983, this volume).

Estimates with current models of possible climatic changes from various causes, both so far as global average effects and the regional variations are concerned, need to be treated with great caution, however, because the models do not yet include many of the important feedback processes. It is because of the importance of understanding the extent to which man's activities can cause climate change, together with our lack of understanding of the mechanisms behind the natural variations of climate that concentrated international effort on the climate problem is being directed.

1.2 The World Climate Research Programme (WCRP)

The WCRP is a component of the World Climate Programme; its objectives are to determine:

(1) to what extent climate can be predicted;
(2) the extent of man's influence on climate.

The programme is not primarily concerned with timescales of millenia or the even longer timescales of the major ice ages, although any understanding of climate must in the end account for the major changes which have occurred in the prehistorical past. The programme's principal aim is the understanding of changes which occur on timescales from several weeks to several decades, a timescale consistent with the feasibility of obtaining comprehensive data sets, the practicalities of numerical modelling, and the major concerns of planners and decision makers.

Although climate is determined by complex interactions on the planetary scale, the WCRP is certainly not limited to the consideration of global average climate; regional climatic anomalies which may develop over periods of months or years are more significant and clearer manifestations of climatic variations. The WCRP is thus concerned with space scales ranging from about 1000 km to global, the lower limit of 1000 km being roughly consistent with the lower limit of timescales of several weeks.

The WCRP is organised internationally by the Joint Scientific Committee (JSC) and the Joint Planning Staff (JPS), both jointly sponsored by the World Meteorological Organisation (WMO) and the International Council of Scientific Unions (ICSU). The JPS is based at the WMO Headquarters in Geneva.

1.3 The Global Atmospheric Research Programme (GARP)

The World Climate Research Programme has developed from the Global Atmospheric Research Programme (GARP) of the 1970s – a programme which has its beginnings in the early 1960s. In an address to the United Nations in 1961, President Kennedy appealed for 'future cooperative efforts between all nations in weather prediction and eventually in weather control'. Stimulated by Kennedy's remarks, following several years of discussion (Ashford, 1982) the GARP was formulated and, in 1967, a joint organising committee (JOC, the predecessor to the JSC) and a joint planning staff (JPS), both jointly sponsored by WMO and ICSU, were set up. The main aim of the programme was to observe the global atmosphere in sufficient detail so as to investigate the way in which different scales of atmospheric motion are organised and how they interact, hence to determine the extent to which the larger-scale motions (1000 km in size and above) can be predicted by numerical models.

After an experiment called the GARP Atlantic Tropical Experiment (GATE) in 1974, whose purpose was to make detailed observations over a period of several months of the particular features of the atmospheric circulation in the tropical Atlantic, the programme culminated in the Global Weather Experiment mounted in 1979. For this experiment, the global system of weather observations was enhanced to cover the globe in a sufficiently comprehensive way (Figs. 1.3 and 1.4) and in a number of institutions around the world a capability in global circulation modelling of the atmosphere was developed based on the largest computers available. First results from the experiments have shown that, with data from the Global Experiments, prediction of the detailed structure of the global circulation can be made up to about five days ahead, and that each improvement either in the data, in its method of assimilation, or in the description of the dynamical or physical processes in the models, leads to a lengthening of the period of prediction.

Fig. 1.3. System of five geostationary and at least two polar orbiting satellites in place for the Global Weather Experiment in 1979.

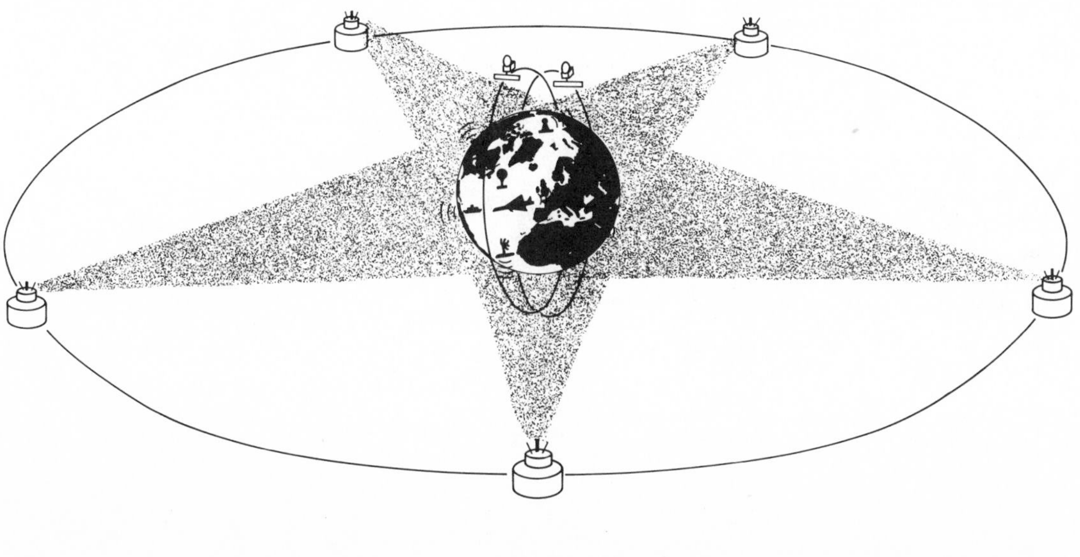

1.4 The climate system

The GARP was concerned with the atmosphere contained between its upper and lower boundaries. At the upper boundary, solar radiation enters the atmospheric system; some solar radiation is reflected out again, infra-red emitted radiation also leaves the top of the atmosphere. The lower boundary is land, ocean or ice, the state of which, for the purpose of GARP models, was considered as defined and fixed. When dealing with the longer periods addressed by the WCRP, dealing with the atmosphere alone is not sufficient; all components of the climate system are involved and have to be considered. We can list them as follows (Fig. 1.5):

(1) the atmosphere, which comprises the earth's gaseous envelope and which is the most variable component of the system. The lower atmosphere possesses a characteristic thermal response time to imposed changes of about one month;

(2) the oceans, which absorb most of the solar radiation incident on their surface and which, because of their high heat capacity, represent a large energy reservoir. The oceans transport about as much heat as does the atmosphere from the equatorial regions to the polar regions (Woods, 1983, this volume). The upper layers of the oceans interact with the overlying atmosphere or ice on timescales of months to years while the deeper ocean waters have thermal adjustment times of the order of centuries;

(3) the cryosphere, which comprises the continental ice, mountain glaciers, surface snow cover and sea ice (Untersteiner, 1983, this volume). Snow cover and the extent of sea ice show large seasonal variations. The glaciers and ice sheets change much more slowly; variations in their volume are closely linked to variations in sea level;

(4) the land surface including the biomass within or above

Fig. 1.4. Data distribution at the European Centre for Medium Range Weather Forecasting for 4 June 1979, 12 GMT ± 3 h. The diagrams show, respectively, the coverage of satellite winds from tracking cloud images from geostationary satellites (top two on left-hand side); surface observations from land stations and ships (bottom left-hand side); satellite temperature soundings (top right-hand side); radiosonde and pilot balloon ascents (upper middle right-hand side); buoys, drop-sondes and automatic stations (lower middle right-hand side); and aircraft observations (bottom right-hand side). (After Bengtsson *et al.* 1982.)

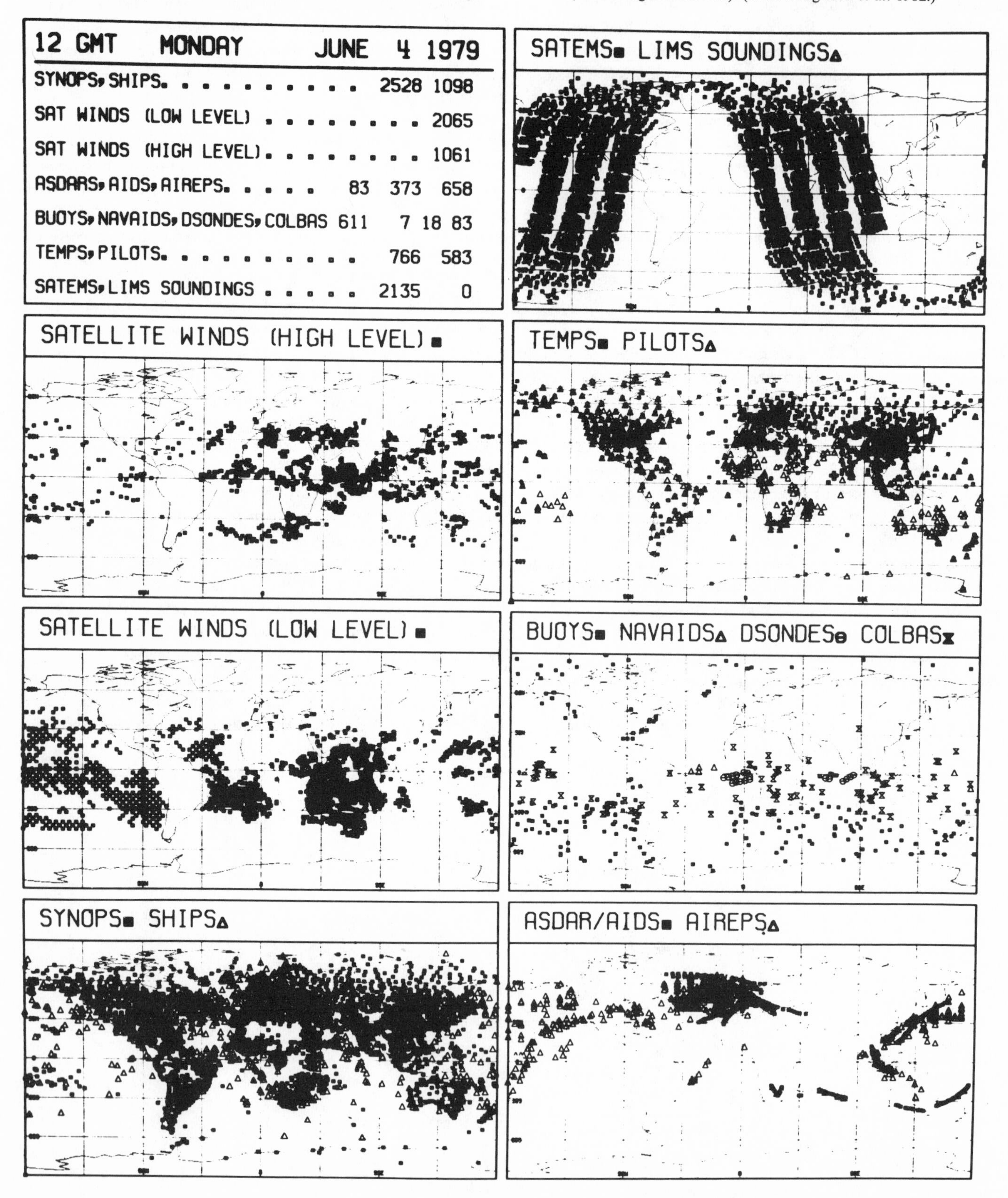

it, and also including those parts such as lakes, rivers and ground water which are important components of the hydrological cycle. Characteristic times for changes in the land surface vary from weeks for agricultural land to decades or centuries for some major forest areas.

Interactions between different components of the climate system occur in many ways. Exchanges of heat (through absorption and emission of radiation, air movement, evaporation of water and precipitation), water and minor chemical constituents (for example, CO_2, Bolin, 1983, this volume) between the land, ice or ocean surface occur on all timescales. The surface is also an important source of airborne particulates, through, for example, volcanoes, dust from deserts or salt particles from sea spray. These in turn, through radiative processes within the atmosphere, can have a significant climatic impact (Kondratyev & Moskalenko, 1983, this volume).

1.5
Climate feedbacks

A large number of feedback processes may be identified within the interactions which occur within the climate system. Some may act to amplify variations within the system (positive feedback) while others act to dampen them (negative feedback). Important examples of simple feedbacks are:

(1) ice–albedo feedback. An ice or snow covered surface reflects away nearly all the solar radiation incident on it, thus leading to further cooling of the surface. If melting of a part of the cover should occur, absorption of solar radiation warming the underlying land or water will lead to further melting – an example of positive feedback;

(2) water vapour–radiation feedback. An increase in temperature at the surface will lead to increased water vapour which, because of its opacity in the infra-red, acts as a radiation blanket over the surface thus further increasing the surface temperature – another example of positive feedback;

(3) cloud–radiation feedback. An increase of temperature at the surface will lead to increased water-vapour content, hence increased cloud cover will be expected. Increased cloud will reduce the amount of solar radiation reaching the surface, which will tend to reduce the surface temperature – an example of negative feedback. Some positive feedback also occurs through the blanketing effect of the cloud.

An illustration of the effect of including the first two of these feedbacks in a numerical model is shown in Fig. 1.6 where the results are shown of experiments on the effect of doubling the carbon dioxide content of the atmosphere with a model similar to that employed in obtaining Fig. 1.2. The effect of ignoring either feedback (1) or (2) above is to reduce the sensitivity of atmospheric temperature to increased carbon dioxide content by about a half.

Fig. 1.6. Model calculations of the sensitivity of the surface temperature to a doubling of the CO_2 content for the cases of full calculation including feedbacks (*a*) fixed ice snow cover (*b*) fixed meridional transport of latent heat (*c*) fixed absolute humidity (*d*) (After Chou *et al.*, 1982.)

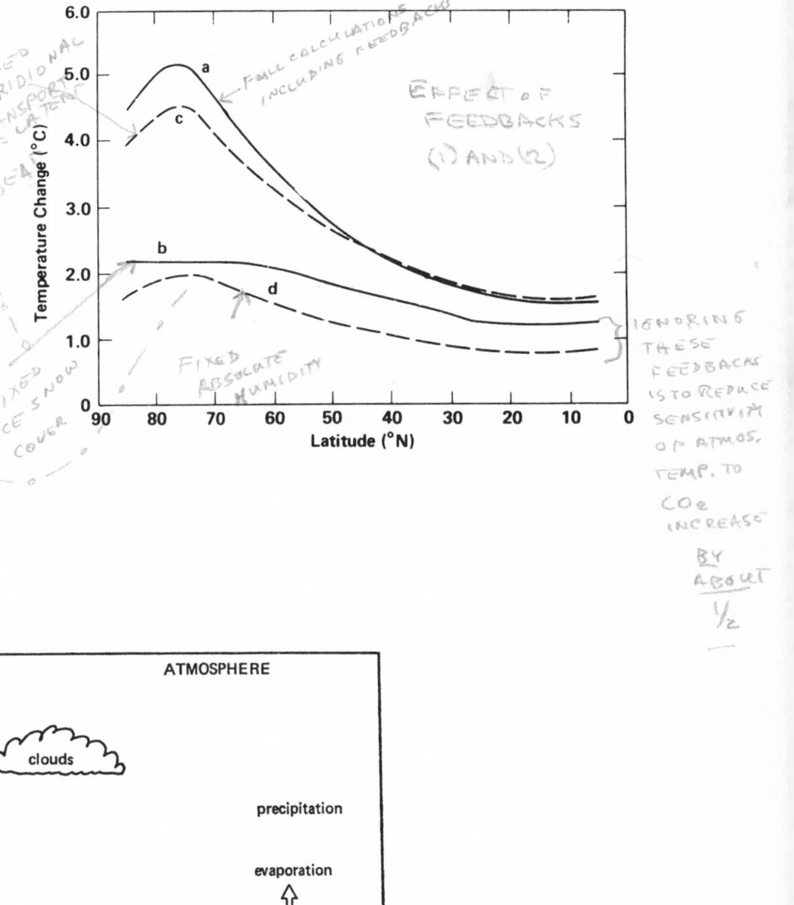

Fig. 1.5. Schematic illustration of the components of the climate system. (From World Meteorological Organisation, 1975.)

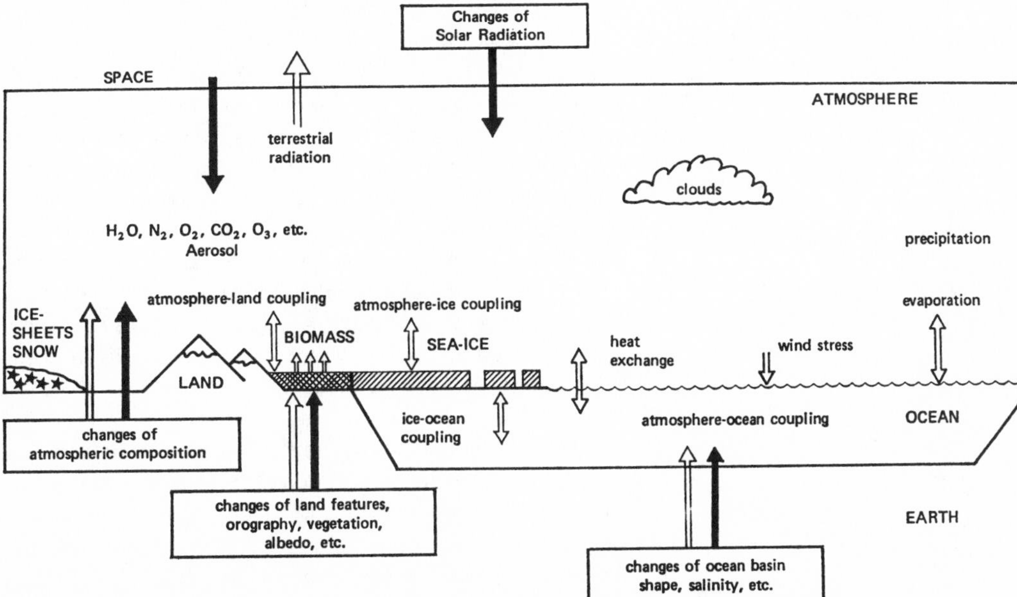

These examples and others are discussed in more detail in the chapters which follow (see especially Webster & Stephens, 1983, and Kondratyev & Moskalenko, 1983, this volume). It must be understood, of course, that the interactions and feedbacks in the whole climate system are very complex; no one process can be considered in isolation. The overall stability of the climate system as we know it must result from a whole variety of internal adjustment processes which as yet we only very partially understand.

1.6
Three research objectives or 'streams'

The knowledge acquired on the space and time structure of climate variations, and the scientific insight gained during recent years of climate research, have led to the formulation of the goals of the WCRP in terms of three objectives or 'streams' of climate research, each corresponding to a different kind of climate predictability and also to a different timescale. The first stream aims at establishing the physical basis for the long-range prediction of weather anomalies over periods of several weeks. The second stream is concerned with the interannual variability of the global atmospheric climate and the tropical oceans over periods of one to several years. Finally, the third stream addresses the problem of long-term variations and the response of the planetary climate to natural or man-made forcing factors over periods of several decades. This third stream of climate research is therefore particularly relevant in the establishment of the physical basis for the sensitivity of climate to external influences such as the increase of the atmospheric carbon dioxide concentration.

The first stream takes into account and aims to predict variability in the atmosphere, variability over short timescales of the land surface, but takes the ocean surface and sea ice to be a fixed defined boundary. The second stream takes into account and aims to predict, in addition, variability over longer timescales in the tropical oceans, and the third stream considers all components of the climate system which vary over timescales of weeks to several decades. Fig. 1.7 illustrates the parts of the system addressed by the three streams.

Fig. 1.7. Illustrating the regimes relevant to the three streams of climate research namely: stream 1: atmosphere alone with fixed lower boundary; stream 2: as stream 1 plus tropical ocean (approximately denoted by heavy dashed line); stream 3: complete system. Atmospheric cross-section shows isotherms for northern hemisphere winter. Ocean cross-section shows isotherms for Pacific at about 160° W.

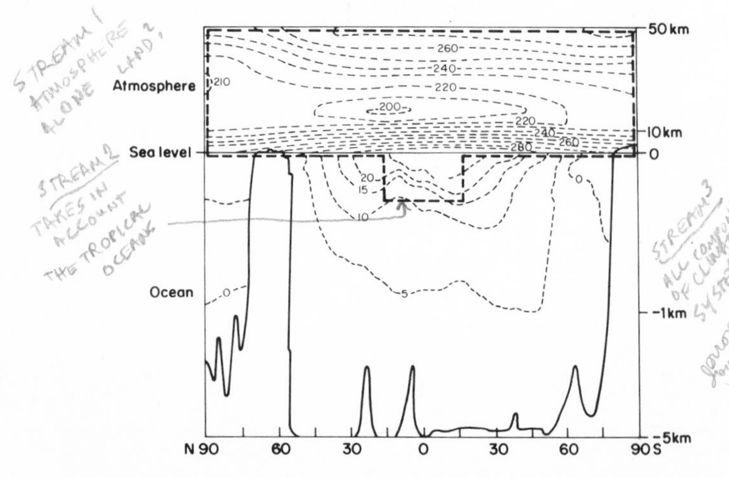

The major tools of the research programme are observations on the one hand and theoretical models on the other. We have seen that a global observing system for the atmosphere was developed during the GARP. For the WCRP, observations of all components of the climate system will be required; indications of the extent of these requirements will be mentioned in Section 1.10; further details will be found in the following chapters of this volume.

Regarding modelling, global circulation models of the atmosphere were developed for the GARP. These need further refinements for the WCRP in ways that are outlined below. Further, for the WCRP, appropriate models for the other components of the climate system and of the processes at the interfaces between these components (for example, processes at the land surface) need to be developed and means of coupling the models together devised. Further discussion of the modelling problems are given in later chapters of this volume, especially those by Leith, and by Simmons & Bengtsson.

1.7
The first stream of climate research: the physical basis for long-range weather forecasting

It is well recognised that there exists a theoretical limit of about two weeks to the predictability of individual weather events; beyond this limit, the relatively fast, non-linear fluid dynamical interactions within the atmospheric flow itself produce unpredictable meteorological variations which may appropriately be called weather noise. Studies based on the observation of the actual behaviour of the atmosphere have indicated, however, that slower variations exist above the level of the weather noise. These variations are generally associated with the larger scales of atmospheric motions and timescales of the order of one month. Thus the persistence of weather patterns observed in the real atmosphere appears to be longer than usually demonstrated by numerical models of the atmospheric circulation, possibly because of the existence of quasi-stable circulation regimes such as the blocking situations which are intermittently observed at mid-latitudes. Longer persistence may also be caused by the slower evolution of land or sea surface boundary conditions, forcing, in turn, a corresponding evolution of the atmospheric flow. For instance, because variations in the fluxes of heat and water vapour at the interface with the oceans influence the atmospheric circulation on timescales of a few weeks, sea surface temperature anomalies which generally show substantial persistence constitute an important predictor for long-range weather forecasts. The first stream of climate research thus aims at understanding those aspects of atmospheric dynamics and thermodynamics which could account for the longer predictability of large-scale atmospheric circulation regimes, and at establishing the physical basis for improved long-range weather predictions on timescales of one to two months.

To achieve this requires the following improvements:

(1) a substantial reduction of systematic errors in numerical weather forecasting models, which cause long-range weather predictions to drift toward unrealistic mean atmospheric flow conditions and away from the actual behaviour of the real atmosphere. This, in turn, requires other improvements namely (2)–(4) below;

(2) a better representation of the effect of terrain roughness and mountains on model atmospheric flows;

(3) better simulation of clouds and the associated radiation fluxes in numerical models (Webster & Stephens, 1983, this volume);

(4) refined estimates of the fluxes of momentum, heat and water vapour from the land surface, taking into account the interplay between soil moisture, vegetation, sea ice and the atmospheric boundary layer (see Mintz, 1983, this volume).

1.8 The second stream of climate research: interannual variability

A growing body of evidence is emerging from observations, indicating that worldwide climatic changes in middle and high latitudes, as well as climatic events on land within the equatorial zone, are related to year-to-year variations in the tropical ocean and the overlying atmosphere. It was in the 1920s that Sir Gilbert Walker first established remote influences in the tropical atmosphere, which were shown to be associated with a large shift of air mass. This phenomenon, which became known as the Southern Oscillation, is manifested by high atmospheric pressure in northern Australia and low pressure on the central South Pacific in one phase, while opposite pressure anomalies develop during the alternate phase. The cause for the persistence of each phase for periods of several months to years was unknown until later work established that the atmospheric pressure differences were related to changes of the sea surface temperature and mean sea level in the islands of the equatorial Pacific Ocean. Thus, the ocean dynamics driven by the atmospheric winds provide the memory which allows the anomalous atmospheric regime to persist.

The effect of this coupled atmospheric and oceanic behaviour is now believed to extend throughout the whole of the tropics. Analysis of 100 years' climatological record has shown that deficient monsoonal rains over India are associated with positive sea surface temperature anomalies in the eastern tropical Pacific Ocean, known as El Nino events (Angell & Gruza, 1983, this volume). Similar connections exist in the tropical Atlantic as far as the Gulf of Guinea where it has been established that warm sea surface temperature in the western tropical Atlantic is associated with droughts in north eastern Brazil. Similarly, warm western Pacific waters are related to droughts in China. These relationships are especially significant because the tropical zone accounts for a large part of the total energy release in the atmosphere. Further, there is also evidence of long-range correlations or teleconnections between the tropical oceans' temperature anomalies and the climate in mid-latitudes, particularly in the winter hemisphere (Fig. 1.8, and Angell & Gruza, 1983, this volume).

In addition to this empirical evidence, theoretical arguments support the idea that the atmosphere and oceans interact more effectively in the tropics. The vanishing of the Coriolis force allows the existence of basin-wide longitudinal oscillation modes trapped in the vicinity of the equator. The tropical oceans therefore exhibit relatively fast dynamic and thermal responses with time constants commensurate with the seasonal variations of the atmospheric circulation regime, thus allowing a strong coupling between the two media. The strongest oceanic signal of this nature has been found in the equatorial Pacific: its most striking manifestation, the El Nino event already mentioned, consists of the appearance of a strong positive sea surface temperature anomaly which develops off the coast of South America over a period of several months. A strong positive anomaly also develops in the central Pacific and persists for periods as long as 18 months. The faster response of tropical oceans has another important scientific implication: it makes it possible to design a first-order model of the tropical oceans interacting with the atmospheric circulation but essentially uncoupled from the mid-latitude ocean circulation.

Fig. 1.8. (*a*) Time series of monthly mean sea surface temperatures (K) in the region near 2.5° S, 130° W in the eastern Pacific Ocean. (*b*) and (*c*) Time series of atmospheric temperature averaged vertically and horizontally over the entire mass of the northern hemisphere (b) and southern hemisphere (*c*). (After Oort, 1982.) The 1963–73 mean annual cycle has been removed.

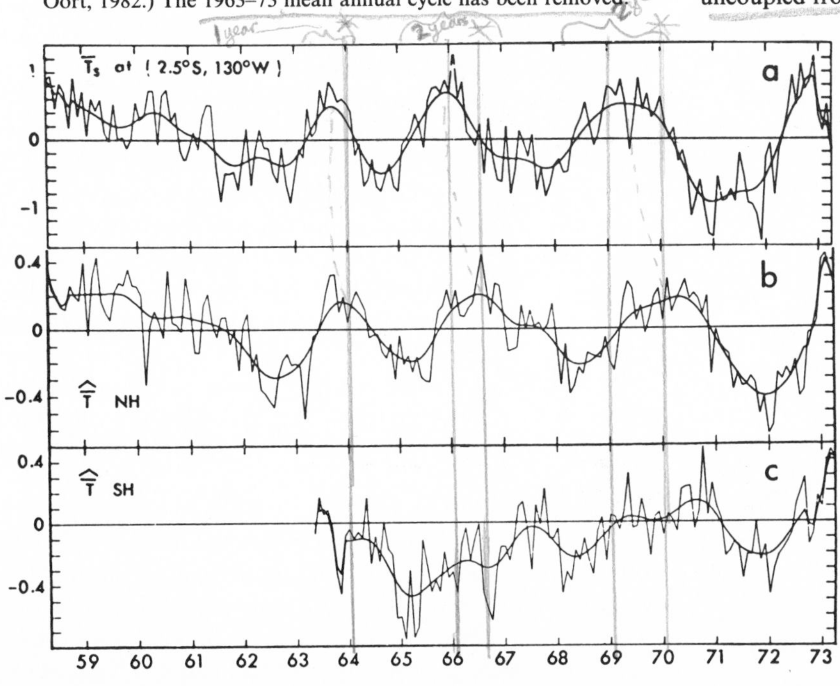

To simulate what occurs in an El Nino event, an efficient and relatively simple model based on the free surface equations, but with active thermodynamics, has been developed by Anderson (1983 and private communication). In the model, wind stress is applied simulating the change of the Walker circulation superimposed on the normal flow in the region of the equatorial Pacific. A decrease in the zonal stress arising from the Walker circulation produces an oceanic response in the form of a Kelvin wave propagating from west to east. Fig. 1.9 shows the sea surface temperature anomaly predicted by the model 100 days after the decreased stress compared with the observed sea surface temperature anomaly at the peak of an El Nino event. Other major features of El Nino events, such as the occurrence of cold anomalies, show up in the model, demonstrating that considerable progress has been made in understanding the basic physics and dynamics underlying the oceanic response to atmospheric forcing in the equatorial regions.

So far as the tropical atmosphere is concerned, its special character makes it particularly sensitive to even modest variations of the ocean surface temperature through a strong modulation of the availability of moisture for the release of latent heat into the atmosphere. Simple models (for instance, Gill, 1980) illustrate the dynamic response of the tropical atmosphere to tropical sea surface temperature anomalies. When such anomalies are introduced into more-elaborate general circulation models (Simmons & Bengtsson, 1983, this volume), features of the circulation are generated which are similar to those shown in observational studies of the connections between sea surface temperature anomalies and atmospheric circulation. It appears that the interaction of the global atmosphere with the tropical oceans alone accounts for the larger part of the predictable variability of the earth's climate over time periods of one season to several years.

The objective of the second stream of climate research is therefore to understand and model the coupled variations of the global atmosphere circulation with the basin-wide transient modes of the tropical Atlantic, Pacific and Indian Oceans, and a programme with the acronym TOGA (Tropical Ocean and Global Atmosphere) has been constituted as the main focus of the second stream of the WCRP.

Because the responses of the three major tropical ocean basins are linked only through their interaction with the global atmosphere, an observation and ocean modelling strategy for each individual basin may be planned independently of the others. This is the path being followed by the international oceanographic community, represented by the Committee on Climatic Changes and the Oceans (a joint committee between the International Oceanographic Commission (IOC) of UNESCO and the Scientific Committee on Ocean Research (SCOR) of the International Council of Scientific Unions), which has joined the JSC in the TOGA programme. The representations of the general atmospheric circulation needed for modelling interannual climate variations do not differ in a major way from the level of refinement required for the long-range weather forecasts as described above for stream one.

1.9 The third stream of climate research: long-term climatic trends and climate sensitivity

The climate of the earth is unique in the solar system in that it involves not only an atmosphere and a land surface, but also a global ocean which absorbs a large fraction of the incoming solar energy, transports this energy in the form of heat on a global scale, supplies a global hydrological cycle with the formation of clouds, precipitation and sea ice and also intervenes strongly in the distribution of radiatively active minor constituents such as CO_2.

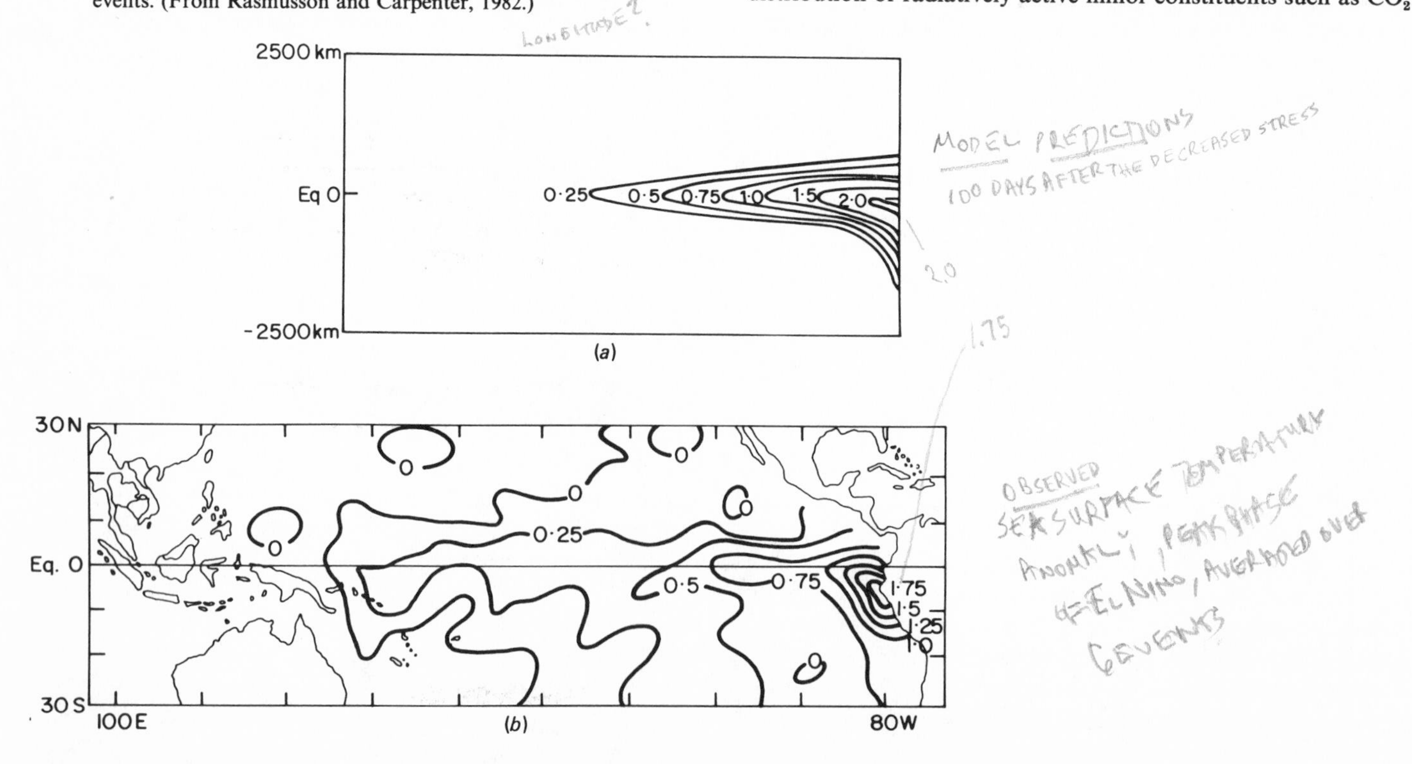

Fig. 1.9. (*a*) Sea surface temperature anomaly (K) as predicted for a simulated El Nino event by an ocean model due to Anderson (private communication). (*b*) Sea surface temperature anomaly (K) for the peak phase of the El Nino averaged over six events. (From Rasmusson and Carpenter, 1982.)

The currents that flow around the world oceans transport heat at a rate which is comparable to the transport by atmospheric winds, so that the seasonal and annual energy and water budgets of the earth can only be balanced globally if account is taken of both atmosphere and oceans. Not only does the thermal inertia of oceans regulate the seasonal temperature excursion of neighbouring land areas, but the meridional transport of excess heat absorbed in the tropical zone by the ocean currents controls the mean meridional temperature profile and thereby regulates the regime of atmospheric winds. The variations of the atmospheric climate over periods of several years to several decades or longer can only be fully understood on the basis of a description of the coupled dynamics of the global atmosphere and the global ocean. Also the oceans will strongly control the response of climate to external stimuli such as the man-induced increase of carbon dioxide concentration in the atmosphere. Because it is a very large heat reservoir, the global ocean could slow down the warming trend induced by excess CO_2 and delay the effect by as much as several decades. In addition, the oceans provide a large storage capacity which now contains about half the total excess CO_2 released recently in the atmosphere: reliable oceanic circulation models are required in order to estimate the rate of penetration of this chemical anomaly from the upper mixed layer of the ocean into the deep water. Consequently, the strategy for understanding the variations of climate over periods of 10 – 100 years and, especially, climate trends which may result from various forcing factors such as the increase of atmospheric CO_2, must include, in addition to the scientific advances already required in the first and second stream of climate research, a substantial improvement of our understanding of the global ocean circulation (Wunsch, 1983, this volume).

Dynamic and thermodynamic processes occurring inside the ocean determine the detailed character of its circulation, including the permanent large-scale currents and gyres, and the transient eddies which may be the oceanic equivalent of 'weather'. A model designed to compute the large-scale ocean heat transport from the surface boundary conditions must accurately describe the surface fluxes of energy, water and momentum which drive the ocean circulation and adequately simulate the circulation regime of the ocean, including both mean and eddy fluxes.

The development of global circulation models of the oceans, which include a good description of the exchanges with the atmosphere at the surface and which also can be coupled to general circulation models of the atmosphere, is one of the main aims of the World Ocean Circulation Experiment (WOCE) which is a primary focus of this third stream of climate research. Resulting from the data acquired during WOCE, estimates will be made of the heat transports by the ocean circulation at different locations and the variability of these transports from year to year.

Plant and animal life on earth depend on the cycling of chemical elements such as nitrogen, oxygen, carbon and sulphur through the soil, air, water and the biomass itself, as well as purely climatic factors such as solar radiation, temperature and precipitation. On the other hand, the extension, height, albedo and water-transfer properties of vegetation influence very significantly the processes at the land–air interface, which determine the fluxes of momentum, heat and water vapour into the atmosphere. Since the variable geographical extension and the nature of vegetation will respond to climatic variations over periods of years to decades, natural or man-made terrestrial ecosystems must be considered as interactive components of the climate system for the study of the long-term sensitivity of climate.

Because of the similarity of the various long-term climate sensitivity problems presented by the changing environment, the third stream of climate research may well be focused on one typical sensitivity problem, for example the quantitative assessment of the time-dependent response of the planetary climate to an increase in atmospheric CO_2. This strategy will not, however, detract from giving attention to the study of other significant factors such as the effect of tropospheric and stratospheric aerosols, the effect of radiatively active minor molecular constituents, the consequence of variations of solar activity and solar radiation, and the secular variation of the earth's orbit. With this in mind, in addition to what has already been included in the first and second streams, the two main components in the third stream of climate research will be the study of the coupled dynamics of the global atmosphere and the global ocean, including the role of sea ice and possibly the continental ice sheets, and the study of the processes involving the interaction of climate with terrestrial ecosystems.

1.10

Observational requirements

In the last sections we have explained the aims of the three 'streams' of the WCRP and detailed some of the major areas of theoretical work which need to be tackled. In this section, we address the observational requirements. Detailed specifications of these in connection with the three streams are being worked out by a number of international working groups. Here we can only describe the requirements in rather general terms, mentioning in particular the satellite observations that are relevant (Houghton, 1981).

For the first stream, that is, for long-range forecasting, the following will be needed:

(1) the existing global network of meteorological observations (see Figs. 1.3 and 1.4), including complete coverage around the equator by geostationary satellites and continuous observations from polar orbiting satellites. The satellite monitoring of the surface temperature of the global oceans is an important requirement for describing the surface forcing anomalies which may account for a good part of the predictable variability of weather on timescales of several weeks. Very high accuracy (better than 0.5 K) is needed for this particular measurement. Current satellite techniques approach this accuracy; new methods under development should enable it to be exceeded (Harries *et al.*, 1983);

(2) more-refined interpretation of observations of clouds from meteorological satellite images. The WCRP has defined the International Satellite–Cloud Climatology Project ISCCP whose purpose is to acquire a 5-year cloud climatology so as to establish the statistics of cloud variability over the period and to enable better descriptions of cloudiness and the associated radiation field to be incorporated into numerical models;

(3) more-refined interpretation of satellite observations of the land surface in terms of quantities which can be employed to describe the surface fluxes of heat and water vapour.

For the second stream, the major requirements additional to those listed above are for observations over tropical regions of the heat content of the upper layer of the ocean, sea surface stress, and sea surface topography (from which surface currents can be derived). Such observations can be made with limited coverage from ships, buoys and island stations, but for adequate coverage over the three tropical ocean basins, satellite observations will be required. The Seasat satellite, in 1978, demonstrated the feasibility of utilising radar scatterometer observations to measure surface stress (Fig. 1.10) and radar altimeter observations to determine the surface topography (Fig. 1.11). All these measurements need to be made not only with good coverage but with high accuracy. Observations of sea ice coverage are also required for stream two; these can be provided by imagery from satellites with passive microwave radiometers.

For the third stream of climate research, global observations of the ocean surface are, of course, required, of the same kind as are necessary for the tropical oceans for stream two. Measurements are also needed of the thermohaline structure and motion within the oceans. In addition to observations from ships, promising new techniques such as acoustic tomography are being developed; further details are given by Wunsch (1983, this volume) and by Munk & Wunsch (1982). A further requirement for stream three is that of accurate measurement of the radiation fluxes from satellites to determine the net energy input and its distribution at the top of the atmosphere, and to provide information required for the estimation of energy fluxes at the ocean surface and of heat transport within the oceans.

For experiments such as TOGA and WOCE, highly co-ordinated sets of observations will be required. For this reason, the JSC is proposing that a five-year period in the late 1980s and early 1990s be denoted an intensive period of ocean observation, during

Fig. 1.10. SASS wind vector comparisons with buoy reports during study in Gulf of Alaska. (After Linwood Jones *et al.*, 1979.)

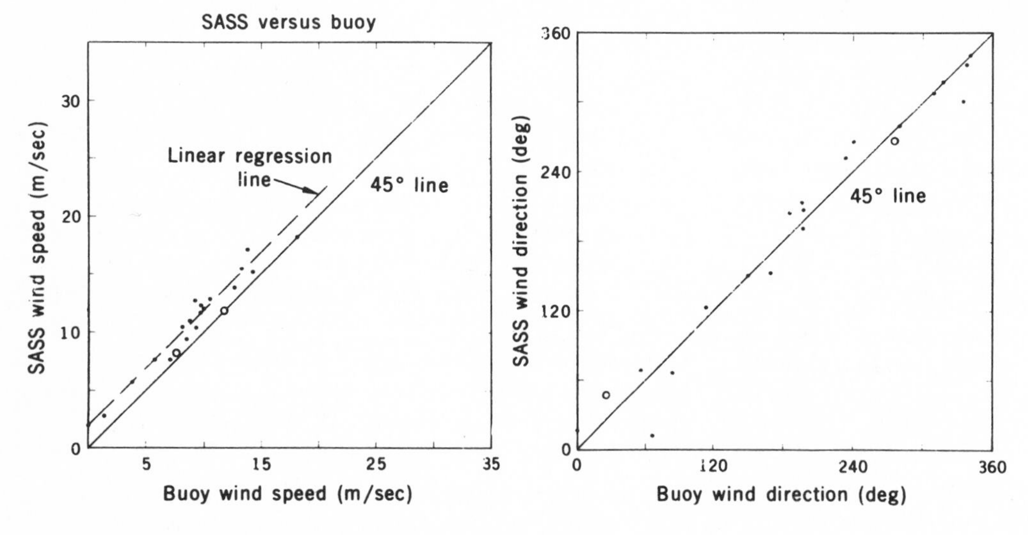

Fig. 1.11. Global circulation of the oceans as observed from radar altimetry from the satellites Seasat and Geos-3 by Cheney & Marsh (1982). Contours show ocean surface topography in cm relative to GEM-L2 Geoid.

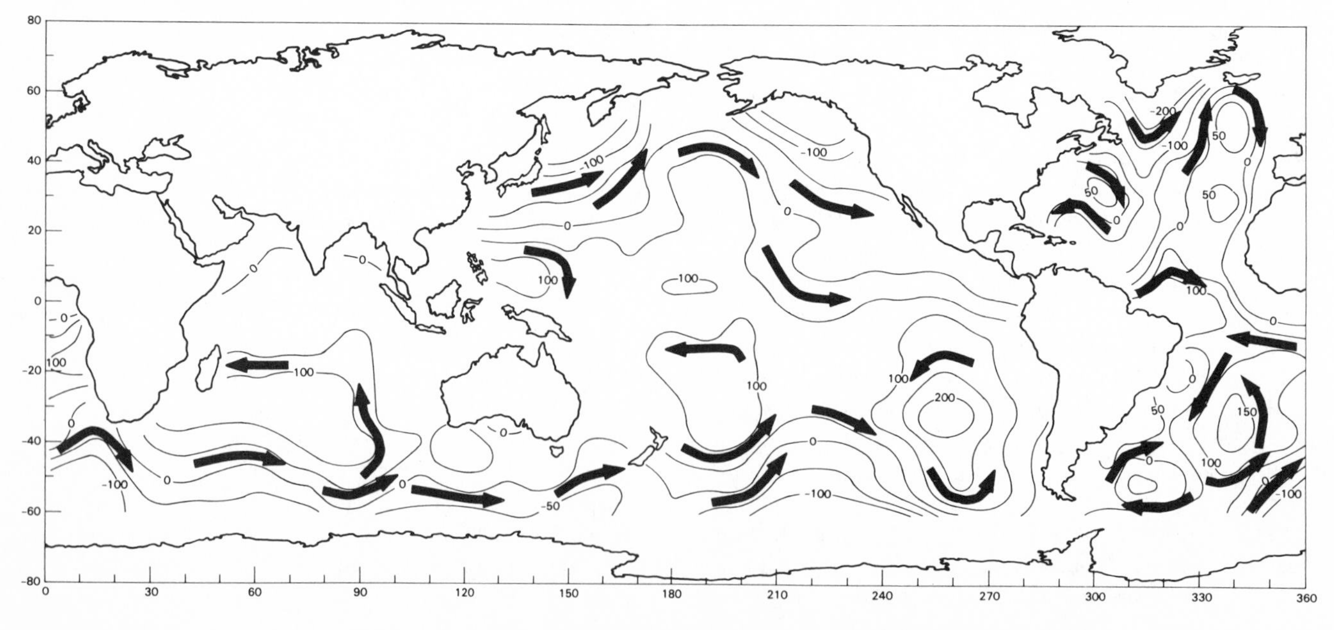

which concentration can be given to the satellite and *in-situ* observations needed for TOGA and WOCE.

1.11

Detection of climate change

However sophisticated and precise climate prediction will become in the course of the World Climate Research Programme, it is nevertheless essential that diagnostic methods be devised to measure even small climate changes and identify, as early as possible, any effect that man's activities or other external factors may have on the global environment.

Climate changes could of course be found through a great many manifestations, the most important of which are changes of distribution of rain and snow, and changes of the mean surface air temperature. But the various environmental quantities which are sensitive to climate shifts may not be equally suitable for climate monitoring and the early detection of a climate drift. For, in addition to being sensitive indicators, the selected climate quantities should also be representative indicators of large-scale changes and be amenable to convenient and accurate measurement. Ideally, one would also like to select climate indicators for which substantial past records are available with the required accuracy and global representativity. No single parameter ideally fulfills these requirements but the near-surface air temperature or, at sea, the surface water temperature, the mean sea level and the areal extent of sea ice are presently thought to be the most suitable indicators of a climate change.

But an early signal of climate drift could not be interpreted unambiguously on the basis of these few measurements alone. It is thought that any scientifically compelling explanation of the cause of a possible observed climate change will require the monitoring of many environmental quantities, including the average radiative fluxes at the top of the atmosphere, the amount and optical properties of aerosols, and the concentration of carbon dioxide and other minor atmospheric constituents. Equally important are global observations of the changing land use and vegetation in natural or managed ecosystems. The detection and characterisation of long-term trends do therefore place stringent demands on the performance and calibration of measuring instruments, as well as data acquisition and analysis methods, particularly with regard to essential satellite observations which are needed to provide the necessary global coverage.

1.12

Organisation of the programme

The Joint Scientific Committee promotes the WCRP objectives by stimulating research in national agencies, institutes and universities, or other multinational institutions, either directly or through international working groups established for a specific scientific purpose.

There is enormous scientific and technical challenge in a programme of research directed towards understanding the climate. Combining as it does the need for the highest performance instrumentation, for data management and organisation on a very large scale, for the development of complex models which demand the largest computing capacity available, together with the glamour of a close association with space research, it is virtually unsurpassed as a field for scientific endeavour. It is also an enterprise about which all of mankind is concerned and in which the whole world can be involved. The climate problem is not one to be solved quickly or easily, but contributing to its solution is enormously worthwhile.

References

Anderson, D. L. T. (1983). *Tellus*, in press.

Ashford, O. M. (1982). 'The launching of GARP', *Weather*, **37**, 265–72.

Bengtsson, L., Konamitsu, M., Kallberg, P. & Uppala, S. (1982). 'FGGE 4-dimensional data assimilation at ECMWF'. *Bull. Amer. Met. Soc.*, **63**, 21–43.

Cheney, R. E. & Marsh, J. G. (1982). 'Global circulation from satellite altimetry'. Presented at the fall AGU Meeting, San Francisco, December 1982.

Chou, M. D., Peng, L. & Arking, A. (1982), 'Climate studies with a multi-layer energy balance model, Part II: The role of feedback mechanisms in the CO_2 problem'. *J. Atmos Sci.*, **39**, 2651–66.

Gill, A. E. (1980). 'Some simple solutions for heat-induced tropical circulation'. *Quart. J. R. Met. Soc.*, **106**, 447–62.

Harries, J. E., Llewellyn-Jones, D. T., Minnett, P. J., Saunders, R. W. & Zavody, A. M. (1983). 'Observations of sea surface temperature for climate research'. *Phil. Trans. Roy. Soc.*, pp. 381–94.

Hays, J. D., Imbrie, J. & Shackleton, M. J. (1976). 'Variations in the earth's orbit; pacemaker of the ice ages'. *Science*, **194**, 1121–32.

Houghton, J. T. (1981). 'Remote sounding of atmosphere and ocean for climate research', *Proc IEE*, **128**, 442–8.

Lamb, H. H. (1966). *The Changing Climate: Selected Papers*. Methuen, London.

Linwood Jones *et al.* (1979), 'Seasat scatterometer: results of the Gulf of Alaska workshop'. *Science*, **204**, 1413–15.

Manabe, S. & Wetherald, R. T. (1980). 'On the distribution of climate change resulting from an increase of CO_2 content of the atmosphere'. *J. Atmos. Sci.*, **37**, 99–118.

Mitchell, J. F. B. (1983). 'The seasonal response of a general circulation model to changes in CO_2 and sea temperatures'. Quart. J. Roy. Met. Soc., **109**, 113–52.

Mitchell, J. M. Jr (1977). 'The changing climate'. *In: Energy and Climate, Studies in Geophysics*. National Academy of Sciences, Washington DC.

Munk, W. & Wunsch, C. (1982). 'Observing the ocean in the 1980s'. *Phil. Trans. Roy. Soc., Lond.*, **A 307**, 439–64.

Oort, A. H. (1982). 'Climate variability – some observational evidence'. Paper presented to the *Study Conference on the Physical Basis of Climate Prediction*, Leningrad USSR, September 1982 (published by WMO, Geneva).

Rasmusson, E. M. & Carpenter, T. H. (1982). 'Variations in tropical sea surface temperatures and surface wind fields associated with the Southern Oscillation/El Nino'. *Monthly Weather Rev.*, **110**, 354–84.

World Meteorological Organisation (1975). *The Physical Basis of Climate and Climate Modelling*, GARP Publications, Series No 16, WMO Geneva.

Global climate research

C. E. Leith
National Center for Atmospheric Research†
Boulder, CO 80307, USA

Abstract

The World Climate Research Program aims to develop dynamical and statistical models of the climate system that will produce reliable estimates of the response of the climate to external or internal influences. Climate is defined to include all the statistical properties of the system consisting of the atmosphere, land surfaces and oceans.

This chapter reviews current activities in three areas of climate research: observation and analysis, climate model development, and theoretical interpretation. The analyses acquired routinely for numerical weather prediction are serving as an increasingly detailed source of atmospheric climate information. Moderate resolution atmospheric models are now generating fairly realistic climate simulations; the next major requirement is for improved ocean circulation models. Theoretical interpretation of model results is greatly aided by analysis of the significance of climate signals as distinct from weather noise.

The clearest signal of interannual climate variability appears to be the Southern Oscillation. The chapter concludes with a review of some relevant observational and modeling research.

2.1 Introduction

Perhaps the greatest stimulus to the current general interest in global climate research was the concern expressed about ten years ago that the introduction of large fleets of supersonic transport (SST) aircraft might seriously damage the stratospheric ozone layer and increase stratospheric aerosols. Research scientists in many nations studied the various chemical, physical, and dynamical aspects of the question. Although their findings were inconclusive their investigations revealed the complexity of stratospheric chemistry and the physics and dynamics of the climate system. As new chemical reactions were considered, the predictions of the impact of SST emissions on ozone changed markedly. Lack of knowledge of the absorptive and scattering properties of aerosols generated by SST led to uncertainty in estimates of their impact on incoming solar radiation. And even for a specified change in incoming solar radiation, estimates of resulting surface temperature change ranged over a factor of about three owing to uncertainties in possible physical and dynamical feedback mechanisms.

Although concern for possible SST impact has receded, at least temporarily, similar new problems have arisen. The impact of artificial chlorofluorocarbon releases to the atmosphere raises the same kind of concerns as did the SST. It is believed that the radiative consequences of measured and projected increases in CO_2 concentration may influence the global climate, as may changes in land surface properties from changing patterns of land use such as by tropical deforestation. It is increasingly important that quantitative estimates of the sensitivity of the global climate to such human impacts be reliable enough to guide public policy.

In turn, variations and fluctuations in climate have an impact on our limited human resources, particularly of food. Any progress

† The National Center for Atmospheric Research is sponsored by the National Science Foundation.

toward prediction of such variations has important consequences for resource planning and management.

The widespread demands for accurate weather prediction have, for many decades, generated detailed observations of the atmospheric component of the climate system. The collection and analysis of these observations have rendered the global atmosphere the most completely measured and recorded fluid in the world. The demands of weather prediction have also led to the development of numerical models of the atmosphere of a complexity and resolution that tax the power of the fastest computers available.

The climate system, however, is larger than the atmosphere alone and involves interactions and physical processes not usually considered important for weather prediction. Thus climate models must include more components and more physics but, unfortunately, they must also sacrifice resolution in order to permit longer simulations with existing computers.

Numerical models are the natural tools for climate sensitivity studies since external influences may be readily changed in them. But confidence in the results depends on the completeness and accuracy of the physical interactions included in the models. Validation of the predicted response to hypothetical influences must be indirect if the projected changes have never been observed in the real atmosphere.

Numerical models are also used to assess the predictability of climate fluctuations on a seasonal and interannual time scale. A large part of such fluctuations is known to arise from sampling of unpredictable weather variations but, for example, the response of the atmosphere to slowly changing tropical ocean surface temperatures provides a potentially predictable climate signal detectable above the weather noise. The most noticeable of these slow climate modes is that associated with the Southern Oscillation centered primarily in the equatorial Pacific Ocean and influencing the North American continent.

Thermal inertia of the ocean introduces much longer time scales into the climate system than would arise from the atmosphere alone. Oceanic time scales range from weeks in the surface layers to millenia in the abyss and are extremely difficult to model in a credible way. An additional ocean modeling difficulty is that spatial scales of importance are an order of magnitude smaller than those in the atmosphere. In order to treat such scales explicitly ocean models require higher spatial resolution at much greater computing expense.

In this chapter I shall begin with a discussion on the definition of climate and of the components of the climate system. I shall then review climate research strategy, which must maintain a balance between three research activities loosely distinguishable as based on observations, models, and theories. Although the final objective of climate research is the development of reliable climate models, confidence in these must rest on comparison with observations and a theoretical understanding of individual components of the system. Finally, I shall review our present state of knowledge and uncertainty in applications of models to studies of the Southern Oscillation as one of the most likely of possibilities for seasonal prediction.

Most attention will be given to recent publications; I shall focus thus on present activities rather than on past history. Many important earlier contributions can be found as references in the cited papers.

2.2
Definition of climate

It is generally recognized that climate is in some sense the average of weather, its fluctuations, and its influence on the surface of the Earth. The main problem in the definition of climate is to make this intuitive idea sufficiently precise to serve as a basis for a theory of climate.

The first step in this direction is to recognize that a climate theory must be probabilistic in nature. Detailed weather fluctuations are treated as a multivariate random process whose statistical properties are a proper subject for climate research. In contrast, the deterministic dynamics of weather are more directly the subject of research on weather prediction. Weather is thus more determined by initial conditions, whereas climate is more determined by boundary conditions.

The next step is to define the nature of the averaging process in terms of which climate statistics are defined. Here, somewhat more subtle problems are encountered. The usual and only operationally feasible way of defining observed climate statistics at a point on the globe is in terms of time averages. Thus we may speak of the mean surface air temperature at a particular weather station for the month of January 1980 or for 30 January months from 1951 to 1980, and consider these to be climate statistics. Clearly then, the time interval or intervals over which the average is defined enters into the specification of such climate statistics. It is, of course, realized that climate statistics for January of one year will differ from those of another year and yet that this may arise as much from statistical sampling fluctuations as from any change in the underlying probabilistic expectations. These expectations are, in fact, what we would prefer as the elements of an ideal climate theory. The statistical sampling fluctuations, which we call weather noise, would then be treated as an unfortunate concomitant to practical observations of the climate.

These distinctions between ideal probabilities and observed statistics are familiar from general theories of probability and statistics. Since the weather noise decreases with increasing sample length, we consider the 30-month January average to be more representative or closer to the true climate than is that for a single January. By extension we might consider the true climate to be defined in terms of an ideal limit of an infinite time average. But this too has technical disadvantages since we wish, in a climate theory, to study the possibility of climate change under slowly changing external forcing influences.

We escape in principle from this final difficulty by a further idealization common to statistical mechanics and turbulence theory. Consider an infinite population or ensemble of earths, each subjected to the same fixed external influences but having different detailed weather patterns. With the passage of time, each earth's weather evolves, but for a suitable distribution the ensemble remains stationary; and if climate is defined in terms of averages over this ensemble it will remained fixed. Note that for this rather elaborate conceptual structure it is now possible to study the time evolution of ensemble average climate for specified changes in external influences, such as incoming solar radiation, imposed on the whole ensemble. If, however, the external influences are assumed to be fixed for all time, then the associated fixed ensemble average climate should be the same as the climate obtained by an infinite time average of any single member of the ensemble such as the Earth. This, of course, requires an ergodic assumption that there is only one stationary ensemble and its climate is approached in the limit of an infinite time average.

In this definition of climate, some influences are considered to be external to the climate system. In practice, what constitutes external influences are chosen in different ways depending on the time scales of interest. On the shortest time scale a climate model could consist of an atmospheric general circulation model with land surface temperature computed by energy balance considerations but with ice cover and ocean surface temperatures specified as external influences. Likewise, for studies of climate sensitivity, specified CO_2 concentration or surface albedo would be treated as external quantities.

On longer time scales the interaction of the atmosphere with the ocean leads to changes in ice cover and ocean surface temperature. The coupling of an atmospheric model to an ocean circulation and ice model produces a more complete climate model with fewer specified external influences. There may still remain, however, a distinction between the ocean and ice component and the atmospheric component of such a model if, as is customary, the former is treated probabilistically and the latter deterministically. In this case we imagine the ensemble of earths to have a common ocean circulation and ice cover determined by ensemble average atmospheric effects.

A complete climate model would, in principle, also treat the ocean and ice probabilistically, but this is difficult to do in practice and, for the time scales of principal interest up to decadal, not appropriate.

Similar considerations apply to relatively slow land surface processes such as soil moisture storage and growth of vegetation.

2.3 Observations and analysis

Research on the global climate can be divided somewhat arbitrarily into three components dealing, respectively, with observations and analysis, with large numerical model simulation, and with theoretical interpretation. It is only through interactions between these three activities that one can develop confidence in predictions of climate response or change.

The primary source of information about the evolving detailed hydrodynamic and thermodynamic state of the atmosphere is the routine observation and analysis cycle carried out by weather services for purposes of global weather prediction. The oldest and most basic observing system for the three-dimensional fields of temperature, water vapor, and winds is the radiosonde network. At each of about 800 stations, of which only about 100 are in the southern hemisphere, rising balloon borne instruments make atmospheric soundings twice each day at noon and midnight Greenwich time. Unfortunately, the radiosonde stations are not uniformly distributed over the globe but, for economic reasons, tend to be concentrated over land areas in midlatitudes in the northern hemisphere. To remedy this situation and to obtain more uniform global coverage, satellite borne sounding systems have been devised, which provide information on temperature and water vapor fields. Compared to radiosondes these have greater resolution and more uniform coverage in horizontal directions, but less vertical resolution and less absolute accuracy. The drift of cloud patterns over oceans as seen by geostationary satellites provides detailed information about wind fields where there are clouds, but with some uncertainty about the height to which the observation applies. There are also available other miscellaneous sources of observations such as winds and temperatures at flight level from aircraft (see Fig. 1.4).

The basic problem of meteorological analysis is to convert this stream of heterogeneous discrete observations into a best estimate of the continuous fields defining the three-dimensional state of the atmosphere at regular intervals in time. These serve as initial states for global weather prediction and provide the basis for verification of prediction models. Records of such analyses go back several decades and provide a data base for statistical analysis of the climatological properties of the three-dimensional atmosphere.

In recent years, available computing power has permitted rapid objective statistical analyses of such records, stored on magnetic tape, to be carried out for the northern hemisphere where data are most complete. The results of such studies have been reported most recently, for example by Blackmon & White (1982), Holopainen, Rontu, & Lau (1982) and Lau & Oort (1981), all of whom provide many references to earlier work. From such studies we know much about the mean fields as well as the three-dimensional spatial distribution of variance and covariance on different time scales. But we can also learn about the relations between these statistical quantities as, for example, in the studies by Holopainen, Rontu & Lau (1982) of the effect of large-scale transient eddies on the time–mean flow.

During the past few years there have been dramatic improvements in the analysis cycle brought about by the progress made by the Global Atmospheric Research Programme (GARP) and by planning for and analysis of the Global Weather Experiment carried out in 1979. Although these developments disrupt to some extent the continuity of climate records, they promise to provide future records of far greater accuracy and detail.

Analysis of observations is now based on optimal statistical interpolation methods (Gandin, 1963) in which the continuous meteorological fields are produced by least square regression methods from pointwise observations. Such methods require a statistical model of the spatial covariances of the meteorological fields based originally on climate statistics. They produce the most likely member of the climate ensemble compatible with the observations and their hypothesized errors. This most likely member can be described as an anomaly from the climatological mean field.

More recently, it has been recognized that model predictions provide sharper *a priori* information at a particular time than do climate statistics. The analysis procedure is easily modified to replace the climate mean by a model prediction and climate statistics by model error statistics to find the most likely correction to a model prediction compatible with new observations. Now we may speak of data assimilation into a model which is providing the mechanism to carry information forward in time.

The statistical models used in data analysis or assimilation typically build in linear dynamical constraints of geostrophic balance between temperature and wind fields. The resulting fields are still not quite satisfactory as initial states for a typical primitive equation prediction model. A further step of nonlinear balancing or initialization is carried out to rid the prediction of spurious gravity mode oscillations. Daley (1981) has recently reviewed developments in the so-called nonlinear normal mode initialization techniques which are becoming standard practice in operational prediction centers.

Rapid progress is being made on two remaining initialization problems. Although nonlinear dynamics has been satisfactorily balanced there is still difficulty with nonlinear physical processes, in particular, latent heat release. The deficiency has been most serious in the treatment of the Hadley circulation arising from heating in the tropics, but Puri & Bourke (1982) have largely solved this specific problem. The other problem is the incompatibility between the linear model of the optimal statistical analysis and the nonlinear model of the initialization. Williamson, Daley & Schlatter (1981) have shown the practical consequences of this discrepancy in producing less than optimal data assimilation. In principle, the problem is solvable by linearization about the predicted state, but in practice this is not computationally feasible with models as complex as those used for operational prediction. Approximations are being sought.

Although these developments have been spurred by the requirements of weather prediction, they have important consequences for future climate diagnostic studies. The nonlinear initialization procedures generate ageostrophic fields, including vertical velocity, that are not directly observable but constitute an important part of the climate mechanism. It has been shown, for example, by Errico (1982) that triple moments involving an ageostrophic wind component can be as important as purely geostrophic triple moments in maintaining eddy fluxes and nonlinear transports in the atmospheric climate system.

The price to be paid for the more detailed and accurate climate records coming out of the full data assimilation process is that they are clearly model dependent. No longer can one separate the 'truth' as observed from the way it is modeled.

Recent analyses of the internal dynamics of the atmospheric climate have sought to determine Eliassen–Palm fluxes in order to understand the mean zonal flow in the atmosphere (Edmon, Hoskins & McIntyre, 1980; Karoly, 1982). This technique borrows from linear wave propagation theory to provide a theoretical basis for analysis. In another related approach, quasi-geostrophic potential vorticity is recognized as the principal quantity of interest in characterizing the dynamics of the system, and analysis of its flux and flux divergence shed light on the maintenance of the mean flow (Holopainen, Rontu & Lau, 1982). Although the potential vorticity flux is less relevant to linear wave theory than is the Eliassen–Palm flux, it is not restricted to a zonally symmetric analysis and thus provides information on regional climate.

It is of great interest, of course, to determine what possible spatially coherent patterns of climate variability may exist. These have been called teleconnection patterns and have been sought in two ways. The older approach, pursued primarily by Rinne and his colleagues (e.g., Rinne & Jarvenoja, 1982), has been to diagonalize the spatial covariance matrix and to extract the principal components of spatial variance often called empirical orthogonal functions (Kutzbach, 1967). As with other statistical properties of climate, there are sampling errors associated with the determination of empirical orthogonal functions, as discussed by North, Bell, Cahalan & Moeng (1982). These severely limit the number of such functions to which physical significance should be ascribed.

The other approach, followed by Wallace & Gutzler (1981) for monthly means, generates teleconnectivity maps giving at each point the maximum absolute value of any remote correlation. The correlation maps centered at the various maxima of the teleconnectivity maps show interesting teleconnection patterns which have been given names according to their geographical location. Statistical significance of these patterns is hard to evaluate, but the procedure only identifies the most important few.

The climate system consists of more than the three-dimensional atmosphere. In particular, observations of surface properties of the Earth form a key part of the climate data base. The relatively slowly changing sea surface temperature fields have traditionally been determined from ships' reports. More recently, satellite based measurements are providing more uniform coverage and identifying more rapidly changing anomalies but with less absolute accuracy. Sea ice cover is relatively easily monitored by satellite techniques and shows considerable interannual variability in addition to the expected annual cycle.

Observations of land surface properties such as temperature and snow cover are, of course, of extremely fine spatial and temporal coverage in regions where people live. They are not, however, well organized as a component of a global climate data base. Even less well organized is information about soil moisture and vegetive cover, both of which are recognized as important to the climate system. Much effort is being made to find satellite based solutions to these observing problems.

There is one kind of special climatological observation for which satellites are particularly suited, and that is the measurement of outgoing radiation from the Earth. This includes reflected solar radiation as observed by broad-band sensors in the visible spectrum and terrestrial radiation as measured in the infrared. The total outgoing radiation suitably integrated over the spherical Earth should balance the incoming solar radiation. But the balance need not be local. Lateral heat transport in both the atmosphere and ocean can compensate for regional imbalance in the radiation budget. In particular, the observed annual average net radiative heating in the equatorial regions and cooling in the polar regions is compensated by poleward heat transport. Oort & Vonder Haar (1976) have shown that atmospheric heat transport as computed from climate statistics is not adequate for this and have estimated the residual oceanic transport required for balance.

2.4 Climate models

The starting point for the development of large-scale climate models has been the development over several decades of numerical models of the general circulation of the atmosphere. Such models explicitly compute the evolution of weather patterns over the globe with as fine a spatial resolution as can be afforded in computing time. In addition to carrying out explicit integrations of the nonlinear dynamical equations, they also compute the influences of physical processes such as cloud formation, precipitation and latent heat release, infrared and visible radiative transfer and heating, and surface heating and evaporation. Until recently, such models usually treated the sea surface temperature as a specified external influence. It is by coupling such an atmospheric model to an interactive ocean model that a full climate model can be produced.

2.4.1

Moderate-resolution atmospheric models

The earlier general circulation models (GCMs) were based on finite difference approximations carried out on a three-dimensional array of grid points. Models developed more recently are based instead on spectral transform methods in which horizontal fields are represented by a truncated expansion in surface spherical harmonics (Bourke, 1974). Vertical structure in either case is resolved with discrete levels and finite difference approximations. The spectral transform methods appear to be slightly more efficient in computing time for a given effective resolution, but are much more robust in the sense that small changes in the model are less likely to lead to numerical instability.

Some of the most realistic climate simulations have been obtained with spectral transform models with nine vertical levels and horizontal truncation at rhomboidal 15 (R15), i.e., with $0 \leqslant |m| \leqslant 15$, $0 \leqslant n-m \leqslant n-m \leqslant 15$, where m and n are, respectively, the zonal and total index of the surface harmonic function $Y(m, n)$. Most of these models were based originally on the spectral transform algorithm of Bourke (1974) and have been developed into GCMs with somewhat differing physical properties in a number of research centers. Results of a 15-year simulation with such a model at the Geophysical Fluid Dynamics Laboratory (GFDL), Princeton, have been reported by Manabe & Hahn (1981), and by N.-C. Lau (1981) who has carried out a study of model teleconnections and empirical orthogonal functions showing some similarity to those of the real atmosphere. Simulation results for a similar model at the Australian Numerical Meteorology Research Centre (ANMRC), Melbourne, have been published by McAvaney, Bourke & Puri (1978). A common GCM problem has been a lower polar stratosphere that is too cold. This appears to have been solved with a modified treatment of clouds and radiation in a variant of the ANMRC model at the National Center for Atmospheric Research (NCAR), Boulder. The climate mean fields for the NCAR model are described by Pitcher, Malone, Ramanathan. Blackmon, Puri & Bourke (1983) and the details of the improved radiation treatment by Ramanathan, Pitcher, Malone & Blackmon (1983). The NCAR model has been run for a 20-year simulation with fixed solar declination angle and sea surface temperature field appropriate for northern winter. It appears to produce realistic mean and transient fields for that season.

These successes with such moderate resolution models raise serious questions about the role of horizontal truncation. Experiments with varying truncation in a spectral transform model show, in fact, a tendency for model mean climate to acquire large-scale biases as resolution is increased (Manabe, Hahn & Holloway, 1979). Similar sensitivity of large-scale behavior to resolution was observed earlier in finite difference models by Manabe, Smagorinsky, Holloway & Stone (1970) and Wellk, Kasahara, Washington & De Santo (1971). It appears that R15 truncation used in spectral transform GCMs may produce the most realistic atmospheric climate. At first this would appear to be good fortune since such models use only about two minutes of CRAY-class computer time for each simulated day. But truncation is an artifact and thus realistic model behavior may be occurring for the wrong reasons.

Frederiksen & Sawford (1980) have suggested that the eddy viscosity prescriptions designed to remove enstrophy from the smallest scales are inadequate and that the system behaves like an inviscid system truncated in wavevector space. These are known to tend toward an equilibrium energy spectrum that depends only on the total energy and enstrophy and on the truncation wavenumber. It happens that for R15 truncation this equilibrium spectrum more or less matches the observed energy spectrum but for higher resolution it has too much energy in the largest scales.

It may also be relevant to this problem that theoretical statistical analysis of the energetics of a two-level quasi-geostrophic atmosphere (Hoyer & Sadourny, 1982) suggests that kinetic energy is trapped in the largest scales but that enstrophy cascades toward small scales where it is lost. The energy trapping is at about the Rossby deformation scale and this is also where baroclinic energy conversion takes place. The consequences of limited resolution on the efficiency of conversion have been studied by Sadourny & Hoyer (1982). They show that truncation at a wavenumber 2.5 times greater than that of the deformation scale diminishes baroclinic conversion by 16%. This corresponds roughly to the R15 truncation of the spectral models described above.

The inclusion of the mean effects of unresolved scales of motion has been one of the oldest problems facing the designers of truncated hydrodynamic models. The present suggestion for climate models is to introduce an artificial mechanism to remove enstrophy but preserve energy. Such a scheme has been described by Sadourny & Basdevant (1981), but it has not yet been fully tested.

An observed feature of the atmospheric circulation is the occurrence of so-called blocking events which appear in preferred geographical locations and have greater persistence than usual midlatitude weather patterns. Owing to their impact on regional weather, the frequency of blocking events is considered to be an important aspect of climate. Any persistent local structure in the atmosphere must survive in spite of the linear dispersive tendency of its wavepacket component Rossby waves. In the modon model of blocking, suggested by McWilliams (1980), the dispersive tendency is balanced by nonlinear effects. For the resolution of such local structures, spectral models, which are free of false numerical linear dispersion, appear to have an innate advantage over grid point models, for which truncation induced artificial dispersion can be important.

2.4.2

Atmospheric model physics

The atmosphere can be thought of as a steam engine deriving much of its kinetic energy from a water evaporation–condensation cycle. It is important therefore that an atmospheric model describe these processes with reasonable accuracy. Evaporation in models is usually based on bulk transfer formulas applied to the ocean, ice and snow, and land surfaces. The greatest uncertainties arise for the last where complex soil moisture physics and vegetive evapotranspiration play a role.

Condensation occurs when water vapor partial pressure reaches the saturation value, which depends on temperature. Water vapor in a model must be transported conservatively from its evaporation source to its condensation sink. The greatest difficulty in doing this arises from the difficulty of resolving the vertical distribution of water vapor and accurately computing its vertical advection. The vertical finite difference methods used may, for example, generate unphysical negative water vapor concentrations through numerical truncation error.

Although some of the condensation may occur on scales large enough to be explicitly resolved in a model, much occurs on a far smaller scale in cumulus convection. Many convective prescriptions have been used in atmospheric models, but none can claim evident superiority, in part because observational data for detailed validation is not yet adequate. The greatest amount of tuning in atmospheric models has been done with parameters in these convective prescriptions in order to simulate the climatologically observed hydrological cycle and cloud distributions. But neither of these are well observed, and a real danger exists that the prescriptions may be too closely tied to the present climate and may interfere with feedback processes important for climate change. There is, in particular, a concern that the static stability of the model troposphere may be artificially determined and may not respond properly to changing external influences.

The atmosphere cannot hold much liquid water in the form of cloud droplets, and in most models it is assumed that any condensed water either precipitates to the surface or evaporates while falling. This avoids the requirement that cloud liquid water be computed in a model as another meteorological field. The presence of clouds, however, is important to the transfer of solar and terrestrial radiation. Many earlier models used for this purpose specified cloud distributions based on climatological information, but some present models are including cloud generating algorithms tied to the convection prescription in order to estimate the fraction of cloud cover as a function of time and three-dimensional position.

It is, of course, the radiative properties of clouds that are important in their definition for models, and these properties depend in complicated ways on their liquid water content, drop size distribution, and morphology. A principal purpose of the International Satellite Cloud Climatology Project (ISCCP) is to provide as much observational information as possible for the validation of model cloud generating algorithms, both in a detailed and a climatological sense.

The greatest source of uncertainty in GCM climate sensitivity experiments at present arises from the crudeness of the cloud–radiation interaction and feedback processes in the models. It is generally recognized that low, warm, white clouds reflect solar radiation with little effect on outgoing infrared radiation, leading to net global cooling, whereas high, cold, thin cirrus reflect little solar radiation but diminish outgoing infrared radiation to produce net heating. Therefore, in any model experiment on the sensitivity of the climate to external specified change, it is important that the model computes credible changes in high and low clouds. Progress is being made on generation of cloud topped boundary layers to produce the climatologically important stratus cloud decks over parts of the ocean. The problem of generating high cirrus clouds is far more difficult owing, in some part, to the difficulty of making sufficiently accurate calculations of water vapor transport to high levels in a model.

This vertical advection problem also renders dubious any model calculation of water vapor in the stratosphere. Unfortunately, radiative balance in the stratosphere is dependent on its water vapor content which is often specified rather than computed on the assumption that the measurements, crude as they are, are more reliable than a model calculation.

In summary, it may be concluded that the most serious problems of atmospheric model physics are related to the hydrological cycle of evaporation, water vapor transport, and precipitation and to the effect of clouds and water vapor on the radiative properties of the atmosphere.

2.4.3 *Ocean circulation models*

The ocean serves as a regulator of the climate system owing to its large heat capacity and ponderous nature. It is not, however, a perfect thermostat, and an important component of any climate sensitivity or predictability question is the determination of how rapidly and how much the ocean surface temperature would change. But surface temperature cannot be isolated from the rest of the physics and dynamics of the ocean so that an ocean circulation model (OCM) is needed as a component of any full climate model. Although the fundamental equations of fluid flow for the ocean are the same as for the atmosphere, there are three major differences which make the development of OCMs more difficult than that of atmospheric GCMs, namely the time scales are longer, the space scales are smaller, and the observations are more sparse.

The time scale problem is made more serious by the range of time scales from weeks to millenia involved in the ocean circulation. Although implicit numerical techniques permit time scales in a model to be determined by the time scales of physical interest, the continuum of important time scales in the ocean presents technical difficulties in devising efficient algorithms.

Perhaps more serious is the problem of oceanic space scales. The static stability of the ocean is such that its Rossby deformation scale is more than an order of magnitude smaller than that of the atmosphere. Since the kinetic energy spectrum in a geophysical fluid tends to be peaked around this scale, the ocean has much of its energy in eddies of around the 50 km scale. An eddy-resolving OCM therefore requires a much higher horizontal resolution than an atmospheric GCM at much higher computing expense. For example, a 30-fold increase in resolution with a 900-fold increase in computing per time step, compensated in part by a 30-fold increase in time scale and time step, leads to a 30-fold increase in arithmetic for a given period of simulation. Clearly a coupled ocean–atmosphere model on this basis would be computationally dominated by the ocean component.

Owing to the great expense of eddy-resolving ocean models, many OCMs have been constructed with coarser resolution closer to that of atmospheric models, and the effects of the unresolved eddies have been introduced through eddy diffusion and viscosity terms. Since such coarse resolution models have produced somewhat realistic behavior, there is considerable controversy over the importance of the eddies to the larger-scale circulation of the ocean.

The resolution of this difficulty is made more difficult by the third special ocean problem, namely the lack of global observations defining the general circulation of the oceans. It is the intent of the World Ocean Circulation Experiment, planned for the late 1980s, to provide this kind of missing information. Satellite altimetric measurements of the height of the ocean surface will provide surface geostrophic flow fields from which, together with bathymetric data, can be deduced a three-dimensional flow field. Satellite scatterometer measurements of surface stress and atmospheric estimates of surface heating and salinity changes provide the forcing mechanisms for ocean circulation models. It is thereby planned to provide the information necessary for OCM validation.

In the meantime, much work is being done with upper ocean models in trying to simulate the annual cycle of mixed layer surface temperature, heat storage, and depth under the influences of surface stress and buoyancy fluxes. It is clear, however, that these models cannot be local but require advection and divergence fields from the underlying general circulation.

The coupling of an atmosphere and ocean circulation model requires a solution to the time scale mismatch problem. If the ocean model is eddy resolving, we have already seen that the atmospheric calculation adds little to the cost. At present, however, coupled models involve an ocean model of the same horizontal resolution as for the atmosphere, thus not eddy resolving, and then the atmospheric calculation becomes dominant. Since it is the statistical time average properties of the atmosphere that are of most importance for driving the ocean, some attempts have been made to deduce such statistics from short sample atmospheric simulations which are then held fixed for a longer ocean simulation, which in turn changes ocean surface temperatures requiring a new atmospheric sample simulation. Dickinson (1981) has pointed out the problems of such asynchronous coupling.

Bretherton (1982) suggests that if the response characteristics of atmospheric surface stress and buoyancy flux properties could be, once and for all, linearly related to patterns of ocean surface temperature change, then the complicated nonlinear atmospheric dynamics model could be replaced by a trivial linear atmospheric response model and all effort could go into the ocean circulation component of the problem. On the other hand, Hasselmann (1976) has shown that variations in ocean surface temperature can be largely a consequence of the random component of atmospheric forcing being integrated by the slow response of the ocean, thus Bretherton's suggestion may have to be supplemented by an appropriately responding random component of atmospheric forcing. Until these problems are resolved, it appears that coupled models will have to pay the price of the arithmetic necessary for synchronous coupling.

2.4.4

Atmospheric macroturbulence models

For time scales of climate variability greater than a year, the detailed initial state of the atmosphere is no longer remembered. The atmosphere on these time scales is a purely stochastic system rapidly responding to a changing surface boundary and to other slow components of the total climate system. The major slow climatic component is the ocean circulation, and a calculation of its evolution requires statistical average information from the atmosphere for surface values of stress, solar and infrared radiative fluxes, evaporation and precipitation. It would clearly be useful, if it were possible, to develop a stochastic model of the atmosphere that would provide such statistical information directly rather than through averaging of a time integration of a detailed atmospheric GCM. Thus we need an atmospheric macroturbulence model that would provide mean fields of temperature, wind, water vapor, clouds and radiation, as well as mean fluxes of heat, momentum, and water vapor. These are, respectively, first and second moment quantities of the atmospheric physics and dynamics. What are the possibilities and problems in developing such a model?

The major technical problem in formulating turbulence equations for a nonlinear dynamical system such as the atmosphere is that each moment evolution equation depends on higher moments. Thus at some level, higher moments must be estimated in terms of lower moments in order to close the system of equations.

The most advanced turbulence models of large-scale atmospheric dynamical processes are still not realistic. They treat the atmosphere as a two-layer quasi-geostrophic fluid in which the turbulence is spatially homogeneous and nearly isotropic. They have, nonetheless, displayed the statistical properties of fully nonlinear baroclinic energy conversion processes (Hoyer & Sadourny, 1982).

The most interesting aspect of the observed global climate is that it is not spatially homogeneous even in longitude. In midlatitudes, eddy-available potential energy is found primarily over continents, conversion occurs at eastern coasts, and the resulting barotropic kinetic energy is found and dissipates over oceans.

Although the turbulence equations may be readily generalized to treat inhomogeneous turbulence, the amount of computation required for their integration becomes prohibitive, exceeding that of a detailed GCM integration. Clearly then, some further simplifications are required, but each such simplification lessens confidence in the validity of the model.

Some progress in modeling inhomogeneous turbulence has been made for studies of the microturbulence of the planetary boundary layer. The simplest closures, which use eddy diffusion coefficients or mixing lengths, have been replaced by closures which include evolution equations for second moment quantities such as turbulent energy. Such models involve a large number of adjustable dimensionless parameters and have succeeded quite well in predicting eddy fluxes through simple boundary layers. These are obviously inhomogeneous in the vertical direction.

Similar approaches should be tried for the macroturbulence of the global atmosphere, but there are two serious complications. The inhomogeneity in this case, as has been noted, is in all three space dimensions, and the geostrophic constraints imposed by the Earth's rotation lead to reverse energy cascade in spatial scale, sometimes referred to as negative viscosity. There is, however, theoretical evidence (Green, 1970) that the eddy transport of potential vorticity may be more easily modeled than that of other quantities and may serve as a basis for a turbulence theory.

It is clear from this discussion that these developments are at a primitive stage compared to the requirements for a stochastic atmospheric component of a full climate model. Thus, general circulation models will continue to be needed to provide statistics through time sample runs and to validate any proposed stochastic atmospheric model.

2.5

Theoretical interpretation

It seems clear that there is no way of avoiding, for the foreseeable future, the computing expense of full coupled atmospheric–ocean models for the most trustworthy estimates of climate sensitivity and predictability. Unfortunately the complexity of such models is a barrier to our understanding of the climate system. We must acquire this understanding instead through theoretical interpretation in terms of simplified models whose properties reflect, at least qualitatively, some of the properties of the total system. Careful analysis of single processes in a simplified framework, such as the statistical analysis of baroclinic energy conversion by Hoyer & Sadourny (1982), can indicate requirements, such as for horizontal resolution, in the full models.

2.5.1

Energy balance models

Much has been learned about the overall energy budget of the Earth through studies with so-called energy balance models (EBMs). Such models were introduced some years ago by Budyko (1969) and Sellers (1969), and have since been considerably extended. In them, all of the complexity of eddy transport in the atmosphere and ocean is reduced to highly simplified, but conservative, energy transport terms. Likewise, the local radiative heating and cooling effects are crudely approximated. In spite of these excessive simplifications, such models have been able to demonstrate the sensitivity of the climate to various formulations of interactive processes such as those between temperature, ice cover, and albedo, or between clouds and radiation. A recent review of EBM studies has been published by North, Cahalan & Coakley (1981).

2.5.2

Weather noise

Energy balance models provide climate mean information that is free of statistical sampling fluctuations. We have noted that in GCMs, such weather noise tends to obscure any computed change in the climate mean. An important part of the theoretical interpretation of observed or modeled climate behavior is therefore an understanding of the nature of this weather noise.

Perhaps the most familiar example of sampling fluctuations is in games of chance. The outcome of each play, of course, is random and unpredictable, but so also with less variance is the average outcome of a sequence of N plays. The theory of probability, which was developed to understand games of chance, provides answers to such sampling problems, which can be applied to the game of wagering on the future climate.

For a sequence of samples, each made up of N values drawn independently from an ensemble, the sample means will fluctuate about the ensemble mean with a variance reduced by a factor of N from that of the ensemble (Cramer, 1946). A natural measure of the magnitude of sampling fluctuations is their standard deviation, which decreases therefore as the square root of the sample size. This is a measure of the rms noise in the sampling process and indicates the extent to which a single sample average is likely to differ from the ensemble mean.

In application to those time sequences of meteorological variables that are time averaged to estimate climate average values, the above result must be generalized to account for serial correlations. For example, if we are computing an average January noon temperature from 31 daily values, we must recognize that the values on two successive days are not independent.

This more-general situation has been analyzed by Leith (1973) and Jones (1975), respectively, for continuous and discrete sampling. Leith introduced the concept of an effectively independent sampling time. He defined this as a correlation time T_0 given by the time integral of the lagged correlation function for the continuous time series analog of the earlier sampling sequence. For time averaging intervals T much longer than T_0, the analog of the discrete independent sample result is valid with $N = T/T_0$. The complete dependence of sampling error on T_0 and T, is given by Leith (1973).

The correlation time T_0 is also a measure of the deterministic predictability of the weather by linear regression techniques or by nonlinear dynamical integration. Note thus that the greater the value of T_0, the greater is the range of deterministic weather predictability, but the higher is the weather noise level, and thus the less predictable is a climate fluctuation.

For a random continuous time series, the frequency power spectrum is given by the Fourier transform of the time-lagged correlation function $R(\tau)$. The correlation time T_0 is thus related to the low-frequency limit of that part of the power spectrum which is attributable to internal weather fluctuations.

Madden (1976, 1977) has estimated T_0 from climate records of surface pressure over the northern hemisphere and temperature over North America. He finds considerable geographical variability with values ranging from two to eight days over the hemisphere. In these papers, Madden has computed the ratio of observed interannual variance of monthly average values to the expected noise variance. Where this ratio is unity, all variability is accounted for as unpredictable noise, and this seems to be the case in midlatitudes, especially for 'continental' climates. But in the polar regions, in the tropics, and for 'maritime' climates, the ratio appears to be significantly greater than one. It is in such regions therefore that useful climate prediction is more likely to be possible.

Weather noise obscures true climate change both in the real atmosphere and in atmospheric general circulation models, and this renders the use of such models for prediction of climate change doubly difficult. If, however, it were possible to devise a statistical dynamical model in which only the statistical properties of the atmosphere were computed, and not its day to day fluctuations, then sampling fluctuations would no longer be a problem for the model although still a problem for the atmosphere.

2.5.3

Climate sensitivity theory

General circulation models have been widely used to estimate the response of climate mean quantities to specified changes in boundary or external conditions, for example of ocean surface temperature, carbon dioxide concentration, or the incoming solar radiation. In abstract terms we may consider the state equation for the unperturbed system to be $\dot{\mathbf{x}} = \mathbf{Q}(\mathbf{x}) + \mathbf{f}$, where the vector $\mathbf{x}$ defines the state of the system at a particular time, the vector function $\mathbf{Q}(\mathbf{x})$ describes the complicated modeled nonlinear dynamics and physics determining the evolution of this state, and the vector $\mathbf{f}$ indicates those external forcing influences that determine the mean statistical properties of the system. We assume that $\langle \dot{\mathbf{x}} \rangle = 0$, i.e., the system is stationary and unchanging in its mean properties. Here the brackets stand for the ensemble average.

Consider an experiment consisting of the imposition at time $t = 0$ of a step change in $\mathbf{f}$ to some new level at which it then remains fixed. It is clear from the equations of motion that $\langle \mathbf{x} \rangle$

will change initially at a rate $\langle \dot{\mathbf{x}} \rangle = \delta\mathbf{f}$, where $\delta\mathbf{f}$ is the imposed step change. This tendency in the mean value $\langle \mathbf{x} \rangle$ will not continue indefinitely but, rather, after some time will level off to a new equilibrium that differs from the original value of $\langle \mathbf{x} \rangle$ by $\delta\langle \mathbf{x} \rangle$. For sufficiently small step changes in the forcing term $\mathbf{f}$, one expects that the final mean response $\delta\langle \mathbf{x} \rangle$ of the climate system will be proportional to $\delta\mathbf{f}$. We may then define a sensitivity matrix $\mathbf{A}$ such that $\delta\langle \mathbf{x} \rangle = \mathbf{A}\delta\mathbf{f}$, and all first-order sensitivity information is contained in the matrix elements of $\mathbf{A}$. The model experiments attempt to determine at least some of these matrix elements by changing some components of the external forcing $\mathbf{f}$.

There is a variant of this procedure which may not be practical with numerical models but is of theoretical interest. Imagine an experiment in which, instead of a step function, $\delta\mathbf{f}$ is imposed as an impulsive forcing which leads to an immediate change $\delta\langle \mathbf{x} \rangle$ in the mean climate. After such a shift in $\langle \mathbf{x} \rangle$, since $\mathbf{f}$ has now been returned to its original value, one expects that $\langle \mathbf{x} \rangle$ will decay back to its original unperturbed value. The nature of this decay can be characterized by a response function for a unit impulse, that is, a matrix-valued Green's function $\mathbf{G}(\tau)$ for mean values of the state vector $\mathbf{x}$, which depends on the elapsed time since the impulse was imposed. It is normalized so that $\mathbf{G}(0) = \mathbf{I}$, the identity matrix, and it is expected to decay towards zero for large τ. The linear response to any history of forcing perturbation may then be written as a convolution integral whose kernel is $\mathbf{G}$.

Using the fluctuation dissipation theorem familiar in statistical mechanics, $\mathbf{G}(\tau)$ may be approximated as being the same, for $\tau > 0$, as the regression matrix function $\mathbf{R}(\tau) = \mathbf{X}(\tau)\,\mathbf{X}^{-1}(0)$, where $\mathbf{X}(\tau) = \langle \delta\mathbf{x}(t)\,\delta\mathbf{x}^{\mathrm{T}}(t+\tau) \rangle$ is the lagged covariance matrix for $\delta\mathbf{x} = \mathbf{x} - \langle \mathbf{x} \rangle$. The fluctuation dissipation relation holds exactly for certain simple statistical mechanical systems of which the climate is not one (Leith, 1975). Experience with turbulence calculations and simple climate models (Bell, 1980) indicates that the approximation $\mathbf{G}(\tau) = \mathbf{R}(\tau)$ is nonetheless relatively good.

The matrix elements of the original response matrix $\mathbf{A}$, which gives the response $\delta\langle \mathbf{x} \rangle$ for step function forcing changes, are easily computed as integrals of the matrix elements of $\mathbf{G}$. Because the diagonal elements of $\mathbf{R}(\tau)$ are qualitatively like the single variable lagged correlation $\mathbf{R}(\tau)$ discussed earlier, the diagonal elements of $\mathbf{A}$ may be estimated as $T_0/2 \approx 3$ days. In the case of step function forcing, we can estimate that the change in level of the mean $\langle \mathbf{x} \rangle$ is given by its initial rate of change $\mathbf{f}$ multiplied by a characteristic time of about three days.

This estimate ignores slow feedback effects which may be included to first order by linear analysis (Leith, 1978). These can either increase or decrease the climate sensitivity. The response time of the external system is assumed to be slow and, for this present discussion, nonrandom. The response properties leading to feedback may be determined therefore by using a numerical model of the external system that includes the appropriate slow physical processes.

Another qualitative estimate arising from consideration of the fluctuation dissipation relation is for the spatial response of climate means to local forcing influences. Inasmuch as the climate response function is similar to the spatial correlation function, we know that the climate mean response to forcing by, say, the local imposition of a change in sea surface temperature should be limited in midlatitudes to a correlation range which is a few thousand kilometers. From this point of view, the climate system is the sum of more-or-less independent regional climate systems having characteristic dimensions of correlation functions. These qualitative relations suggest in general that a careful study should be made of both time and space correlations, as observed in the real atmospheric system, in deciding the most fruitful model climate experiments to be carried out. They also suggest that models used for climate sensitivity studies should be tested for agreement in their time and space correlation properties with the observed present climate as N.-C. Lau (1981) has done.

In the case of observable, slowly changing influences such as ocean surface temperature, the response of the atmospheric climate can be studied by direct statistical analysis of space–time correlations between the observed surface anomaly and anomalies in atmospheric behavior. Such correlations not only reveal the sensitivity of the atmospheric climate to a particular influence but also can serve, through regression analysis, to provide a basis for prediction. The influence of ocean surface temperature anomalies has been widely studied in this way, for example by Barnett & Preisendorfer (1978) for the North Pacific, Rowntree (1972) and Julian & Chervin (1978) for the tropical Pacific, Ratcliffe & Murray (1970) for the North Atlantic and Rowntree (1976) for the tropical Atlantic.

Bretherton (1982) has considered an application of linear sensitivity analysis to ocean circulation studies. The ocean circulation is driven by average radiation flux, surface stress, and evaporation–precipitation difference as determined by the atmosphere. Thus a suitable first-order top boundary condition for a deterministic ocean model need only specify the linear response of these quantities to changing surface temperature. Preliminary studies have shown that, as expected, these responses depend strongly on the spatial scale of the surface temperature anomaly.

2.5.4 *Linear wave analysis*

A common way to gain insight into the workings of a complicated system is to reduce it to a linear problem for which a classical decomposition into independent modes of behavior is possible. Although, in principle, linear analysis is straightforward, in practice there are severe limitations on the size of matrices that can be manipulated with existing computers. The number of degrees of freedom in a nonlinear atmospheric GCM may be of order 100000, but computational linear algebra is only feasible, in general, for fewer than 1000.

Much analysis has been carried out for atmospheric dynamics linearized about specified zonally symmetric mean flows. In this case, the longitudinal degrees of freedom separate out, and for each zonal wavenumber the problem reduces to two-dimensional analysis in the vertical and meridional directions. This is enough, however, to show the existence of absorbing, reflecting, and overreflecting surfaces for wave propagation. These analyses show, for example, that only the longest tropospheric waves should penetrate into the stratosphere and that tropical barriers usually separate equatorial and midlatitude disturbances.

The climatological mean flow is not, however, zonally symmetric, and more recent studies have taken this into account. There are two feasible approaches. One may linearize complicated nonlinear models and observe their time-dependent or stationary linear response to different localized forcing influences (Grose & Hoskins, 1979), or one may reduce the number of model degrees of freedom enough for full linear analysis of eigenvalues and eigenfunctions to be feasible (Frederiksen, 1982).

These studies have shown that many of the observed teleconnection properties of the observed and modeled atmosphere can indeed be qualitatively reproduced by linear analysis.

2.5.5 *Atmospheric blocking theory*

There have been two recent theoretical approaches to a theory of atmospheric blocking events. In one, taken by Charney and his colleagues (Charney & DeVore, 1979; Charney & Strauss, 1980), a Rossby wave forced by topography and damped by friction is recognized to have amplitude-dependent frequency and thus to exhibit nonlinear resonance (Malguzzi & Speranza, 1981). For such a resonance there may for certain conditions, be three stationary solutions of increasing amplitude, the intermediate one being unstable. A blocking event corresponds in this theory to the system being in the high-amplitude stable stationary state. Such a theory explains the preferred geographical locations of blocks but not their local structure.

In a second approach, McWilliams (1980) suggests that in some cases a blocking event may be a modon. Stern (1975) originally derived a modon as a localized dipole solution of the barotropic vorticity equation in which, as with a soliton, linear dispersion is balanced exactly by nonlinear terms. Stern's original modon was a solution confined to a circular domain with a vorticity discontinuity at the boundary. This discontinuity is removed in a more general modon solution given by Larichev & Reznik (1976) in which a rapidly decaying exterior solution is matched to an interior one. Numerical simulations by McWilliams, Flierl, Larichev & Reznik (1981) show modons to be remarkably robust. McWilliams (1980) fits modon parameters to a particular North Atlantic blocking event in January 1963, in an equivalent barotropic model, i.e., one in which a Rossby deformation term accounts for quasi-geostrophic divergence effects. The modon theory explains the local structure of blocks but not their preferred geographical locations. A better theory, perhaps containing elements of these two, is still needed.

A basic issue is whether the climate system displays bimodal behavior. Lorenz (1970) pointed out that this is a theoretical possibility for nonlinear dynamical systems, which he had demonstrated in his famous low-order nonlinear model of convection (Lorenz, 1963). The two blocking theories discussed above are bimodal since they make a clear distinction between the higher and lower resonance or between the presence and absence of a modon.

There are other examples of bimodality observed in the atmosphere. Perhaps the most striking was described many years ago by Yeh, Dao & Li (1959) who noted the rapid shift of the position of the east Asian jet stream between its summer position on the northern side of the Himalayas to its winter position on the southern side. This fluidic switch is shown most graphically by Palmen & Newton (1969) in their Fig. 3.14.

2.6 **Southern Oscillation**

The greatest hope for climate prediction on a seasonal and interannual time scale is to find relatively slow components of climate variability induced by coupling with some ponderous component of the system such as the surface layers of the ocean. An outstanding mode of interannual climate fluctuation has been found to be the Southern Oscillation with its apparent coupling to surface temperature changes in the equatorial Pacific Ocean. Both physics and statistics favor such a mode. The highly nonlinear relation between temperature and water vapor pressure make evaporation and associated latent heat transfer to the atmosphere most sensitive to small changes in temperature where the temperature is highest. The ocean surface mixed layer is relatively thin in the equatorial regions with low heat capacity and less control over temperature excursions. And the weather noise in the tropics is relatively low.

The Southern Oscillation was observed and named by Sir Gilbert Walker in the 1920s. It is a large-scale shift in surface pressure over the tropical regions, most notable in the Indonesian and Pacific regions. A Southern Oscillation Index (SOI) has been defined on the basis of pressure differences as being high when eastern Pacific pressures are high and Indonesian pressures low relative to the long-term average and to the opposite low-index phase. The time period for alternation between phases is not as regular as use of the word 'oscillation' might imply; it varies from about three to seven years.

A number of other phenomena have been associated with the Southern Oscillation through correlation with the SOI. At times of high SOI, the ocean surface temperatures are relatively cold in the equatorial Pacific and along the Peruvian coast, precipitation remains concentrated in Indonesia and is relatively low across the Pacific, the associated longitudinal Walker circulation along the equator is relatively strong, the meridional Hadley circulation is weak, and the southeast trade winds are strong. Bjerknes (1969) suggested that the association of these phenomena was consistent with stronger upwelling of cold water as a consequence of increased surface stress during high-index periods.

During times of low SOI, the southeast trade winds are weak, equatorial and coastal upwelling are diminished, the surface layers of the ocean warm up, intensified El Nino episodes occur along the South American coast, evaporation and precipitation increases over the central equatorial Pacific, the Hadley circulation is strengthened, and the Walker circulation is weakened.

Rasmussen & Carpenter (1982) analyze a number of Southern Oscillation cycles and by compositing techniques display the spatial structure and relative timing of the associated tropical phenomena, in particular the equatorial Pacific Ocean surface temperature anomalies. In recent years an increasing number of global associations have been found, for example by Horel & Wallace (1981), van Loon & Madden (1981), and van Loon & Rogers (1981). Perhaps most remote and surprising is the observation by van Loon, Zerefos & Repapis (1982) of an association between high SOI and low north polar winter stratospheric temperature. At the 100 mb level, the mean temperature difference is about 5 K between SOI extremes. The seasonal predictive value of the SOI for northern hemisphere midlatitudes is evaluated by Chen (1982).

These observations have, of course, provided a challenge to atmospheric modelers. Is it possible to introduce anomalies of the sort observed in the equatorial Pacific Ocean surface temperature field into an atmospheric GCM and reproduce the observed atmospheric responses in pressure, winds and precipitation? The answer is 'yes' with increasing realism in the most recent models. Early studies by Rowntree (1972) and by Julian & Chervin (1978) had shown significant effects. More recently Keshavamurty (1982) reports on the clear and realistic response of the GFDL R15 model to artificial Pacific ocean surface temperature anomalies positioned at three locations along the equator. Similar experiments are underway using the NCAR R15 model (M. L. Blackmon, private communication).

A theoretical analysis of the expected response of the equatorial atmosphere to heating has been carried out by Gill (1980). He showed, with a linearized shallow water model on an equatorial beta plane, that a localized heat source on the equator would tend, through Kelvin wave propagation, to cause a response toward the east three times further than that produced by Rossby wave propagation toward the west. He thus explains qualitatively the nature of the response to the dominant Indonesian heat source to the west in the Indian Ocean and to the east across the Pacific Ocean.

An understanding of the rest of the oscillating system, namely, the response of the ocean to atmospheric changes, is still rather crude. The equatorial surface stress variations with SOI lead to clearly observable east–west changes in sea level and thermocline depth. Simple tropical coupled atmosphere–ocean models have been devised by McWilliams & Gent (1978) and K.-M. Lau (1981), which show roughly how an interactive cycle might occur. Much more detailed and quantitative models of the ocean component are needed. The difficulty of producing these was discussed in an earlier section.

That the Southern Oscillation can have an influence extending from the equatorial Pacific into midlatitudes is theoretically consistent with recent model wave propagation studies by Hoskins & Karoly (1981) and Webster & Holton (1982). These show that tropical westerlies in the Pacific region can open a window for equatorial and cross-equatorial wave propagation into extratropical latitudes.

Most of this chapter has dealt with the problems of developing confidence in model predictions of the climate response to externally imposed influences. In this final section the simulation and possible prediction of an internal slow mode of climate variability has been discussed. These two kinds of prediction are the principal objectives of global climate research, and rapid progress is being made in each.

References

Barnett, T. P. & Preisendorfer, R. W. (1978). 'Multifield analog prediction of short-term climate fluctuations using a climate state vector'. *Journal of the Atmospheric Sciences*, **35**, 1771–87.

Bell, T. L. (1980). 'Climate sensitivity from fluctuation dissipation: some simple model tests'. *Journal of the Atmospheric Sciences*, **37**, 1700–7.

Bjerknes, J. (1969). 'Atmospheric teleconnections from the equatorial Pacific'. *Monthly Weather Review*, **97**, 163–72.

Blackmon, M. L. & White, G. H. (1982). 'Zonal wavenumber characteristics of northern hemisphere transient eddies'. *Journal of the Atmospheric Sciences*, **39**, 1985–98.

Bourke, W. (1974). 'A multi-level spectral model. I. Formulation & hemispheric integrations'. *Monthly Weather Review*, **102**, 687–701.

Bretherton, F. P. (1982). 'Ocean climate modeling'. *Progress in Oceanography*, **11**, 93–129.

Budyko, M. I. (1969). 'The effect of solar radiation variations on the climate of the Earth'. *Tellus*, **21**, 611–19.

Charney, J. G. & DeVore, J. G. (1979). 'Multiple flow equilibria in the atmosphere and blocking'. *Journal of the Atmospheric Sciences*, **36**, 1205–16.

Charney, J. G. & Straus, D. M. (1980). 'Form-drag instability and multiple equilibria in baroclinic, orographically forced planetary wave systems'. *Journal of the Atmospheric Sciences*, **37**, 1157–76.

Chen, W. Y. (1982). 'Fluctuations in northern hemisphere 700 mb height field associated with the Southern Oscillation'. *Monthly Weather Review*, **110**, 808–23.

Cramer, H. (1946) *Mathematical Methods of Statistics*. Princeton University Press, p. 213.

Daley, R. (1981). 'Normal mode initialization'. *Reviews of Geophysics and Space Physics*, **19**, 450–68.

Dickinson, R. E. (1981). 'Convergence rate and stability of ocean–atmosphere coupling schemes with a zero-dimensional climate model. *Journal of the Atmospheric Sciences*, **38**, 2112–20.

Edmon, H. J., Hoskins, B. J. & McIntyre, M. E. (1980). 'Eliassen–Palm cross sections for the troposphere'. *Journal of the Atmospheric Sciences*, **37**, 2600–16.

Errico, R. M. (1982). 'The strong effects of non-geostrophic dynamic processes on atmospheric energy spectra'. *Journal of the Atmospheric Sciences*, **39**, 961–8.

Frederiksen, J. S. (1982). 'A unified three-dimensional instability theory of the onset of blocking and cyclogenesis'. *Journal of the Atmospheric Sciences*, **39**, 969–82.

Frederiksen, J. S. & Sawford, B. L. (1980). 'Statistical dynamics of two-dimensional inviscid flow on a sphere'. *Journal of the Atmospheric Sciences*, **37**, 717–32.

Gandin, L. S. (1963). *Objective Analysis of Meteorological Fields*. Gidrometeorologicheskoe Izdatel'stvo, Leningrad. Translated by Israel Program for Scientific Translations, Jerusalem, 1965, 242pp.

Gill, A. E. (1980). 'Some simple solutions for heat-induced tropical circulation'. *Quarterly Journal of the Royal Meteorological Society*, **106**, 447–62.

Green, J. S. A. (1970). 'Transfer properties of the large-scale eddies and the general circulation of the atmosphere'. *Quarterly Journal of the Royal Meteorological Society*, **96**, 157–85.

Grose, W. L. & Hoskins, B. J. (1979). 'On the influence of orography on large-scale atmospheric flow'. *Journal of the Atmospheric Sciences*, **36**, 223–34.

Hasselmann, K. (1976). 'Stochastic climate models. Part I, theory'. *Tellus*, **28**, 473–85.

Holopainen, E. O., Rontu, L. & Lau, N.-C. (1982). 'The effect of large-scale transient eddies on the time-mean flow in the atmosphere'. *Journal of the Atmospheric Sciences*, **39**, 1972–84.

Horel, J. D. & Wallace, J. M. (1981). 'Planetary-scale atmospheric phenomena associated with the Southern Oscillation'. *Monthly Weather Review*, **109**, 813–29.

Hoskins, B. J. & Karoly, D. J. (1981). 'The steady linear response of a spherical atmosphere to thermal and orographic forcing'. *Journal of the Atmospheric Sciences*, **38**, 1179–96.

Hoyer, J.-M. & Sadourny, R. (1982). 'Closure modeling of fully developed baroclinic instability'. *Journal of the Atmospheric Sciences*, **39**, 707–21.

Jones, R. H. (1975). Estimating the variance of time averages'. *Journal of Applied Meteorology*, **14**, 159–63.

Julian, P. R. & Chervin, R. M. (1978). 'A study of the Southern Oscillation and Walker circulation'. *Monthly Weather Review*, **106**, 1433–51.

Karoly, D. J. (1982). 'Eliassen–Palm cross sections for the Northern and Southern hemispheres'. *Journal of the Atmospheric Sciences*, **39**, 178–82.

Keshavamurty, R. N. (1982). 'Response of the atmosphere to sea surface temperature anomalies over the equatorial Pacific and the teleconnections of the Southern Oscillation'. *Journal of the Atmospheric Sciences*, **39**, 1241–59.

Kutzbach, J. (1967). 'Empirical eigenvectors of sea level pressure, surface temperature, and precipitation complexes over North America'. *Journal of Applied Meteorology*, **6**, 791–802.

Larichev, V. & Reznik, G. (1976). 'Two-dimensional soliton: an exact solution. *Polymode News*,' No. 19 (simultaneous Russian publication in *Reports of USSR Academy of Science*, **231**, No. 5, 1976).

Lau, K.-M. (1981). 'Oscillations in a simple equatorial climate system'. *Journal of the Atmospheric Sciences*, **38**, 248–61.

Lau, N.-C. (1981). 'A diagnostic study of recurrent meteorological anomalies appearing in a 15-year simulation with the GFDL general circulation model'. *Monthly Weather Review*, **109**, 2287–311.

Lau, N.-C. & Oort, A. H. (1981). 'A comparative study of observed northern hemisphere circulation statistics based on GFDL and NMC analyses. Part I: The time mean fields'. *Monthly Weather Review*, **109**, 1380–403.

Leith, C. E. (1973). 'The standard error of time-average estimates of climatic means'. *Journal of Applied Meteorology*, **12**, 1066–9.

Leith, C. E. (1975). 'Climate response and fluctuation dissipation'. *Journal of the Atmospheric Sciences*, **32**, 2022–6.

Leith, C. E. (1978). 'Predictability of climate'. *Nature*, **276**, 352–5.

van Loon, H. & Madden, R. A. (1981). 'The Southern Oscillation. Part I: Global associations with pressure and temperature in northern winter'. *Monthly Weather Review*, **109**, 1150–62.

van Loon, H. & Rogers, J. C. (1981). 'The Southern Oscillation. Part II: Associations with changes in the middle troposphere in northern winter'. *Monthly Weather Review*, **109**, 1163–8.

van Loon, H., Zerefos, C. S. & Repapis, C. C. (1982). 'The Southern Oscillation in the stratosphere'. *Monthly Weather Review*, **110**, 225–9.

Lorenz, E. N. (1963). 'Deterministic nonperiodic flow'. *Journal of the Atmospheric Sciences*, **26**, 131–41.

Lorenz, E. N. (1970). 'Climate change as a mathematical problem'. *Journal of Applied Meteorology*, **9**, 325–9.

Madden, R. A. (1976). 'Estimates of the natural variability of time-averaged sea-level pressure'. *Monthly Weather Review*, **104**, 942–52.

Madden, R. A. (1977). 'Estimates of the autocorrelations and spectra of seasonal mean temperatures over North America'. *Monthly Weather Review*, **105**, 9–18.

Malguzzi, P. & Speranza, A. (1981). 'Local multiple equilibria and regional atmospheric blocking'. *Journal of the Atmospheric Sciences*, **38**, 1939–48.

Manabe, S. & Hahn, D. G. (1981). 'Simulation of atmospheric variability'. *Monthly Weather Review*, **109**, 2260–86.

Manabe, S., Hahn, D. G. & Holloway, J. L., Jr (1979). 'Climate simulations with GFDL spectral models of the atmosphere: Effect of spectral truncation'. *Report of the JOC Study Conference on Climate Models: Performance, Intercomparison and Sensitivity Studies*, Washington DC, 3–7 April 1978. *GARP Publication Series No. 22, Vol. 1*, 41–94 (NTIS N8027917).

Manabe, S., Smagorinsky, J., Holloway, J. L., Jr & Stone, H. M. (1970). 'Simulated climatology of a general circulation model with a hydrologic cycle. III: Effects of increased horizontal computational resolution'. *Monthly Weather Review*, **98**, 175–213.

McAvaney, B. J., Bourke, W. & Puri, K. (1978). 'A global spectral model for simulation of the general circulation'. *Journal of the Atmospheric Sciences*, **35**, 1557–83.

McWilliams, J. C. (1980). 'An application of equivalent modons to atmospheric blocking'. *Dynamics of Atmospheres and Oceans*, **5**, 43–66.

McWilliams, J. C., Flierl, G. R., Larichev, V. D. & Reznik, G. M. (1981). 'Numerical studies of barotropic modons. *Dynamics of Atmospheres and Oceans*, **5**, 219–38.

McWilliams, J. C. & Gent, P. R. (1978). 'A coupled air–sea model for the tropical Pacific'. *Journal of the Atmospheric Sciences*, **35**, 962–89.

North, G. R., Bell, T. L., Cahalan, R. F. & Moeng, F. J. (1982). 'Sampling errors in the estimation of empirical orthogonal functions'. *Monthly Weather Review*, **110**, 699–706.

North, G. R., Cahalan, R. F. & Coakley, J. A., Jr (1981). 'Energy balance climate models'. *Reviews of Geophysics and Space Physics*, **19**, 91–121.

Oort, A. H. & Vonder Haar, T. H. (1976). 'On the observed annual cycle in the ocean–atmosphere heat balance over the northern hemisphere'. *Journal of Physical Oceanography*, **6**, 781–800.

Palmen, E. H. & Newton, C. W. (1969). *Atmospheric Circulation Systems*. Academic Press, New York, xvii, 603 pp.

Pitcher, E. J., Malone, R. C., Ramanathan, V., Blackmon, M. L., Puri, K. & Bourke, W. (1983). 'January and July simulations with a spectral general circulation model'. *Journal of the Atmospheric Science*, in press.

Puri, K. & Bourke, W. (1982). 'A scheme to retain the Hadley circulation during nonlinear normal mode initialization'. *Monthly Weather Review*, **110**, 327–35.

Ramanathan, V., Pitcher, E. J., Malone, R. C. & Blackmon, M. L. (1983). 'The response of a spectral general circulation model to improvements in radiative processes'. *Journal of the Atmospheric Sciences*, in press.

Rasmussen, E. M. & Carpenter, T. H. (1982). 'Variations in tropical sea surface temperature and surface wind fields associated with the Southern Oscillation/El Nino'. *Monthly Weather Review*, **110**, 354–84.

Ratcliffe, R. A. S. & Murray, R. (1970). 'New lag associations between North Atlantic sea temperature and European pressure applied to long-range weather forecasting'. *Quarterly Journal of the Royal Meteorological Society*, **96**, 226–46.

Rinne, J. & Jarvenoja, S. (1982). 'Empirical orthogonal functions of the 500 mb height weighted with respect to the analysis error'. *Monthly Weather Review*, **110**, 907–15.

Rowntree, P. R. (1972). 'The influence of tropical east Pacific Ocean temperature on the atmosphere'. *Quarterly Journal of the Royal Meteorological Society*, **98**, 290–321.

Rowntree, P. R. (1976). 'Response of the atmosphere to a tropical Atlantic Ocean temperature anomaly'. *Quarterly Journal of the Royal Meteorological Society*, **102**, 607–25.

Sadourny, R. & Basdevant, C. (1981). 'Une classe d'opérateurs adaptés à la modélisation de la diffusion turbulente en dimension deux'. *C. R. Acad. Sc. Paris*, **292**, II, 1061–4.

Sadourny, R. & Hoyer, J. M. (1982). 'Inhibition of baroclinic instability in low resolution models'. *Journal of the Atmospheric Sciences*, **39**, 2138–43.

Sellers, W. D. (1969). 'A global climatic model based on the energy balance of the earth–atmosphere system'. *Journal of Applied Meterology*, **8**, 392–400.

Stern, M. E. (1975). 'Minimal properties of planetary eddies'. *Journal of Marine Research*, **33**, 1–13.

Wallace, J. M. & Gutzler, D. S. (1981). 'Teleconnections in the geopotential height field during northern hemisphere winter'. *Monthly Weather Review*, **109**, 784–812.

Webster, P. J. & Holton, J. R. (1982). 'Cross-equatorial response to middle-latitude forcing in a zonally varying basic state'. *Journal of the Atmospheric Sciences*, **39**, 722–33.

Wellk, R. E., Kasahara, A., Washington, W. M. & De Santo, G. (1971) 'Effect of horizontal resolution in a finite-difference model of the general circulation'. *Monthly Weather Review*, **99**, 673–83.

Williamson, D. L., Daley, R. & Schlatter, T. W. (1981). 'The balance between mass and wind fields resulting from multivariate optimal interpolation'. *Monthly Weather Review*, **109**, 2357–76.

Yeh, T.-C., Dao, S.-Y. & Li, M.-T. (1959). 'The abrupt change of circulation over the northern hemisphere during June and October'. *In: The Atmosphere and Sea in Motion.* Rockefeller Inst. Press, New York, pp. 249–67.

J. K. Angell
Air Resources Laboratories NOAA Rockville, MD 20852
G. V. Gruza
All Union Research Institute of Hydrometeorological Information, Obninsk, Kaluga 249020

Climate variability as estimated from atmospheric observations

Abstract

Examined is the variation in mean-annual surface temperature for the last 200 years at two European stations, as well as the variation in mean-annual northern hemisphere surface temperature (NHST) for the last 100 years, obtained by two separate analysis teams. The variation of this temperature with latitude and season is also shown. NHST is indicated to have warmed by 0.4–0.5 K during the past 100 years, with the 1981 value the warmest so far observed. However, NHST cooled by more than 0.5 K between early 1981 and early 1982, so that the warmth of 1981 seems not to be persisting. There has been little evidence of tropospheric warming in north polar latitudes during the last 24 years, but there is evidence of warming in south polar latitudes, as well as in the tropics. In temperate latitudes the cooling during the 1960s has been compensated by a warming during the 1970s. In the northern hemisphere, surface warming began about 1965, but not until about 1975 in the 85–30 kPa layer, and in the 30–10 kPa layer there has been gradual cooling during the last 24 years, resulting in an increase in lapse rate. In the southern hemisphere the low-level lapse rate appears to have decreased, perhaps due to ocean thermal inertia. Rocketsonde data in the western quadrant of the northern hemisphere suggest a 4 K cooling in the middle and high stratosphere between 1970 and 1976, but little temperature change since. Precipitation over the northern hemisphere has been estimated for the last 100 years, and examined by latitude and geographic region. Discussed is the close relation between sea-surface temperature (SST) variations in the equatorial eastern Pacific, and zonally averaged tropical air temperature, as well as Indian summer-monsoon rainfall (ISMR) and tropospheric temperatures in north temperate latitudes.

3.1
Introduction

The traditional concept of weather involves the instantaneous state of the atmosphere, and the evolution of the state as a result of formation, amplification, and decay of individual atmospheric disturbance. The traditional concept of climate involves the generalization of weather over a time period long enough to establish its properties (mean values, variances, space and time correlation functions, probabilities of extreme events, etc.) as those of a statistical ensemble. The climatic system embraces the atmosphere, the hydrosphere (considering the water distributed on and beneath the earth's surface), the cryosphere (comprising the ice and snow on and over the surface), and may be considered as also embracing the lithosphere (comprising the rock, soil and sediment of the earth's crust) and the biosphere (flora, fauna and man himself). Each of these climatic subsystems has different physical characteristics, and is linked to the others by a variety of physical processes.

Since the characteristics of the state of the atmosphere, ocean and land are essentially non-uniform in space, to fully describe them at any given moment, functions of space coordinates (fields) must be used. Climate is a statistical ensemble of such states or 'fields' over several decades. To describe the statistical ensemble it would, in general, be sufficient to specify how often each of the system states occurs. A complete description of global climate, however, is virtually impossible. As a first approximation it is usually sufficient to study only first and second moments, i.e., climatic (long-term) mean values, variances, and the correlations among different physical variables.

It is not completely obvious how one should select material for an empirical assessment of climate; i.e., what is to be used as a climatic sampling for estimating the conceptual ensemble of states of the climatic system. This is not a problem when dealing with computer models where, assuming the external conditions are

constant, one may repeat the experiment any number of times, thus obtaining as big a sampling from the ensemble of states as is necessary for statistical analysis. This cannot be done for real climate since the history of the earth's climate is a unique process and, considering the whole of the earth's history, we deal with only one realization. Consideration of an ensemble of similar realizations is only possible, theoretically or conceptually, as regards certain planets whose climatic evolution was characterized by the same values of external variables.

The topic of climate variability has received increasing attention in recent years largely due to concern regarding the possible greenhouse-warming of the atmosphere by the carbon dioxide released into the atmosphere by the burning of fossil fuels. In view of this concern, as well as to provide reasonable bounds for this study of climate variability as estimated from atmospheric observations, the emphasis in this article is on atmospheric temperature variations and, in particular, recent work which presents estimates of these variations in troposphere and stratosphere, as well as at the earth's surface. Consideration is also given to estimates of long-term variations in precipitation. Accordingly, to be examined in this study of climate variability are variations in:

(1) mean-annual surface temperature at two stations in central Europe during the last 200 years;
(2) mean-annual NHST during the last 100 years obtained by two different analysis teams, as well as the temperature variations at different latitudes (in January and July and the mean for the year) and in different geographical regions;
(3) mean-seasonal NHST during the last 24 years, obtained by two different analysis teams;
(4) mean-seasonal surface and 85–30 kPa temperatures in the five climatic zones during the last 24 years;
(5) mean-seasonal surface, 85–30 kPa, 30–10 kPa, and 10–3 kPa temperatures in the northern and southern hemispheres during the last 24 years;
(6) mean-seasonal temperature in the 26–55 km layer of the northern hemisphere during the last 17 years.
(7) SSTs in the equatorial eastern Pacific, and their relation to tropical air temperatures during the last 24 years and to Indian summer monsoon rainfall (ISMR) during the last 110 years;
(8) northern hemisphere precipitation in January and July during the last 100 years, for latitude zones and different geographical regions.

Mention will also be made of the relation between volcanic eruptions and atmospheric temperature, as well as of significant relations (during the last 24 years) between zonally averaged tropospheric temperature in north temperate latitudes, and the Southern Oscillation and quasi-biennial oscillation.

3.2
Variation in European surface temperature

Some temperature records in Europe extend back 200 years, or to about 1780. Fig. 3.1 shows the variation in mean-annual surface temperature at Vienna, Austria, and Hohenpeissenberg in southern Germany, about 600 km to the west of Vienna, as obtained from *World Weather Records*, a past publication of the Weather Bureau, US Department of Commerce, Washington, DC. In this and subsequent diagrams the temperatures are expressed as deviations from the mean temperature in degrees Kelvin (K). The correlation between these station temperature records, one urban and one non-urban, is 0.86, emphasizing the general representativeness of the temperatures, at least for purposes of regional analysis. During the last 20 years, however, Vienna has been relatively warm compared to Hohenpeissenberg, suggesting the possibility of an urban heat-island effect. Table 3.1 lists the standard deviation of temperature for these two stations, as well as the average year-to-year temperature change over the period of record. These values all equal about 0.8 K.

Even though the interannual variability is large compared to

Fig. 3.1. Variation in mean-annual temperature at Vienna and Hohenpeissenberg. The volcanic eruption of Tambora and Krakatoa are indicated.

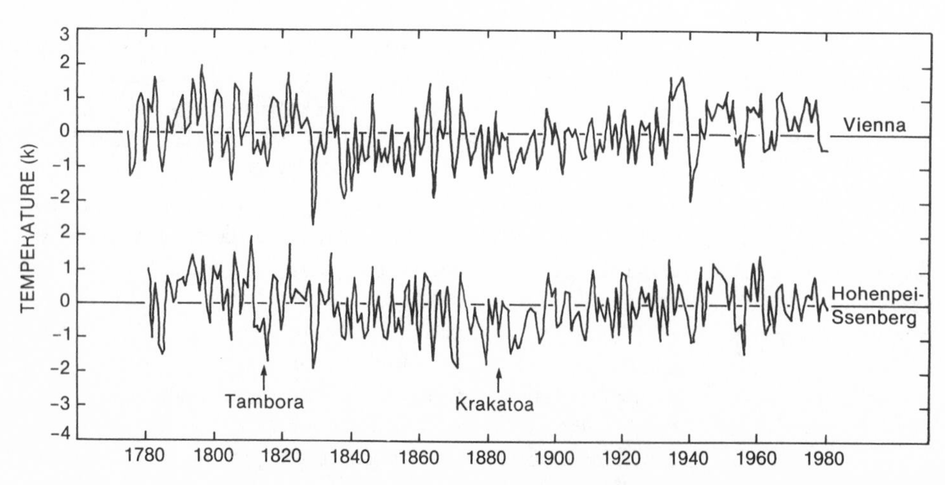

Table 3.1. *Standard deviation of mean-annual surface temperature, and average year-to-year temperature change (K)*

	Years	Standard deviation	Year-to-year change
Vienna	1775–1960	0.81	0.84
Hohenpeissenberg	1781–1960	0.77	0.81
Northern hemisphere			
Vinnikov, *et al.*	1881–1978	0.23	0.13
Jones, *et al.*	1881–1981	0.30	0.16

any long-term trends, a long-term variation in temperature is evident both at Vienna and Hohenpeissenberg, with the temperatures relatively warm around 1800, relatively cool around 1880, and relatively warm again around 1940. Evaluation of the linear regression line for the 200 years of data yields negligible temperature decreases of 0.09 K at Vienna and 0.13 K at Hohenpeissenberg. There are temperature minima following the volcanic eruptions of Tambora in 1815 and Krakatoa in 1883 (both located in what is now Indonesia), but these temperature minima are not unique (see also Taylor *et al.*, 1980). In the case of Tambora there was considerable cooling prior to the eruption.

3.3 Variation in northern hemisphere surface temperature

The subject of climate variability on a hemispheric scale has received ever-increasing attention in recent years. Willett (1950) was one of the first to provide estimates of long-term temperature changes in latitudinal zones, as well as for the whole hemisphere, and this type of work was continued by Mitchell (1961, 1963), Spirina (1969), Yamamoto & Iwashima (1975) and Borzenkova *et al.* (1976), among others. Callender (1961) pointed out possible anthropogenic influences on temperature, and van Loon & Williams (1976*a*, 1976*b*) have attempted to couple temperature changes with heat transport and pressure changes.

Fig. 3.2 shows the variation in mean-annual NHST from 1881 to 1978 (Vinnikov *et al.*, 1980), and from 1881 to 1981 (Jones, Wigley & Kelly, 1982), based on monthly-mean station data interpolated to a 5° latitude by 10° longitude grid. The analysis of Vinnikov *et al.* extends only to 17.5°N, so that it does not represent the entire northern hemisphere. Even so, the correlation between the two temperature records is 0.97, showing that different analysis teams can come to very similar conclusions regarding northern hemisphere temperature variations during the past 100 years. This is perhaps not surprising when it is realized that all analysis teams depend on approximately the same data base.

Over the past 100 years the average year-to-year variation in NHST has been 0.13 K accordingly to Vinnikov *et al.*, and 0.16 K according to Jones, Wigley & Kelly (Table 3.1). Table 3.1 shows that these values are only about half the magnitude of the standard deviations, rather than of similar magnitude as in the case of the Vienna and Hohenpeissenberg data. Thus, as would be anticipated, there is relatively more high-frequency (year-to-year) variation in station data than in hemispheric data. Note that the cooling following the Agung volcanic eruption in 1963 is indicated to be at least as large as the cooling following Krakatoa, a result hardly to be expected in view of the relative magnitudes of the two eruptions.

The temperature increase during the past 100 years (linear regression) is 0.42 K in the case of the Vinnikov data, and 0.48 K in the case of the Jones, Wigley & Kelly (JWK) data. It is this evidence for a long-term temperature increase that has led some (e.g., Hansen *et al.*, 1981) to suggest that a CO_2 greenhouse-warming is already being observed, though there is difficulty in explaining the hemispheric cooling during the interval 1940–60 when there were few volcanic eruptions. It has been shown by Angell & Korshover (1982) that, on the basis of the *Northern Hemisphere Historical Weather Map Series*, extending from 1899 to 1978, hemispheric warming has been associated with northward displacements of the four centers of action (Icelandic Low, Aleutian Low, Azores High and Pacific High), as well as significantly (at the 95% level) lower central pressures of Icelandic and Aleutian Lows.

Fig. 3.3 compares ten-year average values of NHST as obtained by JWK (solid line) with ten-year average values of the mean of Vienna and Hohenpeissenberg temperatures (dashed line). Between 1880 and 1980 there is fairly good agreement between the two traces, and if this were true also during the 1800s there would be little if any evidence for hemispheric warming during the last 200 years. However, the rather extensive analysis of Groveman & Landsberg (1979) suggests that these two stations were not representative of the hemisphere during the 1800s. Further examination of long-term temperature records from a CO_2 warming point of view may be desirable, with special attention to possible urban heat-island effects on the temperature.

Fig. 3.2. Variation in mean-annual NHST according to Vinnikov *et al.* (analysis only to 17.5°N), and Jones, Wigley & Kelly. The volcanic eruptions of Krakatoa and Agung are indicated.

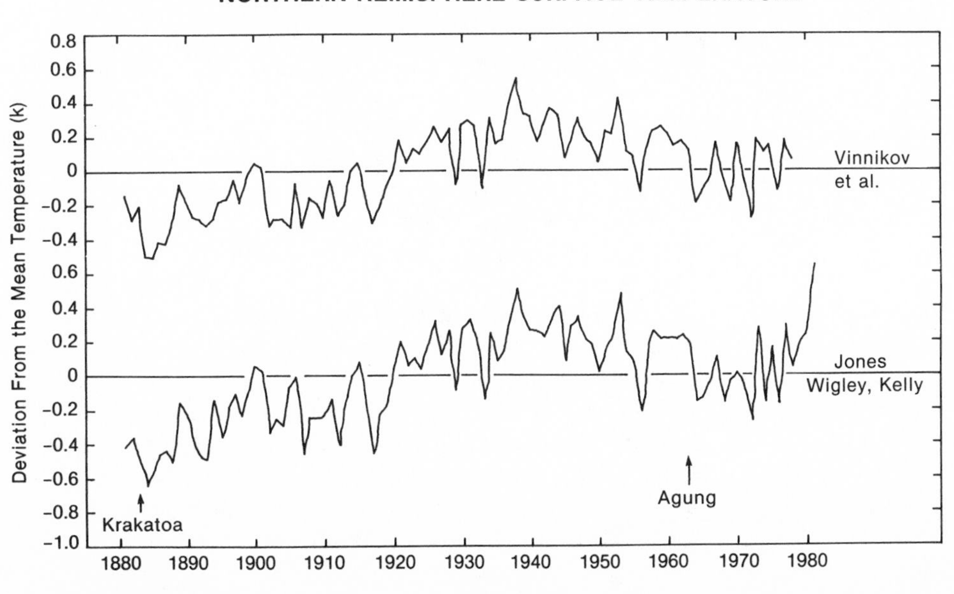

A noteworthy feature of the JWK analysis in Fig. 3.2 is the very warm surface temperature for 1981, more than 0.6 K above the 100-year average, and 0.12 K warmer than the warmest year in the 1930s. Fig. 3.4 shows the recent fluctuations in NHST in more detail, based on seasonal values of JWK and Angell & Korshover (AK). Here the annual temperature cycle has been nearly eliminated by evaluating the seasonal values as deviations from long-term seasonal means. Even though Angell & Korshover (1978) use a sparse data network (42 radiosonde stations in the northern hemisphere, 63 globally), the correlation between the 97 seasonal values is 0.86.

During the past 24 years the average season-to-season variation in NHST (expressed as a deviation from the mean) has been 0.23 K according to JWK, and 0.26 K according to AK. Over the entire interval 1881–1981, the average season-to-season variation in NHST was also 0.23 K according to JWK, so that the seasonal variability during the past 24 years has been typical. The temperature increases over the past 24 years (linear regression) are 0.11 K based on the JWK data and 0.20 K based on the AK data. This difference is mostly the result of the slightly greater cooling indicated to follow the 1963 Agung eruption in the AK analysis. Both analyses suggest a very warm NHST in early 1981 (more than 0.8 K above average), but the AK analysis in particular suggests a rapid cooling thereafter such that in the spring of 1982 the temperature is near average. Thus, the warmth in early 1981 may have been a temporary perturbation, and any cooling due to the El Chichon (Mexico) volcanic eruptions in the spring of 1982 will be superimposed on a cooling already in progress, presumably making detection of the volcanic effect more difficult.

Fig. 3.3. Variation in ten-year average NHST (solid line) and the mean of Vienna and Hohenpeissenberg temperatures (dashed line).

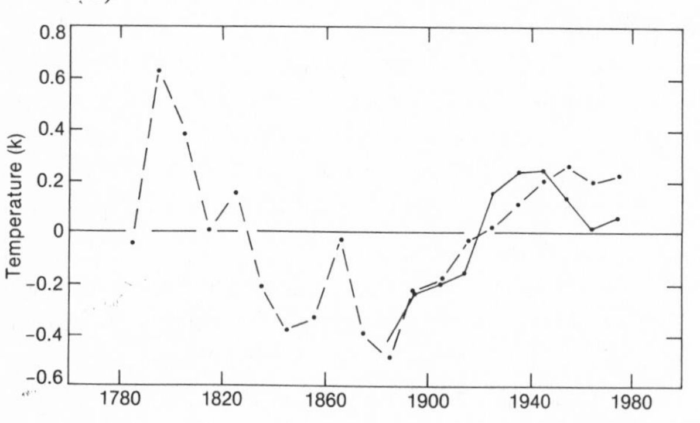

Fig. 3.4. Variation in mean-seasonal NHST according to Jones, Wigley & Kelly and Angell & Korshover. The seasonal values are the deviations from long-term seasonal means, thus almost eliminating the annual temperature cycle. The volcanic eruption of Agung is indicated.

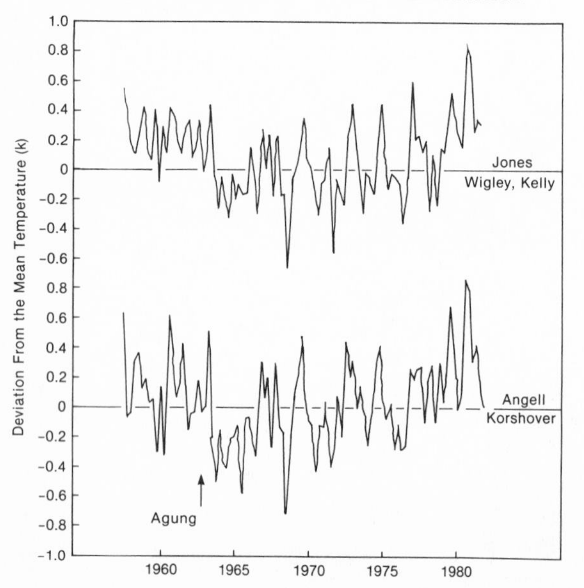

3.4 Temperature variation as a function of latitude and geographic region

The diagram at right in Fig. 3.5 gives the variation in year-average surface temperature at 80, 65, 55, 40, and 20°N based on the analysis of Gruza & Ran'kova (1980). The trend shown by five-year running means suggests that the long-term temperature variations have been greater in polar latitudes than in the tropics. Indeed, at 20°N there is indicated to have been hardly any long-term variation at all. Accordingly, there is evidence for a decrease in meridional temperature gradient between 1890 and 1950, and an increase in this gradient thereafter.

The diagrams at left in Fig. 3.5 show the trends in temperature at these latitudes in January and July. An interesting point is that the tendency for warming during the early 1900s and cooling between 1940 and 1965 shows up strongly in the January data but hardly at all in the July data. Thus, these trends appear to be dominated by winter conditions. At 20°N in July there is even evidence for an overall cooling.

Fig. 3.6 shows the variation in NHST in various geographic regions for January (left) and July, as well as for the year as a whole, again based on the analysis of Gruza & Ran'kova. In January the long-term warming and cooling is greater over the continents of Eurasia and North America than over the Atlantic and Pacific Oceans, but over Africa, unlike the other continents, there is evidence of a gradual long-term warming. In July, only the Atlantic Ocean region shows the tendency for long-term warming followed by cooling, with a tendency for gradual long-term warming particularly apparent over the Pacific Ocean. A more thorough discussion of temperature time-series characteristics as a function of latitude and geographic region is to be found in Gruza *et al.* (1982).

For the more-detailed analysis of recent years, column-mean temperatures have been obtained from the difference in height (thickness) between two constant-pressure surfaces. This has the advantage of giving the mean temperature through a layer. The pressure–height data have been obtained from *Monthly Climatic Data for the World*, a publication of the National Climatic Center, National Oceanic and Atmospheric Administration (NOAA), Asheville, North Carolina.

Fig. 3.7 shows the variation in temperature between 1958 and 1982, at the surface and in the tropospheric 85–30 kPa layer for the five climatic zones, based on the AK analysis. In order to clarify the trends or 1–2–1 weighting (divided by 4) has been applied twice to successive seasonal values, with a conservative 1–1 weighting

Fig. 3.5. Variation in northern hemisphere surface air temperature by latitude for the months of January and July, as well as for the year as a whole (right), based on the analysis of Gruza & Ran'kova (1980), updated to 1981. Five-year running means are also shown.

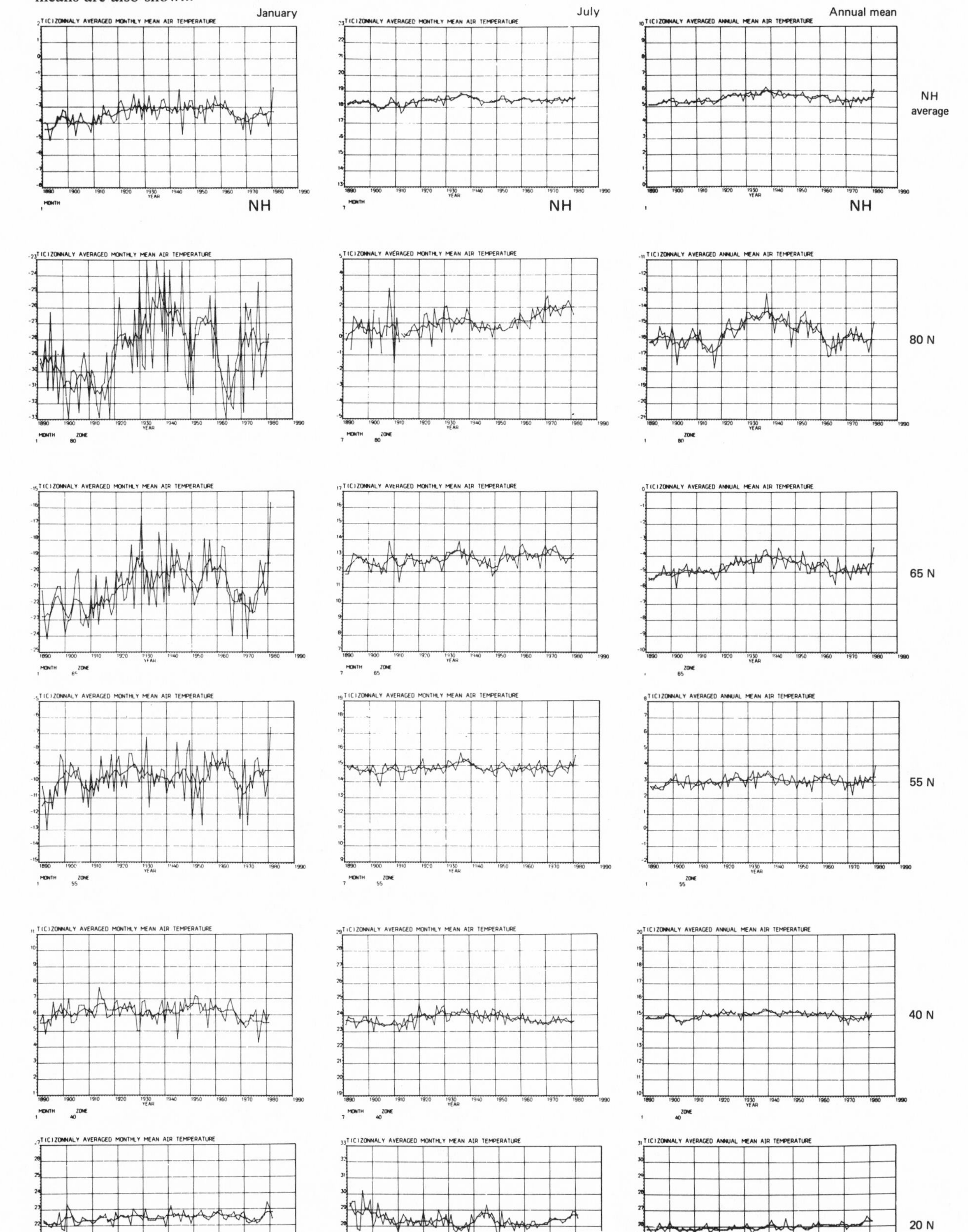

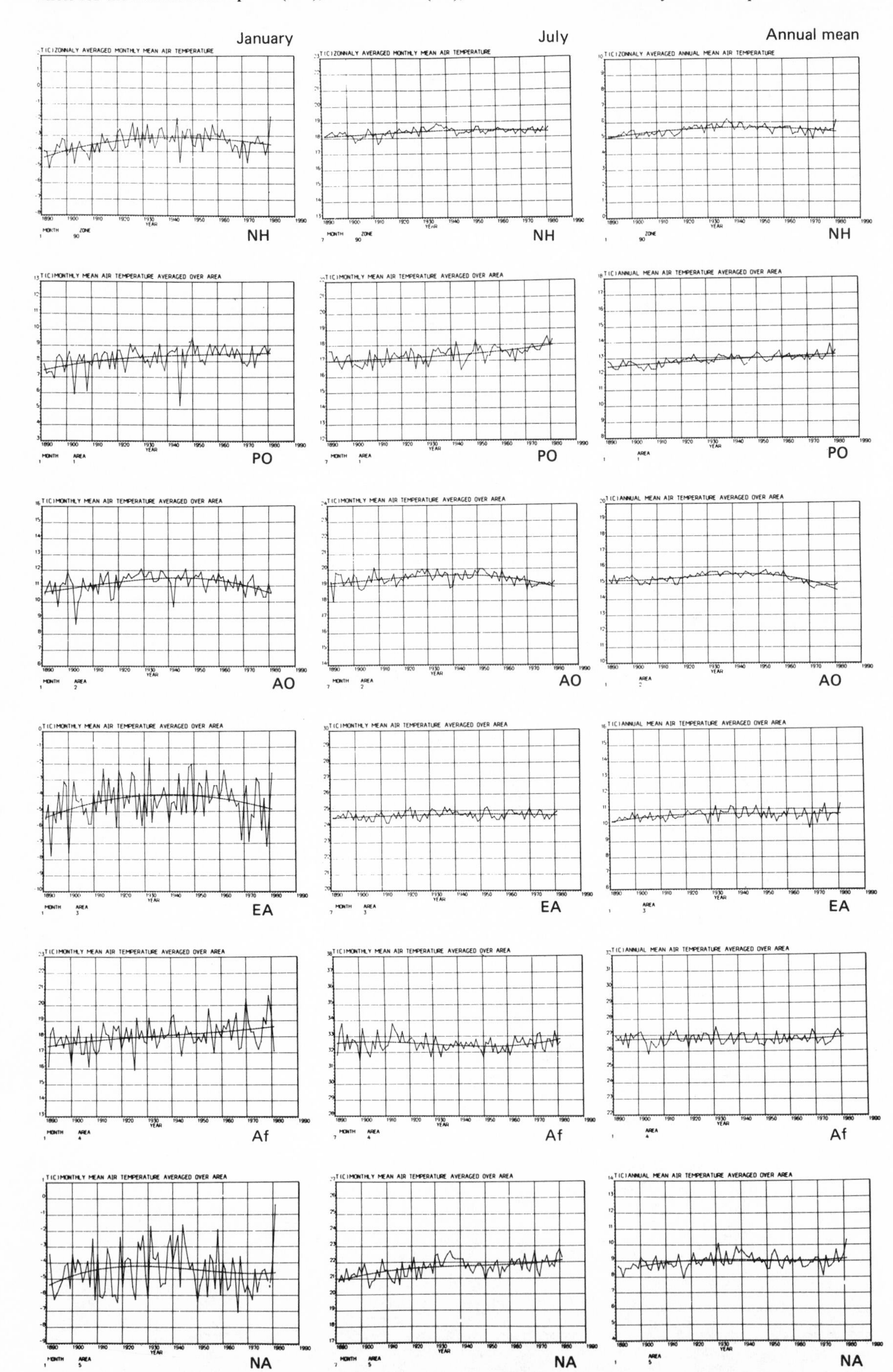

Fig. 3.6. Variation in northern hemisphere surface air temperature for January (left) and July as well as the year as a whole for the northern hemisphere (NH), Pacific Ocean (PO), Atlantic Ocean (AO), Eurasia (EA), Africa (Af) and North America (NA). A cubic polynomial trend (smooth line) has been estimated by the least square method.

(divided by 2) applied twice at the beginning and end of the record (northern spring of 1982).

There is evidence for somewhat different temperature trends in the various climatic zones. The tropics apparently warmed during most of the 24-year interval, whereas in north and south temperate latitudes an early cooling has been compensated by a recent warming. In north temperate latitudes the smoothed surface temperature was indicated to be 1.0 K above average in early 1981, nearly twice the previously observed maximum value. Since then this temperature has decreased rapidly, however, and in the spring of 1982 was near average. In south polar latitudes (Antarctica) there was warming up to about 1975, but little overall temperature change since, whereas in north polar latitudes there is only limited evidence of any long-term temperature change (see also the analysis at right in Fig. 3.5 and Kelly *et al.*, 1982).

Because of the different temperature trends in tropics and temperate latitudes, there was an increase in meridional temperature gradient between these two zones between about 1965 and 1975, but little change thereafter. There has been an increase in the meridional temperature gradient between tropics and north polar latitudes in the interval 1958–82, but a decrease between tropics and and south polar latitudes.

3.5
Temperature variation as a function of height

Fig. 3.8 shows the variation in temperature (smoothing as in Fig. 3.7) at the surface and in 85–30, 30–10 and 10–3 kPa layers

Fig. 3.7. Variation in surface temperature, and mean temperature in the 85–30 kPa layer, in the five climatic zones. A 1–2–1 weighting (divided by 4) has been applied twice to successive seasonal values, except at the beginning and end of the record (northern spring of 1982) where a 1–1 weighting (divided by 2) has been applied twice.

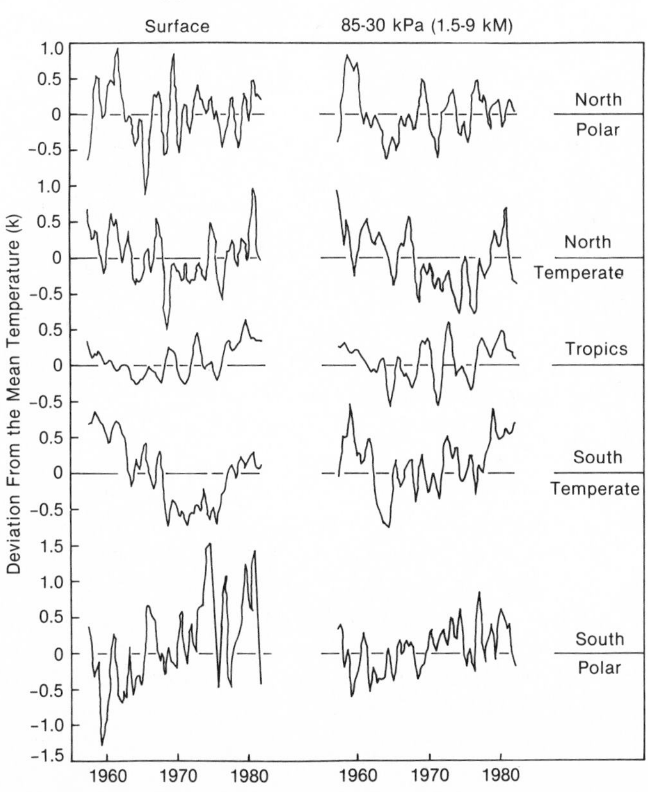

of the northern and southern hemispheres, based on the AK analysis. The hemispheric average is obtained from a 1–2–2–1 weighting of polar, temperate, subtropical, and equatorial zones respectively. Also shown is the temperature variation, between 1965 and 1982, in the 26–55 km layer of the northern hemisphere as obtained from rocketsonde data mostly in the western quadrant.

In the northern hemisphere a surface warming apparently began shortly after the eruption of Agung in1963 (A at bottom of Fig. 3.8), whereas cooling in the 85–30 kPa layer continued until about 1976. This implies an increase in low-level lapse rate between 1964 and 1976 of about 0.01 K per 100 meters, or $\frac{1}{100}$ of the dry adiabatic lapse rate. In the southern hemisphere the opposite variation is indicated, with the tropospheric temperature warming after Agung and the surface temperature warming only in the mid 1970s, with the implication of a decrease in lapse rate. It may be that, in this largely oceanic hemisphere, warming of the atmosphere near the surface is delayed due to ocean thermal inertia (Madden & Ramanathan, 1980).

In the case of the 30–10 kPa layer, the most interesting feature is the absence of the cooling and warming trends observed at the surface and in the 85–30 kPa layer. Rather, particularly in the northern hemisphere, there appears to have been a slight overall cooling during the period of record. This evidence that the cooling and warming at the surface and in the 85–30 kPa layer does not

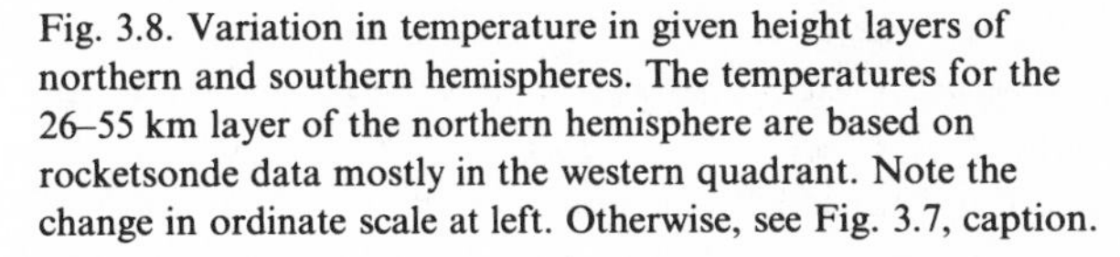
Fig. 3.8. Variation in temperature in given height layers of northern and southern hemispheres. The temperatures for the 26–55 km layer of the northern hemisphere are based on rocketsonde data mostly in the western quadrant. Note the change in ordinate scale at left. Otherwise, see Fig. 3.7, caption.

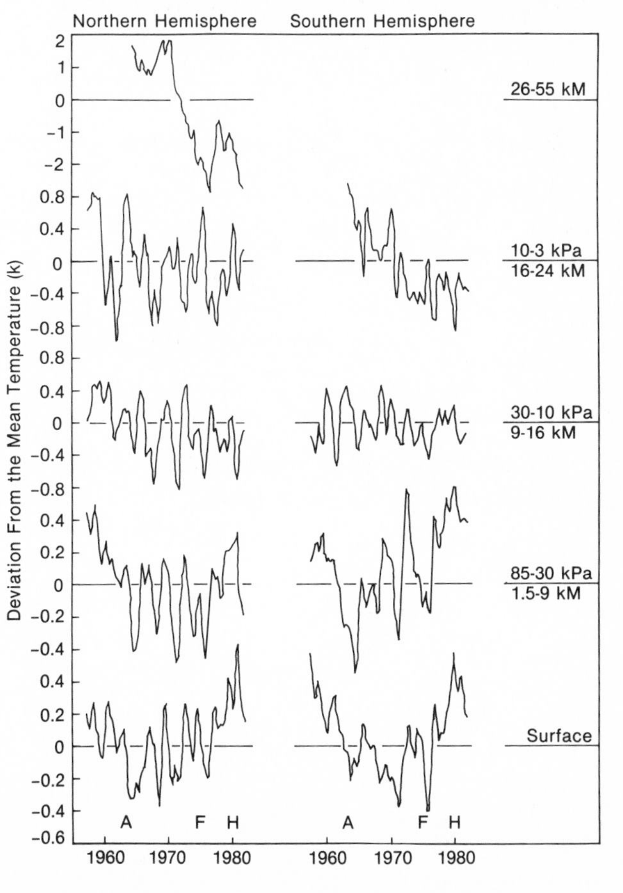

extend to the 30–10 kPa layer may be an important finding since it implies appreciable change in tropospheric lapse rate (an increase of about 0.01 K per 100 meters in the last ten years).

In the 10–3 kPa layer, on the contrary, there is evidence for cooling in the southern hemisphere but not in the northern hemisphere. In both hemispheres there has been an increase in lapse rate between the surface and the low stratosphere (10–3 kPa layer), however. Note the apparent warming of the 10–3 kPa layer of the northern hemisphere following the volcanic eruptions of Agung (A) in Indonesia in 1963, Fuego (F) in Guatemala in 1974, and St Helens (H) in Washington State in 1980.

As a summary of recent temperature values in the troposphere and low stratosphere, Table 3.2 presents the deviation of the mean-annual 1981 temperature from average, for the various climatic zones and various height layers, while Table 3.3 presents a comparison of the tropospheric temperature deviations from the mean for the northern winters of 1980–81 and 1981–82. Table 3.2 shows that at the surface and in the 85–30 kPa layer, all climatic zones were relatively warm in 1981 and, with the exception of south polar latitudes, the 30–10 kPa layer was relatively cool. Table 3.3 shows, on the other hand, that, with the exception of the polar zones, in the troposphere the northern winter (December, January, February) of 1981–82 was cooler than the northern winter of 1980–81. Thus, while tropospheric temperatures during 1981 were unusually warm, this warmth seems not to be persisting.

The trace at upper left in Fig. 3.8 shows the change in mean temperature of the 26–55 km layer (middle and upper stratosphere) of the northern hemisphere as estimated from about ten rocketsonde stations, mostly in the western quadrant. The most striking feature of this trace is the large (4 K) temperature decrease indicated between 1970 and 1976, with evidence for temperature maxima in 1970 and 1978, or near the times of sunspot maxima in 1969 and 1979. The reality of the 4 K decrease has been questioned because of the possibility of instrumental errors in the rocketsondes, and also on the basis of representativeness (rocketsonde data is only available for the western quadrant of the northern hemisphere). A study of the temperature variations in high-level radiosonde data in western and eastern quadrants suggests that limitation of the data to the western quadrant is not a significant factor (similar radiosonde-derived temperature trends in the 10–3 kPa layer of western and eastern quadrants) and, according to Quiroz (1979), there is little evidence for instrumental problems with the rocketsonde data. If the rocketsonde data do turn out to be valid, and representative of the northern hemisphere, then between 1970 and 1976 the average lapse rate between the surface and 50 km increased by $\frac{1}{100}$ of the dry adiabatic lapse rate, the same value obtained for recent tropospheric lapse rate change.

Table 3.2. *Deviation of mean-annual 1981 temperatures (K) from the 1958–77 mean*

	Surface	85–30 kPa	30–10 kPa	10–3 kPa
North polar	0.67	0.10	−1.21	−0.28
North temperate	0.78	0.45	−0.37	0.36
Tropics	0.36	0.21	−0.54	−0.24
South temperate	0.20	0.59	−0.38	−0.51
South polar	1.05	0.39	0.54	0.19
Northern hemisphere	0.61	0.24	−0.65	−0.14
Southern hemisphere	0.36	0.40	−0.25	−0.17
World	0.48	0.32	−0.45	−0.15

Table 3.3 *Comparison of tropospheric temperature deviations from the mean (K) for the northern winters of 1980–81 and 1981–82*

	Surface		85–30 kPa	
	1980–81	1981–82	1980–81	1981–82
North polar	1.74	0.34	−0.35	0.28
North temperate	1.18	0.11	1.05	−0.29
Tropics	0.29	0.22	0.43	0.17
South temperate	0.13	0.00	0.85	0.84
South polar	0.00	0.43	0.42	0.28
Northern hemisphere	0.87	0.20	0.48	−0.18
Southern hemisphere	0.15	0.19	0.59	0.63
World	0.51	0.19	0.54	0.23

On the basis of the large temperature decrease in the 26–55 km layer of the northern hemisphere between 1970 and 1976, it was hypothesized a few years ago that there might be a direct relation between sunspot number and stratospheric temperature and ozone (Callis & Nealy, 1978; Penner & Chang 1978). However, there appears to have been only a relatively slight stratospheric warming between 1976 and 1979 despite the great increase in sunspot number between these years. Thus, it is unlikely that the bulk of the cooling between 1970 and 1976 was due to a decrease in sunspot number and, at most, there has been a small solar-induced temperature perturbation superimposed on a stratospheric cooling trend of unknown cause. It is emphasized that this indicated stratospheric cooling is much too large to be associated, in toto, with any cooling related to the recent CO_2 increase (Manabe & Wetherald, 1975).

3.6 Southern Oscillation

The Southern Oscillation phenomenon was first recognized as a 3–7 year alternation in pressure between the eastern South Pacific and the Indonesian area (Troup, 1965; Trenberth, 1976). More recently it has been found that this pressure alternation is intimately related to SST changes in the equatorial eastern Pacific. Thus, based on seasonal data from 1932 through 1981, there is a highly significant (significant at the 99% level) correlation of −0·64 between SST in the region 0–10°S, 180–80°W, and a normalized pressure difference between Tahiti and Darwin (e.g., Angell, 1981).

The bottom trace of Fig. 3.9 shows the time variation of SST in this particular region of the equatorial eastern Pacific between 1958 and 1982. The range in temperature has approached 3 K, with the changes of largest amplitude in the central portion of the record. The middle trace of Fig. 3.9 shows the time variation in surface–10 kPa (tropospheric) temperature in the tropics, where the pressure-weighted average temperature for this layer has been obtained by letting the surface temperature (expressed as a deviation from the mean) represent the mean temperature in the surface–85 kPa layer (100–85 kPa layer), and then weighting the layer-mean temperature deviations by the pressure differences through the respective layers (15, 55 and 20 kPa).

The variation in zonally averaged surface–10 kPa temperature in the tropics (30°N–30°S) is remarkably similar to the SST

Fig. 3.9. Comparison of SST, in the equatorial eastern Pacific (0–10°S, 180–80°W), and mean tropical air temperatures in the surface–10 and 10–3 kPa layers. The volcanic eruptions of Agung, Fuego, and St Helens are indicated. Note the change in ordinate scale at left.

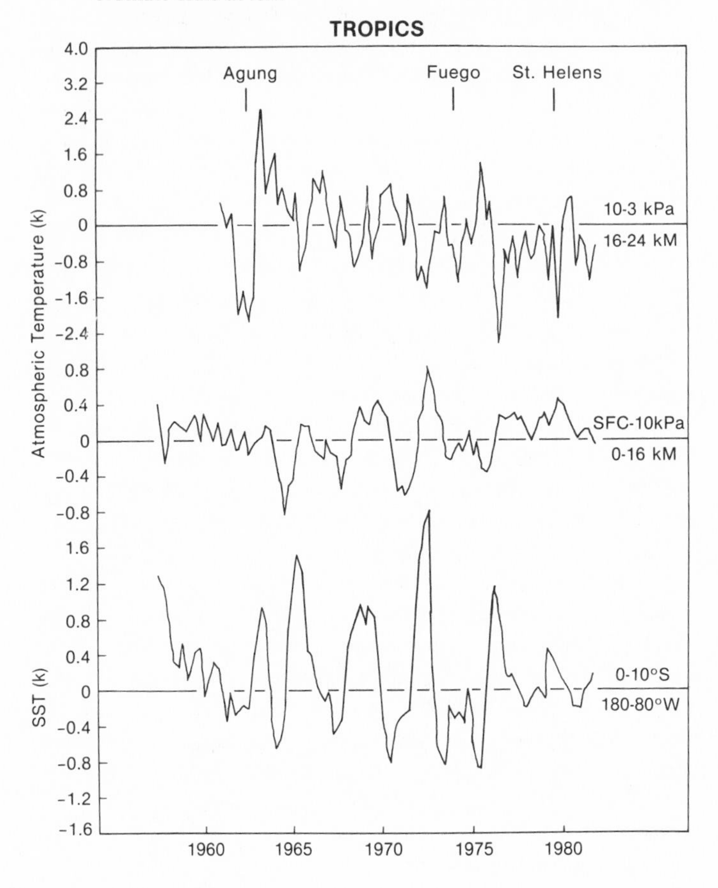

Fig. 3.10. Comparison of the variation in SST in the equatorial eastern Pacific, and ISMR. The volcanic eruptions of Krakatoa and Agung are indicated.

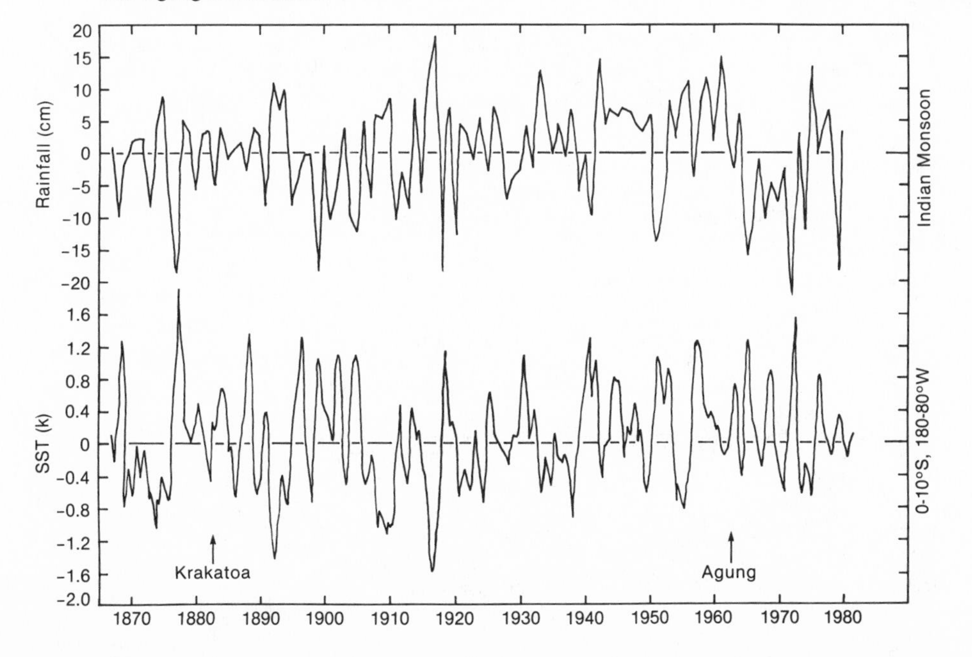

variation, although it is apparent that the atmospheric-temperature variation lags the SST variation. This is one of those very good relationships one does not often find in meteorology. The maximum (and highly significant) correlation of 0.69 between the two sets of data occurs at a lag of two seasons, based on seasonal data from 1958 through 1981. The maximum correlation between this SST and NHST surface–10 kPa temperature is 0.60, and between this SST and global surface–10 kPa temperature is 0.58, again at a lag of two seasons. Thus, when the Southern Oscillation is of large amplitude, as it generally was between 1958 and 1981, it completely dominates the surface–10 kPa temperature variations for hemisphere and world and, indeed, provides an estimate of hemispheric and global temperatures in this layer two seasons in advance.

The top trace of Fig. 3.9 shows the time variation in zonally averaged tropical temperature in the 10–3 kPa layer. In general, there has been cooling in this layer since the Agung eruption, though with evidence that the cooling has almost ceased during the last five years. An increase in lapse rate of about 0.02 K per 100 meters has been associated with the tropical warming in the tropospheric surface–10 kPa layer and cooling in the stratospheric 10–3 kPa layer.

In addition to the close relationship between SST in the equatorial eastern Pacific and zonally averaged tropospheric temperature in the tropics, there has also been a close relationship between this SST and ISMR as estimated by Parthasarathy & Mooley (1978). Fig. 3.10 shows an impressive tendency, particularly since 1950, for ISMR to be relatively small when SST is warm, with a highly significant correlation of −0.62 over the 110 years of record (Angell, 1981). Thus, nearly 40% of the variance in ISMR can be explained in terms of SST variations in the equatorial eastern Pacific, or vice versa. ISMR is most highly correlated with SST in the extended region 0.10°S, 180–80°W two seasons later (the El Nino season), but because of the evidence for progression of the SST warming westward from the South American coast (Rasmusson & Carpenter, 1982), the lag may be in the opposite sense for more localized SST areas, i.e., the SST immediately west of the South American coast may serve as a crude predictor for ISMR. Angell

& Korshover (1982) have shown that, during the interval 1899–1978, there was a significant (at the 95% level) tendency for a warm SST to be associated with southward displacement of the Icelandic Low, relatively high central pressure of the Icelandic Low, but relatively low central pressure of Azores and Pacific Highs and Aleution Low.

While not obvious from Fig. 3.10, the long-term trends in ISMR and SST in the equatorial eastern Pacific are nearly in phase, both being a maximum in about 1950 and a minimum in about 1910. Thus, there is a completely different relation between these two quantities in the long and short term. In the long term they are both also nearly in phase with NHST (Fig. 3.2). Accordingly, over the long term (last 100 years) a warm northern hemisphere has been associated with warm SSTs in the equatorial eastern Pacific and above-average ISMR, as well as northward displacement of the four centers of action (Icelandic Low, Aleutian Low, Azores High, Pacific High).

In view of the evidence for a relation between SST in the equatorial eastern Pacific and global surface–10 kPa temperature, as well as the evidence for a relation between phase of the quasi-biennial oscillation (QBO) and the extent of the 5 kPa north polar vortex (Holton & Tan, 1980), the possible joint influence of SST and QBO on the 85–30 kPa temperature in north temperate latitudes was examined (the 85–30 kPa temperature being more representative than the surface temperature). Based on seasonal values during the interval 1958–81, the combination of east-wind phase of the QBO (at 5 kPa) and cool SST three seasons earlier resulted in a highly significant temperature deviation from average in the north temperate 85–30 kPa layer of −0·4 K. The combination of west-wind phase of the QBO and warm SST three seasons earlier resulted in a significant temperature deviation from average of 0.2 K. Inasmuch as the QBO is regular enough in period for the phase to be forecast several seasons ahead, and the relation depends on a three-season lag with respect to SST, there would seem to be the potential here, on occasion, for the forecasting of zonally averaged north temperate temperatures several seasons in advance.

3.7 Variation in northern hemisphere precipitation

Precipitation estimates for the northern hemisphere are of as much interest as temperature estimates (maybe of more interest from a practical point of view). However, to date there have been serious obstacles in the way of accurately evaluating hemispheric or global precipitation amounts, such as large horizontal gradients in mountain areas and at borders of climatically arid zones where precipitation does not occur every year, as well as lack of data over the oceans. The first problem can partially be solved by smoothing the space non-homogeneity by using 'anomalies' defined by dividing the monthly precipitation amount by the long-term mean amount. In general, though, our confidence with respect to long-term variations in precipitation is not as great as it is with temperature.

Fig. 3.11 shows the estimated variation of such precipitation anomalies for five latitudinal zones in the northern hemisphere for January (left) and July (right), based on the work of Gruza & Apasova (1981). In most latitude bands there tended to be an increase in January precipitation between 1945 and 1960, and a decrease thereafter to 1975. Since the January NHST decreased between 1940 and 1965 (Fig. 3.5), there is some evidence of an out-of-phase relation between the temperature and precipitation,

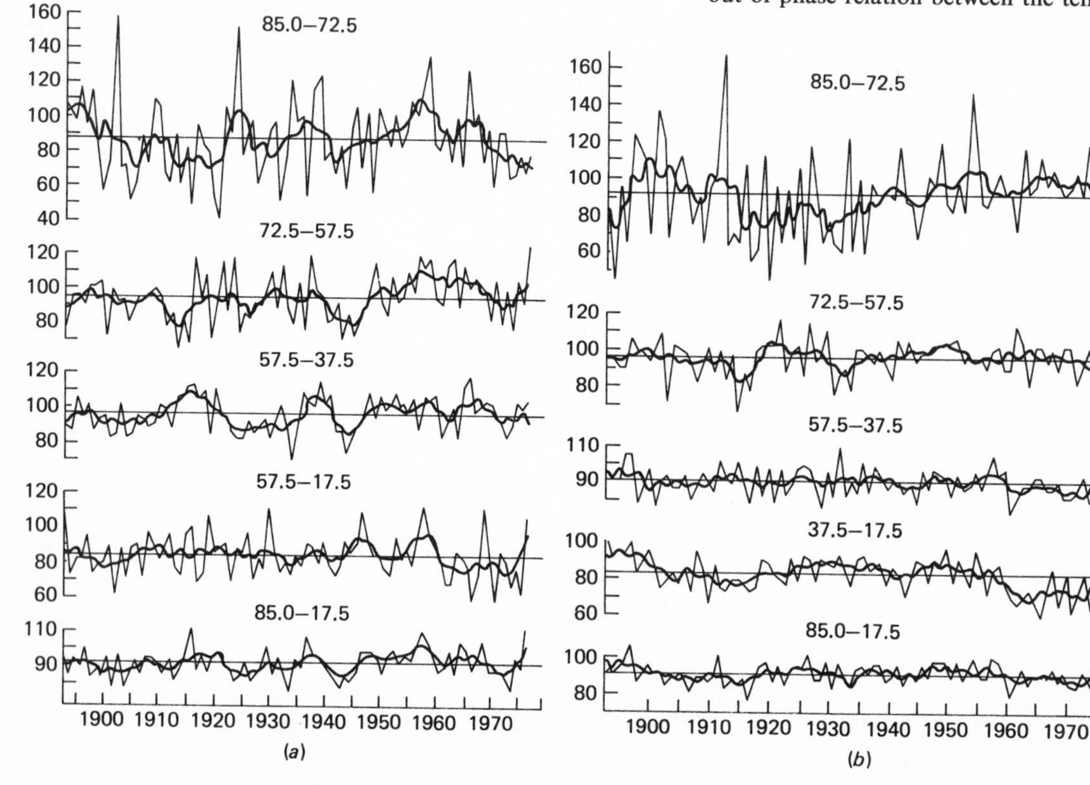

Fig. 3.11. Variation in zonally averaged precipitation anomalies for various latitude bands of the northern hemisphere for January (left) and July (right). The anomalies are expressed as the ratio (in per cent) of monthly amount to long-term mean amount. Five-year running means are also shown.

but it is not a clear-cut one. In July, on the other hand, there has been an in-phase relation between NHST and precipitation in the 17.5–37.5°N latitude band over most of the period of record, i.e., warm temperatures have been associated with an increase in precipitation in this band. This is the same relation found with respect to long-term variations in ISMR and implies a northward displacement of the monsoonal belt as NHST warms.

Fig. 3.12 shows the estimated variation in precipitation in January (left) and July (right) for northern Asia, Africa, America, Eurasia, and the northern hemisphere as a whole. In northern Asia and in Eurasia there is evidence for a long-term increase in January precipitation, but such an increase is not apparent in July. In July the long-term variation in precipitation in America is almost out of phase with that in northern Asia.

3.8 Summary

This summary of results concerning climate variability is keyed to the numerical listing in Section 3.1.

(1) The surface temperature at two stations in central Europe cooled slightly during the 1800s, and warmed slightly during the 1900s but, on the basis of linear regression, exhibited negligible temperature change over the entire 200-year interval.

(2) Based on the results of two separate analysis teams, NHST has warmed by 0.4–0.5 K during the last 100 years (linear regression), with the 1981 temperature the warmest so far observed (0.6 K above the 100-year average). This warming has been greatest in polar latitudes, and relatively large over Africa.

(3) NHST cooled by more than 0.5 K between early 1981 and early 1982, so that the warmth of 1981 does not appear to be persisting. This may make the detection of cooling due to the El Chichon (Mexico) volcanic eruption in the spring of 1982 more difficult.

(4) At the surface and in the tropospheric 85–30 kPa layer there has been a general warming in the tropics and south polar latitudes during the last 24 years, and in temperate latitudes the recent warming has compensated the earlier cooling, but in north polar latitudes there is little evidence of a temperature change during this interval.

(5) In the northern hemisphere the surface temperature began warming in about 1965, the 85–30 kPa temperature not until about 1975, and the 30–10 kPa temperature has tended to cool over the entire 24-year interval, resulting in an increase in lapse rate. In the southern hemisphere, however, the 85–30 kPa temperature apparently began to warm in about 1965, and the surface temperature not until about 1975, yielding a decrease in low-level lapse rate, perhaps due to ocean thermal inertia.

(6) There was a rocketsonde-derived temperature decrease of about 4 K between 1970 and 1976 in the 26–55 km layer of the western quadrant of the northern hemisphere, but little temperature change since, so there is unlikely to be a basic association between this temperature and sunspot number.

(7) During the past 24 years there has been a highly significant relation between SST in the equatorial eastern Pacific and zonally averaged tropospheric temperature in the tropics, with air–temperature changes lagging SST changes by about two seasons. This relation has been so strong it has dominated the global temperature variations as well. During the past 110 years there has also been a highly significant tendency for warm SST in this region to be associated with below-average monsoon rainfall in India.

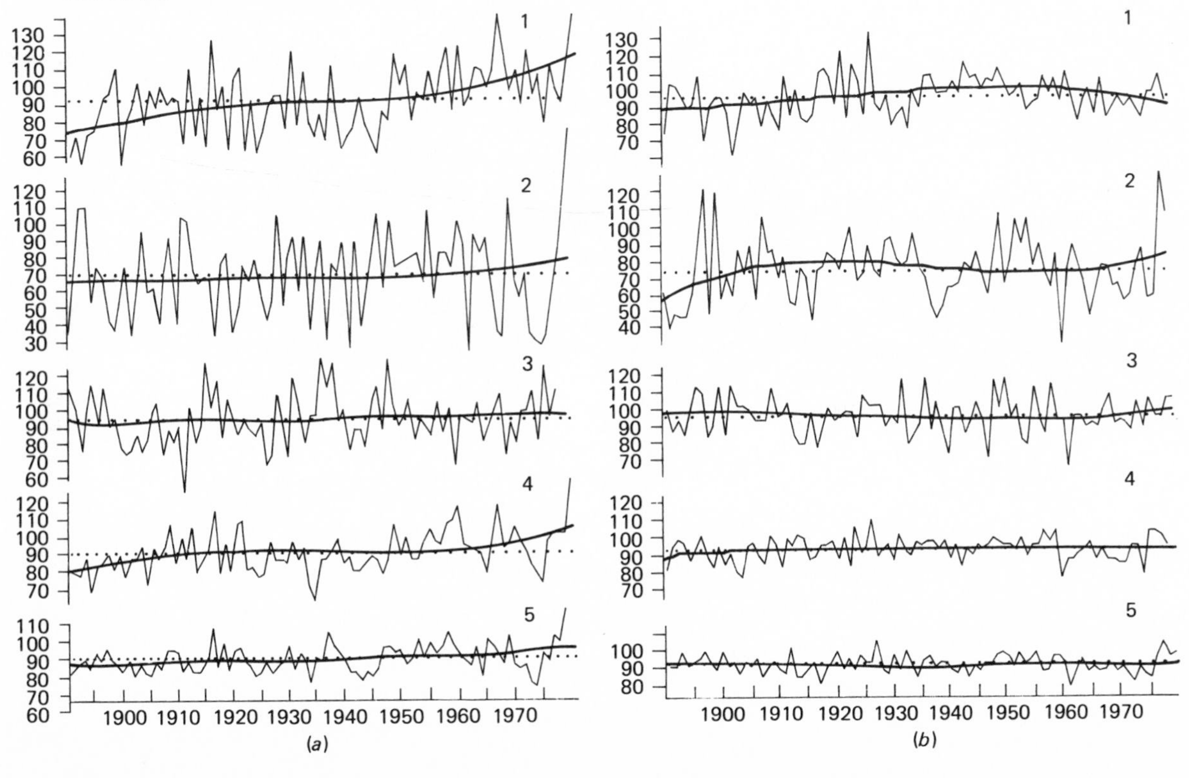

Fig. 3.12. Variation in precipitation anomalies for January (left) and July (right) for Northern Asia (1), Africa (2), America (3), Eurasia (4), and the northern hemisphere as a whole (5), where the anomalies are expressed as the ratio (in per cent) of monthly amount to long-term mean amount. A cubic polynomial trend is also shown.

(8) In July there has been an in-phase relation between NHST and precipitation in the 17.5–37.5°N band during the last 100 years, implying a northward displacement of the monsoonal belt as NHST warms. In northern Asia and Eurasia there is evidence for a long-term increase in January precipitation.

Finally, based on the interval 1958–81, in the east-wind phase of the quasi-biennial oscillation (QBO) at 5 kPa, the mean 85–30 kPa temperature in north temperate latitudes has been very significantly below average (−0.4 K) three seasons after cool SST, and in the west-wind phase of the QBO this temperature has been significantly above average (0.2 K) three seasons after warm SST.

References

Angell, J. K. (1981). 'Comparison of variations in atmospheric quantities with sea surface temperature variations in the equatorial eastern Pacific'. *Monthly Weather Review*, **109**, 230–43.

Angell, J. K. & Korshover, J. (1978). 'Global temperature variations, surface-100 mb: An update into 1977'. *Monthly Weather Review*, **106**, 755–70.

Angell, J. K. & Korshover, J. (1982). 'Comparison of year-average latitude, longitude and pressure of the four centers of action with air and sea temperature, 1899–1978'. *Monthly Weather Review*, **110**, 300–3.

Borzenkova, I. I., Vinnikov, K. Ya., Spirina, L. P. & Stekhnovsky, D. I. (1976). 'Air temperature variations in the Northern Hemisphere for the period 1881–1975'. *Meteorologiya i Gidrologiya*, **7**, 27–35.

Callender, G. S. (1961). 'Temperature fluctuations and trends over the earth'. *Quarterly Journal of the Royal Meteorological Society*, **87**, 1–12.

Callis, L. B. & Nealy, J. E. (1978). 'Solar UV variability and its effect on stratospheric thermal structure and trace constituents'. *Geophysical Research Letters*, **5**, 249–52.

Groveman, B. S. & Landsberg, H. E. (1979). 'Simulated northern hemisphere temperature departures, 1579–1980'. *Geophysical Research Letters*, **6**, 767–69.

Gruza, G. V. & Apasova, Ye, G. (1981). 'Climatic variability of the northern hemisphere monthly precipitation amounts'. *Meteorologiya i Gidrologiya*, **5**, 5–16.

Gruza, G. V. & Ran'kova, E. Ya (1980). *Structure and Variability of Observed Climate, Northern Hemisphere Air Temperature*. Gidrometeoizdat, Leningrad, 72 pp.

Gruza, G. V., Kleshchenko, L. K. & Timofeyeva, T. P. (1982). 'On the variability of the northern hemisphere atmospheric temperature and circulation regimes'. *Meteorologiya i Gidrologiya*, **3**, 8–20.

Hansen, J., Johnson, D., Lacis, A., Lebedeff, P., Lee, P., Rind, D. & Russel, G. (1981). 'Climate impact of increasing atmospheric carbon dioxide'. *Science*, **213**, 957–66.

Holton, J. R. & Tan, H. C. (1980). 'The influence of the equatorial quasi-biennial oscillation on the global circulation at 50 mb'. *Journal of Atmospheric Sciences*, **37**, 2200–08.

Jones, P. D., Wigley, T. M. L. & Kelly, P. M. (1982). 'Variations in surface air temperature: Part 1, northern hemisphere, 1881–1980. *Monthly Weather Review*, **110**, 59–70.

Kelly, P. M., Jones, P. D., Sear, C. B., Cherry, B. S. G. & Tavakol, R. K. (1982). 'Variations in surface air temperatures: Part 2. Arctic regions, 1881–1980'. *Monthly Weather Review*, **110**, 71–83.

Madden, R. A. & Ramanathan, V. (1980). 'Detecting climate change due to increasing carbon dioxide'. *Science*, **209**, 763–8.

Manabe, S. & Wetherald, R. T. (1975). 'The effects of doubling the CO_2 concentration on the climate of a general circulation model'. *Journal of Atmospheric Sciences*, **32**, 3–15.

Mitchell, J. M. (1961). 'Recent secular changes of global temperature.' *Annals of the New York Academy of Sciences*, **95**, 245–50.

Mitchell, J. M. (1963). 'On the worldwide pattern of secular temperature change'. *In: Changes of Climate*, Arid Zone Research XX, UNESCO, Paris, pp. 161–81.

Parthasarathy, B. & Mooley, D. A. (1978). 'Some features of a long homogeneous series of Indian summer monsoon rainfall'. *Monthly Weather Review*, **106**, 771–81.

Penner, J. E. & Chang, J. S. (1978). 'Possible variations in atmospheric ozone related to the eleven-year solar cycle'. *Geophysical Research Letters*, **5**, 817–20.

Quiroz, R. S. (1979) 'Stratospheric temperatures during solar cycle 20'. *Journal of Geophysical Research*, **84**, 2415–20.

Rasmusson, E. M. & Carpenter, T. H. (1982). 'Variations in tropical sea surface temperature and surface wind fields associated with the Southern Oscillation/El Nino'. *Monthly Weather Review*, **110**, 354–84.

Spirina, L. P. (1969). 'On the secular course of mean air temperature in the Northern Hemisphere'. *Meteorologiya i Gidrologiya*, **1**, 85–9.

Taylor, B. L., Gal-Chen, T. & Schneider, S. H. (1980). 'Volcanic eruptions and long-term temperature records: an empirical search for cause and effect'. *Quarterly Journal of the Royal Meteorological Society*, **106**, 175–99.

Trenberth, K. E. (1976). 'Spatial and temporal variation of the Southern Oscillation'. *Quarterly Journal of the Royal Meteorological Society*, **102**, 639–53.

Troup, A. J. (1965). 'The "southern oscillation"'. *Quarterly Journal of the Royal Meteorological Society*, **91**, 490–506.

van Loon, H. & Williams, J. (1976*a*). 'The connection between trends of mean temperature and circulation at the surface. Part I, winter'. *Monthly Weather Review*, **104**, 365–80.

van Loon, H. & Williams, J. (1976b). 'The connection between trends of mean temperature and circulation at the surface. Part II, summer'. *Monthly Weather Review*, **104**, 1003–11.

Vinnikov, K. Ya., Gruza, G. V., Zakharov, V. F., Kirillov, A. A., Kovyneva, N. P. & Ran'kova, E. Ya. (1980). 'Present-day climatic changes in the northern hemisphere'. *Meteorologiya i Gidrologia*, **6**, 5–17.

Willett, H. C. (1950). 'On the present climatic variation'. *Centenary Proceedings of the Royal Meteorological Society*, 195–206.

Yamamoto, R. & Iwashima, T. (1975). 'Change of the surface air temperature averaged over the northern hemisphere and large volcanic eruptions during the years 1951–1972'. *Journal of the Meteorological Society of Japan*, **53**, 482–6.

4

Atmospheric general circulation models: their design and use for climate studies

A. J. Simmons and L. Bengtsson
European Centre for Medium Range Weather Forecasts, Shinfield Park, Reading UK

Abstract

A general review of atmospheric general circulation modelling is presented, with emphasis on the use of models for climate studies. The opening section briefly sets this type of modelling in the context of climate modelling as a whole. It is followed by two further sections of an introductory nature, these outlining the historical development of general circulation models and the processes of primary importance that must be taken into account in constructing them. Sections 4.4 and 4.5 then review the design of the models themselves, discussing both their numerical formulations for the solution of the basic adiabatic equations, and their parameterizations of a number of processes that cannot be explicitly described by the numerical form of the equations. Discussion of the use and performance of these models is prefaced by Section 4.6 which gives an account of the general predictability of the atmosphere. The following section describes the actual use to which models have been put in a number of climate studies, these ranging from the experimental prediction of monthly means to evaluation of the possible response of the longer-term climate to an increase in the carbon dioxide content of the atmosphere. The results of such studies must be judged in the light of the ability of the models to simulate the present climate, and this is discussed in Section 4.8. Some concluding remarks, including brief comment on the interpretation of model results, is given in Section 4.9.

4.1 Introduction

The purpose of this article is to discuss the design and use of atmospheric models for the study of climate. Climate is generally understood as referring to the average weather characteristics over a particular period of time for a particular point or region of the earth's surface. Traditionally, 30-year means have been chosen for its definition, but here we shall adopt a more broad interpretation, and consider averaging periods ranging upward from a few weeks. The time span of interest thus begins at the limit of deterministic weather prediction imposed by the inherent instability of large-scale atmospheric flow, and we shall consider an upper limit of several decades imposed by the computational requirements of comprehensive atmospheric models and by the response time of the deep oceanic circulation. We shall thus discuss the models and modelling problems associated with the prediction of monthly and seasonal means, with the simulation of the longer-term climatological state, including its shorter-term variability, and with the prediction of the influence of changes in the external forcing, either natural or anthropogenic, on the longer-term climate.

Models of a wide variety qualify for the general description of 'climate model', but here we shall restrict attention to the type generally referred to as atmospheric general circulation models (AGCMs). These are numerical models which explicitly simulate the day-to-day evolution of the large-scale weather systems that are an essential component of the climate, and which include parameterizations of the predominant smaller-scale dynamical and physical processes such as moist convection, turbulent mixing and radiation. They generally include equations for the relatively rapid variation in time of land-surface parameters, but have for the most part been used with either fixed or climatologically varying sea-surface temperatures. A two-way interaction between atmosphere and ocean cannot, however, be neglected for comprehensive modelling of longer-term climatic change (as discussed elsewhere in this publication), and the coupling of atmospheric and oceanic circulation models has been the goal of a number of studies, among them those of Manabe (1969), Bryan (1969), Manabe *et al.* (1975, 1979*a*), Washington *et al.* (1980), Wells (1979*a*), and Pollard (1982).

The AGCMs are but one of a hierarchy of atmospheric models which can be used in the study of climate. Analytical or simpler numerical models of the individual processes which together account for the earth's climate play an invaluable role in clarifying the nature of these processes. They thereby help in the design of the more comprehensive models, and they may also suggest experimental studies to be performed with the latter models and aid the interpretation of results. The various processes involved in climate are, however, highly interactive, and it is the more complete models that offer the major hope for obtaining reliable quantitative results. The simpler models are inevitably specific to the process in question, and will not be reviewed here.

Another class of model must also be mentioned. It comprises what are known as the statistical–dynamical atmospheric models. This type includes models of the zonally averaged surface heat balance, models of the time evolution of the zonally averaged state of the atmosphere, and models which resolve the very largest scales of atmospheric wave motion. All such models parameterize the active synoptic ($\gtrsim$ 1000 km) scales of motion. The parameterizations may be based on the statistics of observed atmospheric behaviour (and are thus biased towards the current climatic state), or on (imperfect) theoretical knowledge of the behaviour of synoptic-scale weather systems. They may alternatively use statistics derived from more-comprehensive models. The statistical–dynamical models are less demanding of computational resources than the AGCMs and thus are the potential tool for studies of climate on time scales longer than can be accomplished with the more-expensive models. Their use is appealing in that the time scales of weather systems are short compared with those of interest in climate studies, but their design involves an additional, and difficult, degree of parameterization, not only for the actual working of the model, but also for the interpretation of results in terms of the climates of local regions. Reviews of this type of model have been published by Schneider & Dickinson (1974) and in contribution No. 16 to the *GARP Publication Series* (1975). In addition, results from a number of models have been described in *GARP Publication Series No.* 22 (1979). No further discussion will be included in this article.

4.2

The development of general circulation modelling

The origins of climate modelling as considered here follow close behind the first application of the computer for short-range weather prediction. The original barotropic forecast experiments by Charney *et al.* (1950) were soon followed by the development of the baroclinic models capable of describing both the growth and decay of the predominant middle-latitude weather systems (Charney & Phillips, 1953). It was soon recognized that such models could be of use not only for weather forecasting, but also for studying the mean long-term behaviour of the atmosphere. A fascinating account of early thinking on this subject may be found in papers by von Neumann, Charney, Phillips, Mintz, Smagorinsky and others, presented in the proceedings of a conference held at Princeton in 1955 to discuss the application of numerical techniques to the study of the general circulation (Pfeffer, 1960). One year later, Phillips's (1956) account of the first general circulation experiment was published in the open literature.

The following two decades witnessed very substantial advances in general circulation modelling. Phillips's original model used the approximate 'quasi-geostrophic' equations and a minimal (two-layer) vertical resolution of the atmosphere, with a latitudinally varying but otherwise constant distribution of heating and cooling. Subsequent developments in modelling technique and computer power led to integrations using the more complete 'primitive' equations and more realistic energy sources and sinks (Smagorinsky, 1963), global integrations including orographic forcing (Mintz, 1965), enhancements in vertical resolution to include a coarse representation of the planetary boundary layer and stratosphere (Smagorinsky *et al.*, 1965), inclusion of moist processes (Manabe *et al.*, 1965; Manabe, 1969), and other such milestones. Model descriptions were published by other centres (Kasahara & Washington, 1967; Corby *et al.*, 1972; Somerville *et al.*, 1974) and, in 1974, eight different groups provided descriptions for a catalogue of AGCMs (*GARP Publication Series No. 14*), while others were also active. Four years later, nine groups described simulated model climates at a JOC study conference, and their results may be found in *GARP Publication Series No.* 22 (1979).

Some of the progress in recent years will be discussed in detail in the remainder of this article. Technical developments have continued, but there has also been a growing application of models in a variety of climate studies. At one end of the time range, first (and rather encouraging) experimental studies of predictability on the monthly time scale have taken place (Shukla, 1981*a*; Miyakoda *et al.*, 1983). At the other end, a model run for a period of almost two decades has succeeded in reproducing much of the interannual variability of the atmosphere (Manabe & Hahn, 1981; Lau, 1981), and similar conclusions from integrations over periods ranging from several years to a decade have been reported by Cubasch (1981*a*), Schlesinger & Gates (1981), and Volmer *et al.* (1983*a*, b). Studies of the response to changes in various forcing factors or initial conditions have increased and, coupled with simpler model results and data studies, have led to an increased awareness of the important role of such quantities as sea-surface temperature, albedo and soil wetness.

One further recent development is perhaps worthy of special mention. Numerical weather prediction and general circulation modelling grew from similar origins in the early 1950s, but modelling for the two applications differed widely. Weather forecasting was concentrated on a time range of up to, at most, a few days ahead, over which period close attention to the climatological balance between resolved and parameterized processes was not of paramount importance. It was also subject to operational time constraints which necessitated the use of less than hemispheric domains. However, as was recognized at the outset of numerical modelling, and more specifically by Miyakoda *et al.* (1972) in their experimental study of medium-range predictability, the climatological balance of the forecast model may become of importance after several days of prediction. This has been confirmed in practice (Hollingsworth *et al.*, 1980; Bengtsson & Simmons, 1983), and has lead to an emphasis being placed on achieving such a balance in the development of the global operational forecast model at ECMWF, where medium-range predictions are carried out daily to ten days ahead. As a consequence, short-term (50-day) climate simulations play a role in the testing of alternative formulations for this model, and a form of it has been used (in collaboration with the meteorological services of France and the Federal Republic of Germany) for some longer-term climate studies (Cubasch, 1981*a*; Volmer *et al.*, 1983*a*,*b*). Thus a clear distinction can no longer be drawn between numerical models for climate studies and weather forecasting, and we may anticipate that future improvements in model design will result not only from comparisons of the climate simulations produced by different models, but also from studies of the growth of climate error in forecast experiments.

4.3
Processes of primary importance

The fundamental process driving the earth's climatic system is the heating by incoming short-wave solar radiation and the cooling by long-wave radiation to space. The heating is strongest at tropical latitudes, while cooling predominates at the polar latitudes of at least the winter hemisphere (Vonder Haar & Suomi, 1971). The latitudinal gradient of heating drives currents in the atmosphere and ocean which provide the heat transfer required to balance the system. Estimates for the northern hemisphere indicate that transfer by the atmosphere and ocean are of comparable magnitude (Vonder Haar & Oort, 1973; Oort & Vonder Haar, 1976).

The bulk of the net incoming solar radiation is absorbed not by the atmosphere but by the underlying surface. Evaporation of moisture and the heating of the surface lead, however, to much of this energy being transferred to the atmosphere as latent and, to

Fig. 4.1. The mean annual radiation and heat balance of the atmosphere, relative to 100 units of incoming solar radiation. (From US National Academy of Sciences, 1975.)

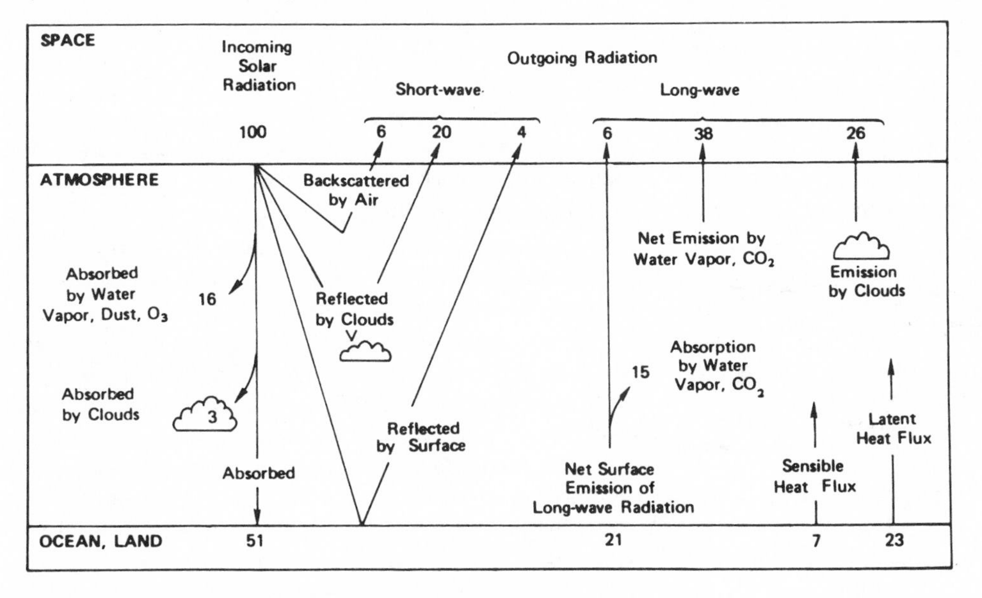

Fig. 4.2. Schematic illustration of the earth's climatic system, with some examples of the physical processes responsible for climate and climatic change. (From Gates, 1979.)

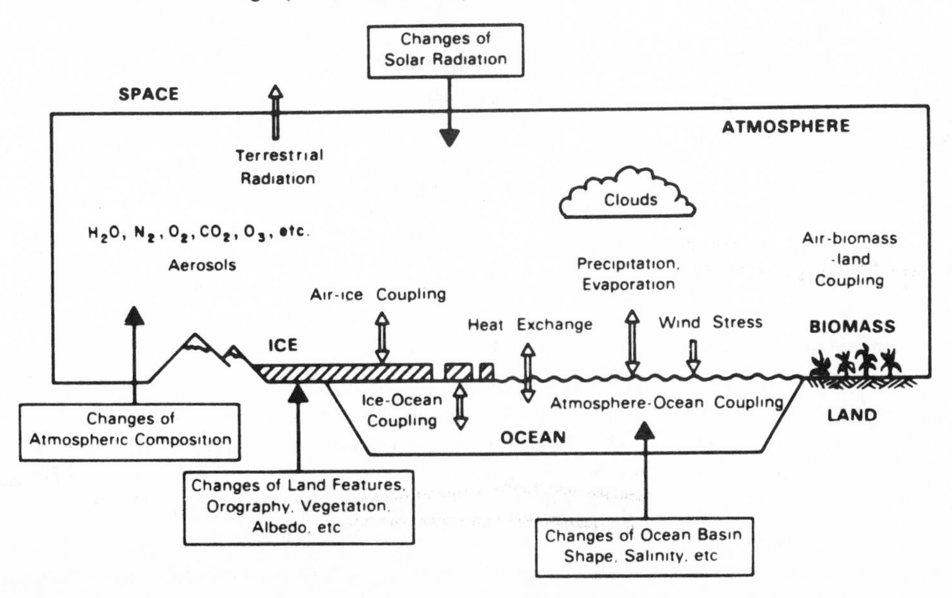

a lesser extent, sensible heat. Thus the dominant direct heating of the atmosphere is found to be the latent heat release associated with deep tropical convection. Fig. 4.1 summarizes the overall radiation and heat balance.

The atmosphere's meridional energy transfer across middle latitudes is accomplished largely by transient weather systems with a time scale of the order of days, which themselves develop due to the so-called 'baroclinic' instability (Charney, 1947; Eady, 1949) of the predominantly zonal flow set up by the differential radiative heating. This flow is further perturbed, especially in the northern hemisphere, by orographic effects and land–sea contrasts (Charney & Eliassen, 1949; Smagorinsky, 1953). These strongly influence local climate, and on a larger scale lead to a concentration of baroclinic transfer along the storm tracks of the North Pacific and Atlantic Oceans (Blackmon *et al.*, 1977). Lower-frequency variability, of which the quasi-stationary 'blocking' anticyclones are just one manifestation, also occurs both in the extratropics and in tropical circulation systems. It may arise either from slowly varying boundary conditions such as sea-surface temperature (e.g. Bjerknes, 1966, 1969; Namias, 1969) or from variability in the dynamical response to a fixed external forcing (e.g. Charney & DeVore, 1979; Simmons *et al.*, 1983). Its prediction is a goal of short-term climate forecasting, and its statistics form a part of the longer-term climate that must be simulated in any comprehensive study.

Many other processes are important in determining the detailed behaviour of the atmospheric component of the climatic system, as illustrated schematically in Fig. 4.2. Firstly, turbulent vertical transfers of heat, moisture and momentum influence all larger scales of motion, and are determined both by the nature of the underlying surface and by the nature of the larger-scale flow itself. The type of surface, as characterized by its albedo, also determines the proportion of incoming solar radiation that is reflected back towards space. Latent heat release not only helps drive the mean circulation, but can also be a significant component of synoptic-scale systems, in middle latitudes as well as the tropics. The related clouds play an important role in reflecting incoming short-wave radiation and in absorbing and emitting long-wave components.

The radiative heating and cooling also depend on the gaseous constituents of the atmosphere, notably carbon dioxide, water vapour and ozone. The distribution of water vapour is strongly influenced by the overall temperature and by dynamical transfers, and similar considerations apply to the stratospheric distribution of ozone. The carbon dioxide content may be regarded as constant for many studies, although there is widespread interest in the climatic impact of higher future levels likely to arise from man's burning of fossil fuels, from deforestation, and from changing agricultural practices. In this case, interchange with the ocean and land (through vegetation) may vary as components of the climatic system change. Other factors that can influence the radiation budget and thereby the climate are aerosols such as volcanic dust, and the orbital parameters of the earth. Different, though fixed, values of the latter have been used in model studies of past climate (Mason, 1976; Kutzbach & Otto-Bliesner, 1982).

Some of the processes outlined in preceding paragraphs are such that climate is critically sensitive to the characteristics of the underlying surface. Some surface properties, such as the distribution of sea, land and mountains may be regarded as fixed in time for all practical purposes, but other properties change substantially and relatively rapidly in response to the state of the atmosphere. Land-surface temperature is an obvious example of a particularly fast response, while soil moisture and snow and ice cover are examples of more slowly varying but important quantities. The type of vegetation also influences surface transfers. In general, parameterization of the processes which determine such surface characteristics form part of what is understood by an AGCM.

Sea-surface temperature is another boundary condition which not only influences but is also influenced by the atmospheric circulation. Modelling of the ocean has traditionally been regarded as a task separate from that of atmospheric modelling, although its importance for climate modelling has long been recognized, and reference has been made in Section 4.1 to a number of studies in which oceanic and atmospheric models have been coupled. In general, the complexity of the oceanic model required depends on the time scale of the problem under consideration. The response time of models of the deep oceanic circulation to changes in surface forcing is of the order of centuries, which in itself poses problems in the coupling with atmospheric models to determine an equilibrium state. Simpler models of the surface 'mixed layer' may be satisfactory for time scales of months and years.

Further discussion of many of the processes mentioned here, together with detailed references, will be found in other contributions to this publication. Although our main aim has been to outline those processes which must be included in models for the quantitative simulation and prediction of climate, the nature of the processes also has implications for the experimental strategy to be followed when using climate models. The varying nature of synoptic-scale weather systems introduces an inevitable element of uncertainty into time means over periods which can be substantially longer than the characteristic lifetime of these systems. Moreover, the prevailing synoptic situation (or weather type) may strongly influence the response of a model atmosphere to a particular boundary forcing or change in parameterization scheme, say. Obtaining statistically significant model results thus becomes a matter for careful experimental design, and may demand a substantial computational effort. Such matters will not be considered in detail in this article, and the interested reader is referred to papers by Leith, Chervin and Hasselmann in *GARP Publication Series No. 22* (1979).

4.4 Numerical formulations

4.4.1 *Basic considerations*

Reviews of the numerical techniques used in forecast and general circulation models have been given in the two volumes of *GARP Publication Series No. 17* (1976, 1979). In these models, the spatial variation of the atmospheric state is approximated by a number of discrete values, for example values at points on a three-dimensional grid covering the globe, and the models then predict these values at regular time intervals. In general, the computation required to advance the model by one time step is of the order of the number of discrete values chosen to represent the predicted variables. In addition, the time-step itself cannot be freely chosen, but must be less than a value which is typically linearly dependent on the spatial resolution of the model. The determining factor is usually the horizontal resolution, so that a halving of the smallest resolved horizontal-length scale implies an increase by a factor of about 8 in the computational cost of a simulation of a fixed time period. Since use of higher resolution should lead, in principle at least, to a more realistic model climate, the modeller with fixed computational resources is faced with balancing the potential benefits of higher resolution against the cost in terms of the number of experiments that can be performed. In practice, techniques and parameterizations developed and thoroughly tested for one resolution may not work satisfactorily at higher resolution, leading to a further choice as to the extent to which attention should be concentrated on model development, or on performing specific climate studies with a well-tried, if limited, model.

The models in regular use differ much in the detail of their numerical formulation, and it is beyond the scope of this article to provide a comprehensive catalogue. A general discussion is given in the following paragraphs, and some specific details will be given for the model with which we are most familiar, namely the model used for the first phase of operational medium-range prediction at ECMWF.

4.4.2

The primitive equations

The majority of models are based on a set of equations known as the primitive equations. As discussed by Phillips (1973), the governing dynamic, thermodynamic and conservation equations are mapped to a spherical geometry, and the reduced set of primitive equations is obtained by assuming the height scale of the motion to be small compared with its horizontal-length scale, an acceptable approximation for horizontal scales upwards of tens of kilometres. The basic predicted variables are the discretized horizontal wind components, u and v, temperature, T, water vapour as represented usually by the specific humidity q, and (in most formulations) surface pressure, p_s.

We illustrate the form of the primitive equations using the vertical coordinate system in most common use, namely the 'sigma' coordinate system proposed by Phillips (1957), for which the vertical coordinate, σ, is given by

$$\sigma = p/p_s, \tag{4.1}$$

where p is pressure. In this case the equations become

Momentum

$$\frac{D\mathbf{v}}{Dt} + f\mathbf{k}\times\mathbf{v} + \nabla\phi + R_d T\nabla \ln p_s = \mathbf{P}_v + \mathbf{K}_v. \tag{4.2}$$

Thermodynamic

$$\frac{DT}{Dt} - \frac{R_d T\omega}{c_{pd}\, p_s \sigma} = P_T + K_T. \tag{4.3}$$

Moisture conservation

$$\frac{Dq}{Dt} = P_q + K_q. \tag{4.4}$$

Mass conservation

$$\frac{Dp_s}{Dt} + p_s\left(\nabla\cdot\mathbf{v} + \frac{\partial\dot\sigma}{\partial\sigma}\right) = 0. \tag{4.5}$$

Hydrostatic

$$\frac{\partial\phi}{\partial\sigma} = -\frac{R_d T}{\sigma}. \tag{4.6}$$

Here t is time and D/Dt denotes the rate of change moving with a fluid particle, which in σ coordinates takes the form

$$\frac{D}{Dt} = \frac{\partial}{\partial t} + \mathbf{v}\cdot\nabla + \dot\sigma\frac{\partial}{\partial\sigma};$$

$\mathbf{v}$ is the horizontal velocity vector, $\mathbf{v} = (u, v, 0)$, and ∇ is the two-dimensional gradient operator on a surface of constant σ; f is the Coriolis parameter (twice the earth's angular rotation rate multiplied by the sine of latitude); $\mathbf{k}$ is the unit vertical vector; ϕ is the geopotential (the acceleration due to gravity multiplied by the height of a surface of constant pressure); R_d is the gas constant for dry air, and C_{pd} the specific heat of dry air at constant pressure; P_X denotes the rate of change of variable X due to the parameterized processes of radiation, convection, turbulent vertical mixing and large-scale precipitation, and further detail concerning the inclusion of these will be given in Section 4.5. Finally, K_X represents the rate of change of X due to the explicit horizontal smoothing that is usually included in models to prevent an unrealistic growth of the smallest resolved scales. The latter term would ideally be regarded as representing the influence of unresolved scales of motion on the explicitly predicted scales, and treated as part of the parameterization. In practice, since the smallest scales in a model are inevitably subject to numerical misrepresentation, it is common to choose empirically a computationally convenient form of smoothing, and to adjust it so that contour plots of the predicted variables do not appear excessively rough.

A predictive equation for surface pressure is obtained by integrating Eq. (4.5) from $\sigma = 0$ to $\sigma = 1$, using the boundary conditions $\dot\sigma = 0$ at $\sigma = 0$ and $\sigma = 1$:

$$\frac{\partial p_s}{\partial t} = -\int_0^1 \nabla\cdot(p_s\mathbf{v})\,d\sigma. \tag{4.7}$$

Vertical velocities are not explicitly predicted, but they too can be deduced from Eq. (4.5). The sigma- and pressure-coordinate forms, $\dot\sigma$ and ω, are given by

$$p_s\dot\sigma = \sigma\int_0^1 \nabla\cdot(p_s\mathbf{v})\,d\sigma - \int_0^\sigma \nabla\cdot(p_s\mathbf{v})\,d\sigma \tag{4.8}$$

and

$$\omega \equiv \frac{Dp}{Dt} = \sigma\mathbf{v}\cdot\nabla p_s - \int_0^\sigma \nabla\cdot(p_s\mathbf{v})\,d\sigma. \tag{4.9}$$

The form of the primitive equations given above neglects the local mass of water vapour compared with that of dry air, an approximation that can introduce a small, but not always negligible, error in moist tropical regions. It is thus common for models to use a more accurate form in which the temperature appearing in Eqs. (4.2) and (4.6) is replaced by the virtual temperature, T_v, which is defined by

$$T_v = T\left\{1 + \left(\frac{R_v}{R_d} - 1\right)q\right\}, \tag{4.10}$$

where R_v is the gas constant for water vapour. In addition, the second term on the left-hand side of Eq. (4.3) may be multiplied by a factor

$$\left\{1 + \left(\frac{R_v}{R_d} - 1\right)q\right\}\Big/\left\{1 + \left(\frac{C_{pv}}{C_{pd}} - 1\right)q\right\},$$

where C_{pv} is the specific heat of water vapour at constant pressure.

In addition to representing the evolution of the basic atmospheric variables, models generally include predictive equations for several surface fields, as will be discussed later. Furthermore, other predicted variables may be introduced for particular applications.

Ozone and other trace constituents have been included for some stratospheric studies (e.g. Cunnold *et al.*, 1975; Schlesinger & Mintz, 1979; Allam *et al.*, 1981), and a formulation for a separate prediction of liquid water has been developed by Sundqvist (1981). A type of boundary-layer parameterization which uses the turbulent kinetic energy as a predicted variable has been tested by Miyakoda & Sirutis (1977).

4.4.3

Vertical discretization

The vertical structure of model variables is most commonly represented by values defined at a number of levels in the vertical. For the usual sigma coordinate, the pressure of a particular level is proportional to the surface pressure, and coordinate surfaces thus rise over rather than intersect mountains. Alternative coordinates which are also terrain-following at low levels, but for which upper levels are at constant pressures, have also been adopted, notably for stratospheric studies (Schlesinger & Mintz, 1979; Fels *et al.*, 1980) but also for more-general application (Simmons & Burridge, 1981; Simmons & Strüfing, 1983). Fig. 4.3 illustrates the usual case in which all predicted variables are defined at the same levels.

A minimum of two levels is required to describe the baroclinic growth of middle-latitude disturbances, and has continued to be used (e.g. Schlesinger & Gates, 1980) ever since the first general circulation experiments. A variety of higher resolutions has become common, however, both to ensure a more accurate dynamical representation, and to facilitate parameterization. Many studies have used nine levels (e.g. Manabe & Hahn, 1981; Shukla, 1981*a*; Miyakoda *et al.*, 1983; Ramanathan *et al.*, 1982), and a maximum number of around 18 levels has generally been adopted for tropospheric studies (e.g. Miyakoda & Sirutis, 1977; Hunt 1978), although considerably more levels are used in some stratospheric and mesospheric models (Fels *et al.*, 1980; Hunt, 1981). Some parameterizations of the planetary boundary layer require that at least two or three model levels lie within the lowest kilometre of the atmosphere, while a similar number of stratospheric levels may be chosen as the minimum to avoid distortion of tropospheric planetary-wave structures, although the precise influence of the stratospheric representation on tropospheric simulations has been a matter of debate (e.g. Lindzen *et al.*, 1968; Nakamura, 1976; Kirkwood & Derome, 1977; Bates, 1977; Mechoso *et al.*, 1982; Simmons & Strüfing, 1983). The distribution of 15 levels used in the first operational ECMWF model is included in Fig. 4.3.

The vertical finite-difference schemes in common use are generally characterized by their second-order accuracy (for slowly varying distributions of levels) and by their satisfying certain integral constraints in common with the continuous equations. Notable are the conservation of mass and energy (e.g. Corby *et al.*, 1972; Arakawa & Lamb, 1981; Burridge & Haseler, 1977), while angular-momentum conserving schemes have also been proposed (Arakawa & Lamb, *loc. cit.*; Simmons & Burridge, 1981). In addition, special care is generally taken to avoid inaccuracy in the calculation of the horizontal pressure gradient (Kurihara, 1968; Corby *et al.*, 1972; Gary, 1973; Sundqvist, 1976; Nakamura, 1978; Mesinger, 1981*a*; Simmons & Burridge, 1981).

4.4.4

Horizontal discretization

Two different techniques, finite-difference and spectral, have been widely adopted for the horizontal, although the use of finite-element methods has also been examined (Cullen, 1974; Staniforth & Mitchell, 1978), just as indeed it has in the vertical (Staniforth & Daley, 1977). In the finite-difference method, variables are represented at one of a wide variety of grids, which vary according to the relative locations of wind and temperature components (Arakawa & Lamb, 1981), and according to whether the grid separation is regular (or approximately regular) in longitude or physical distance as the poles are approached (Kurihara, 1965). The grid of the ECMWF model is shown in Fig. 4.4. Following work such as that of Lilly (1964), Arakawa (1966) and Sadourny (1975), attention is increasingly directed towards ensuring conservation properties of the finite-difference schemes (in particular, conservation of energy and enstrophy) as a means of achieving computational stability and realistic simulations without recourse to excessive horizontal smoothing (Arakawa & Lamb,

Fig. 4.3. An example of the vertical distribution of variables in a model with N layers. The column of values on the right-hand side denotes the pressures at which variables are represented in the 15-level resolution used operationally at ECMWF, assuming a surface pressure of 1000 mb.

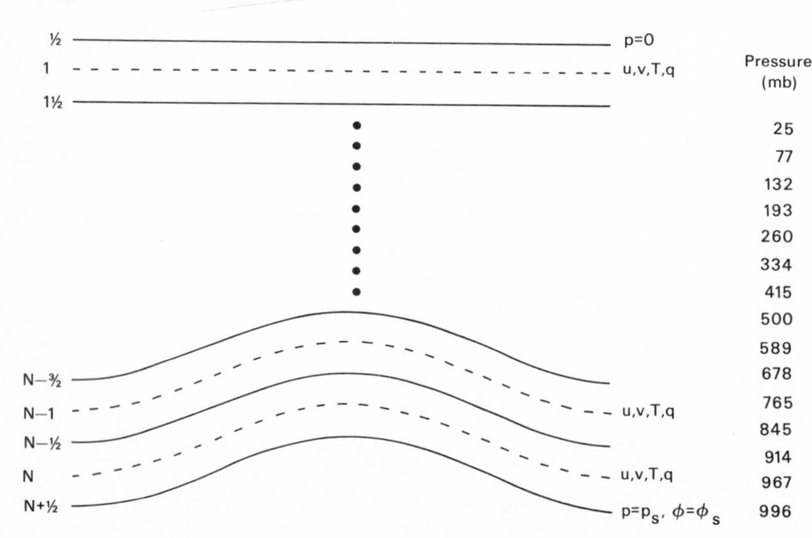

1981; Mesinger, 1981*b*; Janjić, 1983). Schemes are commonly second-order accurate, although fourth-order schemes have also been investigated by a number of authors (Arakawa, 1966; Gerrity *et al.*, 1972; Kalnay-Rivas *et al.*, 1977; Williamson, 1978; Mesinger, 1981*b*). Resolutions for the most part lie in the range from two to five degrees of latitude and longitude.

In the spectral method the predicted variables, which generally include vorticity and divergence rather than the horizontal wind components (Bourke, 1972), are represented in terms of truncated expansions of spherical harmonics:

$$X(\lambda,\theta,\sigma,t) = \sum_{m=-M}^{M} \sum_{n=|m|}^{N} X_n^m(\sigma,t)\, P_n^m(\sin\theta)\, e^{im\lambda} \tag{4.11}$$

Fig. 4.4. The horizontal distribution of variables in the ECMWF finite-difference model. Operationally, $\Delta\lambda = \Delta\theta = 1.875°$.

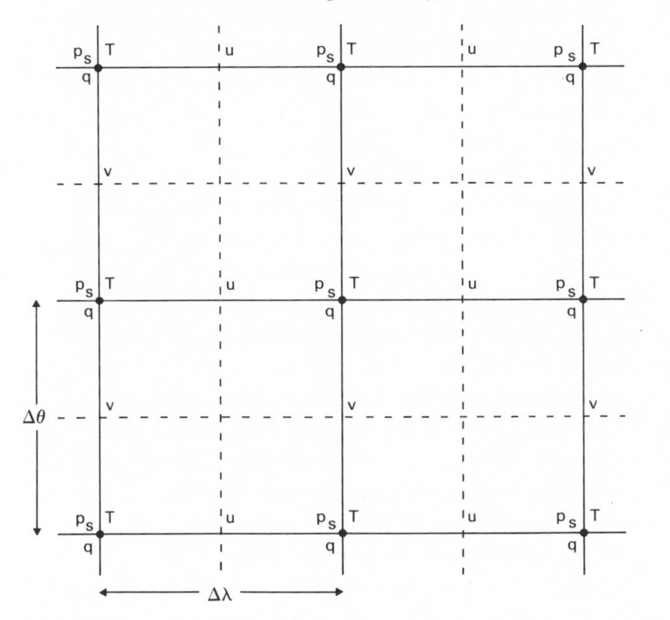

where X is any variable; θ is latitude; λ is longitude and the P_n^m are associated Legendre functions. Although used by Silberman as early as 1954, it was not until the development of the transform technique (Eliasen *et al.*, 1970; Orszag, 1970; Machenhauer & Rasmussen, 1972) that spectral models became fully competitive with finite-difference models. In this technique, nonlinear terms (including parameterized contributions) are evaluated on an almost regular latitude–longitude grid, and the spectral tendencies ($\partial X_n^m/\partial t$) calculated by quadrature. Descriptions of early multi-level models and experiments were published by Bourke (1974), Hoskins & Simmons (1975) and Daley *et al.* (1976), and spectral general circulation experiments have been reported by McAvaney *et al.* (1978), Manabe & Hahn (1981), Cubasch (1981*a*), Pitcher *et al.* (1983), and others. A general review has been given by Machenhauer (1979).

Spectral models, in general, lack the variety (caused by the use of various grids and difference schemes) of their finite-difference counterparts, but truncations of the two types illustrated in Fig. 4.5 are in common use. Rhomboidal truncation (which has $N = |m| + M$ in Eq. (4.11), and which we shall denote by *RM*) has been generally more popular than triangular truncation ($N = M$, denoted *TM*), although comparisons indicate that the two give short-range forecasts of equivalent accuracy (Daley & Bourassa, 1978), with a tendency if anything for triangular truncation to be preferred in medium-range predictions (Jarraud, personal communication). A comparison of general circulation simulations using spectral and finite-difference models of several different resolutions (R15, R21 and R30, and 500 and 250 km grids) has been made by Manabe *et al.* (1979*b*), who report in general favour of the spectral method. Similar conclusions were reached by Girard & Jarraud (1982) in the case of medium-range forecast experiments at relatively high resolution (T63 and 1.875°), although at such resolutions differences become generally smaller than those which are found to result from a substantial reduction in resolution. The generally satisfactory behaviour of efficient low-resolution (R15 or T21) spectral models has led to them being adopted for the integrations over many years discussed by Manabe & Hahn (1981), Lau (1981), Cubasch (1981*a*) and Volmer *et al.* (1983*a*, *b*).

Fig. 4.5. The regions of wavenumber space (shaded) for which spectral components are retained in rhomboidal truncation (*RM*), left, and triangular truncation (*TM*), right.

Rhomboidal

n

M

|m|

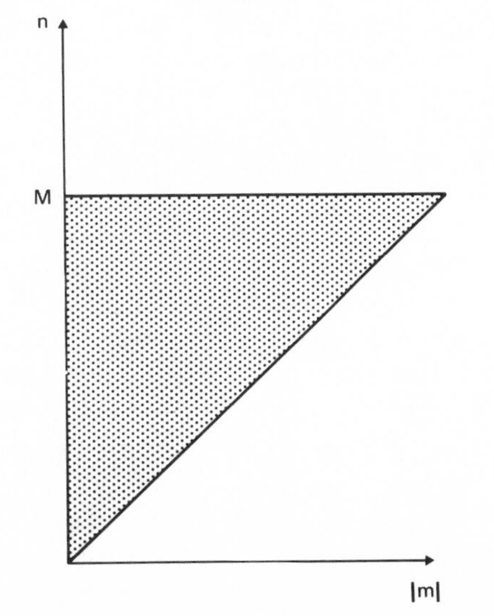

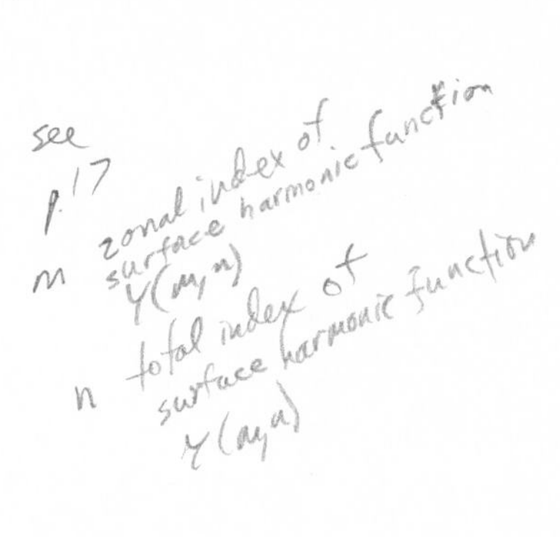

4.4.5
Efficient time schemes

The increasing use of spectral models in general circulation studies has also arisen from the ease with which the semi-implicit method of stepping models forward in time may be incorporated in them. In this method, pioneered by Marchuk (1965) and Robert *et al.* (1972), linearized terms describing small-amplitude gravity-wave motion are treated in such a way as to allow a significantly longer time step than would otherwise be possible, at least provided that care is taken in the linearization (Simmons *et al.*, 1978; Simmons & Burridge, 1981). Characteristic time steps then vary from the order of one hour for a T21 spectral truncation to 20 minutes for T63, and are such that truncation error due to temporal finite-differencing generally remains negligible. Semi-implicit or alternative economical schemes have been widely and quite quickly adopted in finite-difference models for weather forecasting (e.g. Burridge, 1975; Gauntlett *et al.*, 1976; Gadd, 1978; Burridge, 1979), but apart from the work of Marchuk *et al.* (1979) this has not been the case in general circulation modelling. The semi-implicit spectral models have thus been left with a clear lead in terms of computational efficiency. The extent to which this lead will be maintained in the light of the development of Lagrangian techniques (Robert, 1981, 1982; Bates & McDonald, 1982) remains to be seen.

4.4.6
Computational requirements

Computational requirements in terms of elapsed times on a CRAY-1 computer are presented in Table 4.1 for a variety of spectral and finite-difference resolutions. Values are for a one-month simulation using a 15-level vertical resolution, and they may be scaled linearly for alternative numbers of levels or alternative simulation periods. They are based on ECMWF models optimized for high-resolution use on a 1M-word machine. Although differences in timing can result from differences in technique and in the complexity of parameterizations, the dominant sensitivity is to the resolution of the model.

4.5
Parameterizations

The parameterizations required in models are naturally determined by the processes of primary importance for climate, as outlined in Section 4.3. Involving as it does the need to represent the effect of unresolved small-scale processes on the larger, explicitly resolved scales, the development of parameterization schemes requires fundamental understanding of the working of the atmosphere on scales smaller than about 200 km in the horizontal and 1 km in the vertical. Since model performance and observational and theoretical knowledge are far from perfect in many instances, parameterizations tend to vary substantially from model to model, at least in questions of detail, and schemes for a particular model are typically adjusted more frequently than the basic numerical formulation. As such, a comprehensive review is difficult to give in a limited space, and only a brief summary of the types of parameterization currently used will be given below. Reference should be made to *GARP Publication Series No. 8* (1972) for an earlier, but very much more complete review. Although the following discussion will be divided according to individual processes, it must be recognized that, in practice, different components of an overall parameterization are highly interactive, just as the corresponding processes are in the real atmosphere. Particular examples include the treatments of tropical convection, cloud cover and surface heating by radiation, and the treatments of boundary-layer cloud, radiation and turbulence. The general extent of interactions is illustrated in Fig. 4.6.

Table 4.1. *Elapsed time on a CRAY-1 computer for a one-month simulation using ECMWF models*

Model	Resolution	Elapsed time (hours)
Spectral	T21	0.5
Spectral	T40	4.5
Spectral	T63	12
Finite-difference	3.75°	1.5
Finite-difference	1.875°	12

4.5.1
Radiation and clouds

The theoretical basis of radiative transfer is reasonably well understood, but a comprehensive calculation is highly complex. The predominant problems are thus of a practical nature, and they may be broadly divided into two categories. The first concerns the choices of radiative processes to be included, and the approximations to be adopted in their calculation. This inevitably includes the question of the computational time that can be devoted to this particular (and usually rather expensive) component of the model. The second category of problem concerns the determination of appropriate input parameters for the radiative calculations

A general account of the calculation of radiative transfer has been given by Paltridge & Platt (1976). The parameterizations in actual use in various models have in the main been developed over a number of years, and we shall not attempt a comparison of them, but instead refer the interested reader to the various model descriptions referenced elsewhere in this article. Schemes are typically specific to particular models which differ in many other respects, and it is in general difficult to gain a clear impression of the importance of particular aspects of the radiative calculation, which differ from model to model. Comparisons of two schemes in otherwise identical forecast models have been reported in ECMWF (1981) and by Geleyn *et al.* (1982*a*), while Ramanathan *et al.* (1983) have shown that the careful treatment of a number of specific processes can have a pronounced influence on general circulation simulations for January, especially at stratospheric levels.

The input provided to the radiative calculation in general comprises the model's predicted temperature and moisture fields (and possibly also ozone in some applications), together with fixed (climatological or perturbed) distributions of other active gases and aerosols. The distribution of surface albedo may be in part climatological and in part predicted, for example using the model snow cover. In addition, a prescription of cloud cover is required, and this is perhaps the most uncertain element of the whole radiative parameterization. The simplest approach is to utilize a climatological distribution, as adopted in several early models, but this evidently restricts the usefulness of a model for those climate studies where cloud feedback processes are important, and more generally inhibits the development of parameterizations of other processes for which the interaction of radiation with the local cloud distribution is significant. Thus there has been an increasing tendency for cloud distributions to be parameterized in terms of model variables. Cloud cover is typically chosen to depend on the predicted relative humidity, and may also be related to the occurrence of convection and to the thermal structure of the boundary layer (e.g. Slingo, 1980). Particular problems worthy of note relate to the radiative properties of cirrus clouds and the interaction between boundary-layer cloud and radiation. Examples of general circulation experiments concerned with cloud processes include those of Schneider *et al.* (1978), Hunt (1978), Herman *et al.* (1980) and Wetherald & Manabe (1980).

4.5.2

The planetary boundary layer

Boundary-layer parameterization schemes may be divided into two classes according to the vertical resolution of the model close to the ground. If the resolution is such that, at most, one level lies within the boundary layer, then the so-called 'bulk' parameterization schemes must be used to represent the boundary layer as a whole. Alternatively, the boundary-layer structure may be explicitly, albeit crudely, resolved by locating several levels within the lowest 2 km of the model atmosphere. In both cases various degrees of sophistication may be adopted, and the merits of the two approaches are a matter of current debate. A general review has been given recently by Driedonks & Tennekes (1981).

In the simplest of the bulk approaches, the surface fluxes are calculated from a basic surface drag law using either the wind at the lowest model level, or a wind extrapolated from more than one level. The drag coefficients and turning of the wind through the boundary layer may exhibit a dependence on the underlying surface and the stability of the lowest model layer. Turbulent fluxes either vanish in the free atmosphere or are treated by simple eddy diffusivities. A review of the use of such schemes has been given by Bhumralkar (1976), while Arya (1977) examines them in the light of additional observational and theoretical results. Washington & Williamson (1977) describe a scheme in which model variables are predicted at an 'anemometer level' in order to compute the surface fluxes. In a different approach, equations for the boundary-

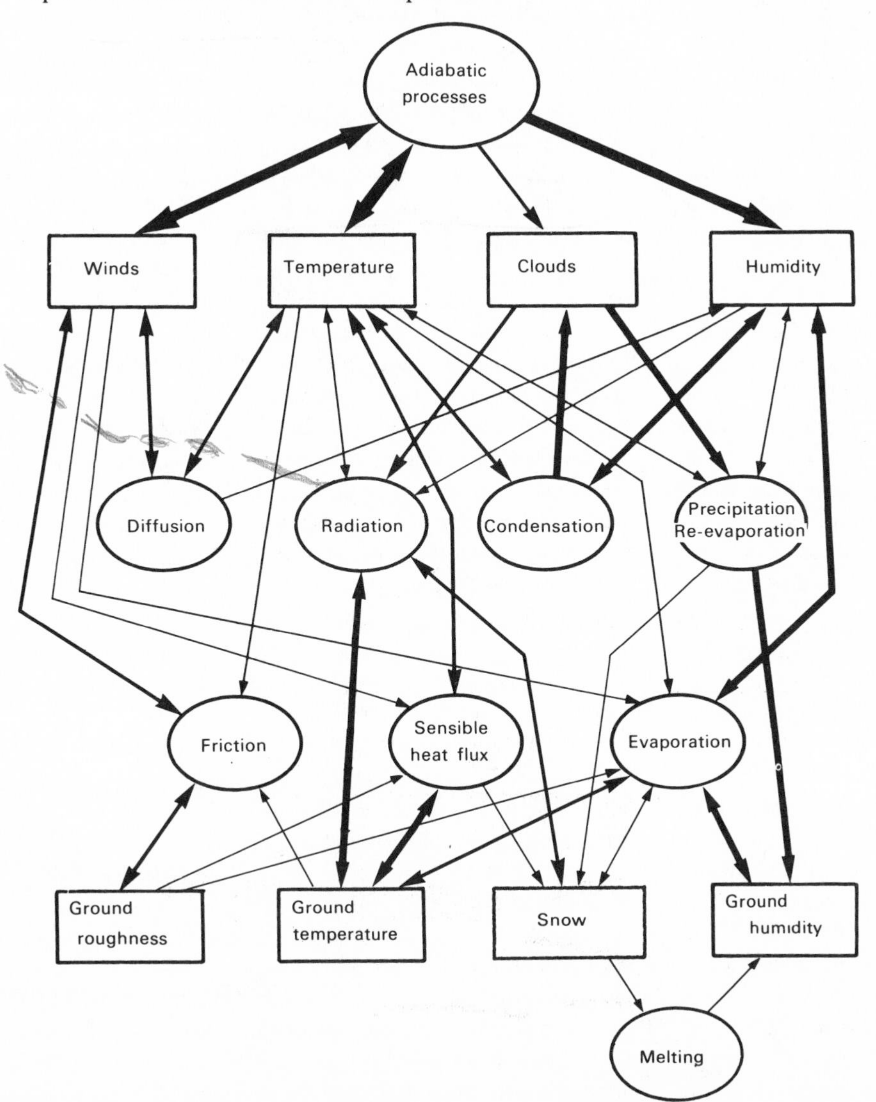

Fig. 4.6. Schematic illustration of the processes commonly included in atmospheric general circulation models. The thickness of a particular arrow gives a qualitative indication of the importance of the interaction the arrow represents.

layer height and layer-mean temperature, horizontal wind and moisture were incorporated into an AGCM by Randall (1976), and Mechoso *et al.* (1982) describe results from a version in which a generalized vertical coordinate is used so that the top of the boundary layer becomes a coordinate surface.

In the explicit boundary-layer models the lowest level is chosen to be within a few tens of metres of the ground, and surface fluxes are determined using either a logarithmic wind profile or, more generally, Monin–Obukhov similarity theory. Above the lowest level, turbulent vertical fluxes are represented as the product of eddy diffusivities and the vertical gradients of the explicitly resolved fields. In the much-used scheme originally described by Smagorinsky *et al.* (1965), the diffusivity is computed from the local wind shear and a mixing length which decreases linearly with height from the lowest model level to become zero at 2.5 km above the surface. The diffusivity depends also on the local stability, measured by the Richardson number, and uses a mixing length which does not vanish in the free atmosphere in the schemes discussed by Delsol *et al.* (1971) and Louis (1979). A 'higher-order closure' scheme in which diffusivities are related to an additional predicted variable, the local turbulent energy, has been developed from the work of Mellor & Yamada (1974) and tested by Miyakoda & Sirutis (1977). This scheme has been successfully used in the prediction experiments reported by Miyakoda *et al.* (1983).

4.5.3
Moist convection

The parameterizations of moist convection in regular use vary substantially in complexity, and thus in computational cost. Of the three types we shall consider, the simplest is the moist convective adjustment scheme introduced by Manabe *et al.* (1965). In this scheme, convection occurs when a model layer becomes saturated and the lapse rate exceeds the moist adiabatic rate. The temperature and moisture fields are then instantly adjusted to a moist adiabat in such a way that moist static energy is conserved. The amount of precipitation is determined by the moisture that must be removed to ensure that the relative humidity does not exceed 100%, or some lower threshold value.

In parameterizations based on proposals by Kuo (1965, 1974), the occurrence of convection depends not only on the existence of an unstable lapse rate, but also on there being a net convergence of moisture by the large-scale flow and surface fluxes. A fraction of the converged moisture is precipitated, and the remainder moistens the atmospheric column, with vertical profiles of heating and moistening derived from differences in temperature and moisture between the cloud air and the larger-scale environment. Detailed descriptions of specific schemes of this type are given, for example, by Anthes (1977) and Tiedtke *et al.* (1979).

A different approach has been taken by Arakawa & Schubert (1974). A number of clouds extend from a common base to different model levels, and each cloud entrains environmental (cloud-free) air through its sides and detains cloud air at its top. The grid-square mean heating and moistening are determined from the net effect on the environment of entrainment, detrainment and the convectively induced subsidence of environmental air. The computation is made possible (or 'closed') by assuming that the convection acts instantaneously to counter any tendency of other processes to destabilize each cloud type. In this sense it resembles the moist adiabatic adjustment scheme.

The performance of different parameterizations of convection has been tested by Krishnamurti *et al.* (1980) and Lord (1982) using observed data, with encouraging results found for both the Kuo (1974) and Arakawa–Schubert schemes. Comparisons using comprehensive numerical models exhibit, however, a general sensitivity of the simulated convective heating to the choice of parameterization scheme (e.g. Miyakoda & Sirutis, 1977; Hollingsworth *et al.*, 1980; Geleyn *et al.*, 1982*b*). More fundamentally, Moncrieff (1981) has pointed out that schemes such as all of those mentioned here do not model the crucial role of downdraughts in deep convection and, in particular, do not represent the generally counter-gradient momentum transfers found in models of intense organized convective systems. As a result of these and other considerations, the parameterization of convection remains a major research topic in atmospheric modelling.

4.5.4
Nonconvective precipitation

The treatment of nonconvective precipitation is typically one of the simpler elements of an overall parameterization scheme. It is computed after other dynamical and physical processes which change the temperature and water vapour content, and generally comprises the condensation, with associate latent-heat release, of sufficient vapour to keep the relative humidity below a fixed threshold value. This value varies from model to model in the general range of 80 to 100%. Values lower than 100% are thought of as representing the small-scale nature of much precipitation, which may occur at points within a grid square, even if in the mean the square is less than saturated. Too low a threshold in the boundary layer may, however, lead to unrealistic amounts of precipitation from the lowest level above a wet surface, as suggested by the forecast experiments of Hollingsworth *et al.* (1980).

In the simplest schemes, all condensed moisture falls instantly to the ground, but the evaporation of precipitation may also be taken into account, as indeed it may also be in the convective case. For example, Somerville *et al.* (1974) and Washington *et al.* (1979) use a scheme in which evaporation occurs until lower layers reach the threshold value for precipitation, while a dependence on the density of precipitation is included by Tiedtke *et al.* (1979), following a proposal of Kessler (1969). A range of further refinements may also be adopted, including an explicit representation of liquid water (Sundqvist, 1981).

4.5.5
Surface values

Changes in a number of the land-surface characteristics used for the calculation of surface heat, moisture and momentum fluxes are generally computed. The surface temperature is either determined diagnostically assuming a local radiative balance (e.g. Smagorinsky *et al.*, 1965), or computed using a prognostic equation for one or more layers of soil (e.g. Corby *et al.*, 1972). A diurnal cycle of incoming radiation is often, though not always, included. The hydrology of the land surface is generally represented following the approach originally adopted by Manabe (1969), and this comprises the prediction of soil moisture (again using one or more layers) and snow depth. Precipitation is classed as rain or snow according to a low-level temperature criterion, and snow-melt (which also influences the surface temperature) and run-off are included, as well as changes due to evaporation. The surface roughness may vary, as in the scheme of Tiedtke *et al.* (1979), according to the nature of the orography, vegetation and urbanization, and over sea according to the surface wind stress. A more-general influence of the vegetation is included by Washington *et al.* (1979), following Deardorff (1978). Attention to further such detail in the treatment of surface characteristics may be expected in the future if current indications (discussed in Section 4.7) concerning the role of surface processes are confirmed by additional studies.

4.6 Predictability

Three types of atmospheric predictability may be distinguished. One is what is known as 'deterministic' predictability, and relates to the prediction of the instantaneous state of the atmosphere for as many days or weeks ahead as may be possible. Such predictions are generally referred to as 'short-range' or 'medium-range' weather forecasts. The other two types are what may be regarded as climate predictability of the *first* and *second* kinds (Lorenz, 1975). The first kind is associated with the prediction of the statistical properties of the atmosphere, for example temporal or spatial means, for time ranges beyond the limit of deterministic predictability. Initial conditions in the atmosphere and at the earth's surface are as important in this case as they are in deterministic prediction, although different aspects of the initial state may be emphasized. This type of prediction may be referred to as 'long-range' forecasting. Climate predictability of the second kind concerns the prediction of the impact of changes in the external forcing. Statistics which are essentially independent of any experimental initial conditions are sought.

Although not the prime concern of this article, the deterministic predictability of the atmosphere must be briefly discussed as it provides a lower bound for climate predictability of the first kind. The first point to stress is that the extent to which the atmosphere may be deterministically predictable depends very much on what particular aspect of the atmospheric state is being considered since, for example, even if the larger synoptic-scale disturbances are accurately forecast, the same may not be true of smaller-scale rain-bearing systems embedded in them. Moreover, from a modelling viewpoint, distinctions between different types of predictability can be artificial. High-resolution limited-area models are the appropriate tool for the very short-range deterministic prediction of the smaller-scale systems, while we have already discussed how a clear distinction cannot be drawn between the models used for longer-range deterministic forecasting and those used for climate prediction.

It is appropriate to distinguish between practical and theoretical limits of deterministic predictability. The practical limit is simply that actually achieved by the best operational weather forecasting systems, or by their research counterparts. For extratropical synoptic scales of motion, this limit at present lies, on average, in the range of about four to seven days depending on the accuracy demanded of a particular forecast, although individual cases of potentially useful forecasts over a range of ten days or more may be found (Bengtsson & Simmons, 1983). The theoretical limit exists because small differences in initial conditions (representative of unavoidable uncertainty in the observation and modelling of the smaller scales of motion) inevitably grow ultimately to dominate the synoptic scales in idealized forecast experiments. A number of past studies have suggested a theoretical predictability limit of the order of two to three weeks (e.g. Charney *et al.*, 1966; Smagorinsky, 1969), a figure confirmed by a study of the actual performance of the ECMWF system (Lorenz, 1982). The clear difference between the practical and theoretical limits is a measure of the improvements in weather forecasts that can be expected to follow from better model design and a better determination of the initial atmospheric (and surface) state. Such improvements are unlikely to occur independently of an improved ability to make climate predictions of one or the other kind.

Selected examples from studies of climate predictability will be given in the following section. Some of these relate directly to predictability of the first kind, attempting to determine what potential there is for making useful predictions of some aspects of atmospheric behaviour for a month or season ahead. This predictability may be considered as fundamentally internal to the atmosphere if it depends strongly on the initial atmospheric state rather than on anomalous surface conditions, and we shall follow Shukla (1981*a*) and refer to this as 'dynamical' predictability. Such predictability would occur if, for example, the forcing of an anomalous circulation pattern by transient weather systems were to be correctly represented in the time mean, even though the instantaneous intensity and position of the individual systems could not be predicted. Alternatively, predictability of the first kind may be strongly influenced by forcing external to (though influenced by) the atmosphere itself, for example forcing due to an anomalous distribution of sea-surface temperature.

Sensitivity to external forcing has commonly been studied not in long-range forecast experiments, but by performing predictability experiments of the second kind. For some applications the motivation has been to understand the working of past climatic regimes (e.g. Williams *et al.*, 1974; Gates, 1976; Manabe & Hahn, 1977), but in other cases (some of which will be discussed below) the sensitivity of climate simulations to changed boundary conditions has been investigated as a guide to the processes that might be important for forecasting on the monthly or seasonal time scale. Insofar as the model used provides a realistic representation of the unperturbed climate, the latter studies give an overall indication of the importance of a particular boundary forcing. However, they may underestimate the potential impact of that forcing on particular forecasts, since the response to anomalous forcing may depend strongly on the synoptic situation (e.g. Shukla & Wallace, 1983; Arpe, personal communication).

4.7 Some examples of climate studies

4.7.1 *Numerical forecasts of monthly means*

Forecasts for a month or season ahead are currently carried out on a routine operational basis using statistical methods, but results exhibit, at best, only a marginal skill (Nichols, 1980; Gilchrist, 1981*a*; Nap *et al.*, 1981). Consideration has thus been given to the use of comprehensive general circulation models for such forecasting, although intermediate approaches based on simpler models have also been adopted (e.g. Marchuk & Skiba, 1976; Chao *et al.*, 1982; Miyakoda & Chao, 1982). Concerning the use of comprehensive models, a first attempt by Gilchrist (1977*a*) gave a suggestion of success in the prediction of the larger scales of motion, and recent work by Shukla (1981*a*) and Miyakoda *et al.* (1983) has provided more substantial evidence of the possible usefulness of this approach. Some results from operational medium-range prediction are also relevant in this context (Bengtsson & Simmons, 1983).

The study reported by Shukla (1981*a*) represents an extension of the type of experiment used to determine the theoretical limit of deterministic predictability. Monthly mean forecasts from initial conditions randomly perturbed by amounts typical of observational errors were found to vary less than monthly means from markedly different (observed) initial conditions, despite the use of the same climatological boundary conditions for each forecast. This result was found for the average over the first month of a two-month forecast period, and suggests a potential for dynamical predictability on at least a monthly scale, since in these experiments the lack of deterministic predictability of synoptic-scale systems was apparently not such that different forecasts of these systems gave rise to substantially different means over the first month. Means taken over the second month were, however, found to be unpredictable, at least given the fixed boundary conditions used in the experiments.

A specific example of a remarkably accurate monthly forecast has been presented by Miyakoda *et al.* (1983). Success was achieved in simulating the events of January 1977 and, in particular, a record-breaking cold spell over eastern North America, as shown in Fig. 4.7. Results were found to be sensitive to the model adopted, and the best performance was obtained by using the highest resolution and most-sophisticated parameterization scheme. Climatological initial conditions were used at the surface. Other forecasts performed for this month have been of mixed accuracy (Strüfing, 1982*a*,*b*; Cubasch, personal communication), and the modelling factors which yield success in this case remain to be precisely determined.

Further evidence of the ability of models to represent actual climatic anomalies may be found from the study of the performance of medium-range forecast models. Bengtsson & Simmons (1983) illustrate how the mean of all ECMWF forecasts for ten days ahead, produced daily during July 1980, succeeded in representing the temperature anomaly associated with a major heatwave and drought over southeastern North America. Such an anomaly was, of course, present in the initial conditions for the predictions, but the forecast model (which used climatological sea-surface temperatures) was evidently capable of maintaining the strength of the anomaly over at least a ten-day period.

Such isolated studies as those mentioned here cannot yield a definitive picture of the predictability of the atmosphere on a monthly time scale. They nevertheless provide encouraging first results and, taken with evidence from studies of sensitivity to boundary forcing (to be discussed later), they justify a substantial research effort to establish the feasibility of making routine monthly predictions using comprehensive numerical models.

4.7.2 *The simulation of low-frequency variability*

As comprehensive numerical models have been developed, attention has naturally been placed first on examining their representation of the time mean circulation of the atmosphere. In addition, their successful use for weather prediction has confirmed that they can accurately describe the day-to-day evolution of individual weather systems and the development of new disturbances. However, with the exception of the simulation of the atmosphere's annual cycle, it is only recently that attention has been concentrated on the ability of models to reproduce the lower-frequency variability of the atmosphere, for example as characterized by the differences between monthly means from different years, or by differences in seasonal transitions from year to year.

Results from an 18-year simulation using an annual variation of solar radiation have been reported by Manabe & Hahn (1981) and Lau (1981), and similar experiments have been performed for six- and ten-year periods using a different model (Cubasch, 1981*a*; Volmer *et al.*, 1983*a*,*b*). In both models the sea-surface temperature varied only according to a climatological annual cycle, but soil moisture and snow depth evolved as predicted variables. The cloud

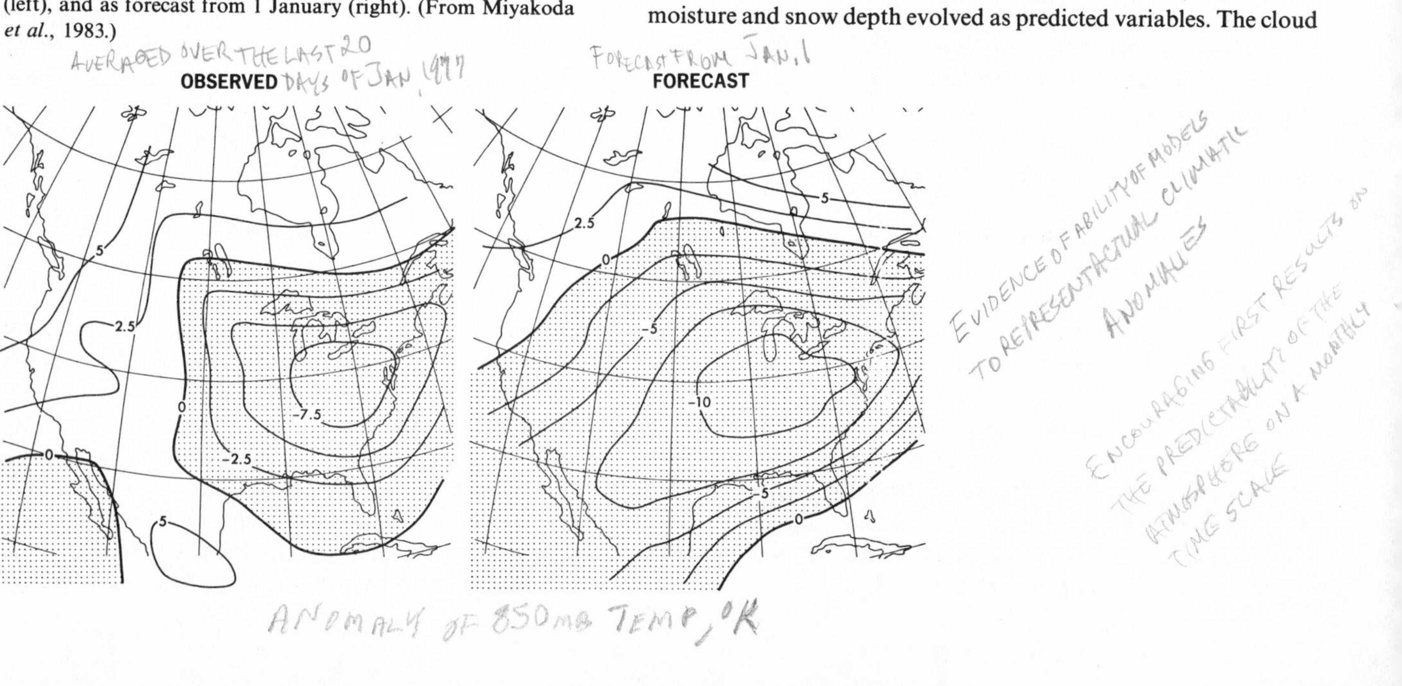

Fig. 4.7. The observed anomaly of 850 mb temperature (K) for North America averaged over the last 20 days of January 1977 (left), and as forecast from 1 January (right). (From Miyakoda *et al.*, 1983.)

cover was held constant in time in the 18-year experiment, and varied interactively in the shorter simulations, although no differences in results have been ascribed to this. The simulations, in fact, each exhibited a significant interannual variability, with the occurrence of substantial and persistent anomalies, including both local perturbations in the northern hemisphere, resembling the observed teleconnection patterns (e.g. Wallace & Gutzler, 1981), and spells of unusually strong (or weak) zonal flow in the southern hemisphere. The latter type of anomaly also occurs in nature (e.g. Trenberth, 1979), and a simulated example is shown in Fig. 4.8. Overall, the amount of low-frequency variability was found to be somewhat less than observed in middle latitudes (at least for the northern hemisphere), and substantially less in the tropics.

The above results are in accord with the view that surface anomalies are a more important source of low-frequency variability in the tropics than in middle latitudes (Charney & Shukla, 1981). It is, however, difficult to draw quantitative conclusions concerning the importance of the various processes that can cause variability on time scales of a month and upwards, since discrepancies between the simulations and reality may be due either to the omission of some external forcings or to an inadequate treatment of those processes included in the models. An ability to simulate long-lasting anomalies also does not immediately imply that the development, persistence and decay of actual anomalies are predictable for more than a few days or weeks ahead. Nevertheless, the results from the extended simulations are important as they imply a capability for studying the impact of external changes on the variability of climate, as well as its mean. They also provide a basis for carefully controlled numerical experimentation which may help elucidate the mechanisms and predictability of the low-frequency components of the general circulation.

4.7.3
The response to sea-surface temperature anomalies

One of the most common applications of AGCMs has been to investigate the response to anomalies in sea-surface temperature. The first experiment, by Rowntree (1972), was conducted to test suggestions by Bjerknes (1966, 1969) that fluctuations of ocean temperatures in the tropical Pacific could be responsible for significant variations in the atmospheric circulation over and downstream of the extratropical North Pacific. Rowntree's results, and those subsequently obtained for heating from both the tropical Pacific (Julian & Chervin, 1978; Wells, 1979*b*; Rowntree, 1979;

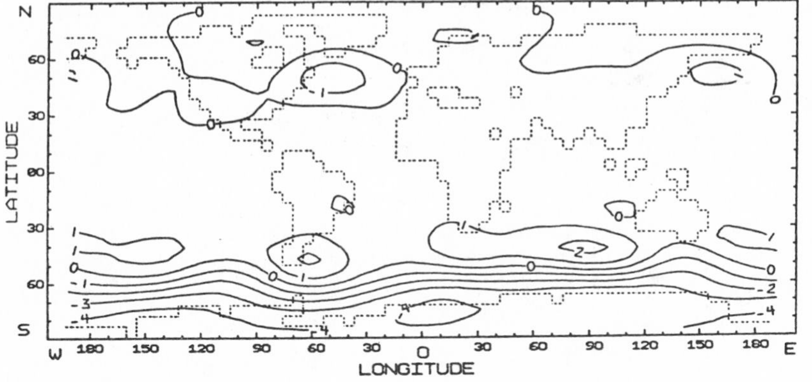

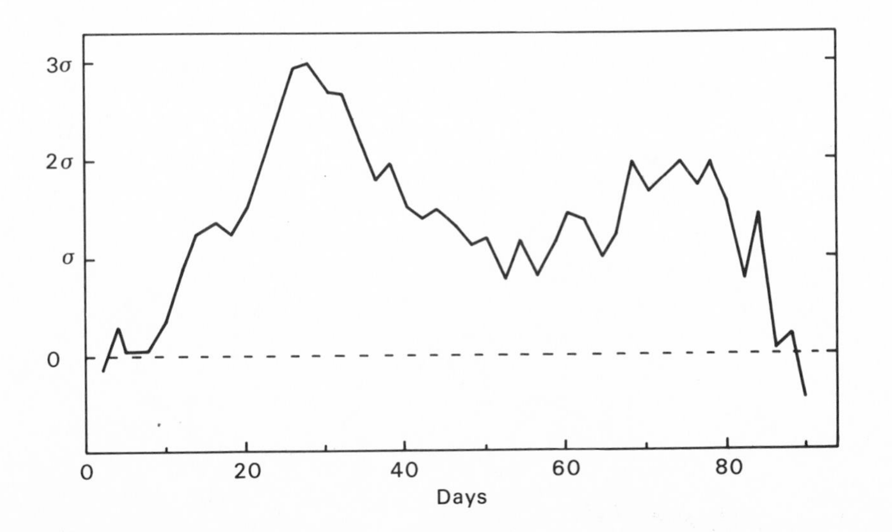

Fig. 4.8. An example of a persistent 500 mb height anomaly. Upper: The pattern of the first principal component of non-seasonal variability. Lower: The coefficient of this component (expressed in terms of its rms value σ) during a period in which its sign did not change for almost 90 days. At the peak of the anomaly, the zonal mean 500 mb flow was more than 60% stronger than normal. (From Volmer *et al.* 1983*a*.)

Keshavamurty, 1982; Blackmon *et al.* 1983; Shukla & Wallace, 1983) and the tropical Atlantic (Rowntree, 1976), indeed indicate some influence of tropical anomalies on the middle-latitude circulation of the winter hemisphere, and they are in at least qualitative agreement with observational studies (e.g. Horel & Wallace, 1981). The picture deduced from comprehensive models and observations has also been confirmed by a number of studies using idealized models (e.g. Egger, 1977; Opsteegh & van den Dool, 1980; Hoskins & Karoly, 1981; Webster, 1981; Simmons, 1982*a*; Hendon & Hartmann, 1982; Simmons *et al.*, 1983), and these models have in turn indicated a potential sensitivity of the response in comprehensive models, suggesting that results may depend strongly on the synoptic situation in short-term anomaly experiments, and that systematic deficiencies in the model climate (for example in the meridional and zonal variations of the middle-latitude jet streams) may bias the response in longer-range simulations. Confirmation of some of this sensitivity may be found in the numerical experiments of Shukla & Wallace (1983).

Other studies have investigated the more local tropical response to anomalous sea-surface temperatures. Moura & Shukla (1981) report experiments which confirm observational and theoretical results in associating drought over northeast Brazil with temperature anomalies in the tropical Atlantic, and an example of their results is shown in Fig. 4.9. Shukla (1975) has also described experiments relating a decrease in monsoon rainfall over India to relatively cold Arabian-Sea temperatures. These latter results were not confirmed using a different model (Washington *et al.*, 1977), but Shukla (1981*b*) reports similar conclusions from use of a third model, and ascribes the discrepancy to a less realistic simulation of the unperturbed monsoon flow in the model used by Washington *et al.* If so, this serves as another example of the importance of achieving an accurate climate simulation by models that are to be used for anomaly experiments.

A third type of study has concerned the response to middle-

Fig. 4.9. The response of rainfall to a sea-surface temperature anomaly in the tropical Atlantic. The upper map shows the anomaly, and the lower graphs the rainfall from an unperturbed control simulation and a simulation with the anomaly included. Rainfall is plotted as 15-day running means for the two areas, A and B, shown in the upper map. (From Moura & Shukla, 1981.)

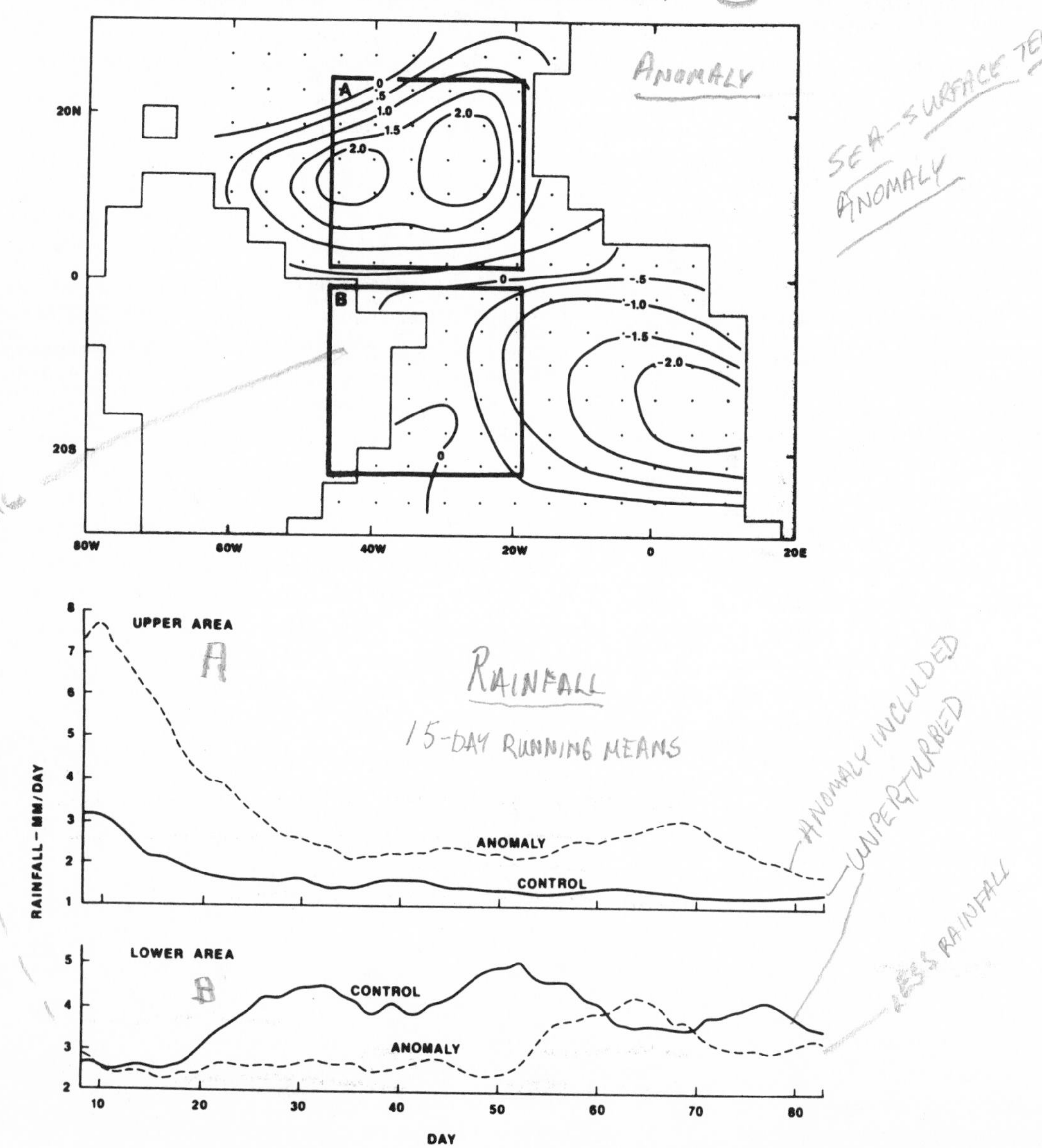

latitude anomalies. Numerous experiments have been reported (e.g. Spar, 1973; Houghton *et al.*, 1974; Simpson & Downey, 1975; Kutzbach *et al.*, 1977; Rowntree, 1979; Shukla & Bangaru, 1979; Chervin *et al.*, 1980) but, for the most part, they have not shown a clearly significant response distant from the anomaly. Only the results of Shukla & Bangaru are indicative of the pronounced downstream influence suggested by a number of observationally based studies (e.g. Namias, 1969, 1978; Ratcliffe & Murray, 1970), although the lack of a definitive response from comprehensive models is consistent with simpler model experiments discussed by Webster (1981).

Overall, a reasonably clear qualitative picture of the role of sea-surface temperature anomalies has arisen from the combined use of observational data and both comprehensive and idealized models, at least for tropical anomalies. Sensitivity to the synoptic situation and to model deficiencies make it difficult to draw quantitative conclusions, and the topic thus remains one which is the subject of much continuing research.

4.7.4

Sensitivity to other surface conditions

The influence of several surface conditions other than sea-surface temperature has been investigated in a number of studies. Sensitivity to the initial distribution of soil moisture was examined for 20-day tropical simulations by Walker & Rowntree (1977), using a model in which the subsequent amount of soil moisture was predicted in the manner outlined earlier. An idealized 'desert' zone which was initially dry remained so throughout, in striking contrast to the balance of evaporation and precipitation set up over the same region when it was initially moist. Miyakoda *et al.* (1979) and Miyakoda & Strickler (1981) found several pronounced differences in summer forecasts for the northern hemisphere using different fixed distributions of soil moisture, and Shaw (1981) illustrated how forecasts of monsoon flow can be markedly sensitive to the initial prescription of (a subsequently predicted) soil moisture. Further examples can be found in Mintz's contribution to this volume.

The role of the surface albedo is central to a mechanism for drought proposed by Charney (1975). He argued that an increase in albedo could result in increased radiative cooling of the atmosphere, enhanced sinking, decreased rainfall, and reduced vegetation. Since the albedo increase could itself result from reduced vegetation due either to incipient drought or to other natural or anthropogenic influences, a biogeophysical feedback giving rise to or enhancing drought in sub-tropical desert margins was possible. Subsequent experiments using an AGCM confirmed the feasibility of part of this mechanism (Charney *et al.*, 1977). Higher prescribed albedos over semi-arid regions were indeed found to result in reduced local precipitation. Although results were sensitive to relatively crude parameterizations of evapotranspiration, they have largely been confirmed by more recent experimentation (Sud & Fennessy, 1982).

Sensitivity to the location of the sea-ice limit in the Arctic has been examined by Herman & Johnson (1978). A change in sea-ice cover comparable with the difference between maximum and minimum observed winter limits over a 17-year period was found not only to affect the local atmospheric circulation, but also to produce apparently significant changes in middle and sub-tropical latitudes. Snow cover is another factor which it has been suggested might influence atmospheric anomalies on the seasonal time scale (e.g. Hahn & Shukla, 1976), but numerical experiments relating to this topic have yet to be undertaken.

In general, studies such as those discussed here place heavy demands for accuracy in the parameterizations describing both the evolution of surface fields and the effect of these fields on the resolved atmospheric circulation. Results contribute towards an improved understanding of the potential for seasonal predictability and the mechanisms of drought in the tropics and sub-tropics and, as such, the processes involved may be expected to be the subject of repeated attention as parameterization schemes are refined, and tropical simulations become closer to reality.

4.7.5

The response to an increase in carbon dioxide content

There have been numerous studies of the possible climatic impact of an increased concentration of atmospheric carbon dioxide, but these have, for the most part, used models simpler than those considered here. There is, however, a need for the application of the more-comprehensive models to this important problem, particularly as they are the only tool for the consistent prediction of local, geographically dependent changes. The experiments reported to date with these models have used some quite different oceanic representations, and substantial quantitative differences in atmospheric response have been found. Quite apart from their relevance to the carbon dioxide problem, the results discussed here are indicative of some of the interactive processes that can be of importance in climatic change, and provide examples of the possible sensitivity of model results to the type of parameterization adopted.

Experiments performed by Manabe & Wetherald (1975, 1980) used an annual mean insolation and an idealized distribution of land and ocean, with the latter represented as a region with an unbounded supply of moisture for evaporation, but with no heat capacity and no heat transfer by currents. A mean surface temperature rise of 3 K was found in response to a doubling of carbon dioxide, with a larger mean rise of 6 K and local rises of 9 K or more in response to a quadrupling. The local maxima resulted from a drier soil (and consequently reduced evaporation) in the continental interior at middle latitudes, and from a poleward retreat of snow cover and sea ice (and consequently reduced albedo) at high latitudes. The mean latitudinal temperature gradient was decreased, and there was reduced middle-latitude eddy activity and a poleward shift of the associated rainbelt.

A zero heat capacity ocean has a temperature which adjusts immediately to changes in the surface fluxes of radiation. At the other extreme, fixed climatological temperatures were used in experiments (with realistic geography) reported by Gates *et al.* (1981) and Mitchell (1983). Doubling carbon dioxide in these models resulted in a much smaller change, although summer surface temperature rises of over 2 K were found for parts of North America and Eurasia. Mitchell also carried out an experiment in which the carbon dioxide doubling was accompanied by a fixed increase of 2 K in ocean temperature. Atmospheric temperature changes were in this case similar to those found by Manabe & Wetherald, except near the surface at high latitudes where a fixed sea-ice distribution was used.

Further experiments, these using a simple mixed-layer ocean

model, have been described by Manabe & Stouffer (1980) and Manabe *et al.* (1981). Results were obtained for both idealized and realistic distributions of land and ocean. A quadrupling of carbon dioxide gave a maximum increase of 7 K in the summer surface temperature at middle latitudes of the idealized model, and similar increases were found with realistic geography, as is shown in Fig 4.10. Winter surface temperature rises (also shown in Fig. 4.10) were as high as 15 K or more, and were generally largest at high latitudes and in the northern hemisphere, where the reduction in sea ice and snow cover was largest. The summer soil moisture was significantly reduced both at middle latitudes, where the reduction in rainfall from spring to summer occurred earlier than in the unperturbed case, and at high latitudes where there was an earlier and smaller total snow-melt, allied with enhanced evaporation due to a lower albedo once the snow had disappeared.

Additional studies are currently being carried out both by modelling groups whose work has been mentioned here, and by others. Attention is being directed towards a more comprehensive treatment of the ocean, and first results have been presented by Bryan *et al.* (1982).

4.8

Some problems in climate simulation

Some of the specific problems that arise in the design and use of AGCMs for climate studies have already been mentioned where appropriate in preceding sections. Here we shall take a different view, and discuss a number of problems as seen by examining some features of the actual performance of these models in the simulation of the present climate. Some aspects of their performance in weather prediction will also be considered since many of the errors found in the 'climate' of a forecast model, as revealed by simulations over extended time ranges, may also be clearly seen in the 'systematic' forecast errors that are identified

Fig. 4.10. An example of a computed increase in surface air temperature (K) resulting from a quadrupling of the carbon dioxide content of a model atmosphere. (*a*) Annual mean. (*b*) December–February mean. (*c*) June–August mean. Shading denotes regions where the increase exceeds 5 K. (From Manabe & Stouffer, 1980.)

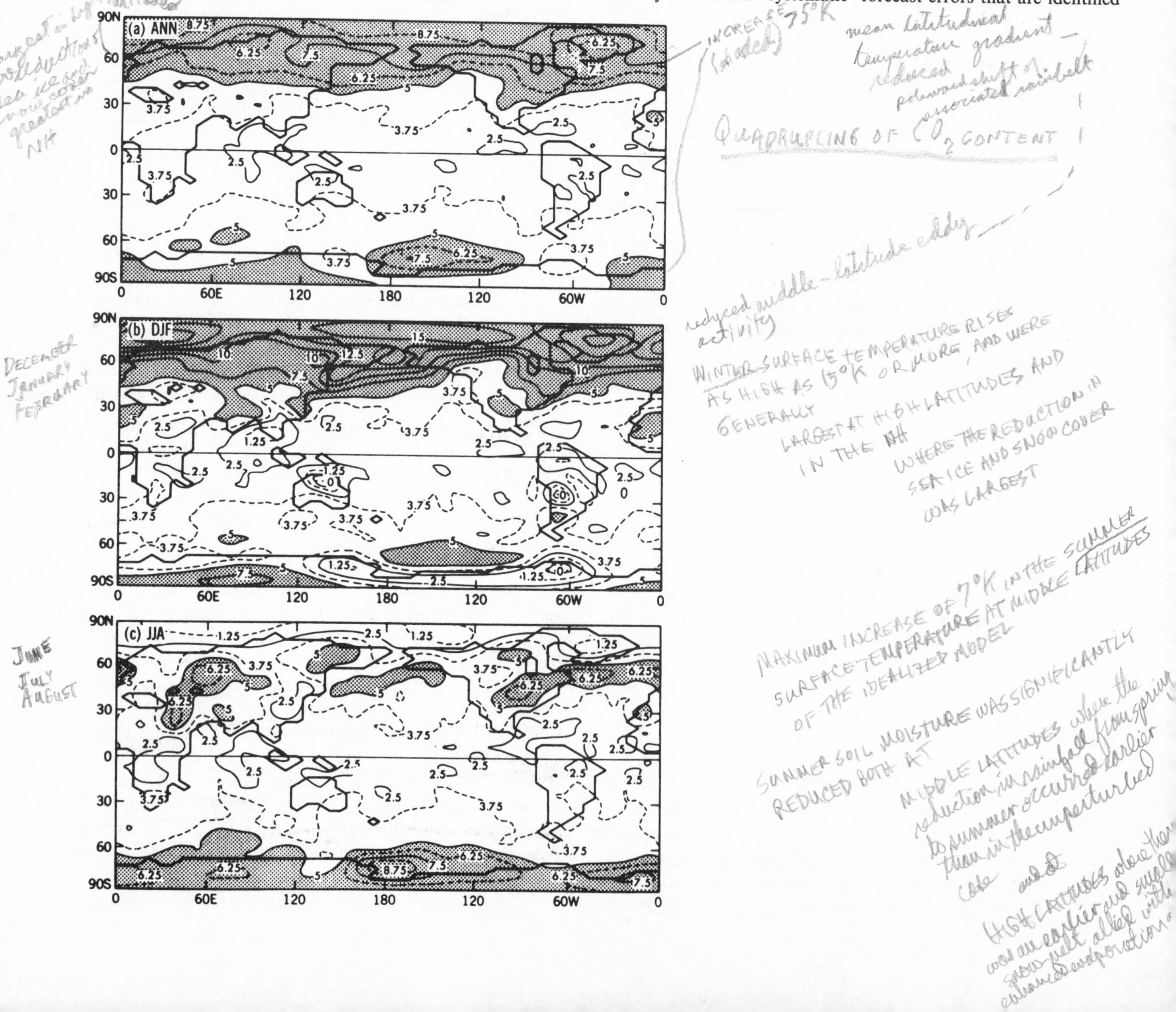

by averaging over a representative sample of individual, shorter-range forecasts. These systematic errors grow in amplitude as the time range of the forecast increases, and their spatial pattern may change. Their behaviour is of particular interest because the manner in which they evolve may itself be suggestive of their source, and thus of the source of corresponding errors in the simulation of climate (Wallace *et al.*, 1983).

4.8.1
The mean temperature structure

Climate simulations by a number of AGCMs are presented in *GARP Publication Series No. 22* (1979). They show that these models achieve a general success in representing the mean thermal structure of the troposphere, and any overall biases are difficult to discern from the published latitude–height cross-sections of zonal mean temperature. Cross-sections of mean temperature error presented for some series of medium-range forecasts do show a tendency for tropospheric cooling by up to several degrees (Miyakoda *et al.*, 1972; Bengtsson & Simmons, 1983), and a similar deficiency has been noted by Druyan *et al.* (1975). Such errors can, however, be sensitive to a number of the detailed aspects of model design, including not only particular features of the parameterizations of diabatic processes (e.g. Hollingsworth *et al.*, 1980), but also such features as the prescription of the orography (Wallace *et al.*, 1983) and the amount of horizontal smoothing of model fields (Girard & Jarraud, 1982), both of which may influence precipitation and the associated latent-heat release.

Temperature errors become more pronounced at upper levels, where there is a general tendency for too cold a stratosphere, particularly near the winter pole. Associated with this the stratospheric polar-night jet is generally too strong, and often insufficiently separated from the main sub-tropical jet. We have already remarked on the possible contribution of the radiative parameterization to these errors (Ramanathan *et al.*, 1983), but there may also be dynamically induced biases due to inadequate stratospheric resolution (Fels *et al.*, 1980) or inadequate tropospheric forcing of quasi-stationary stratospheric disturbances (O'Neill *et al.*, 1982). Heat transfer by the latter is an important component of the thermal balance of the extratropical winter stratosphere, and underestimation of the tropospheric stationary-wave component is a model fault which will be discussed further below.

4.8.2
The zonal circulation

A number of common, though by no means universal, deficiencies in the representation of the zonally averaged flow may be noted. At the surface, the middle-latitude westerlies are generally found to be a few metres per second stronger than observed in the northern hemisphere, and this discrepancy can become more pronounced at higher resolution (Manabe *et al.*, 1979*b*; Cubasch, 1981*b*). Conversely, higher resolution generally yields a better simulation of the stronger surface westerlies of the southern hemisphere. Surface easterlies in the tropics and near the poles are, for the most part, well captured, although excessive polar pressures may arise from use of a model grid that is irregular in longitude (Dey, 1969; Holloway and Manabe, 1971; Holloway *et al.*, 1973). In the upper troposphere, westerly maxima are usually found to be too strong (again by a few metres per second), and displaced slightly poleward and upward. Tropical easterlies tend to be underestimated near the tropopause, and overestimated at higher stratospheric levels.

4.8.3
Spatial variability

Simulations of the time-averaged sea-level pressure field, for the northern hemisphere winter, commonly exhibit Icelandic and Aleutian Lows which are too deep and shifted towards the east, and these deficiencies have also been found to become more marked at higher resolution (Manabe *et al.*, 1979*b*; Cubasch, 1981*b*), as illustrated in Figure 4.11. In addition, an erroneous westerly surface flow over North America has been noted by Blackmon & Lau (1980), Lau (1981) and Wallace *et al.* (1983), and this may also be seen in Fig. 4.11. Evidence that the simulation of the surface low-pressure areas may be sensitive to detail in the formulation of the boundary-layer parameterization has been presented by Louis *et al.* (1981), while the diagnostic and experimental results described by Wallace *et al.* (1983) suggest that some deficiencies of the surface pressure field may be due to an inadequate representation of orographic effects. For example, underestimation of the barrier presented to the low-level zonal flow by the Rocky Mountains may account for the erroneous model flow over North America.

The study by Wallace *et al.* (*loc. cit.*) was motivated by a systematic tendency of the ECMWF model to gradually weaken the time-averaged wave component of the extratropical flow at middle and upper tropospheric levels. For example, Fig. 4.12(*a*) shows mean forecast errors in 500 mb height for 100 consecutive winter forecasts for ten days ahead, superimposed on the mean 500 mb height for the same 100-day period. The correspondence between negative errors and ridges and between positive errors and troughs is clear evidence of the failure of the model to maintain the mean wave pattern. Similar forecast errors have been reported by Miyakoda *et al.* (1972), and a general lack of stationary-wave amplitude has also been noted for climate simulations (e.g. Pratt, 1979; Straus & Shukla, 1981; O'Neill *et al.*, 1982), although Pratt (*loc. cit.*) and Manabe *et al.* (1979*b*) show that this is not invariably the case. The results obtained by Wallace *et al.* are highly indicative of an insufficient orographic forcing in the ECMWF model at least, as suggested by Fig. 4.12(*b*) which shows how mean negative errors in one-day forecasts are located over mountainous regions. An inadequate forcing of the extratropics originating from erroneous diabatic heating in tropical regions may not be ruled out as an additional source of stationary-wave error.

4.8.4
Temporal variability

Detailed studies of the temporal variability exhibited by several different models have been published by Pratt (1979), Blackmon & Lau (1980), Straus & Shukla (1981), Manabe & Hahn (1981), and others. Results generally indicate that this variability is smaller than observed, particularly at upper levels, for larger scales, and for lower frequencies. Conversely, in medium-range forecasts with the ECMWF system, the variability associated with transient disturbances tends to be overestimated (Hollingsworth *et al.*, 1980; Wallace *et al.*, 1983) and the overall cycle of energy conversions is also generally stronger than observed (Bengtsson, 1981*a*). Both synoptic- and planetary-scale eddy energy has been found to increase in finite-difference models when the horizontal resolution is increased and the associated horizontal smoothing decreased (Manabe *et al.*, 1970; Wellck *et al.*, 1971; Williamson, 1978), but the component associated with transient waves did not increase at higher resolution in experiments reported by Manabe *et al.* (1979*b*) using both finite-difference and spectral models.

The prediction of a particular case of blocking has been found to be sensitive to horizontal resolution (Bengtsson, 1981*b*), but it has yet to be established whether the longer-term simulation of low-frequency variability is similarly sensitive. This could, for example, be the case if higher-frequency transient eddies, themselves directly sensitive to model resolution, were to play an important role in maintaining lower-frequency disturbances, as discussed by Green (1977) and Hoskins *et al.* (1983). As we have already mentioned, an underestimation of low-frequency variability may occur due to the use of some fixed boundary conditions.

Fig. 4.11. The January mean sea level pressure (mb) from R15, R21 and R30 spectral simulations, and as observed. (From Manabe *et al.*, 1979*b*.)

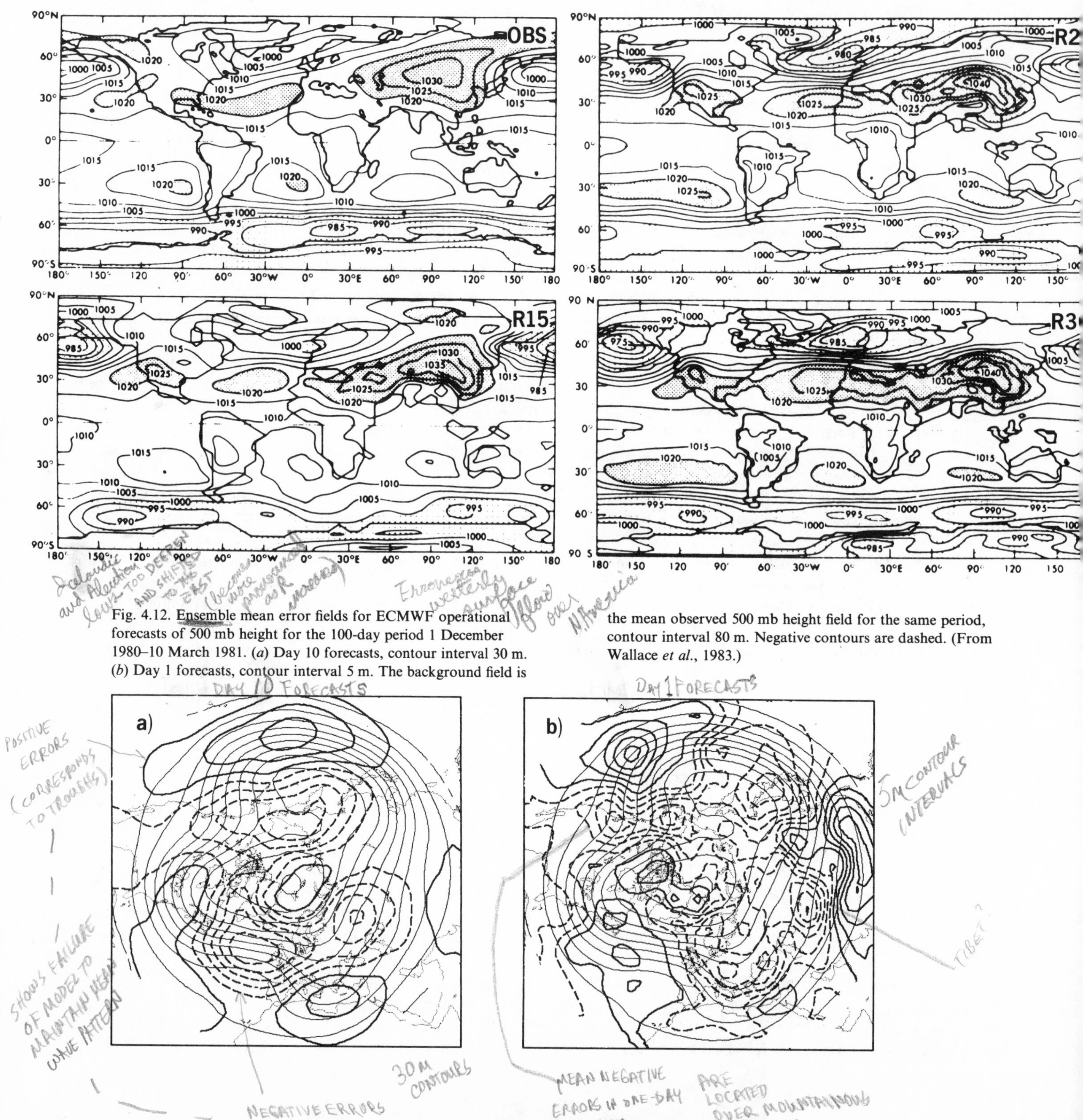

Fig. 4.12. Ensemble mean error fields for ECMWF operational forecasts of 500 mb height for the 100-day period 1 December 1980–10 March 1981. (*a*) Day 10 forecasts, contour interval 30 m. (*b*) Day 1 forecasts, contour interval 5 m. The background field is the mean observed 500 mb height field for the same period, contour interval 80 m. Negative contours are dashed. (From Wallace *et al.*, 1983.)

Fig. 4.13. (*a*) A simulation of net precipitation for February (upper), and the climatology for the December–February period (lower), from Newell *et al.* (1974), after Möller (1951). The diagonal and solid shadings correspond to similar precipitation rates for the two maps. Model grid squares in which no precipitation occurred are indicated by dots. More detail of the simulation is given by Simmons (1982*b*). (*b*) As (*a*), but for a July simulation and the June–August climatology.

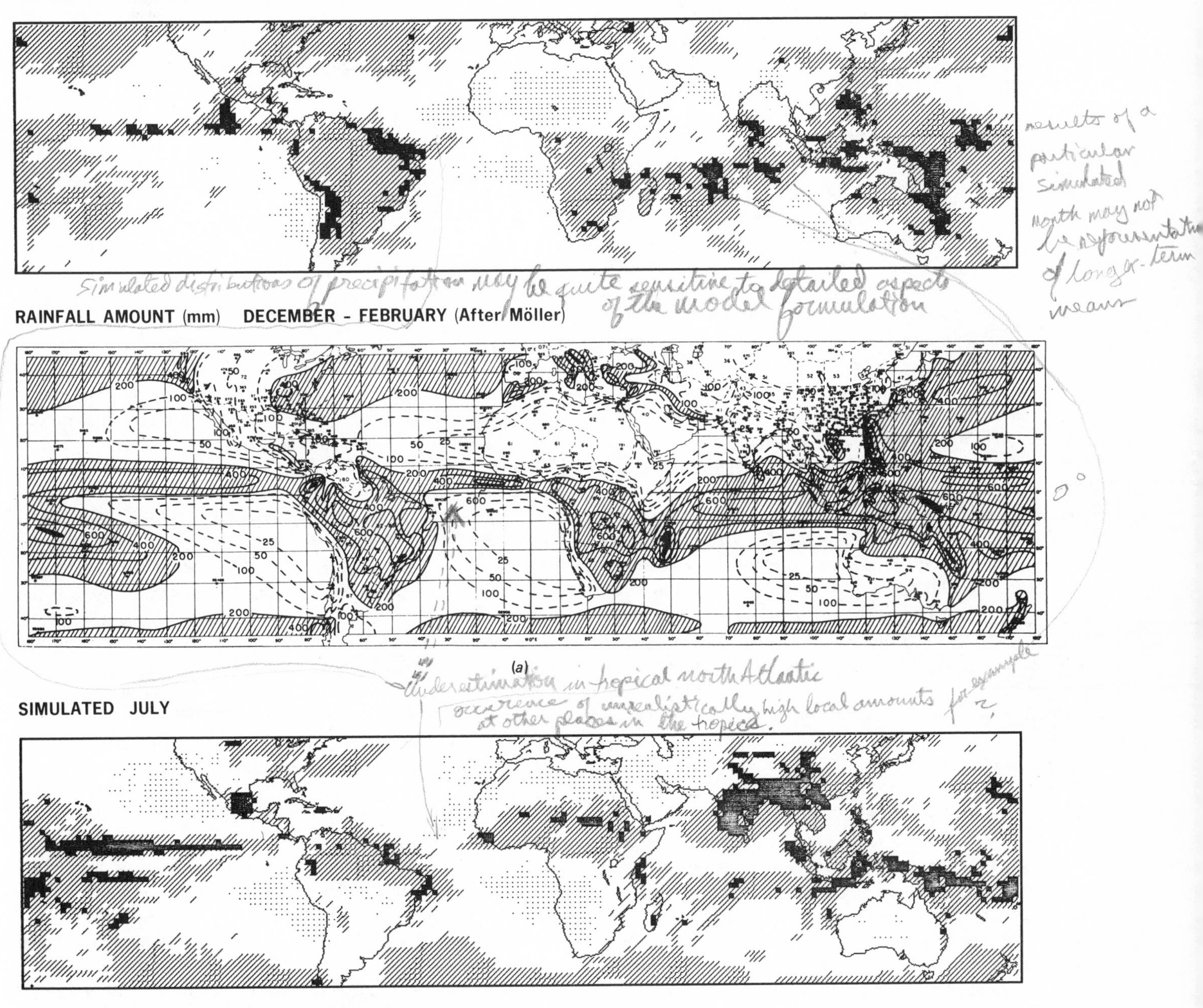

RAINFALL AMOUNT (mm) JUNE - AUGUST

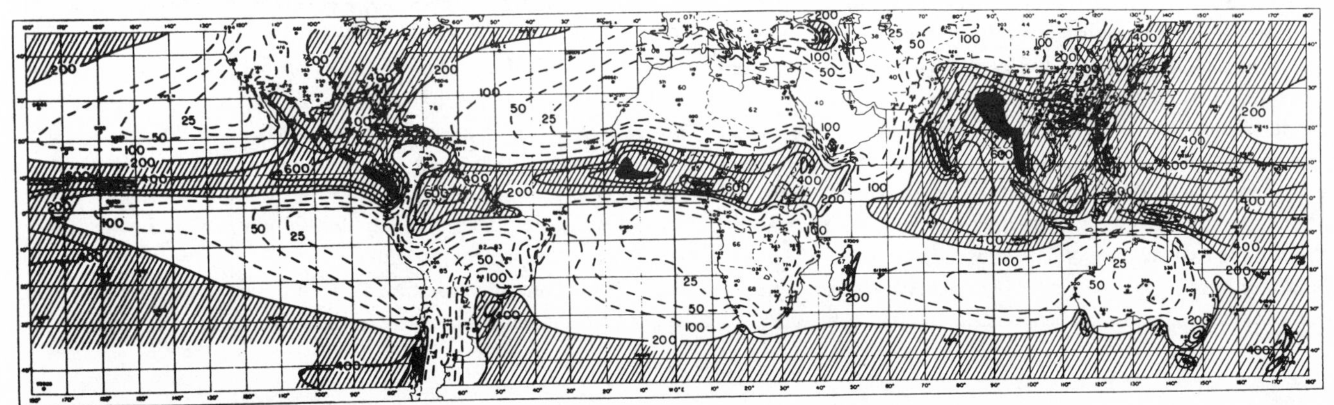

(*b*)

4.8.5
Precipitation

Simulated global precipitation maps presented in *GARP Publication Series No.* 22 (1979) and elsewhere show that despite low levels of transient eddy activity, the amounts of precipitation that occur in the storm tracks of the North Atlantic and Pacific Oceans are, for the most part, realistic, although cases of either excessive or deficient rainfall may also be found. Elsewhere, the overall distribution of wet and dry regions is quite well simulated, as are seasonal changes in precipitation. Examples are presented in Fig 4.13, although it should be noted that in matters of detail, such results for particular simulated months may not be representative of longer-term means, and that, in general, simulated distributions of precipitation may be quite sensitive to detailed aspects of the model formulation. Deficiencies which are common to several models include a general underestimation of rainfall in the tropical North Atlantic, and the occurrence of unrealistically high local amounts at other places in the tropics.

Local rainfall distributions are, in general, highly sensitive to model resolution (e.g. Manabe *et al.*, *1979b*) and to the prescription of the model orography. Considering as a particular example the Asian Summer Monsoon, articles by Gilchrist (1977*b*, 1981*b*) and Washington (1981), which include reviews of earlier studies by Hahn & Manabe (1975) and Washington & Daggupaty (1975), discuss a general lack of detail in the simulation of precipitation in the vicinity of the Indian sub-continent. A tendency for models to produce excessive amounts of rainfall was also noted. Some more-realistic features have been found in a higher-resolution simulation using a relatively detailed orography (Simmons, 1982*b*), but local rainfall maxima were again overestimated. Thus, even for a country as large as India, models are some way from reproducing local detail in simulated rainfall patterns, and sensitivity or predictability studies can only be expected at present to provide results averaged over quite large areas, for some parts of the world at least.

4.8.6
The tropical climate

The numerical modelling of the tropical atmosphere is discussed more generally in *GARP Publication Series No. 20* (1978). Accounts of the overall simulation of the Asian Summer Monsoon may be found in the papers referred to in the preceding paragraph, while Rowntree (1978) has given a review of tropical prediction and simulation with emphasis on west Africa and the Atlantic. In general, the persistent components of the tropical climate, such as the monsoons, trade winds and desert heat lows, can be strongly influenced by atmospheric and surface processes that have to be parameterized in models, and there may be substantial interactions between these processes and the larger-scale circulations. In a model, erroneous latent heating that is associated with deficiencies in precipitation may drive erroneous local circulations, and these circulations may further change the distribution of precipitation. An example is discussed briefly by Simmons (1982*b*). We have already mentioned how different models may differ in their representation of monsoon flow and, just as in reality changes in tropical heating associated with sea-surface temperature anomalies may influence the extratropics, so also unrealistic model distributions of diabatic heating may affect the representation of stationary waves in middle latitudes. Improved accuracy in the simulation of the tropical climate is thus a major goal in model development, and its attainment will benefit not only a wide range of climate studies, but also numerical weather prediction, since the rapid growth of systematic errors can contaminate forecasts for the tropics much more rapidly than those for middle latitudes (Bengtsson & Simmons, 1983).

4.9
Concluding remarks

The past two-and-a-half decades have witnessed an immense development in our ability to model climate. Twenty-five years ago, pioneering attempts were being made to reproduce the gross features of the working of the atmospheric ‘heat engine’. Today, we have reached a stage at which we seek to verify the performance of models in their representation of natural atmospheric variability on time scales of up to a year, in their simulation of the detailed structure of monsoon flows, in their description of the many processes involved in the hydrological cycle, and in numerous other respects. Indeed, the discussions of model performance given in this article have concentrated on what are in many cases quite minor deficiencies in the simulation of major features of the atmospheric general circulation. Whilst models do tend to share common defects, this is becoming less so than in the past, and it has proved difficult to generalize about some specific aspects of model behaviour.

The development of increasingly realistic models has been accompanied by their increasing use for a wide range of purposes. AGCMs now have a regular, operational application in deterministic weather prediction, and they are being applied experimentally to investigate the potential for routine prediction of some of the statistics of weather on the monthly and seasonal time scales. They are being used to evaluate the importance of a variety of surface processes during anomalous climatic regimes, both those which are natural properties of the overall climatic system and those which might result from man’s activity. They are giving insight into the nature of past climates, and they are beginning to provide quantitative predictions of some possible future courses of climatic change.

Although comprehensive models are capable of representing the major features of the earth’s climate, they are all to some extent or other subject to deficiencies in their description of detail. For example, errors may occur in the intensity and position of a larger-scale component, or there may be a more-complete misrepresentation of a local feature. Such defects manifest themselves in the systematic errors of numerical weather prediction models, which can produce significant distortions of individual forecasts at a time range when, on average, useful predictions can still be made. Moreover, studies of the sensitivity of climate may be biased by deficiencies in the simulation of the unperturbed atmospheric state. Climate models thus cannot be used properly without a thorough appreciation of their limitations, and their use is at its most powerful when it is possible for conclusions to be reinforced by observational studies and results from more idealized models. Similarly, more-refined applications of climate models are likely to require what might be a rather gradual refinement in our ability to simulate the present climate.

For a climate prediction to be of direct practical use, it must apply to the meteorological variables of interest to specific types of user at what are often quite specific locations. Variables of particular importance include the surface temperature, precipitation and cloudiness, and attempts to summarize model performance in terms of practical measures based on these parameters, for example measures which relate to crop growth are in their infancy (Gates, personal communication). An example of a first attempt is a verification by Manabe *et al.* (1979*b*) in terms of the geographical climate classification devised by Köppen (1931). Some of the difficulties in providing forecasts for specific locations are similar to those encountered in interpreting the output from numerical weather prediction models, and even if aspects of the large-scale climate are predictable, further increases in model resolution or the use of statistical or dynamical interpretation techniques, such as developed for weather forecasting, may be required before local climate forecasts of substantial practical benefit can be provided.

In conclusion, it must be remarked that the progress that has been achieved over the past 25 years has been made possible not only by the developments in modelling technique that have been reviewed here, but also by the very substantial increases in computing capacity that have occurred with developments in technology over this period. These developments are continuing, and further advances in climate modelling should follow. The accurate prediction of both short-term climatic variations and potential longer-term climatic change is the central goal of climate research, and comprehensive atmospheric modelling has a dominant role to play in the progress towards this goal.

Acknowledgement

In preparing this review we have in part drawn on material from a number of lecture notes and workshop proceedings published by ECMWF. We express our gratitude to the many staff members and visiting scientists who have contributed to these publications.

References

Allam, R. J., K. S. Groves & A. F. Tuck (1981). 'Global OH distribution derived from general circulation model fields of ozone and water vapour'. *J. Geophys. Res.*, **86**, 5303–20.

Anthes, R. A. (1977). 'Hurricane model experiments with a new cumulus parameterisation scheme'. *Mon. Wea. Rev.*, **105**, 287–300.

Arakawa, A. (1966). 'Computational design for long-term numerical integration of the equations of fluid motion: two-dimensional incompressible flow. Part 1'. *J. Comp. Phys.*, **1**, 119–43.

Arakawa, A. & W. H. Schubert (1974). 'Interaction of a cumulus cloud ensemble with the large-scale environment'. *J. Atmos. Sci.*, **31**, 674–701.

Arakawa, A. & V. R. Lamb (1981). 'A potential enstrophy and energy conserving scheme for the shallow water equations'. *Mon. Wea. Rev.*, **109**, 18–36.

Arya, S. P. S. (1977). 'Suggested revisions to certain boundary layer parameterization schemes used in atmospheric circulation models'. *Mon. Wea. Rev.*, **105**, 215–27.

Bates, J. R. (1977). 'Dynamics of stationary, ultra-long waves in middle latitudes'. *Quart. J. Roy. Met. Soc.*, **103**, 397–430.

Bates, J. R. & A. McDonald (1982). 'Multiple-upstream semi-Lagrangian advective schemes: analysis and application to a multi-level primitive-equation model'. *Mon. Wea. Rev.*, **110**, 1831–42.

Bengtsson, L. (1981*a*). 'Review of recent progress made in medium range weather forecasting'. *Proceedings of the ECMWF Seminar on Problems and Prospects in Long and Medium Range Weather Forecasting*, pp. 91–112.

Bengtsson, L. (1981*b*). 'Numerical prediction of atmospheric blocking – a case study'. *Tellus*, **33**, 19–42.

Bengtsson, L. & A. J. Simmons (1983). 'Medium range weather prediction – operational experience at ECMWF'. *Large-Scale Dynamical Processes in the Atmosphere*, eds. B. J. Hoskins & R. P. Pearce, Academic Press, pp. 337–63.

Bhumralkar, C. M. (1976). 'Parameterization of the planetary boundary layer in atmospheric general circulation models'. *Rev. Geophys. Space Phys.*, **14**, 2, 215–26.

Bjerknes, J. (1966). 'A possible response of the atmospheric Hadley circulation to equatorial anomalies of ocean temperature'. *Tellus*, **18**, 820–9.

Bjerknes, J. (1969). 'Atmospheric teleconnections from the equatorial Pacific'. *Mon. Wea. Rev.*, **97**, 163–72.

Blackmon, M. L. & N.-C. Lau (1980). 'Regional characteristics of the Northern hemisphere wintertime circulation: a comparison of the simulation of a GFDL general circulation model with observations'. *J. Atmos. Sci.*, **37**, 497–514.

Blackmon, M. L., J. M. Wallace, N-C. Lau & S. L. Mullen (1977). 'An observational study of the northern hemisphere wintertime circulation'. *J. Atmos. Sci.*, **34**, 1040–53.

Blackmon, M. L., J. E. Geisler & E. J. Pitcher (1983). 'A general circulation model study of January climate anomaly patterns associated with interannual variation of equatorial Pacific sea surface temperatures'. *J. Atmos. Sci.*, **40**, 1410–25.

Bourke, W. (1972). 'An efficient one-level primitive-equation spectral model'. *Mon. Wea. Rev.*, **100**, 683–9.

Bourke, W. (1974). 'A multi-level spectral model: I Formulation and hemispheric integrations'. *Mon. Wea. Rev.*, **102**, 688–701.

Bryan, K. (1969). 'Climate and ocean circulation: III Ocean model'. *Mon. Wea. Rev.*, **97**, 806–27.

Bryan, K., F. G. Komro, S. Manabe & M. J. Spelman (1982). 'Transient climate response to increasing atmospheric carbon dioxide'. *Science*, **215**, 56–9.

Burridge, D. M. (1975). 'A split semi-implicit reformulation of the Bushby–Timpson 10-level model'. *Quart. J. Roy. Met. Soc.*, **101**, 777–92.

Burridge, D. M. (1979). 'Some aspects of large scale numerical modelling of the atmosphere'. *Proceedings of ECMWF Seminar on Dynamical Meteorology and Numerical Weather Prediction*, Vol. 2, pp. 1–78.

Burridge, D. M. & J. Haseler (1977). 'A model for medium range weather forecasting – adiabatic formulation'. *ECMWF Technical Report No. 4*, 46 pp.

Chao, J-P., Guo, Y-F. & Xin, R-U. (1982). 'A theory and method of long-range numerical weather forecasts'. *J. Met. Soc. Japan*, **60**, 282–91.

Charney, J. G., (1947). 'The dynamics of long waves in a baroclinic westerly current'. *J. Meteor.*, **4**, 135–62.

Charney, J. G. (1975). 'Dynamics of deserts and droughts in the Sahel'. *Quart. J. Roy. Met. Soc.*, **101** 193–202.

Charney, J. G. & A. Eliassen (1949). 'A numerical method for predicting the perturbations of middle latitude westerlies'. *Tellus*, **1**, 38–54.

Charney, J. G. & N. A. Phillips (1953). 'Numerical integration of the quasi-geostrophic equations for barotropic and simple baroclinic flows'. *J. Meteor.*, **10**, 71–99.

Charney, J. G. & J. G. DeVore (1979). 'Multiple flow equilibria and blocking'. *J. Atmos. Sci.*, **36**, 1205–16.

Charney, J. G. & J. Shukla (1981). 'Predictability of monsoons'. *In: Monsoon Dynamics*, eds. M. J. Lighthill & R. P. Pearce, Cambridge University Press, pp. 99–109.

Charney, J. G., R. Fjörtoft & J. von Neumann (1950). 'Numerical integration of the barotropic vorticity equation'. *Tellus*, **2**, 237–54.

Charney, J. G., R. G. Fleagle, V. E. Lally, H. Riehl & D. Q. Wark (1966). 'The feasibility of a global observation and analysis experiment'. *Bull. Amer. Met. Soc.*, **47**, 200–20.

Charney, J. G., W. J. Quirk, S. Chow & J. Kornfield (1977). 'A comparative study of the effects of albedo change on drought in semi-arid regions'. *J. Atmos. Sci.*, **34**, 1366–85.

Chervin, R. M., J. E. Kutzbach, D. D. Houghton & R. G. Gallimore (1980). 'Response of the NCAR general circulation model to prescribed changes in ocean surface temperature. Part II: Midlatitude and subtropical changes'. *J. Atmos. Sci.*, **37**, 308–32.

Corby, G. A., A. Gilchrist & R. L. Newson (1972). 'A general circulation model of the atmosphere suitable for long period integrations'. *Quart. J. Roy. Met. Soc.*, **98**, 809–32.

Cubasch, U. (1981*a*). 'Preliminary assessment of long-range integrations done with the ECMWF global model'. *ECMWF Technical Memorandum No. 28*, 21 pp.

Cubasch, U. (1981*b*). 'The performance of the ECMWF model in 50-day integrations'. *ECMWF Technical Memorandum No. 32*, 74 pp.

Cullen, M. J. P. (1974). 'Integrations of the primitive equations on a sphere using the finite element method'. *Quart. J. Roy. Met. Soc.*, **100**, 555–62.

Cunnold, D., F. Alyea, N. A. Phillips & R. Prinn (1975). 'A three-dimensional dynamical–chemical model of atmospheric ozone'. *J. Atmos. Sci.*, **32**, 172–94.

Daley, R. & Y. Bourassa (1978). 'Rhomboidal versus triangular spherical harmonic truncation: some verification statistics'. *Atmosphere-Ocean*, **16**, 187–96.

Daley, R., C. Girard, J. Henderson & I. Simmonds (1976). 'Short term forecasting with a multi-level spectral primitive-equation model'. *Atmosphere*, **14**, 98–134.

Deardorff, J. (1978). 'Efficient prediction of ground surface temperature and moisture, with inclusion of a layer of vegetation'. *J. Geophys. Res.*, **83**, 1889–1903.

Delsol, F., K. Miyakoda & R. H. Clarke (1971). 'Parameterized processes in the surface boundary layer of an atmospheric circulation model'. *Quart. J. Roy. Met. Soc.*, **97**, 181–208.

Dey, C. H. (1969). 'A note on global forecasting with the Kurihara grid'. *Mon. Wea. Rev.*, **97**, 597–601.

Driedonks, A. G. M. & H. Tennekes (1981). 'Parameterization of the atmospheric boundary layer in large-scale models'. *Bull. Amer. Met. Soc.*, **62**, 594–8.

Druyan, L. M., R. C. J. Somerville & W. J. Quirk (1975). 'Extended-range forecasts with the GISS model of the global atmosphere'. *Mon. Wea. Rev.*, **103**, 779–95.

Eady, E. T. (1949). 'Long waves and cyclone waves'. *Tellus*, **1**, 33–52.

ECMWF (1981). *Proceedings of a Workshop on Radiation and Cloud-Radiation Interaction in Numerical Modelling*, 209 pp.

Egger, J. (1977). 'On the linear theory of the atmospheric response to sea surface temperature anomalies'. *J. Atmos. Sci.*, **34**, 603–14.

Eliasen, E., B. Machenhauer & E. Rasmussen (1970). 'On a numerical method for integration of the hydrodynamical equations with a spectral representation of the horizontal fields'. *Inst. of Theor. Met., Univ. of Copenhagen, Report No. 2.*

Fels, S. B., D. J. Mahlman, M. D. Schwarzkopf & R. W. Sinclair (1980). 'Stratospheric sensitivity to perturbations in ozone and carbon dioxide: Radiative and dynamical response'. *J. Atmos. Sci.*, **37**, 2265–97.

Gadd, A. J. (1978). 'A split explicit integration scheme for numerical weather prediction'. *Quart. J. Roy. Met. Soc.*, **104**, 569–82.

'Parameterization of sub-grid scale processes'. *GARP Publication Series No. 8*, WMO/ICSU, Geneva.

'Modelling for the first GARP global experiment'. *GARP Publication Series No. 14*, WMO/ICSU, Geneva.

'The physical basis of climate and climate modelling'. *GARP Publication Series No. 16*, WMO/ICSU, Geneva.

'Numerical methods used in atmospheric models'. *GARP Publication Series No. 17*, WMO/ICSU, Geneva.

'Numerical modelling of the tropical atmosphere'. *GARP Publication Series No. 20*, WMO/ICSU, Geneva.

'Report of the JOC Study Conference on climate models: performance, intercomparison and sensitivity studies'. *GARP Publication Series No. 22.*

Gary, J. M. (1973). 'Estimate of truncation error in transformed coordinate, primitive equation atmospheric models'. *J. Atmos. Sci.*, **30**, 223–33.

Gates, W. L. (1976). 'The numerical simulation of ice-age climate with a global general circulation model'. *J. Atmos. Sci.*, **33**, 1844–73.

Gates, W. L. (1979). 'The physical basis of climate'. *Proceedings of the World Climate Conference.* WMO, Geneva, pp. 112–31.

Gates, W. L., K. H. Cook & M. E. Schlesinger (1981). 'Preliminary analysis of experiments on the climatic effects of increased CO_2 with an atmospheric general circulation model and a climatological ocean'. *J. Geophys. Res.*, **86**, 6385–93.

Gauntlett, D. J., L. M. Leslie & D. R. Hincksman (1976). 'A semi-implicit forecast model using the flux form of the primitive equations'. *Quart. J. Roy. Met. Soc.*, **12**, 203–17.

Geleyn, J-F., A. Hense & H. J. Preuss (1982*a*). 'A comparison of model generated radiation fields with satellite measurements. *Beitr. Phys. Atmosph.*, **55**, 253–86.

Geleyn, J-F., C. Girard & J-F. Louis (1982*b*). 'A simple parameterization of moist convection for large-scale atmospheric models'. *Beitr. Phys. Atmosph.*, **55**, 325–34.

Gerrity, J. P., R. D. McPherson & P. D. Polger (1972). 'On the efficient reduction of truncation error in numerical weather prediction models'. *Mon. Wea. Rev.*, **100**, 637–43.

Gilchrist, A. (1977*a*). 'An experiment in extended range prediction using a general circulation model and including the influence of sea-surface temperature anomalies'. *Beitr. Phys. Atmos.*, **50**, 25–40.

Gilchrist, A. (1977*b*). 'The simulation of the Asian Summer Monsoon'. *Pure and Applied Geophys.*, **115**, 1431–8.

Gilchrist, A. (1981*a*). 'Long-range forecasting in the Meteorological Office'. *Proceedings of the ECMWF Seminar on Problems and Prospects in Long and Medium Range Weather Forecasting*, pp. 21–90.

Gilchrist, A. (1981*b*). 'Simulation of the Asian summer monsoon by an 11-layer general circulation model'. *In: Monsoon Dynamics*, eds. M. J. Lighthill & R. P. Pearce, Cambridge University Press, pp. 131–45.

Girard, C. & M. Jarraud (1982). 'Short and medium range forecast differences between a spectral and a grid-point model. An extensive quasi-operational comparison'. *ECMWF Technical Report No. 32.*, 178 pp.

Green, J. S. A. (1977). 'The weather during July 1976: some dynamical considerations of the drought'. *Weather*, **32**, 120–5.

Hahn, D. G. & S. Manabe (1975). 'The role of mountains in the south Asian monsoon circulation'. *J. Atmos. Sci.*, **32**, 1515–41.

Hahn, D. G. & J. Shukla (1976). 'An apparent relationship between Eurasian snow cover and Indian monsoon rainfall'. *J. Atmos. Sci.*, **33**, 2461–2.

Hendon, H. H. & D. L. Hartmann (1982). 'Stationary waves on a sphere: Sensitivity to thermal feedback'. *J. Atmos. Sci.*, **39**, 1906–20.

Herman, G. F. & W. T. Johnson (1978). 'The sensitivity of the general circulation to Arctic sea ice boundaries: a numerical experiment'. *Mon. Wea. Rev.*, **106**, 1649–64.

Herman, G. F., M-L. C. Wu & W. T. Johnson (1980). 'The effect of clouds on the Earth's solar and infra-red radiation budgets'. *J. Atmos. Sci.*, **37**, 1251–61.

Hollingsworth, A., K. Arpe, M. Tiedtke, M. Capaldo & H. Savijärvi (1980). 'The performance of a medium-range forecast model in winter – impact of physical parameterizations'. *Mon. Wea. Rev.*, **108**, 1736–73.

Holloway, J. L. & S. Manabe (1971). 'Simulation of climate by a global general circulation model: I Hydrologic cycle and heat balance'. *Mon. Wea. Rev.*, **99**, 335–70.

Holloway, J. L., M. J. Spelman & S. Manabe (1973). 'Latitude–longitude grid suitable for numerical time integration of a global atmospheric model'. *Mon. Wea. Rev.*, **101**, 69–78.

Horel, J. D. & J. M. Wallace (1981). 'Planetary scale atmospheric phenomena associated with the interannual variability of sea-surface temperature in the equatorial Pacific'. *Mon. Wea. Rev.*, **109**, 813–29.

Hoskins, B. J. & A. J. Simmons (1975). 'A multi-layer spectral model and the semi-implicit method'. *Quart. J. Roy. Met. Soc.*, **101**, 637–55.

Hoskins, B. J. & D. J. Karoly (1981). 'The steady linear response of a spherical atmosphere to thermal and orographic forcing'. *J. Atmos. Sci.*, **38**, 1179–96.

Hoskins, B. J., I. James & G. H. White (1983). 'The shape, propagation and mean-flow interaction of large-scale weather systems'. *J. Atmos. Sci.*, **40**, 1595–612.

Houghton, D. D., J. E. Kutzbach, M. McClintock & D. Suchman (1974). 'Response of a general circulation model to a sea-surface temperature perturbation'. *J. Atmos. Sci.*, **31**, 857–68.

Hunt, B. G. (1978). 'On the general circulation of the atmosphere without clouds'. *Quart. J. Roy. Met. Soc.*, **104**, 91–102.

Hunt, B. G. (1981). 'The maintenance of the zonal mean state of the upper atmosphere as represented in a three-dimensional general circulation model extending to 100 km'. *J. Atmos. Sci.*, **38**, 2172–86.

Janjić, Z. I. (1983). 'Non-linear advection schemes and energy cascade on semi-staggered grids'. Submitted for publication.

Julian, P. R. & R. M. Chervin (1978). 'A study of the southern oscillation and Walker circulation phenomena'. *Mon. Wea. Rev.*, **106**, 1433–51.

Kasahara, A. & W. M. Washington (1967). 'NCAR global general circulation model of the atmosphere'. *Mon. Wea. Rev.*, **95**, 389–402.

Kalnay-Rivas, E., A. Bayliss & J. Storch (1977). 'The 4th order GISS model of the global atmosphere'. *Beitr. Phys. Atmos.*, **50**, 299–311.

Keshavamurty, R. N. (1982). 'Response of the atmosphere to sea surface temperature anomalies over the equatorial Pacific and the teleconnections of the Southern Oscillation'. *J. Atmos. Sci.*, **39**, 1241–59.

Kessler, E. (1969). 'On the distribution and continuity of water substance in atmospheric circulations'. *Met. Monogr.*, **10**, 1–84.

Kirkwood, E. & J. Derome (1977). 'Some effects of the upper boundary condition and vertical resolution on modelling forced stationary planetary waves'. *Mon. Wea. Rev.*, **105**, 1239–51.

Köppen, W. (1931). *Grundriss der Klimakunde*, Walter de Gruyter, Berlin.

Krishnamurti, T. N., Y. Ramanathan, H-L. Pan, R. J. Pasch & J. Molinari (1980). 'Cumulus parameterization and rainfall rates I'. *Mon. Wea. Rev.*, **108**, 465–72.

Kuo, H. L. (1965). 'On formation and intensification of tropical cyclones through latent heat release by cumulus convection'. *J. Atmos. Sci.*, **22**, 40–63.

Kuo, H. L. (1974). 'Further studies of the parameterization of the influence of cumulus convection on large-scale flow'. *J. Atmos. Sci.*, **31**, 1232–40.

Kurihara, Y. (1965). 'Numerical integration of the primitive equations on a spherical grid'. *Mon. Wea. Rev.*, **93**, 399–415.

Kurihara, Y. (1968). 'Note on finite difference expressions for the hydrostatic relation and pressure gradient force'. *Mon. Wea. Rev.*, **96**, 654–6.

Kutzbach, J. E. & B. L. Otto-Bliesner (1982). 'The sensitivity of the African–Asian monsoonal climate to orbital parameter changes for 9000 years B.P. in a low-resolution general circulation model'. *J. Atmos. Sci.*, **39**, 1177–88.

Kutzbach, J. E., R. M. Chervin & D. D. Houghton (1977). 'Response of the NCAR general circulation model to prescribed changes in ocean surface temperature. Part I: Mid-latitude changes'. *J. Atmos. Sci.*, **34**, 1200–13.

Lau, N.-C. (1981). 'A diagnostic study of recurrent meteorological anomalies appearing in a 15-year simulation with a GFDL general circulation model'. *Mon. Wea. Rev.*, **109**, 2287–311.

Lilly, D. K. (1964). 'Numerical solutions for the shape-preserving two-dimensional thermal convection element'. *J. Atmos. Sci.*, **21**, 83–98.

Lindzen, R. S., E. S. Batten & J. W. Kim (1968). 'Oscillations in atmospheres with tops'. *Mon. Wea. Rev.*, **96**, 133–40.

Lord, S. J. (1982). 'Interaction of a cumulus cloud ensemble with the large-scale environment. Part III: Semi-prognostic test of the Arakawa-Schubert cumulus parameterization'. *J. Atmos. Sci.*, **39**, 88–103.

Lorenz, E. N. (1975). 'Climate Predictability'. *GARP Publication Series No. 16*, pp. 132–6.

Lorenz, E. N. (1982). 'Atmospheric predictability experiments with a large numerical model'. *Tellus*, **34**, 505–13.

Louis, J-F. (1979). 'A parametric model of vertical eddy fluxes in the atmosphere'. *Boundary-Layer Meteorol.*, **17**, 187–202.

Louis, J-F., M. Tiedtke & J-F. Geleyn (1981). 'A short history of the PBL parameterization at ECMWF'. *Proceedings of ECMWF Workshop on Planetary Boundary Layer Parameterization*, pp. 59–80.

Machenhauer, B. (1979). 'The spectral method'. *GARP Publication Series No. 17*, II, pp. 121–275.

Machenhauer, B. & E. Rasmussen (1972). 'On the integration of the spectral hydrodynamical equations by a transform method.' *Inst. of Theor. Met., Univ. of Copenhagen, Report No. 4.*

Manabe, S. (1969). 'Climate and the ocean circulation: I. The atmospheric circulation and the hydrology of the earth's surface'. *Mon. Wea. Rev.*, **97**, 739–74.

Manabe, S. & R. T. Wetherald (1975). 'The effects of doubling the CO_2 concentration on the climate of a general circulation model.' *J. Atmos. Sci.*, **32**, 3–15.

Manabe, S. & D. G. Hahn (1977). 'Simulation of the tropical climate of an ice-age'. *J. Geophys. Res.*, **82**, 3889–911.

Manabe, S. & R. T. Wetherald (1980). 'On the distribution of climate change resulting from an increase in CO_2 content of the atmosphere'. *J. Atmos. Sci.*, **37**, 99–118.

Manabe, S. & R. J. Stouffer (1980). 'Sensitivity of a global climate to an increase of CO_2 concentration in the atmosphere'. *J. Geophys. Res.*, **85**, 5529–54.

Manabe, S. & D. G. Hahn (1981). 'Simulation of atmospheric variability'. *Mon. Wea. Rev.*, **109**, 2260–86.

Manabe, S., J. Smagorinsky & R. F. Strickler (1965). 'Simulated climatology of a general circulation model with a hydrologic cycle'. *Mon. Wea. Rev.*, **93**, 769–98.

Manabe, S., J. L. Holloway, Jr & H. M. Stone (1970). 'Simulated climatology of a general circulation model with a hydrologic cycle, III. Effects of increased horizontal computational resolution'. *Mon. Wea. Rev.*, **98**, 175–212.

Manabe, S., K. Bryan & M. J. Spelman (1975). 'A global ocean–atmosphere climate model, Part I. The atmospheric circulation'. *J. Phys. Ocean*, **5**, 3–29.

Manabe, S., K. Bryan & M. J. Spelman (1979*a*). 'A global ocean–atmosphere climate model with seasonal variation for future studies of climate sensitivity'. *Dyn. Atmos. Oceans*, **3**, 393–426.

Manabe, S., D. G. Hahn & J. L. Holloway (1979*b*). 'Climate simulation with GFDL spectral models of the atmosphere: effect of spectral truncation'. *GARP Publication Series No. 22*, pp. 41–94.

Manabe, S., R. T. Wetherald & R. J. Stouffer (1981). 'Summer dryness due to an increase of atmospheric CO_2 concentration'. *Climate Change*, **3**, 347–85.

Marchuk, G. I. (1965). 'A new approach to the numerical solution of differential equations of atmospheric processs'. *WMO Tech. Note No. 66*, Geneva, pp. 286–94.

Marchuk, G. I. & Yu. N. Skiba (1976). 'Numerical calculation of the conjugate problem for a model of the thermal interaction of the atmosphere with the oceans and continents'. *Atmos. Ocean Phys.*, **12**, 279–84.

Marchuk, G. I., V. P. Dymnikov, V. N. Lykosov, V. Ya. Galin, I. M. Bobyleva & V. L. Perov (1979). 'A global model of the general atmospheric circulation'. *Atmos. Ocean. Phys.*, **15**, 321–31.

Mason, B. J. (1976). 'Towards the understanding and prediction of climatic variations'. *Quart. J. Roy. Met. Soc.*, **102**, 473–98.

McAvaney, B., W. Bourke & K. Puri (1978). 'A global spectral model for simulation of the general circulation'. *J. Atmos. Sci.*, **35**, 1557–83.

Mechoso, C. R., M. J. Suarez, K. Yamazaki, J. A. Spahr and A. Arakawa (1982). 'A study of the sensitivity of numerical forecasts to an upper boundary in the lower stratosphere'. *Mon. Wea. Rev.*, **110**, 1984–93.

Mellor, G. L. & T. Yamada (1974). 'A hierarchy of turbulence closure models for planetary boundary layers'. *J. Atmos. Sci.*, **31**, 1791–806.

Mesinger, F. (1981*a*). 'On the convergence and error problems of the calculation of the pressure gradient force in sigma coordinate models'. *Geophys. Astrophys. Fluid Dynam.*, **19**, 105–17.

Mesinger, F. (1981*b*). 'Horizontal advection schemes of a staggered grid – An enstrophy and energy-conserving model'. *Mon. Wea. Rev.*, **109**, 467–78.

Mintz, Y. (1965). 'Very long-term global integration of the primitive equations of atmospheric motion'. *WMO Tech. Note No. 66*, Geneva, pp. 141–55.

Mitchell, J. F. B. (1983). 'The seasonal response of a general circulation model to changes in CO_2 and sea temperatures'. *Quart. J. Roy. Met. Soc.*, **109**, 113–52.

Miyakoda, K. & J. Sirutis (1977). 'Comparative integrations of global models with various parameterized processes of subgrid-scale vertical transports: description of the parameterizations'. *Beitr. Phys. Atmosph.*, **50**, 445–87.

Miyakoda, K. & R. J. Strickler (1981). 'Cumulative results of extended forecast experiment. Part III: Precipitation'. *Mon. Wea. Rev.*, **109**, 830–42.

Miyakoda, K. & Chao J-P. (1982). 'Essay on dynamical long-range forecasts of atmospheric circulation'. *J. Met. Soc. Japan*, **60**, 292–308.

Miyakoda, K., G. D. Hembree, R. F. Strickler & I. Shulman (1972). 'Cumulative results of extended forecast experiments. I: Model performance for winter cases'. *Mon. Wea. Rev.*, **100**, 836–55.

Miyakoda, K., G. D. Hembree & R. F. Strickler (1974). 'Cumulative results of extended forecast experiments. II: Model performance for summer cases'. *Mon. Wea. Rev.*, **107**, 395–420.

Miyakoda, K., T. Gordon, R. Caverly, W. Stern, J. Sirutis & W. Bourke (1983). 'Simulation of a blocking event in January 1977'. *Mon. Wea. Rev.*, **111**, 846–69.

Möller, F. (1951). 'Vierteljahrskarten des Niederschlags für die Ganze Erde'. *Petermanns Geogr. Mitt.*, **95**, 1–7.

Moncrieff, M. W. (1981). 'A theory of organized steady convection and its transport properties'. *Quart. J. Roy. Met. Soc.*, **107**, 29–50.

Moura, A. D. & J. Shukla (1981). 'On the dynamics of droughts in northeast Brazil: observations, theory and numerical experiments with a general circulation model'. *J. Atmos. Sci.*, **38**, 2653–75.

Nakamura, H. (1976). 'Some problems in reproducing planetary waves by numerical models of the atmosphere'. *J. Met. Soc. Japan*, **54**, 129–46.

Nakamura, H. (1978). 'Dynamical effects of mountains on the general circulation of the atmosphere: I. Development of finite-difference schemes suitable for incorporating mountains'. *J. Met. Soc. Japan*, **56**, 317–39.

Namias, J. (1969). 'Seasonal interactions between the North Pacific Ocean and the atmosphere during the 1960's'. *Mon. Wea. Rev.*, **97**, 173–92.

Namias, J. (1978). 'Multiple causes of the North American abnormal winter of 1976–77'. *Mon. Wea. Rev.*, **106**, 279–95.

Nap, J. L., H. M. van den Dool & J. Oerlemans (1981). 'A verification of monthly weather forecasts in the seventies'. *Mon. Wea. Rev.*, **109**, 306–12.

Newell, R. E., J. W. Kidson, D. G. Vincent & G. J. Boer (1974). *The General Circulation of the Tropical Atmosphere*, Vol. 2, *MIT Press, Cambridge, Massachusetts*, 371 pp.

Nichols, N. (1980). 'Long-range weather forecasting: value, status and prospects'. *Rev. Geophys. Space Phys.*, **18**, 771–88.

O'Neill, A., R. L. Newson & R. J. Murgatroyd (1982). 'An analysis of the large-scale features of the upper troposphere and the stratosphere in a global three-dimensional general circulation model'. *Quart. J. Roy. Met. Soc.*, **108**, 25–53.

Oort, A. H. & T. H. Vonder Haar (1976). 'On the observed annual cycle in the ocean–atmosphere heat balance over the northern hemisphere'. *J. Phys. Oceanog.*, **6**, 781–800.

Opsteegh, J. D. & H. M. van den Dool (1980). 'Seasonal differences in the stationary response of a linearized model: prospects for long-range weather forecasting?' *J. Atmos. Sci.*, **37**, 2169–85.

Orszag, S. A. (1970). 'Transform method for calculation of vector coupled sums: application to the spectral form of the vorticity equation'. *J. Atmos. Sci.*, **27**, 890–5.

Paltridge, G. W. & C. M. R. Platt (1976). *Radiative Processes in Meteorology and Climatology*, Elsevier, 318 pp.

Pfeffer, R. L. (ed.) (1960). *Dynamics of Climate*, Pergamon Press, 137 pp.

Phillips, N. A. (1956). 'The general circulation of the atmosphere: a numerical experiment'. *Quart. J. Roy. Met. Soc.*, **82**, 123–64.

Phillips, N. A. (1957). 'A coordinate system having some special advantages for numerical forecasting'. *J. Met.*, **14**, 184–5.

Phillips, N. A. (1973). 'Principles of large scale numerical weather prediction'. *In: Dynamical Meteorology*, ed. P. Morel, Reidel, 1–95.

Pitcher, E. J., R. C. Malone, V. Ramanathan, M. L. Blackmon, K. Puri & W. Bourke (1983). 'January and July simulations with a spectral general circulation model'. *J. Atmos. Sci.*, **40**, 580–604.

Pollard, D. (1982). The performance of an upper-ocean model coupled to an atmospheric GCM: preliminary results'. *Climate Research Institute, Oregon State Univ., Report No. 31*, 33 pp.

Pratt, R. W. (1979). 'A space-time spectral comparison of the NCAR and GFDL general circulation models of the atmosphere'. *J. Atmos. Sci.*, **36**, 1681–91.

Ramanathan, V., E. J. Pitcher, R. C. Malone & M. L. Blackmon (1983). 'The response of a spectral general circulation model to refinements in radiative processes'. *J. Atmos. Sci.*, **40**, 605–30.

Randall, D. A. (1976). 'The interaction of the planetary boundary layer with large-scale circulations'. Ph.D. thesis, University of California, Los Angeles, 247 pp.

Ratcliffe, R. A. S. & R. Murray (1970). 'New lag associations between North Atlantic sea temperature and European pressure applied to long-range weather forecasting'. *Quart. J. Roy. Met. Soc.*, **96**, 226–46.

Robert, A. J. (1981). 'A stable numerical integration scheme for the primitive meteorological equations'. *Atmos. Ocean*, **19**, 35–46.

Robert, A. J. (1982). 'A semi-Lagrangian and semi-implicit, numerical integration scheme for baroclinic models of the atmosphere'. *Met. Soc. Japan*, **60**, 319–25.

Robert, A. J., J. Henderson & C. Turnbull (1972). 'An implicit time intergration scheme for baroclinic models of the atmosphere'. *Mon. Wea. Rev.*, **100**, 329–35.

Rowntree, P. R. (1972). 'The influence of tropical east Pacific Ocean temperatures on the atmosphere'. *Quart. J. Roy. Met. Soc.*, **98**, 290–321.

Rowntree, P. R. (1976). 'Response of the atmosphere to a tropical Atlantic Ocean temperature anomaly'. *Quart. J. Roy. Met. Soc.*, **102**, 607–26.

Rowntree, P. R. (1978). 'Numerical prediction and simulation of the tropical atmosphere'. *In: Meteorology over the Tropical Oceans*, Roy. Met. Soc., 278 pp.

Rowntree, P. R. (1979). 'Statistical assessments of sea temperature anomaly experiments'. *GARP Publication Series No. 22*, pp. 482–500.

Sadourny, R. (1975). 'The dynamics of finite difference models of the shallow-water equations'. *J. Atmos. Sci.*, **32**, 680–9.

Schlesinger, M. E. & Y. Mintz (1979). 'Numerical simulation of ozone production, transport and distribution with a global atmospheric general circulation model'. *J. Atmos. Sci.*, **36**, 1325–61.

Schlesinger, M. E. & W. L. Gates (1980). 'The January and July performance of the OSU two-level atmospheric general circulation model'. *J. Atmos. Sci.*, **37**, 1914–43.

Schlesinger, M. E. & W. L. Gates (1981). 'Preliminary analysis of the mean annual cycle and inter-annual variability simulated by the OSU two-level atmospheric general circulation model'. *Climatic Research Institute, Oregon State Univ., Report No. 23*, 47 pp.

Schneider, S. H. & R. E. Dickinson (1974). 'Climate modelling'. *Rev. Geophys. Space Phys.*, **12**, 447–93.

Schneider, S. H., W. M. Washington & R. M. Chervin (1978). 'Cloudiness as a climatic feedback mechanism: effects on cloud amounts of prescribed global and regional surface temperature changes in the NCAR GCM'. *J. Atmos. Sci.*, **35**, 2207–21.

Shaw, D. B. (1981). 'ECMWF operational forecasts in the SW and NE monsoon regions'. *Proceedings of ECMWF Workshop on Tropical Meterology and its Effects on Medium-Range Prediction at Middle Latitudes*, pp. 53–86.

Shukla, J. (1975). 'Effect of Arabian sea surface temperature anomaly on Indian summer monsoon: a numerical experiment with the GFDL model'. *J. Atmos. Sci.*, **32**, 503–11.

Shukla, J. (1981*a*). 'Dynamical predictability of monthly means'. *J. Atmos. Sci.*, **38**, 2547–72.

Shukla, J. (1981*b*). 'Predictability of monthly means: Part II – Influence of the boundary forcings'. *Proceedings of ECMWF Seminar on Problems and Prospects in Long and Medium Range Weather Forecasting*, pp. 261–312.

Shukla, J. & B. Bangaru (1979). 'Effect of a Pacific sea-surface temperature anomaly on the circulation over North America: a numerical experiment with the GLAS model'. *GARP Publication Series No. 22*, pp. 501–18.

Shukla, J. & J. M. Wallace (1983). 'Numerical simulation of the atmospheric response to equatorial Pacific sea surface temperature anomalies'. *J. Atmos. Sci.*, **40**, 1613–30.

Silberman, I. (1954). 'Planetary waves in the atmosphere'. *J. Met.*, **11**, 27–34.

Simmons, A. J. (1982*a*). 'The forcing of stationary wave motion by tropical diabatic heating'. *Quart. J. Roy. Met. Soc.*, **108**, 503–34.

Simmons, A. J. (1982*b*). 'The numerical prediction and simulation of the tropical atmosphere – a sample of results from operational forecasting and extended integrations at ECMWF'. *Tropical Droughts – Meteorological Aspects and Implications for Agriculture.* WMO, Geneva, pp. 81–103.

Simmons, A. J. & D. M. Burridge (1981). 'An energy and angular-momentum conserving vertical finite-difference scheme and hybrid vertical coordinates'. *Mon. Wea. Rev.*, **109**, 758–66.

Simmons, A. J. & R. Strüfing (1983). 'Numerical forecasts of stratospheric warming events using a model with a hybrid vertical coordinate'. *Quart. J. Roy. Met. Soc.*, **109**, 81–111.

Simmons, A. J., B. J. Hoskins & D. M. Burridge (1978). 'Stability of the semi-implicit time scheme'. *Mon. Wea. Rev.*, **106**, 405–12.

Simmons, A. J., J. M. Wallace & G. W. Branstator (1983). 'Barotropic wave propagation and instability, and atmospheric teleconnection patterns'. *J. Atmos. Sci.*, **40**, 1363–92.

Simpson, R. W. & W. K. Downey (1975). 'The effect of a warm mid-latitude sea-surface temperature anomaly on a numerical simulation of the general circulation of the Southern Hemisphere'. *Quart. J. Roy. Met. Soc.*, **101**, 847–67.

Slingo, J. M. (1980). 'A cloud parameterization scheme derived from GATE data for use with a numerical model'. *Quart. J. Roy. Met. Soc.*, **106**, 747–70.

Smagorinsky, J. (1953). 'The dynamical influences of large scale heat sources and sinks on the quasi-stationary mean motions of the atmosphere'. *Quart. J. Roy. Met. Soc.*, **79**, 342–66.

Smagorinsky, J. (1963). 'General circulation experiments with the primitive equations, I. The basic experiment'. *Mon. Wea. Rev.*, **93**, 99–164.

Smagorinsky, J. (1969). 'Problems and promises of deterministic extended range forecasting'. *Bull. Amer. Met. Soc.*, **50**, 286–311.

Smagorinsky, J., S. Manabe & J. L. Holloway Jr (1965). 'Numerical results from a 9-level general circulation model of the atmosphere'. *Mon. Wea. Rev.*, **93**, 727–68.

Somerville, R. C. J., P. H. Stone, M. Halem, J. E. Hansen, J. S. Hogan, L. M. Druyan, G. Russell, A. A. Lacis, W. J. Quirk & J. Tenenbaum (1974). 'The GISS model of the global atmosphere'. *J. Atmos. Sci.*, **31**, 84–117.

Spar, J. (1973). 'Some effects of surface anomalies in a global general circulation model'. *Mon. Wea. Rev.*, **101**, 91–100.

Staniforth, A. N. & R. W. Daley (1977). 'A finite-element formulation for the vertical discretization of sigma-coordinate primitive-equation models'. *Mon. Wea. Rev.*, **105**, 1108–18.

Staniforth, A. N. & H. L. Mitchell (1978). 'A variable-resolution finite-element technique for regional forecasting with the primitive equations'. *Mon. Wea. Rev.*, **106**, 439–47.

Straus, D. M. & J. Shukla (1981). 'Space-time spectral structure of a GLAS general circulation model and a comparison with observations'. *J. Atmos. Sci.*, **38**, 902–17.

Strüfing, R. (1982*a*). 'On the effect of energy/enstrophy conservation in the finite difference scheme of the ECMWF gridpoint model'. *ECMWF Tech. Memo. No. 49*, 34 pp.

Strüfing, R. (1982*b*). 'Some comparisons between linear and non-linear horizontal diffusion schemes for the ECMWF grid-point model'. *ECMWF Tech. Memo. No. 60*, 42 pp.

Sud, Y. C. & M. Fennessy (1982). 'A study of the influence of surface albedo on July circulation in semi-arid regions using the GLAS GCM'. *J. Climatology*, **2**, 105–25.

Sundqvist, H. (1976). 'On vertical interpolation and truncation in connexion with use of sigma system models'. *Atmosphere*, **14**, 37–52.

Sundqvist, H. (1981). 'Prediction of stratiform clouds: results from a 5-day forecast with a global model'. *Tellus*, **33**, 242–53.

Tiedtke, M., J-F. Geleyn, A. Hollingsworth & J-F. Louis (1979). 'ECMWF model-parameterization of sub-grid scale processes'. *ECMWF Tech. Rep. No. 10*, 46 pp.

Trenberth, K. (1979). 'Interannual variability of the 500 mb zonal flow in the southern hemisphere'. *Mon. Wea. Rev.*, **107**, 1515–24.

U.S. National Academy of Sciences (1975). *Understanding Climatic Change*, Washington, DC, 239 pp.

Volmer, J-P., M. Deque & M. Jarraud (1983*a*). 'Large scale fluctuations in a long range integration of the ECMWF spectral model'. *Tellus*, **35**, 173–88.

Volmer, J-P., M. Deque & D. Rousselet (1983*b*). 'Long range behaviour of the atmosphere: a comparison between simulation and reality'. Submitted for publication.

Vonder Haar, T. H. & V. E. Suomi (1971). 'Measurements of the earth's radiation budget from satellites during a five-year period. Part I. Extended time and space means'. *J. Atmos. Sci.*, **28**, 305–14.

Vonder Haar, T. H. & A. H. Oort (1973). 'New estimate of annual poleward energy transport by Northern Hemisphere oceans'. *J. Phys. Oceanog.*, **3**, 169–72.

Walker, J. M. & P. R. Rowntree (1977). 'The effect of soil moisture on circulation and rainfall in a tropical model'. *Quart. J. Roy. Met. Soc.*, **103**, 29–46.

Wallace, J. M. & D. S. Gutzler (1981). 'Teleconnections in the Geopotential Height Field during the northern hemisphere winter'. *Mon. Wea. Rev.*, **109**, 784–812.

Wallace, J. M., S. Tibaldi & A. J. Simmons (1983). 'Reduction of systematic forecast errors in the ECMWF model through the introduction of an envelope orography'. *Quart. J. Roy. Met. Soc.*, **109**, in press.

Washington, W. M. (1981). 'A review of general-circulation model experiments on the Indian monsoon'. *In Monsoon Dynamics*, eds.

M. J. Lighthill & R. P. Pearce, Cambridge University Press, pp. 111–30.

Washington, W. M. & S. M. Daggupaty (1975). 'Numerical simulation with the NCAR global circulation model of the mean conditions during the Asian-African summer monsoon'. *Mon. Wea. Rev.*, **103**, 105–14.

Washington, W. M. & D. L. Williamson (1977). 'A description of the NCAR global circulation models'. *In: Methods in Comp. Physics*, Vol. 17, *General Circulation Models of the Atmosphere*, J. Chang, ed., Academic Press, pp. 111–72.

Washington, W. M., R. M. Chervin & G. V. Rao (1977). 'Effects of a variety of Indian Ocean surface temperature anomaly patterns on the summer monsoon circulation: experiments with the NCAR general circulation model'. *Pure and Applied Geophysics*, **115**, 1335–56.

Washington, W. M., R. E. Dickinson, V. Ramanathan, T. Mayer, D. L. Williamson, G. Williamson & R. Wolski (1979). 'Preliminary atmospheric simulation with the third-generation NCAR general circulation model'. *GARP Publication Series No. 22*, pp. 95–138.

Washington, W. M., A. J. Semtner, G. A. Meehl, D. J. Knight & T. A. Mayer (1980). 'A general circulation experiment with a coupled atmosphere, ocean and sea ice model'. *J. Phys. Oceanog.*, **10**, 1887–08.

Webster, P. J. (1981). 'Mechanisms determinining the atmospheric response to sea surface temperature anomalies'. *J. Atmos. Sci.*, **38**, 554–71.

Wellck, R. E., A. Kasahara, W. M. Washington & G. de Santo (1971). 'Effect of horizontal resolution in a finite-difference model of the general circulation'. *Mon. Wea. Rev.*, **99**, 673–83.

Wells, N. C. (1979*a*). 'A coupled ocean-atmosphere experiment: The ocean response'. *Quart. J. Roy. Met. Soc.*, **105**, 355–70.

Wells, N. C. (1979*b*). 'The effect of a tropical sea-surface temperature anomaly in a coupled ocean-atmosphere model'. *J. Geophys. Res.*, **84**, 4985–97.

Wetherald, R. T. & S. Manabe (1980). 'Cloud cover and climate sensitivity'. *J. Atmos. Sci.*, **37**, 1485–510.

Williams, J., R. G. Barry & W. M. Washington (1974). 'Simulation of the atmospheric circulation using the NCAR global circulation model with ice age boundary conditions'. *J. Appl. Met.*, **13**, 305–17.

Williamson, D. L. (1978). 'The relative importance of resolution, accuracy and diffusion in short-range forecasts with the NCAR global circulation model'. *Mon. Wea. Rev.*, **106**, 69–88.

Cloud–radiation interaction and the climate problem

Peter J. Webster† and Graeme L. Stephens

CSIRO, Division of Atmospheric Physics

PO Box 77, Mordialloc, Australia, 3195

Abstract

Observational evidence of the timescales over which radiation–cloud interaction occurs is presented. Variations, shown in terms of net outgoing longwave radiation, are found to exist with almost equal magnitude on timescales ranging from synoptic to interannual. The association of these variations with changes in boundary forcing (e.g. equatorial Pacific Ocean warmings) and subsequent changes in heating distributions is discussed.

The manner in which the long-term dynamics of the oceans and atmosphere may be influenced by cloudiness variability is investigated using both open and closed structures. In open structures, the changes in field variables relative to variations of cloud types, heights, etc. are calculated. With closed model structures, the feedback between the changes in the field variables and an induced cloud feedback is included. The results of these two model types and the parallel observational studies, as they pertain to climate systems, are discussed at length.

5.1 Introduction

There is little quarrel that the primary energy source of the earth's climate system is the incoming solar stream and that the major sink is the outgoing longwave stream. Controversy stems from another quarter: the manner in which the radiative streams (i.e. $F(z)$) are modulated. As the shortwave gain and the longwave loss are not usually in local balance, a three-dimensional redistribution of energy occurs as the system seeks equilibrium. Within a climate system, clouds of varying distribution, character, thickness and height appear as intrinsic features. Their roles as important modifiers of the radiation budget, on either regional or planetary scales or as integral elements of feedback loops which enhance or buffer changes in a climate system, is a question of concern and debate.

Of all climate constituents, clouds have the potential of exercising maximum impact on both the longwave and shortwave streams. The controversial question is whether or not clouds affect the *net* radiation stream. Cess (1976), for example, argues that the shortwave loss of energy due to cloud reflection is exactly balanced by longwave enhancement by cloud absorption. Recent studies by Ohring & Clapp (1980) suggest that the net radiation balance at the top of the atmosphere is sensitive to changes in cloud amount. Hartmann & Short (1980) have shown that the quantities discussed by Cess and Ohring & Clapp are fundamentally different. For example, Cess calculated a quantity like the total derivative of net flux with cloud (i.e., $\mathrm{d}F/\mathrm{d}Q$), while Ohring & Clapp computed the sensitivity of the radiation balance to changes in cloudiness (i.e., $\partial F/\partial Q$).

However, though aimed principally at the planetary-scale radiation budgets, neither the Ohring–Clapp nor Cess study addresses the problem of *regional* or *vertical* energy redistributions

† Permanent affiliation: Department of Meteorology, Pennsylvania State University, University Park, Pennsylvania, 16802, USA.

by cloud. This is important because energy conversions and dynamic structures critically depend on the vertical distribution of total diabatic heating to which radiational forcing is a considerable contributor. Consequently, the question of the impact of clouds on the net outgoing radiation flux at the top of the atmosphere may have only a partial relevance to the impact of clouds on the climatic state of the earth system. Conceivably, the net flux at the top of the atmosphere is the radiative quantity which is least sensitive to changes in cloud amount. That is, a particular value of the net flux need not represent a unique state of the column below.

Thus, a definitive study of the cloud–climate problem has remained an elusive attainment. Observational studies have been hampered by the need to establish control situations. Similarly, because of the multi-faceted interdependencies of cloud and climate, investigations by theoretical and numerical models are equally difficult, and possibly premature, with current models (Hunt *et al.*, 1980). The utilization of simple energy balance models tends to be constrained by the simplicity of the adjunct radiation models and cloud parameterizations, which often depend only on a single atmospheric temperature that is a prescribed and tuned function of the surface temperature (Stephens & Webster, 1979), and the neglect of all but the simplest dynamic transports.

The problems facing the larger and more sophisticated models are underlined by the study of Hunt (1977) who found the earth climate system, at least as simulated by a general circulation model (GCM), to be relatively insensitive to cloudiness. However, Hunt *et al.* (1980) questioned this lack of sensitivity on the basis that Hunt's model ignored certain critical feedback mechanisms in common with other general circulation models. In particular, the Hunt study neglected the relative humidity feedback in the radiation calculations.

A number of groups employing model generated cloud fields have overcome some of the restrictions Hunt (1977) imposed upon his experiments. For example, Gates & Schlesinger (1977) have made exhaustive and critical comparisons of their model generated cloud fields with observation in a manner similar to Herman (1981) for the NASA–GLAS model. Although only zonally averaged aspects of the model products were compared with observation, many of the gross features of the observed mean cloud climatologies were present.

Considering the impact of clouds upon climate structures, a number of groups have conducted studies in the spirit of Hunt (1977). For example, Shukla & Sud (1981) conducted studies in which clouds were successively held constant or allowed to vary. Their results indicate significant differences between experiments. This is a somewhat different conclusion to that reached by Wetherald & Manabe (1980), although the make-up of the experiments was entirely different and the question of whether the latter paper was testing a partial or total derivative change of climate variable to cloud remains to be resolved. We will discuss this point later.

Other problems await investigation which may be approached using much simpler models. For example, Stephens & Webster (1979) discussed the sensitivity of radiative forcing to both variations of cloud amount and of cloudy- and clear-sky optical properties. Within the confines of an extremely simple model they found the radiative balance to be a strong function of both cloud character and cloud height. The results allowed them to make specific recommendations regarding the treatment of cloud parameterization and radiative transfer in both simple and complex climate models. However, the major simplification of the Stephens–Webster model was a fixed state constraint. That is, the radiation balance was calculated relative to a fixed atmospheric state for a given cloud structure such that the results could only be interpreted as an initial tendency of the atmosphere to the imposition of cloud.

From the above paragraphs it is apparent that there is a growing literature describing the three principal modes of investigation of the cloud–radiation climate problem. The main purpose of this paper is to assess the progress made to date and to determine the climate sensitivity to cloud amount, cloud type and cloud character. In order to establish that cloud and climate are in some manner linked, we will first show that cloudiness varies as a function of climate variability. In the third section we will establish a more formal link between cloud and climate by indicating how, through the alteration of the magnitude or the distribution (and often both) of the diabatic heating, clouds may directly interact with the dynamics of the atmospheric, oceanic and cryospheric components of the climate system, as well as the dynamics associated with induced changes of the continental surfaces. In later sections, we will review results obtained from both modelling and observational studies and discuss how they relate to the speculations listed in Section 5.3.

5.2 **Cloud distributions and associations**

Cloud distribution possesses a substantial variability with respect to both space and time. Probably the degree of variability was not appreciated until the advent of the satellite and the archiving of the data product. Although little comment was made regarding the interannual variability as sensed by the satellite, mean monthly and seasonal cloud brightness charts spanning a number of years were available over a decade ago.

Cloud variability may be associated with variations in the dynamics of the atmosphere on a number of scales. This may be seen for the lower end of the temporal scale from the study of Webster & Stephens (1980*a*) for the Winter Monsoon Experiment region. Fig. 5.1 shows a series of five-day average distributions of upper-level cloudiness as determined from the Geostationary Meteorological Satellite (GMS). The difference in the mean infrared (IR) radiances indicate modulation of the synoptic scale variability. However, the important aspect of Fig. 5.1 is that high-level cloudiness in the tropics can be shown to be associated with variations in precipitation. Thus variations in cloudiness can be considered as indicators of variations in the large-scale latent heating of the atmosphere. What Webster & Stephens (1980*a*) sought to argue was that clouds, by their very existence, were partially responsible for the large-scale latent heating distribution both in the horizontal and in the vertical.

The mutual dependency of the components of the total diabatic heating function proposed by Webster & Stephens (1980*a*) depended upon the radiative destabilization of the extended middle- and upper-level cloud layers; themselves debris products of the cumulus-tower convection, and the subsequent release of latent heat within the extended cloud. Such clouds had already been shown by detailed radar measurements during GATE (GARP

Fig. 5.1. Variation of the upper troposphere cloud amount over the Indonesian region determined from GMS IR radiance data for six five-day periods in December 1978. (From Webster & Stephens, 1980*a*.)

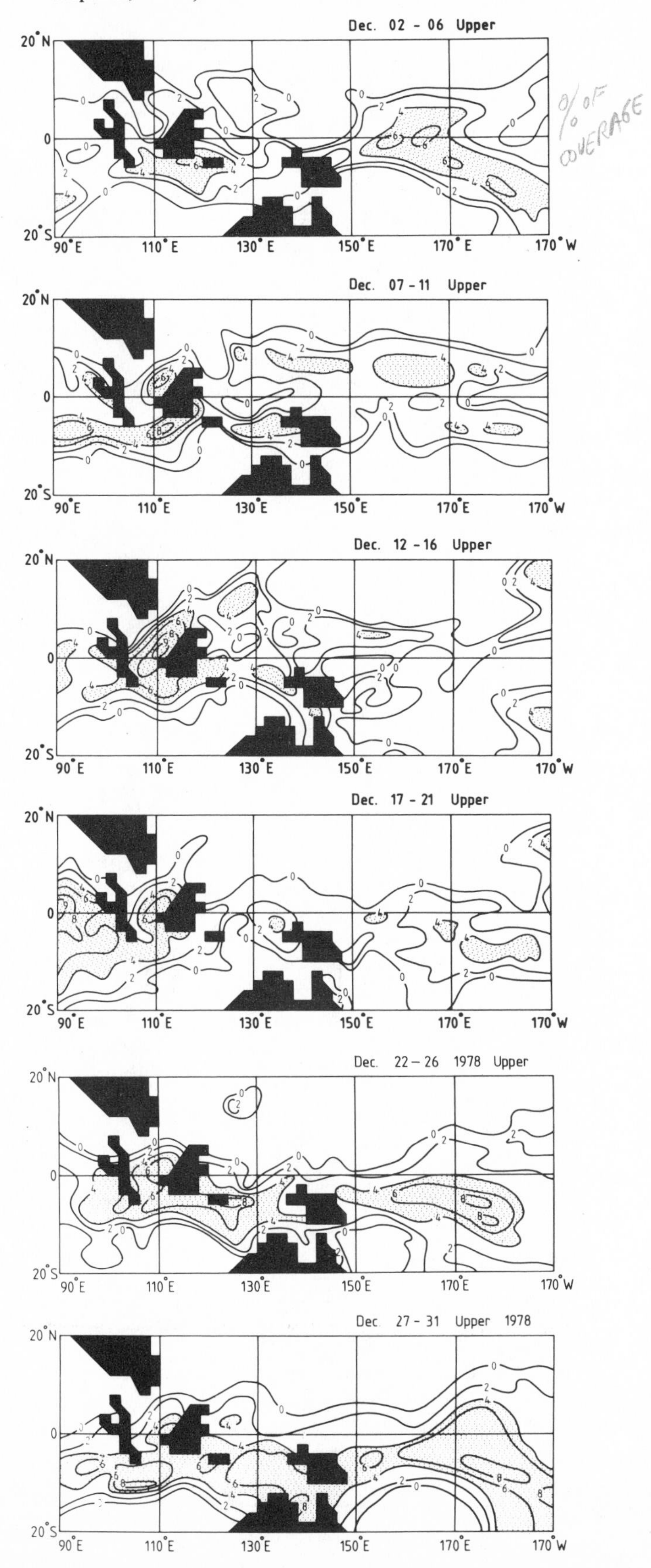

Atlantic Tropical Experiment) to supply some 40% of the total disturbance precipitation (see Houze, 1982, for summary). Houze (1982) has suggested, on the basis of the cloud-modified radiative heating profile plus the resultant upper-level release of latent heat, that the total disturbance scale diabatic heating distribution is substantially altered. In particular, the heating takes place at a substantially higher level and occurs over a broader scale.

The region shown in Fig. 5.1 is the most persistent and intensive region of latent heating in the tropics. Thus variability shown on the space- and timescales of the diagram are obviously important on the shorter timescale. However if we extend both the area of interest to the entire tropical belt and the observational period to a number of years, we find that this same region, the western Pacific Ocean, is involved in much stronger and longer period variations. We can best attest to this variability from the diagram taken from Liebmann & Hartmann (1982) and shown in Fig. 5.2 (*a*) and (*b*). To derive these figures, Liebmann & Hartmann compiled the mean winter and summer IR category flux (i.e. December, January and February, DJF; and June, July and August, JJA) for the years 1974–78 and obtained the seasonal deviations for each of these years. Plotted in contours of 5 Wm^{-2}, the eight panels of Fig. 5.2 illustrate the variability. It should be noted that an increase in outgoing flux (solid contours) may be interpreted as a decrease in cloudiness. The dashed contours should be interpreted in the opposite sense. Quite obvious from the diagram is that, over the four-year period, not only is there a large change in the IR radiative flux at the top of the atmospheric column in the tropics, but there is an accompanying large change in cloudiness. For example, a 60 Wm^{-2} difference exists in IR emittance between DJF 1976 and DJF 1978.

Liebmann & Hartmann chose the four-year study period (1974–78) because it spanned the 1976 El Nino phenomena in the central and eastern Pacific Ocean. The relative warming of this oceanic region may be deduced from the sea-surface temperature curve in Fig. 5.3.

Other studies (e.g. Ramage, 1975; Horel & Wallace, 1981; and many others) have shown that significant circulation changes accompany the strong boundary forcing events such as El Nino. Such changes, and the manner in which their effect may set up climatically significant teleconnection patterns between low and high latitudes have been the subject of a number of theoretical studies (e.g. Hoskins & Karoly, 1981; and Webster, 1981*a*,*b*), 1982). In a similar mode Webster & Holton (1982) have speculated on the effect the changes in the tropical circulation during these events may have on wave propagation between the hemispheres.

In the light of the above, it is an important goal to establish the role of cloud in these critical climate events, beyond, of course, accompanying large-scale changes in latent heating. Remembering that Webster & Stephens (1980*a*) and, later, Houze (1982), have hinted at the important cooperative role between radiative destabilization in cloud layers and latent heat release on the synoptic and supersynoptic timescales, we can immediately identify the effect of the same physical process on the interannual timescale. Principally, the total diabatic heating distribution on climate timescales is an aggregate of successive synoptic scale heating events and, in that sense, the cooperative role of latent and radiative heating is quite important. In fact, in using a general circulation model with a

Fig. 5.2. Seasonal deviations from the four mean outgoing IR radiance distributions for (*a*) DJF and (*b*) JJA for the years 1975–78. Positive anomalies are dashed and may be interpreted as decreases in cloudiness. Solid lines denote negative radiance anomalies and hence cloudiness increases. Contours are 5 Wm^{-2} intervals. The magnitude of some of the extremes are marked (From Liebmann & Hartmann, 1982.)

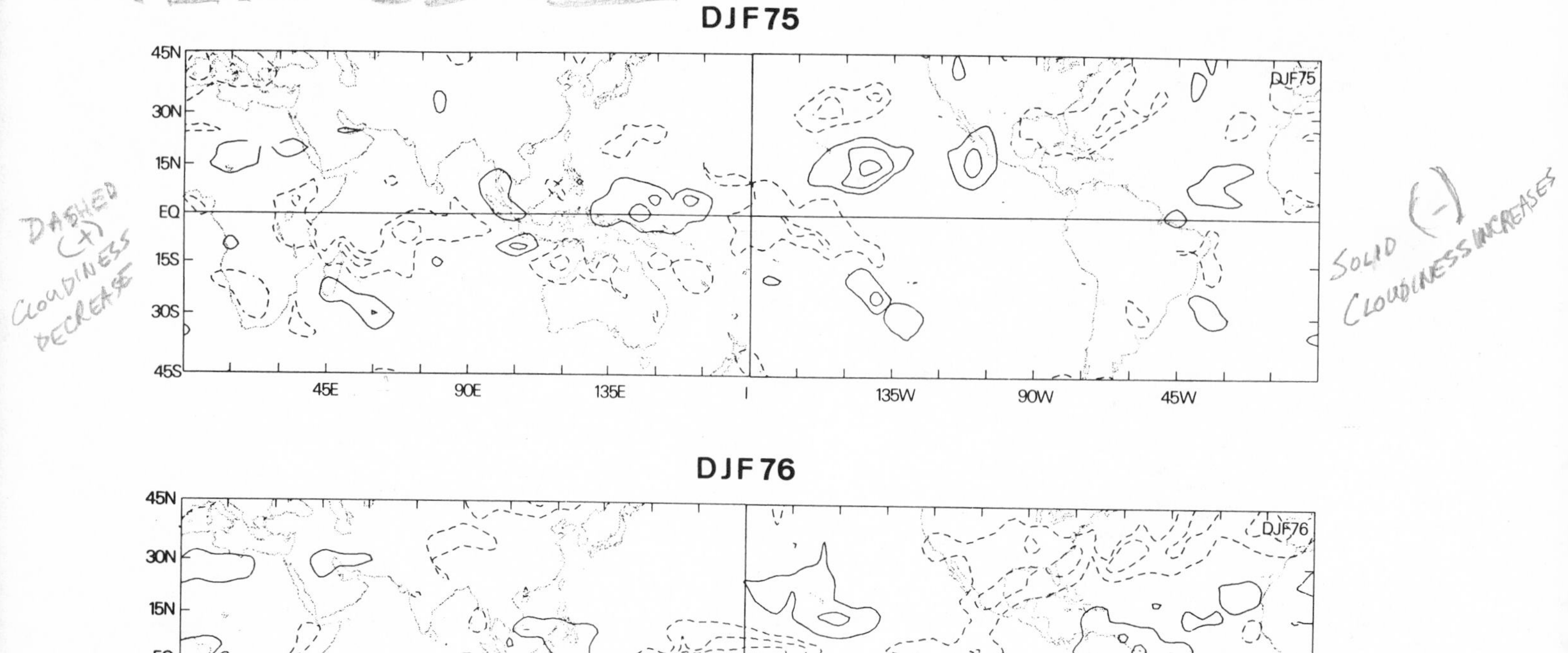

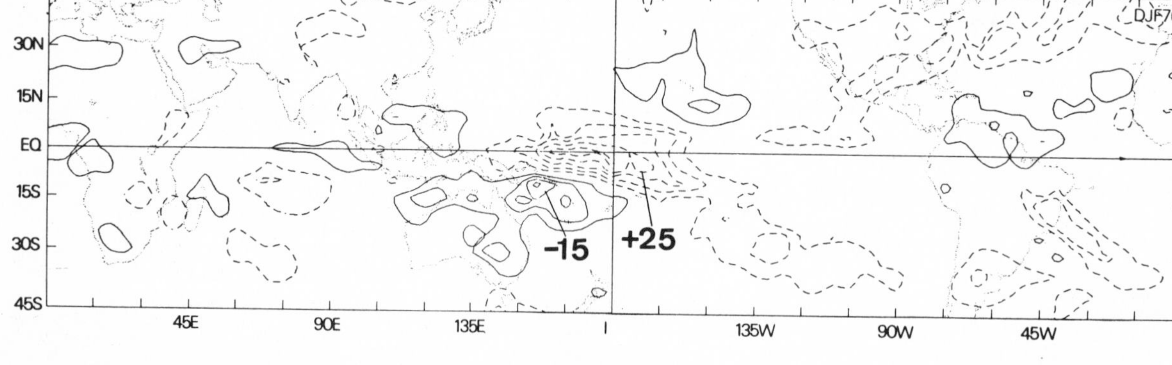

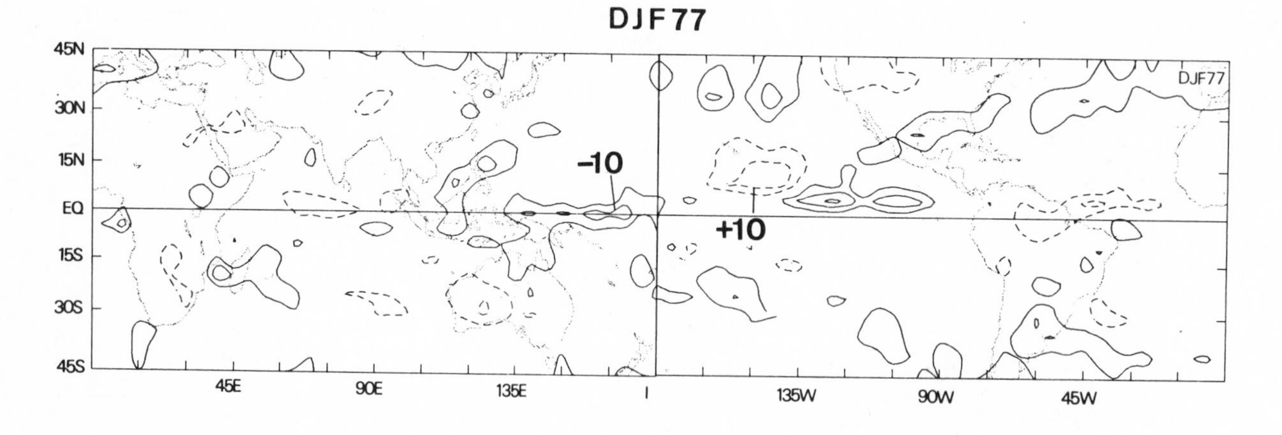

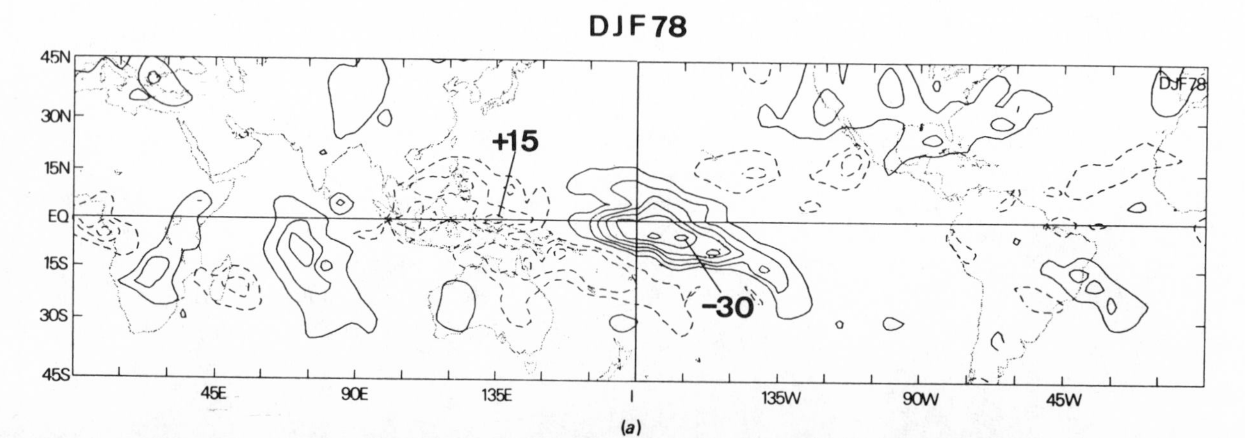

(*a*)

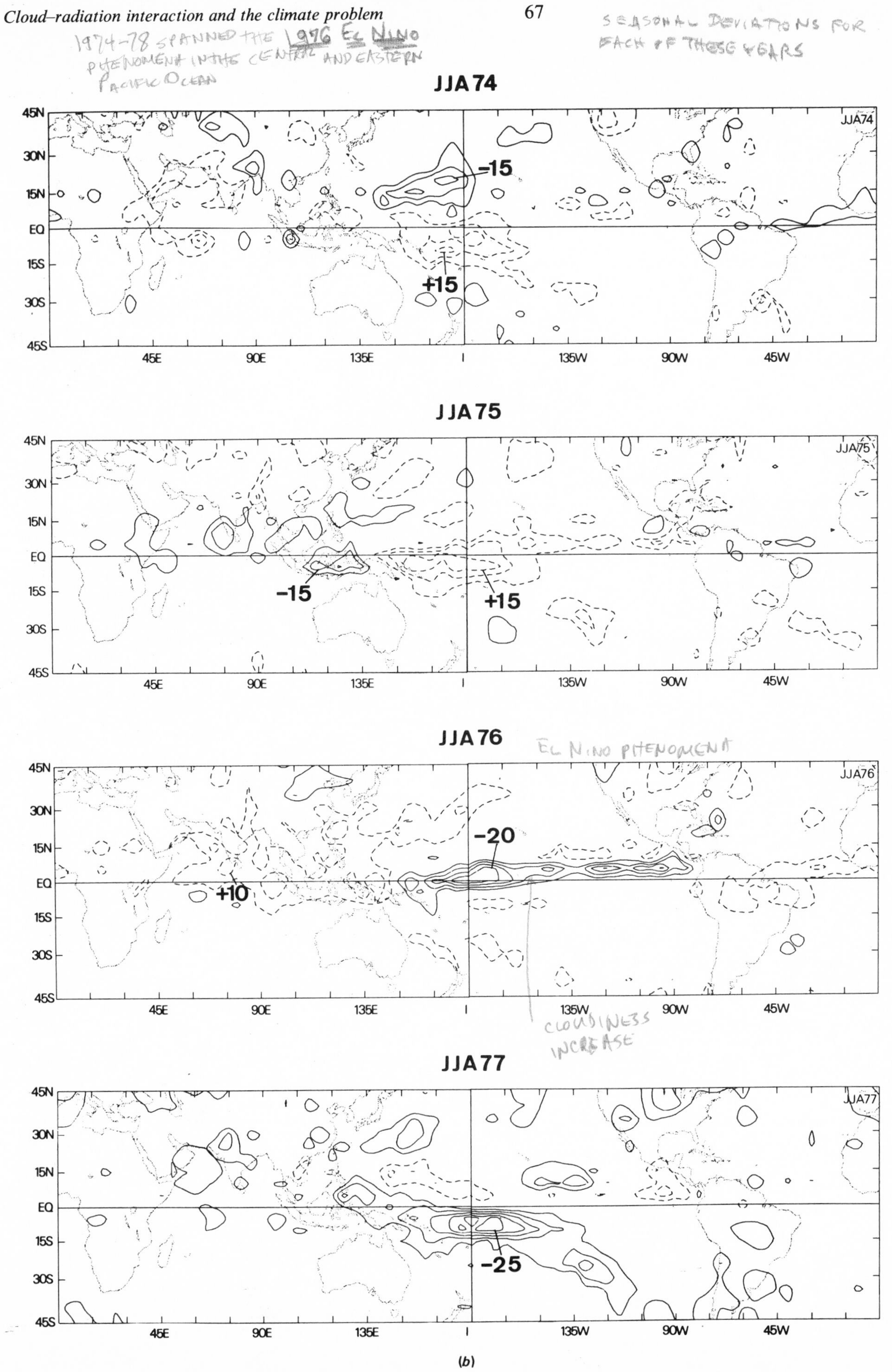
JJA 74
JJA74
−15
+15
JJA 75
JJA75
−15
+15
JJA 76
JJA76
−20
+10
JJA 77
JJA77
−25
45N
30N
15N
EQ
15S
30S
45S
45E
90E
135E
135W
90W
45W

(b)

heating profile which takes into account the cooperative heating processes, Hartmann, *et al.* (1982) have been able to produce a dynamic response to equatorial heating, which more keenly matches the observed climate response.

Even though there exists some observational evidence for the existence of radiative heating–latent heating cooperative effects, the substantiation is basically theoretical. The same observational problems (i.e. the difficulty in establishing a control) do not allow a thorough assessment of the degree of interaction of clouds with the physical processes which *produce* climate change. There is even some doubt from observations that, given a change of cloudiness associated with a particular dynamic event, the clouds will act to enhance or reduce the climate change.

Despite the observational difficulties, some studies using satellite data have made significant progress towards this goal. More usually, the effect of clouds on climate has been approached from the development of physical and mathematical analogues of the climate system.

5.3 Clouds and the total diabatic heating field

Clouds appear to be intrinsic features of processes associated with the modification and redistribution of the components of the total diabatic heating field of the climate system. Such associations are identifiable in the list of Arakawa (1975) which shows how clouds may effect climate. Arakawa's basic processes are:

(i) the coupling of dynamic and hydrological processes through the release of latent heat and by evaporation, and by the redistribution of sensible and latent heat and momentum;
(ii) the coupling of radiative and dynamical–hydrological processes in the atmosphere through the reflection, absorption and emission of radiation;
(iii) the coupling of hydrological processes in the atmosphere and in the ground via precipitation, and;
(iv) the influencing of couplings between the atmosphere and the ground through the modification of radiation and the turbulent transfers through the surface.

Arakawa's list refers to a non-interactive boundary; other processes may be added, referring specifically to the hydrosphere and cryosphere:

(v) the influence of the energy balance at the ocean surface through modification of the incident radiant flux, thus modifying evaporation by heat transfers at the surface;
(vi) the alteration of the visible longwave radiation ratio of the incident flux at the surface of the ocean by the clouds and the consequent alteration of the absorption of the radiation within the ocean itself;
(vii) the modification of the cryospheric heat balance by cloud.

Modification and redistribution of the total diabatic heating field within the climate system is apparent in all of the seven processes listed above. For example (ii) describes the modification by cloud of the incoming and outgoing radiative streams, while (iii) describes the manner in which the radiative heating enters the hydrology cycle by providing heat for evaporation, tending to the eventual release of latent heat in cloud structures. The influences of cloud variation on ocean dynamics may be subtle. Although not receiving much attention in the literature, (vi) allows for clouds to influence ocean dynamics even if the total net radiation flux received at the surface remains constant! This is because clouds may decrease the incident visible flux but increase the longwave flux of the surface. As short- and longwave radiation have different absorption characteristics in liquid water, the final energy absorption distribution within the mixed layer, and possibly even at greater depth, may be quite cloud dependent.

5.3.1 *Clouds and diabatic processes in the atmosphere*

The role of diabatic processes in the atmosphere may be seen by considering the global energy budget. If K represents the total kinetic energy of the system and P the available potential energy, we can write an expression for the total energy of the atmosphere $(K+P)$ as:

$$\frac{\mathrm{d}}{\mathrm{d}t}(K+P) = G - E, \tag{5.1}$$

where G is a generation term of available potential energy and E is the dissipation of kinetic energy. For a quasi-geostrophic system, G may be written as:

$$G = \int_v \frac{f_0^2}{\sigma} R \frac{\partial \psi}{\partial p}\, \mathrm{d}v, \tag{5.2}$$

where f_0 and σ are Coriolis and stability parameters; R is the diabatic heating and $\partial\psi/\partial p$ is a measure of the temperature of the column; $\int \mathrm{d}v$ indicates a volume integration. Eq. (5.1) states that the rate of change of the total energy is given by the difference between the generation of available potential energy due to the correlation of diabatic heating (R) and the temperature $(-\partial\psi/\partial p)$ and the kinetic energy dissipation by surface friction. The implication of Eqs. 5.1 and 5.2 to climate research is obvious. If R is large then a sound knowledge of both G and E will be necessary in order to accommodate an adequate energy conservation.

Fig. 5.4 shows the vertical distribution of the zonally averaged heating rates attributable to transport mechanisms as a function of latitude. Calculations were made using data from Oort & Rasmussen (1971) and Newell *et al.* (1972) for the northern hemisphere summer (i.e. June–August). At all latitudes the radiative cooling to space ($\dot{Q}_{\mathrm{RAD}}$) and the condensational heating are major processes. Dynamic transports appear as small residuals. The profiles suggest that both the radiative effects and the condensational heating must be well known in order to calculate the residual with some accuracy. What is not apparent from Fig. 5.4 (*a*) or from the data from which it evolved is the role clouds play in the establishment of the form or magnitude of the components of the diabatic heating fields.

Fig. 5.3. Representation of the sea surface temperature variation in the eastern Tropical Pacific Ocean for 1974–77. (From Liebmann & Hartmann, 1982.)

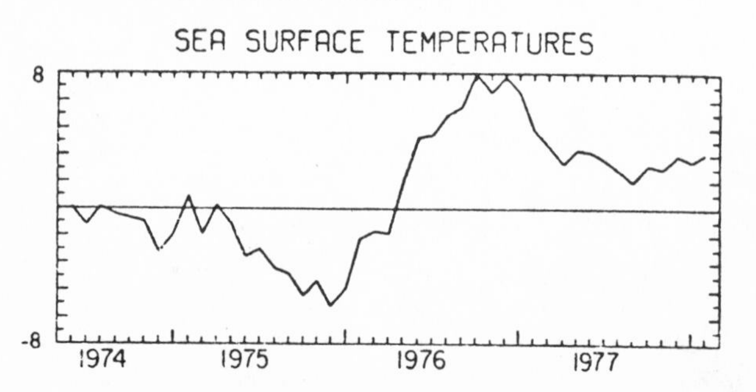

The problem is compounded when the longitudinal or eddy structure of the diabatic heating is considered. If a zonal mean is defined ($\bar{\ }$) and a deviation from that zonal mean ($'$), we may write down a mean available potential energy ($\bar{P}$) equation and an eddy available potential energy (P') equation. These are

$$\frac{d\bar{P}}{dt} = -\{\bar{P}\cdot P'\} - \{\bar{P}\cdot\bar{K}\} + \{\bar{R}\cdot\bar{P}\} \tag{5.3}$$

and

$$\frac{dP'}{dt} = \{\bar{P}\cdot P'\} - \{P'\cdot K'\} - \{P'\cdot R'\}. \tag{5.4}$$

The first two terms on the right-hand sides of Eqs. 5.3 and 5.4 refer to the energy conversions between the indicated energy forms. For our discussion the most important terms are underlined and refer (respectively) to the generation of mean zonal available potential energy and of eddy zonal available potential energy. The two terms are defined as:

$$\{\bar{R}\cdot\bar{P}\} = -\int_v \frac{f_0^2}{\sigma}\,\bar{R}\,\frac{\partial\bar{\psi}}{\partial p}\,dv \tag{5.5}$$

and

$$\{R'\cdot P'\} = \int_v \frac{f_0^2}{\sigma}\,R'\,\frac{\partial\psi'}{\partial p}\,dv. \tag{5.6}$$

Eq. (5.5) simply states that $\bar{P}$ is generated if mean zonally averaged diabatic heating correlates positively with the zonal mean temperature. As the net diabatic heating *abundance* occurs in the *warm* equatorial regions (i.e. latent, sensible and radiative effects) and the diabatic *deficit* occurs in the *cool* higher latitudes, the $\{\bar{R}\cdot\bar{P}\}$ is positive. In a similar manner, if eddy heating correlates with longitudinal variations in temperature, then P' is generated (i.e. $-\{P'\cdot R'\} < 0$). Oort & Rasmussen (1971) estimate by residual methods that $\{P'\cdot R'\}$ is roughly 25% of $\{\bar{R}\cdot\bar{P}\}$ in magnitude and 25% of either $\{\bar{P}\cdot P'\}$ or $\{P'\cdot K'\}$. Consequently the term is important from large-scale energetic considerations.

Fig. 5.4(*b*) shows a partial representation of energy transports and heating rates for three locations (I, the arid regions of Saudi Arabia; II, the Arabian Sea; and III the Bay of Bengal). Radiational cooling is important at all the three adjacent locations but with a variation of form from the relatively dry and cloudless Arabian region to the moist, convective and cloudy Bay of Bengal.

Convective heating is a maximum over the Bay of Bengal and radiative cooling over the Arabian desert region. Large local imbalances between the radiational and condensational heating are apparent and the compensatory (longitudinal and latitudinal) dynamic transports are considerably larger than is evident in Fig. 5.4 (*b*). What emerges are strong zonal and meridional transports

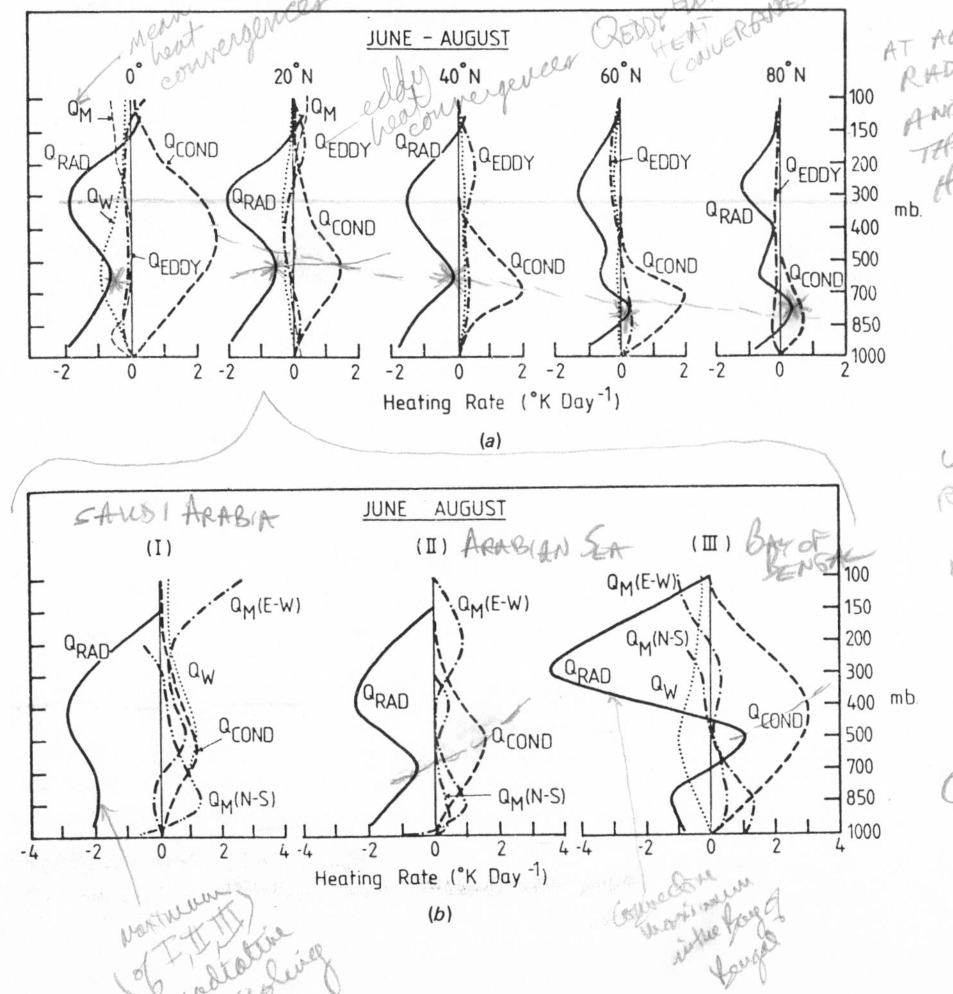

Fig. 5.4. (*a*) Vertical distribution of heating rates attributable to zonally averaged transport mechanisms at various latitudes in northern hemisphere summer. Q_{RAD}, Q_{COND}, Q_M, Q_{EDDY}, and Q_W refer to heating from radiation, condensation, mean heat convergences, eddy heat convergences and adiabatic motions.

(*b*) Same as (*a*) except along 25°N during July. $Q_M(E-W)$ and $Q_M(N-S)$ refer to heat convergences by zonal and meridional mean motions respectively. Locations I, II and III refer to Saudi Arabia, the Arabian Sea and the Bay of Bengal. (From Webster, 1981*a*.)

($\dot{Q}_M(N-S)$ and $\dot{Q}_M(E-W)$) or, in other words, vigorous thermally forced dynamic modes.

5.3.2

Clouds and heating in the ocean

Radiation incident at the surface of the ocean is dispersed in a number of ways. Downward directed terrestrial radiation from the atmosphere is absorbed very close to the surface. The shortwave radiation is absorbed through the ocean column with a strong spectral dependence. The 'red' end of the spectrum is absorbed within a few metres whereas the 'blue–green' radiation is absorbed at considerably greater depths. With the spectral dependency in mind, the absorption of visible radiation is given by (Simpson & Dickey, 1981):

$$\frac{I}{I_0} = R \exp\left(\frac{z}{\psi_1}\right) + (1-R) \exp\left(\frac{z}{\psi_2}\right), \tag{5.7}$$

where R is the proportion following the 'red-end' absorption characteristics with attenuation depth ψ_1; ψ_2 represents the attenuation of the blue–green component. R and both ψ_1 and ψ_2 may also be strong functions of turbidity. An example of the three-scale distribution is given in Fig. 5.5. The left-hand panel shows the disposition of the IR radiation and the right-hand panel the visible. Attenuation depths ψ_1 and ψ_2 are shown for the visible beam. The lettering denotes the thermocline depths for a number of locations. Note the importance of the positions ψ_1 and ψ_2 relative to these depths. If $\psi_2 > A_1, \ldots, D$, then horizontal advection must remove the heating or cooling anomaly, rather than mixed layer dynamics.

The incident solar beam, the downward longwave radiation and the upward longwave radiation from the surface are all integral components of the surface energy balance and the mixed layer structure itself. Whether or not the incident solar beam plays a *direct* role in climate timescales which are longer than those associated with the mixed layer depends upon the relative scale of the mixed layer depth h and the attenuation depth ψ_2. If $h \gg \psi_2$, then variations in the radiation balance will be constrained to effects within the mixed layer. If $h < \psi_2$, solar radiation may be absorbed below the thermocline and may be unaffected by the 'rapid' dynamic and thermodynamic adjustments of the mixed layer.

The factors which will ascribe the greatest change to I_0 and the downward directed terrestrial IR are clouds. For a given location, the incident solar beam and the downward IR will vary by factors which are proportional to the total liquid water path of the atmospheric column above (Stephens & Webster, 1981). Generally, as cloud amount changes, the *total* radiation reaching the surface (visible plus IR) does not change as much as the individual components of the total radiation. For example, the solar beam is depleted with an increase in cloud amount and as the downward IR from the cloud increases. Which component is largest (thus producing either a net radiation gain or loss at the surface) depends on many factors which will be discussed in Section 5.4.2. However the change in the proportion of the long- and shortwave radiation produced by cloud is important since the vertical distribution of radiant energy deposited in the ocean column will be a strong function of cloud amount, cloud height and cloud type.

A number of studies have attempted to test the sensitivity of the ocean mixed layer to variations in attenuation depth or the magnitude of I_0. Kraus & Turner (1967) point out that the predicted minimum depth of the summer thermocline was 80 m if the attenuation depth was 20 m, but 57 m if it were changed to 10 m. Similar sensitivity occurs with the variation in the intensity of the incoming beam. From these considerations, it seems plausible that the variation in cloudiness noted from year to year about the 1976 El Nino, shown in Fig. 5.2, may produce considerable interannual variation in the equatorial mid-Pacific Ocean.

If the structure of the deeper, sub-thermocline ocean is to be influenced in the manner described above, it will only occur in those geographical regions where $h < \psi_2$. Such conditions are apparent only in the summer hemisphere and also in the low latitudes at all times of the year. Thus we may speculate that when prolonged anomalies occur in cloudiness over a prolonged period in time, such as those shown in Fig. 5.2, cloudiness may enter *directly* into the long-term variability of the climate system by causing regions of sub-thermocline ocean to be anomalously heated or cooled.

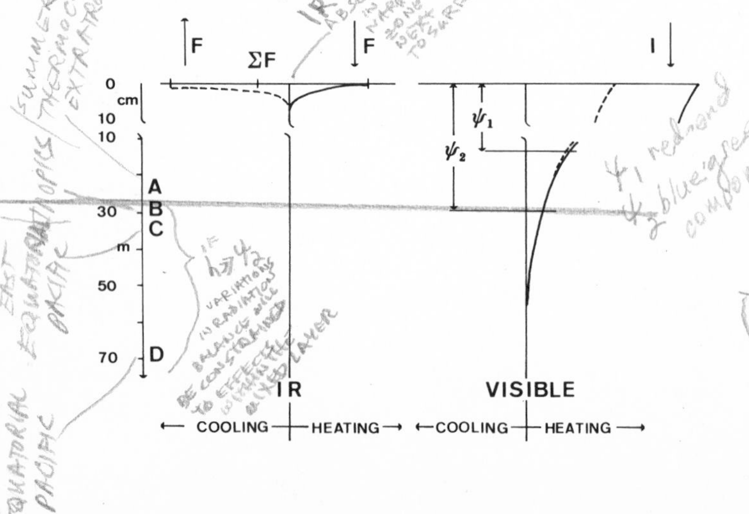

Fig. 5.5. The disposition of the incoming radiation with depth in the ocean column. Downward directed terrestial IR is absorbed in a narrow zone close to the surface. The 'blue–green' visible stream penetrates furthest with an attenuation depth of ψ_1 m. The rear end of the visible stream is attenuated at ψ_2 m. Arrow labelled A shows depth of the average summer thermocline of the extratropical ocean. B shows the average depth for the tropics. C and D denote the east and west equatorial Pacific Ocean mean thermocline depths.

5.4

Studies of cloud–radiation interaction and climate

We may express a climate system in the following manner:

$$L = L(X) = L(X_1, X_2, \ldots; C, \ldots), \tag{5.8}$$

where

$$C = C(L, X) = C(L, X_1, X_2, \ldots). \tag{5.9}$$

In such a non-linear system, L describes a family of climate variables which depend on X_i, one of which is the cloudiness of the atmosphere, C. In a fully interactive system C may also be a function of L_i and X_i.

Studies which attempt to assess the sensitivity of a climate system to variations in cloud top height, cloud base height or cloud composition fall into two main groups. The first group of studies, both observational and theoretical, allow the climate–cloud feedback to remain *open* (i.e. non-interactive) such that in Eq. (5.8), C is constant (for example) and in Eq. (5.9), $C \neq C(L, X_i)$. That is, changes to the climate system, which are produced by a change in a cloudiness specification, are not allowed to feedback into the cloud structure. The second group of studies allows the climate system to be fully coupled as described by Eqs. (5.8) and (5.9). In contrast to the first group of studies (the open-loop studies), the second class are referred to as *closed-loop*. Fig. 5.6 shows a schematic of both groups.

5.4.1

Observational studies

Few observational studies exist which seek a specific assessment of climate sensitivity to cloud amount. The scarcity of studies underlines the difficulty in experimental design and establishment of a meaningful control.

As radiation data is basically satellite derived, the few studies revolve around the radiation budget at the top of the atmosphere

$$R_{\mathrm{NET}} = I_0\,(1-\alpha)-F, \tag{5.10}$$

where I_0 is the available insolation, α is the albedo of the climate system and F is the longwave radiation emitted to space; α is used as a measure of cloud amount. Noting that $R_{\mathrm{NET}} = R_{\mathrm{NET}}\,(A_c, C_T, q, \alpha_s, \ldots)$, where the dependences are, respectively, cloud amount, cloud property, temperature and moisture distribution and surface albedo, the question of sensitivity of climate to cloud means evaluating partial deviations such as:

$$\left.\frac{\partial R_{\mathrm{NET}}}{\partial A_c}\right|_{C,T,q,\alpha_s,\ldots} \tag{5.11}$$

In a pioneering study, Cess (1976) found a value of Eq. (5.9) to be effectively zero, specifically 2·5 Wm^{-2}, which suggested an almost complete cancellation between the shortwave and longwave effects. If Cess's estimate was correct, the radiation budget at the top of the atmosphere would be nearly insensitive to cloud amount. However, Hartmann & Short (1980) point out that rather than calculating a partial derivative, Cess was probably calculating the *total derivative* $\mathrm{d}R_{\mathrm{NET}}/\mathrm{d}A_c$. The study of Cess can properly be classified as a closed-loop observational study.

Ellis (1978) went further toward calculating the partial derivative (Eq. (5.11). He compared the outgoing radiation between clear skies and for average cloudiness. The method requires an accurate determination of a clear-sky period and of the fractional cloud cover at other times. However, because of this, A_c, as well as C and α_s, were not precisely invariant. $\partial R_{\mathrm{NET}}/\partial A_c$ was found to be -35 Wm^{-2}, indicating, as interpreted by Hartmann & Short (1980), that the distribution of cloud sensed by Ellis was about twice as effective in reflecting solar radiation as in trapping terrestrial radiation. Ohring & Clapp (1980), Hartmann & Short (1980) and Ohring *et al.* (1981) obtained estimates of $\partial F/\partial A_c$ and $\partial R_{\mathrm{NET}}/\partial A_c$ at specific points using month-to-month variations (Ohring & Clapp) and day-to-day variations (Hartmann & Short) between cloudy and clear skies. Calculation of the quantities at the same point eliminated the variability of the ground albedo, α_s. All studies found similar estimates of Eq. (5.11) to those of Ellis (1978) and these are reproduced in Fig. 5.7 from Hartmann & Short (1980). The triangular section can be seen to contain all estimates, except that of Cess (1976), and indicates the dominance of increase in the shortwave reflection over longwave loss by clouds.

Both Ohring *et al.* (1981) and Hartmann & Short (1980) have produced global distributions of the sensitivity factor given by Eq. (5.11). Ohring *et al.*'s distribution produced from a 45-month set of monthly mean radiation data is shown in Fig. 5.8. The entire domain is negative, which indicates the dominance at the top of the atmosphere of the shortwave effect of clouds over their longwave effects. Besides the universality of the sign of the factor,

Fig. 5.6. Schematic representation of the open- and closed-loop model system described in the text relative to Eqs. (5.8) and (5.9).

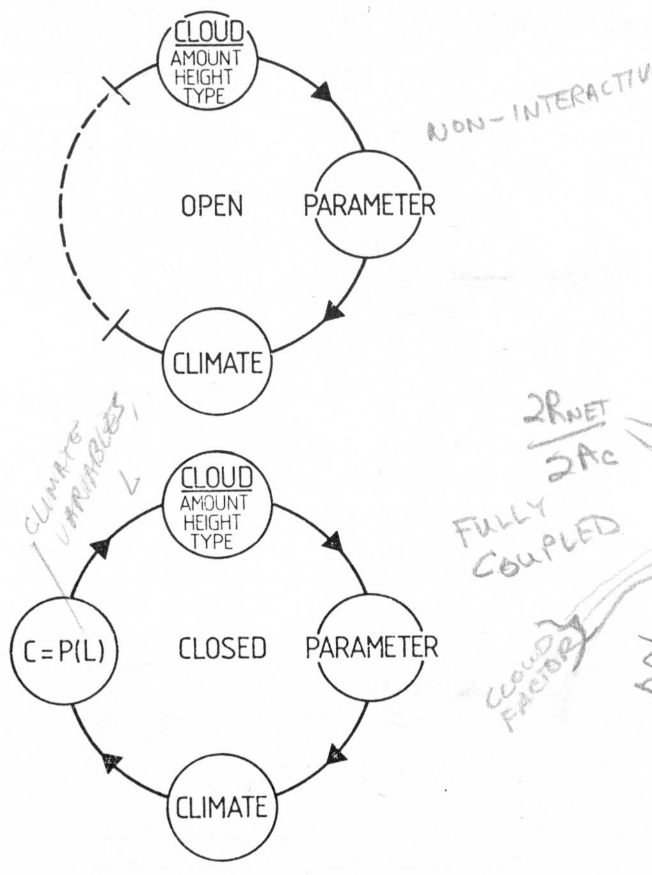

Fig. 5.7. Graph comparing the various estimates of the cloud–climate sensitivity compiled by Hartmann & Short (1980). Coordinates are the differences between overcast and clear skies of the albedo (ordinate) and the outgoing IR (abscissa). Radial lines refer to the Hartmann–Short cloud factor. Heavy dashed lines show $\partial R_{\mathrm{NET}}/\partial A_c$ of Eq. (5.11). A, S, C, E and O refer to the earlier estimates of Adem (1967), Schneider (1972), Cess (1976), Ellis (1978) and Ohring & Clapp (1980). The triangle encloses the range of probable mean values of cloud–climate sensitivity for the globe.

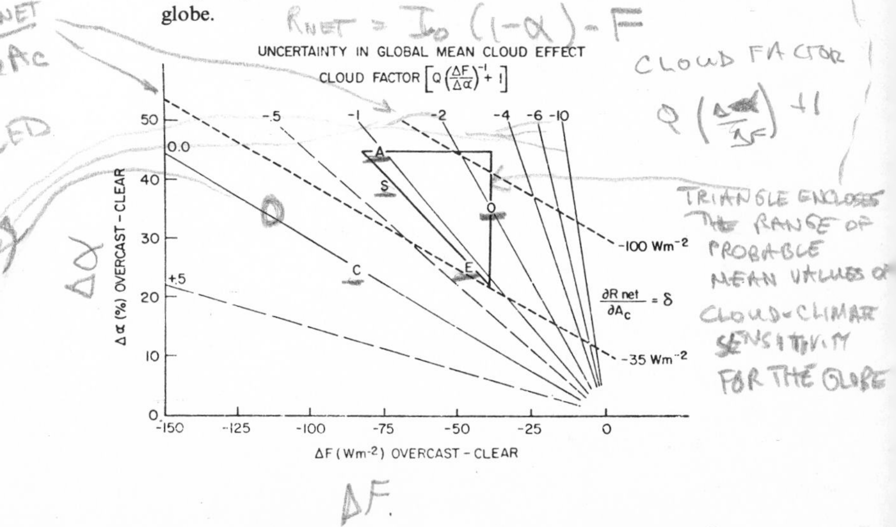

the distribution shows considerable character. By relating this variation to the known features of the climate system (such as deserts, zones of precipitation, subsidence, etc.), it may be possible to obtain additional information which may not be readily extractable from the usual satellite product such as a visible brightness or IR radiance distribution. The potential use in calculating such fields may be seen from Hartmann & Short (1980) who show two fields which are similar to Fig. 5.8, except for two solstitial seasons. The distributions are markedly different.

5.4.2

Theoretical and modelling studies

There are a number of model types which comprise the open- and closed-loop models.

(1) *One-parameter energy balance models*: the parameter is usually surface temperature and the atmospheric structure is expressed as a functional form of that one parameter. Basic physics involve the surface energy budget and horizontal heat diffusion.

(2) (a) *One-parameter radiative–convective models*: the parameter used is the vertically dependent atmospheric temperature. Radiative equilibrium is achieved throughout the atmospheric column, subject to the constraint of gravitational or convective stability. Energy balance calculations are made at the surface with parameterized flux forms relative to the atmospheric temperature structure above.

(b) *Radiative–convective atmosphere with a simple ocean*: as above except the surface energy balance includes heat fluxes from below the ocean subject to an adjusting mixed layer.

(3) *Multi-parameter radiative dynamic models*: such models are usually multi-dimensional, often global in extent and contain full dynamics and (occasionally) an interactive ocean.

All of the model types listed above may be used for experimental open-loop studies. Types (1) and (2) are restricted to open-loop studies simply because, for example, they lack the parametric breadth to fully express the production of clouds. Type (3), the general circulation model, alone has the capability of closed-loop studies.

Fig. 5.8. Global distribution of the net sensitivity $\partial R_{NET}/\partial A_c$ in units of Wm^{-2}. Sensitivity values greater than $-80\ Wm^{-2}$ are shaded. Note that the entire domain is negative. (From Ohring *et al.*, 1981.)

In summary, we may summarize the model attributes and deficiencies as follows:

The open-loop model group possesses the following attributes:

(1) Although it is not necessary, the open-loop model usually possesses only a few degrees of freedom. This simplicity allows the possibility of a fine grid resolution, especially in the vertical. The overall simplicity of the model renders the results relatively easy to understand.

(2) An extremely detailed cloud parameterization may be attached to the fine grid resolution. The amount of cloud, the cloud height and their optical properties may be carefully specified.

(3) The open-loop model allows careful controlled experimentation.

However, open-loop studies have the following drawbacks:

(1) The very simplicity of the model makes it difficult to compare the model results with the real climate. Basic model structure in the vertical is parameterized by the assumption of equilibrium lapse rate.

(2) The results may only be interpreted as the initial tendencies of the real climate system to the change in a particular cloud property. Because of this, open-loop models often appear to be considerably more sensitive than closed-loop models.

(3) It is difficult to develop parallel observational studies with open-loop models.

The closed-loop model group possesses a number of attractive features. These are:

(1) The physical complexity of the parent climate model allows for a greater confidence in the similitude between the model and the real climate system, which allows meaningful comparisons between observational studies and the model climate.

(2) The system may feed back on to the cloudiness, thus allowing an assessment of sensitivity *beyond* the initial tendencies noted in the open-loop studies.

On the negative side:

(1) Because of the more general physical domain, the grid resolution of the system is rather coarse and cloud structures are often sub-grid scale. The lack of vertical resolution may be especially critical in calculating cloud–radiation interaction.

(2) The success of the climate simulation depends on how well

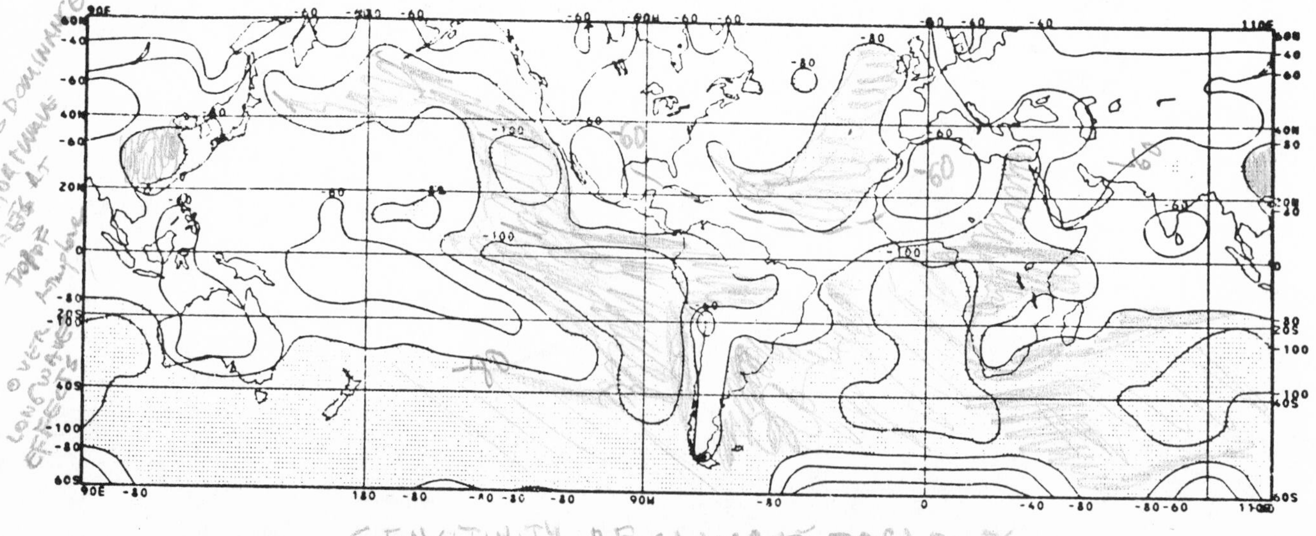

cloud production and dissipation is parameterized, as well as the formulation of the cloud interaction with the radiative field as in the open-loop model.

Thus both groups have their various attributes and deficiencies. The two modelling philosophies are complementary, however, with the more restrictive open-loop studies suggesting experiments and possible climate sensitivities to be studied more comprehensively by the closed-loop models. Furthermore it should be noted that the general circulation models are in a state of evolution and it is hoped that their deficiencies will decrease with time.

In the following paragraphs, we will summarize some of the studies which utilize the open- and closed-loop sensitivities.

5.4.3
Open-loop model studies

A fairly extensive literature exists on open-loop model studies of climate sensitivity. For example, studies by Budyko (1969), Sellers (1969), Schneider (1972) and North (1975) have employed simple energy balance models to questions of climate stability and the effect on climate changing atmospheric composition. Radiative–convective models were originally introduced by Manabe & Strickler (1964) and Manabe & Wetherald (1967) and have been utilized in many studies for a variety of climate problems. In the following paragraphs we will concentrate on a series of studies which utilize the same radiative convective model and the same cloud property parameterization. Specifically we will summarize the results of Stephens & Webster (1979, 1981 and 1982) and Webster and Stephens (1980*a* and 1980*b*).

Fig. 5.9 shows the cloud albedo–cloud emissivity relationships used in the four studies. Both cloud albedo and cloud emissivity are correlated to the liquid water path in the atmospheric column, allowing the expression of both the longwave and shortwave optical properties of the clouds as functions of one parameter. The parameterization originated in the study of Stephens (1978).

Fig. 5.9. Albedo and emissivity relationships as a function of the liquid water path (gm^{-2}). Curves represent the basic cloud optical property parameterization of Stephens (1978). The effective emittance is invariant with latitude. However, the cloud albedo is distinctly zenith angle dependent. Clouds which fall into the particular liquid water path range are shown on the upper abscissa. (From Stephens & Webster, 1981.)

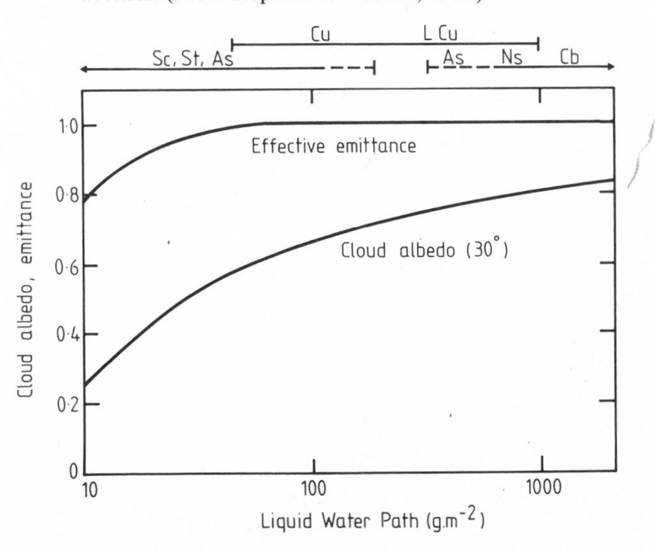

In the studies described below, the radiative scheme described by Stephens & Webster (1979) was coupled to a convective adjustment scheme similar to the Manabe & Wetherald (1967) model with a fixed climatological relative humidity distribution. Clouds of a specified distribution, amount and type, relative to the parameterization shown in Fig. 5.7, were 'inserted' into the radiative–convective model and an equilibrium 'climate' produced. All results pertain to the change of temperature at a land surface.

It should be remembered that the large differences in equilibrium temperatures reported below and said to occur between clear and overcast conditions are really initial tendencies. If the model were fully 'closed', other factors which are purportedly omitted from the Stephens & Webster (1981) model may minimize the difference.

5.4.3.1
Sensitivity to cloud amount

Fig. 5.10 shows the equilibrium surface temperature distribution as a function of cloud amount for the three cloud layers indicated, using both summer and winter solstice insolation values at 35°N. Each cloud species shows a different variation of temperature with changing cloud amount. Middle and low clouds tend to decrease the surface temperature as cloud amount is increased, whereas high clouds show the opposite trend. The reason for the difference may be seen in Fig. 5.10. Low and middle cloud of the thicknesses shown here possess water paths that are greater than 100 gm^{-2}, which typifies an emissivity of unity and a high albedo. The high clouds have smaller liquid water or ice paths (perhaps $< 15\ gm^{-2}$), which corresponds to a reduction in cloud albedo by a factor of four and only 20% in emissivity. Thus, in the case of low and middle clouds, the albedo effect dominates over emissivity effect and the net energy input to the surface is decreased as cloud

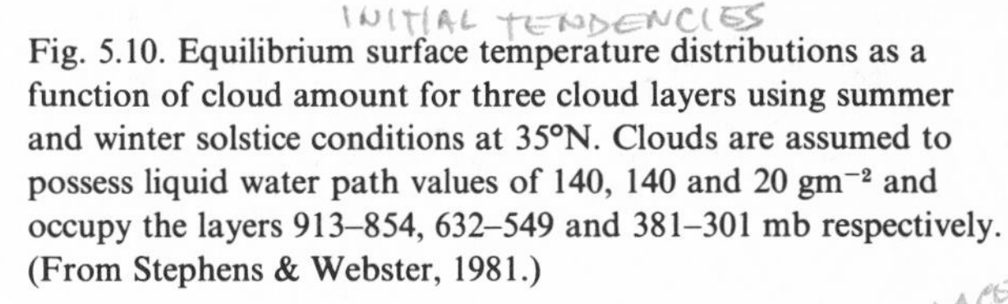

Fig. 5.10. Equilibrium surface temperature distributions as a function of cloud amount for three cloud layers using summer and winter solstice conditions at 35°N. Clouds are assumed to possess liquid water path values of 140, 140 and 20 gm^{-2} and occupy the layers 913–854, 632–549 and 381–301 mb respectively. (From Stephens & Webster, 1981.)

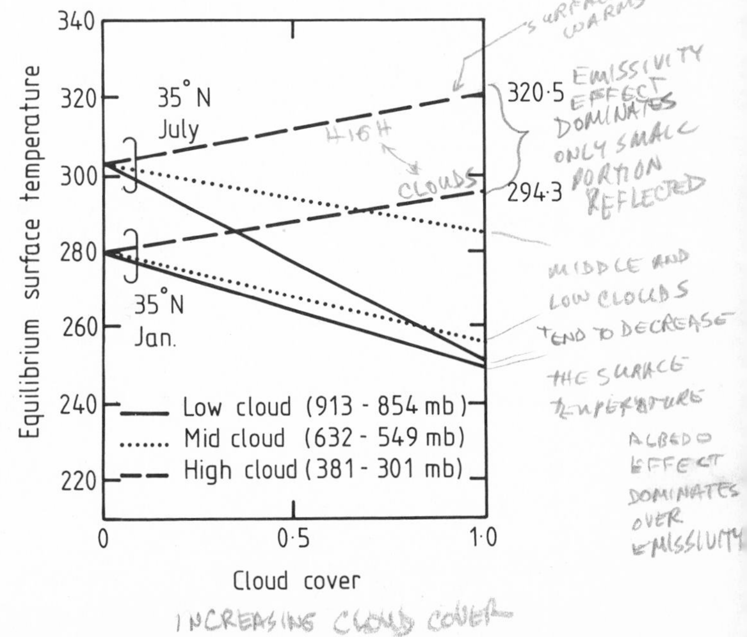

amount is increased. However, with high clouds the emissivity effect dominates and only a small percentage of the incident solar beam is reflected. Consequently, an increase in high cloud amount tends to raise the surface temperature.

5.4.3.2

Sensitivity to cloud height

(i) *Layer clouds*: Fig. 5.11 (from Stephens & Webster, 1981) shows the surface temperature difference between clear and overcast skies for three layer clouds as a function of water or ice–water path (gm^{-2}). High (H), middle (M) and low (L) clouds are shown and are assumed to occupy the slabs 318–301, 632–549 and 913–854 mb. Winter insolation values for 5, 35 and 65°N are chosen. Generally, for a given liquid or ice–water path, the effect of an increase in cloud height is to decrease the cooling effect of the cloud or to increase the warming effect. Except for slightly different absorption characteristics, the shortwave effect of cloud is much the same irrespective of height. However, as the cloud temperature is assumed to be the same as ambient temperature, the longwave loss to space by the system is reduced as the cloud top is raised.

Considered in terms of the cloud *base* height, the reasoning regarding the character of Fig. 5.9 assumes a paradoxical nature. Besides the solar radiation reaching the ground, the downward flux of IR is the principal determinant of the surface temperature. In a cloudy atmosphere the downward flux of IR originates in the cloud layer and in the background water vapour distribution which exists whether the atmosphere is cloudy or not. Thus it would seem that the lower the base of the cloud the greater the downward IR flux simply because the base temperature would be warmer. However, Fig. 5.9 shows the opposite effect. We will return to this paradox later.

(ii) *Overlapping clouds*: in reality clouds possess considerably more compound structures than the layer clouds assumed above. In order to assess the effect of overlapping cloud structures (Stephens & Webster, 1983), Fig. 5.12 shows the four cloud distribution functions used in the study. Fig. 5.13 shows the distribution of ΔT_g (T_g is equivalent to T_s) as a function of vertically integrated liquid water content for the distribution functions 1 and 2 of Fig. 5.12, which can be referred to as the 'high' and 'low' cloud profiles respectively.

In essence, the character of the overlapping cloud distributions is the same as for the larger cloud shown in Fig. 5.11. As before, the cooling of the surface by low clouds exceeds that of the higher cloud even though the vertically integrated water path is identical.

(iii) *The cloud base effect*: the paradoxical surface temperature response may be explained in the following manner. In order to assess the effect of a cloud, the contribution of the *clear* atmosphere to the surface balance must first be understood. As the background water vapour distribution is concentrated near the ground, the lower atmosphere of the clear sky is already opaque when viewed from the ground. Thus a cloud placed in the lower atmosphere may

Fig. 5.11. Surface temperature difference between clear and overcast conditions as a function of liquid water or ice–water path (gm^{-2}) for three cloud layers. Results are shown for 5, 35 and 65°N in winter with a surface albedo of 0·102. (From Stephens & Webster, 1981.)

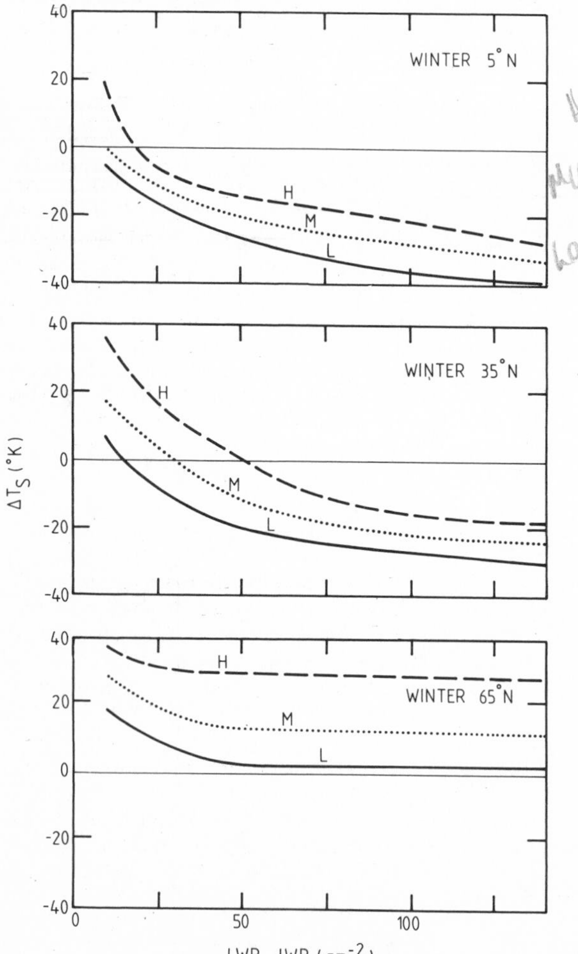

Fig. 5.12. Four cloud distribution functions used in study. (From Stephens & Webster, 1983.)

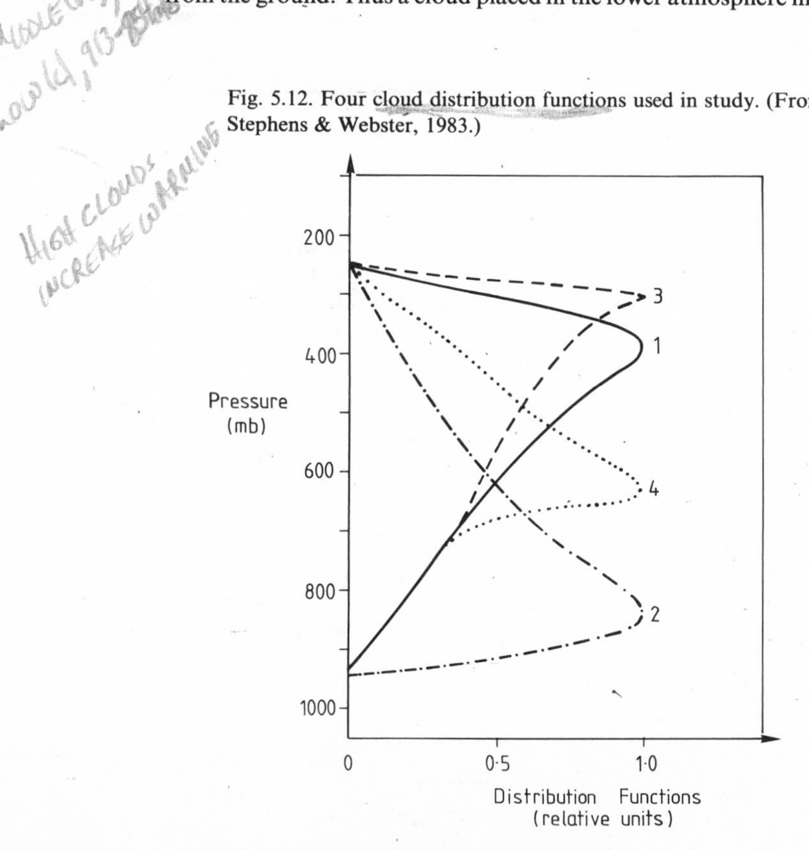

not have as large an impact as a cloud placed in the relatively dry upper troposphere. Fig. 5.14 shows the relative increase of layer emissivity with respect to height by the placement of a cloud of given emissivity at different levels. In the tropical atmosphere, the ratio of the emissivities is hardly changed in the lower atmosphere but may increase by a factor of ten in the upper atmosphere. This effect is compensated somewhat by the decrease of temperature with height, but, as can be seen from the right-hand panel of Fig. 5.4, the compensation is only partial.

5.4.3.3

Cloud–radiation surface albedo sensitivities

In a cloud-free atmosphere the effect of surface albedo on surface temperature is dramatic. Increases in surface albedo cause substantial cooling at the surface. However, when clouds are included in the system, the effect is changed considerably.

Fig. 5.15 shows the difference in equilibrium surface temperature between a clear and overcast sky of the indicated cloud type as a function of surface albedo. Because of the albedo effect, a cloud with a specified liquid water path (and therefore albedo and emittance) *can change from being a net cooler to being a net warmer of the surface*. For example, at 35°N, middle- and low-level clouds of 140 gm^{-2} are usually net surface coolers. However, with surface

Fig. 5.13. The change in the surface temperature between clear skies and an overcast sky of vertical distribution function 1 and 2 of Fig. 5.10. ΔT_g is plotted against the vertically integrated liquid water content. (From Stephens and Webster, 1983.)

Fig. 5.14. The relative increase in layer emissivity for clear to overcast sky for the three given cloud emissivities as a function of mid-pressure of the cloud situated in a 100 mb slab. Also shown is the relative change in black body longwave flux as a function of pressure arising from changing cloud temperature. Curves are deduced for a model tropical atmosphere. (From Stephens & Webster, 1983.)

Fig. 5.15. Surface temperature difference between clear and overcast skies as a function of surface albedo. Results are for 5, 35 and 65°N in winter. Clouds possess the indicated liquid water path and the low (L), medium (M), and high (H) clouds occupy the same layers as in Fig. 5.8.

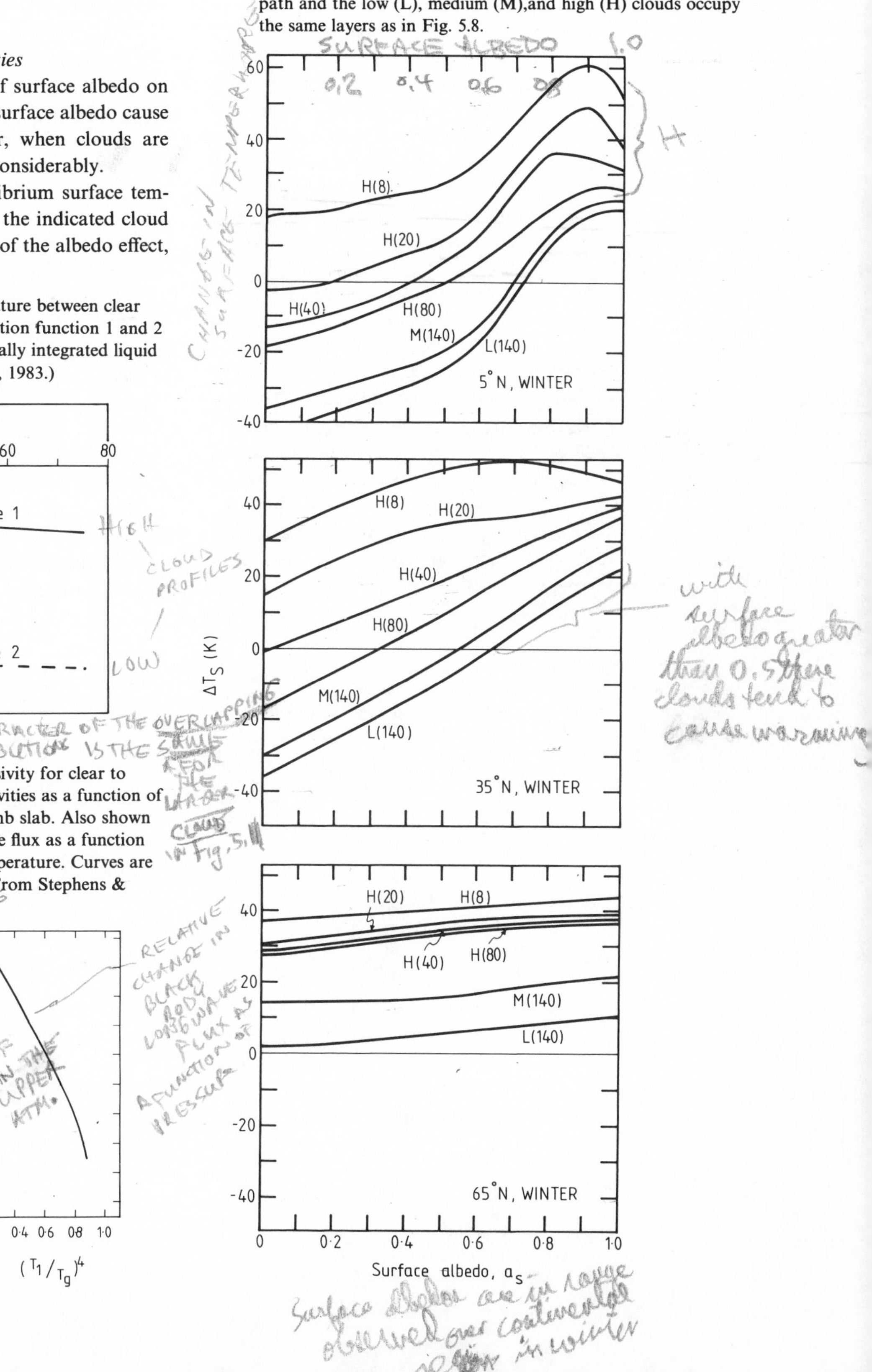

albedos greater than about 0.5 the clouds tend to warm the surface. This is important because these surface albedos are in the range observed over continental regions in winter. The effect probably has palaeoclimatological implications.

5.4.3.4

Clouds and atmospheric sensitivity to composition changes

Fig. 5.16 shows the vertical structure of the difference between the equilibrium temperature profiles for atmospheres possessing 300 and 600 ppm CO_2 concentrations for clear skies and for selected overcast conditions. The result for the clear sky (heavy curve), a surface warming of about 2.5 °C and a substantially larger cooling of the stratosphere, is very similar to that found by Manabe & Wetherald (1967). With the inclusion of clouds, surface temperature changes decrease from about 2.6 K (clear case) to about 0.5–1.0 °C for optically thick clouds and to even smaller values for high thin clouds. It appears that the effect of cloud is to *reduce* the sensitivity of the lower part of the model atmosphere to CO_2 effects. The physical mechanism which accomplishes the reduction is the effective partitioning of the stratosphere and the troposphere by the cloud layer. With the enhanced stratospheric IR emission impeded by the cloud layer, the surface temperature will only increase by an amount which is determined by increased IR emission originating in the *sub-cloud layer*.

The various surface temperature sensitivities discussed in the preceding paragraphs are summarized in Table 5.1. Sensitivities are expressed as the surface temperature change (K) which would be produced by a 1% change in some parameter. The table indicates the fact that the model climate appears most sensitive to cloud amount changes. In fact, a 5% decrease in cloudiness would be equivalent to a 1% change in solar constant. Likewise, the temperature increase produced by the doubling of CO_2 could be offset by a 10% increase in low cloud amount or a 10% in high cloud. The latter figures are listed in Table 5.2.

Arguments about compensatory climatic parameters are difficult to develop when it is realized that we have little idea whether cloudiness would tend to increase or decrease with an induced column temperature rise. The problem is compounded further when it is realized that high cloud and low cloud work in opposite directions, a feature which originates from the different albedo-emittance relationships of high and low clouds. Of course, this conclusion says nothing of the *sign* of the cloud amount variation with a changing atmospheric temperature. These are the problems the *closed-loop models* must tackle.

5.4.4

Closed-loop studies

A successful closed-loop study is an extremely difficult problem and remains one of the major challenges in climate dynamics. As indicated in the open-loop studies, the radiative structure of the climate system is a strong function of cloud height, amount, type and composition. Furthermore, the very existence of a cloud structure and form depends on the moisture distribution of a particular region, which, in turn, depends upon the dynamic and thermal state of the atmosphere. It is the faithful representation of each of the interlocking states which is the major concern of the closed-loop model.

A number of large-scale models have self-generating and interactive cloud components within the model climate. In an exhaustive and critical comparison between model results and observed cloud statistics Gates & Schlesinger (1977) found the general sense of both the zonally averaged and the zonally asymmetric distributions of cloudiness to be reproduced, but found some fairly large differences in magnitude. For example, zonally

Fig. 5.16. Vertical structure of the *difference* between equilibrium temperature profiles for atmosphere pinening 300 and 600 ppm CO_2 concentrations for clear skies and overcast conditions. Cloud slab heights as in Fig. 5.8 and liquid water path as indicated. (From Webster & Stephens, 1983.)

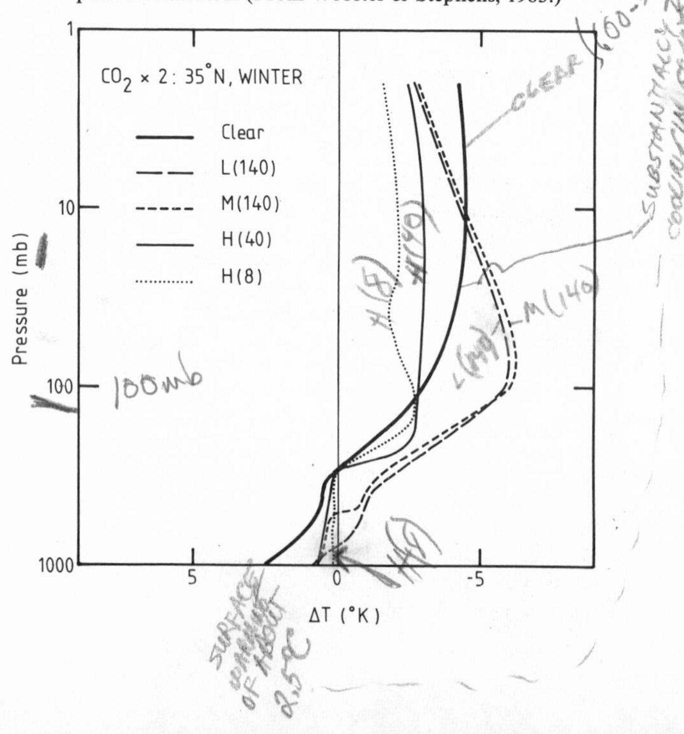

Table 5.1. *Surface temperature sensitivities to changes in various parameters at 35°N in winter. Implications from open-loop studies*

	Sensitivity $\frac{\alpha}{100}\frac{dT_s}{d\alpha}$				
	Clear	L(140)	M(140)	H(40)	H(8)
Parameter α					
Solar constant	+1.26	+1.20	+1.20	+1.25	+1.27
Cloud amount	—	−0.28	−0.22	+0.04	+0.26
Surface albedo					
$\alpha_s < 0.5$	−0.8	−0.13	−0.17	−0.30	−0.46
$\alpha_s > 0.7$	−0.9	−0.14	−0.18	−0.40	−1.20
CO_2	+0.026	+0.006	+0.005	+0.004	+0.004

Table 5.2. *The per cent variation in cloudiness which would be required in order to match a temperature change of (A) 2.6 K and (B) 0·5 K by a CO_2 atmospheric concentration doubling. Implications from open-loop studies*

	(A) 2.6 K increase			(B) 0.5 K increase		
	L(140)	M(140)	H(8)	L(140)	M(140)	H(8)
Cloud amount (%)	+9.3	+11.8	−10.0	+1.9	+2.3	−1.9

averaged cloud simulations are underestimated by a factor of two in the winter southern hemisphere but are fairly accurately represented in the summer northern hemisphere. A similar study has been carried out by Herman (1981) for the NASA–GLAS (Goddard Laboratory for Atmospheric Science) model. Although many of the features of the observed cloudiness are apparent in the simulation, considerable differences still occur at higher latitudes.

Wetherald & Manabe (1980) used a truncated version of the Geophysical Fluid Dynamics Laboratory (GFDL) model with idealized geography to seek the sensitivity of the model climate to changes in insolation for both fixed and variable cloud amount. They found little difference between the fixed and variable cloud experiments; and attributed the lack of sensitivity to compensatory changes in cloud amount and height modifying the solar radiation and the outgoing longwave radiation. Shukla & Sud (1981) point out that Wetherald & Manabe examined only the zonally and globally averaged climate sensitivity in the mean climate to variable and fixed clouds assumptions.

Herman *et al.* (1980) and Shukla & Sud (1981) conducted a sensitivity experiment with the NASA–GLAS model with the aim of comparing the zonally asymmetric structures produced by variable and fixed cloud assumptions. Significant differences were produced in both studies. For example, Shukla & Sud found that in the fixed cloud run, the generation of P' (in Eq. (5.4)), the conversion of P' to K' and the variance of the transient and stationary components were found to be larger. Furthermore, evaporation and precipitation were enhanced in the fixed cloud experiment. Shukla & Sud and Herman *et al.* are among the first to suggest, via closed-loop experimentation, that climate is sensitive to interactive cloud variability. In their studies, cloud varies in a manner which tends to reduce the amplitude of various factors in the atmosphere; i.e., clouds produce a negative feedback effect on climate.

Shukla & Sud and Herman *et al.* have hinted at important relationships between cloud and climate. However, as they imply, it is difficult to generalize beyond the climate system they studied (the NASA–GLAS model), with its own particular parameterization, attributes and deficiencies, to a physically significant relationship. Probably a more definite statement concerning the magnitude of the sensitivity of the real climate system must await improvements of model structures including the cloud–radiation and cloud–dynamic–thermodynamic parameterization.

An example of the challenge facing the closed-loop models may be obtained from important studies aimed at assessing the impact of an increasing CO_2 concentration in the atmosphere. We might ask what are the 'requirements' of the Manabe & Wetherald (1980) paper which would allow a statement from the model results about the real climate system? Closed-loop studies (Tables 5.1 and 5.2 summarize the sensitivities) suggest that an increase in low cloud of 10% or a decrease in high cloud of the same amount may compensate for a CO_2 doubling. If the open-loop sensitivities are correct (and, paradoxically, it will require a closed-loop model to test the magnitude as the open-loop studies only offer an initial tendency), then a general circulation model may need to simulate the amount of cloudiness at a number of levels to within a few per cent.

5.5 Concluding remarks

Possibly the most exciting development in the study of climate–cloud relationships has been the common progress in all three modes of research; observational studies, simple open-loop model development and the continual improvement of large-scale general circulation models. By intercomparing the results of all three modes, a general conclusion has emerged that cloudiness may be a significant climate constituent.

It is interesting that probably none of the three can individually approach a definitive and absolute answer to the question of cloud and climate without reference to the other. Despite the exciting advances of Ohring & Clapp (1980), Hartmann & Short (1980) and Ohring *et al.* (1981), it is doubtful that the determination of a partial derivative (for example, $\partial R_{NET}/\partial A_c$) can ever be completely approached. Also, being tied to satellite data, it will be difficult to obtain an assessment of the surface sensitivity. At the same time, the simpler open-loop models can only indicate initial tendencies of the complete system. However, they can serve as useful indicators of areas of sensitivity that must be taken into account by the more sophisticated general circulation models.

The final answers will probably emerge from the closed-loop model studies. The degree of definitiveness will depend upon the ability of the model to simulate, very accurately, the actual climate. With the establishment of an excellent control, a variety of cloud–climate related questions may be resolved by careful experimentation.

References

Adem, J. (1967). 'On the relations between outgoing longwave radiation, albedo and cloudiness'. *Mon. Wea. Rev.*, **95**, 257–260.

Arakawa, A. (1975). 'Modelling clouds and cloud processes for use in climate models'. *GARP Publication Series, No. 16* (ICSU/WMO), pp.183–97.

Budyko, M. (1969). 'The effect of solar radiation variations on the climate of the Earth'. *Tellus*, **21**, 61–9.

Cess, R. D. (1976). 'Climate change: An appraisal of atmospheric feedback mechanisms employing zonal climatology'. *J. Atmos. Sci.*, **33**, 1831–43.

Ellis, J. S. (1978). 'Cloudiness, the planetary radiation budget and climate'. Ph.D. Thesis, Department of Atmospheric Sciences, Colorado State University, 129 pp.

Gates, W. L. & M. Schlesinger (1977). 'Numerical simulation of the January and July global climate with a two-level atmospheric model'. *J. Atmos. Sci.*, **34**, 36–76.

Hartmann, D. & D. Short (1980). 'On the use of Earth radiation budget statistics for studies of clouds and climate'. *J. Atmos. Sci.*, **37**, 1233–50.

Hartmann, D., H. H. Henden & R. Houze (1984). 'Some implications of the meso-scale circulations in tropical cloud clusters for large scale dynamics and climate. *J. Atmos. Sci.*, **41**.

Herman, G. (1981). 'Cloud–radiation experiments conducted with GLAS general circulation models'. *In: Clouds in Climate, Report of Workshop at NASA–GISS*, New York, 29–31 Oct., 1980.

Herman, G. F., M. C. Wu & W. Johnson (1980). 'The effect of clouds on the Earth's solar and IR budgets'. *J. Atmos. Sci.*, **37**, 1251–61.

Horel, J. D. & J. M. Wallace (1981). 'Planetary scale atmospheric phenomenon associated with the Southern Oscillation'. *Mon. Wea. Rev.*, **109**, 813–29.

Hoskins, B. & D. Karoly (1981). 'The steady linear response of a spherical atmosphere to thermal and orographic forcing'. *J. Atmos. Sci.*, **38**, 1179–96.

Houze, R. (1982). 'Cloud clusters and large scale vertical motions in the tropics'. *J. Met. Soc. Japan*, **60**, 396–410.
Hunt, B. G. (1977). 'On the general circulation of the atmosphere without clouds'. *Quar. J. Roy. Meteor. Soc.*, **104**, 91–102.
Hunt, G. E., V. Ramanathan & R. M. Chervin (1980). 'On the role of clouds in the general circulation of the atmosphere'. *Quar. J. Roy. Meteor. Soc.*, **106**, 213–15.
Kraus, E. B. & J. S. Turner (1967). 'A one-dimensional model of the seasonal thermocline: II. The general theory and its consequences'. *Tellus*, **19**, 98–106.
Liebmann, B. & D. Hartmann (1982). 'Interannual variation of outgoing IR and association with tropical circulation changes during 1974–78'. *J. Atmos. Sci.*, **39**, 1153–62.
Manabe, S. & R. Strickler (1964). 'Thermal equilibrium of the atmosphere with a convective adjustment'. *J. Atmos. Sci.*, **21**, 361–85.
Manabe, S. & R. T. Wetherald (1967). 'Thermal equilibrium of the atmosphere with a given distribution of relative humidity'. *J. Atmos. Sci.*, **24**, 241–59.
Manabe, S. & J. L. Hollaway, Jr (1971). 'Simulation of the climate by a global general circulation model 1, hydrology cycle and heat balance'. *Mon. Wea. Rev.*, **99**, 335–69.
Newell, R., D. Vincent, J. Kidson & G. Boer (1972). *The General Circulation of the Tropical Atmosphere and the Interaction with Extratropical Latitudes*, Vol. 1., MIT Press, 258 pp.
North, G. R. (1975). 'Analytic solutions to a simple climate model with diffusive heat transport'. *J. Atmos. Sci.*, **32**, 1301–7.
Ohring, G. & P. F. Clapp (1980). 'The effect of changes in cloud amount on the net radiation at the top of the atmosphere'. *J. Atmos. Sci.*, **37**, 447–54.
Ohring, G., P. F. Clapp, T. R. Heddingham & A. F. Krueger (1981). 'The quasi-global distribution of the sensitivity of the earth–atmosphere radiation budget to clouds'. *J. Atmos. Sci.*, **38**, 2539–41.
Oort, A. & E. Rasmussen (1971). *Atmospheric Circulation Statistics.* NOAA Prof. Paper. No. 4., US Dept. of Commerce, 323 pp.
Ramage, C. (1975). 'Preliminary discussion of the meteorology of the 1972–73 El Nino'. *BAMS*, **46**, 4–15.
Schneider, S. (1972). 'Cloudiness as a global feedback mechanism. The effect of the radiation balance and surface temperature of variations in cloudiness'. *J. Atmos. Sci.*, **29**, 1413–22.
Sellers, W. D. (1969). 'A global climate model based on the energy balance of the earth–atmosphere system'. *J. Appl. Meteor.*, **8**, 392–400.
Shukla, J. & Y. Sud (1981). 'Effect of cloud–radiation feedback on the climate of a general circulation model'. *J. Atmos. Sci..*, **38**, 2337–53.
Simpson, J. J. & T. D. Dickey (1981). 'The relationship between downward irradiance and upper ocean structure'. *J. Phys. Ocean*, **11**, 309–23.
Stephens, G. L. (1978). 'Radiative properties of extended water clouds: II Parameterisations'. *J. Atmos. Sci.*, **35**, 2133–32.
Stephens, G. L. & P. J. Webster (1979). 'Sensitivity of radiative forcing to variable cloud and moisture'. *J. Atmos. Sci.*, **36**, 1542–56.
Stephens, G. L. & P. J. Webster (1981). 'Clouds and climate: sensitivity of simple systems'. *J. Atmos. Sci.*, **38**, 235–47.
Stephens, G. L. & P. J. Webster (1983). 'Cloud decoupling of surface and upper radiation balances', to appear in *J. Atmos. Sci.*, **40**.
Webster, P. J. (1981*a*). 'Review of cloud interaction with other climate elements'. *In: Clouds in Climate, Report of Workshop at NASA–GISS*, New York, 29–31 Oct., 1980.
Webster, P. J. (1981*b*). 'Mechanisms affecting the atmospheric response to sea-surface temperature anomalies'. *J. Atmos. Sci.*, **38**, 554–71.
Webster, P. J. (1982). 'Seasonality in the atmospheric response to sea-surface temperature anomalies'. *J. Atmos. Sci.*, **39**, 24–40.
Webster, P. J. & G. L. Stephens (1980*a*). 'Tropical upper tropospheric extended cloud: Inferences from winter MONEX'. *J. Atmos. Sci.*, **37**, 1521–41.
Webster, P. J. & G. L. Stephens (1980*b*). 'Gleaning CO_2-climate. relationships from model calculations'. *In: Carbon Dioxde and Climate: Australian Research.* Editor, G. I. Pearman, Australian Academy of Science, pp. 185–95.
Webster, P. J. & J. R. Holton (1982). 'Low latitude and cross-equatorial response to middle-latitude forcing in a zonally varying basic state'. *J. Atmos. Sci.*, **39**, 722–33.
Wetherald, R. T. & S. Manabe (1980). 'Cloud cover and climate sensitivity'. *J. Atmos. Sci.*, **37**, 1485–1510.

The sensitivity of numerically simulated climates to land-surface boundary conditions†

Yale Mintz
Department of Meteorology, University of Maryland, College Park, MD 20742 and Laboratory for Atmospheric Sciences, NASA Goddard Space Flight Center, Greenbelt, MD 20771

Abstract

This review describes, interprets, and compares 11 sensitivity experiments that have been made with general circulation models to see how land-surface boundary conditions can influence the rainfall, temperature, and motion fields of the atmosphere. In one group of experiments, different soil moistures or albedos are prescribed as time-invariant boundary conditions. In a second group, different soil moistures or different albedos are initially prescribed, and the soil moisture (but not the albedo) is allowed to change with time according to the governing equations for soil moisture. In a third group, the results of constant versus time-dependent soil moistures are compared.

All of the experiments show that the atmosphere is sensitive to the land-surface evapotranspiration: so that changes in the available soil moisture or changes in the albedo (which affects the energy available for evapotranspiration) produce large changes in the numerically simulated climates.

6.1 Introduction

6.1.1 *Some observational and theoretical considerations*

Averaged for the globe and for the year, the measured river water drainage from the continents is about a third as large as the measured precipitation (Baumgartner & Reichel, 1975, Table 9; Korzun, 1978, Table 150). This means that on the average the land-surface evapotranspiration is about two-thirds as large as the precipitation.

In some continental regions, during part of the year, the evapotranspiration is larger than the precipitation. This cannot be known from measurements of river flow, but can be derived from measurements of the transport of water vapor by the atmosphere. An example of this for the central and eastern United States, in July, is shown in Fig. 6.1.

On the left in the diagram is the vertical distribution of the water vapor transport divergence, as given by twice daily rawinsonde measurements for two July months, and averaged for the region 80–100°W, 30–47.5°N, which is an area of about (2000 km)^2 (Rasmusson, 1968, Table 1 and Fig. 2). From the surface to the 930 mb level there is a water vapor transport convergence of 1.4 gm/cm^2 per month: or 14 mm/month equivalent water depth. Above the 930 mb level there is divergence of 36 mm/month. Integrated over the entire depth of the atmosphere there is a net divergence (a net removal of water from the region) of 22 mm/month.

From the beginning to the end of July the change in the water vapor content of the atmosphere is very small. Therefore, the 22 mm of water that are removed from the region must come from the water stored in the soil, which means that the evapotranspiration

† Review paper presented at the *JSC Study Conference on Land Surface Processes in Atmospheric General Circulation Models*, Greenbelt, USA, 5–10 January 1981.

is 22 mm/month larger than the precipitation. Inasmuch as the measured average July precipitation in this region is about 94 mm/month, the average July evapotranspiration must be about 116 mm/month (3.7 mm/day). (A comparable analysis for the central and eastern United States by Benton *et al.* (1953, Figs. 24, 26) gave a July evapotranspiration of 121 mm/month (3.9 mm/day).) We can interpret this water budget as follows:

The net radiational heating of the ground in this region, in July, is about 140 watt/m² (Budyko, 1963, Plate 21), which if used entirely for evapotranspiration would put about 150 mm/month (4.8 mm/day) of water into the air. But if we accept the aerologically derived evapotranspiration of 3.7 mm/day ($LE = 107$ watt/m²), there will be a sensible heat transfer from the ground to the atmosphere, $H = (R_N - LE) = (140 - 107) = 33$ watt/m²; and a Bowen ratio, (H/LE), equal to 0.31. Here, L = latent heat of evapotranspiration, E = evapotranspiration rate.

The 116 mm/month of water vapor, forced into the atmospheric planetary boundary layer by the radiational heating of the surface, combines with the 14 mm/month brought into the region by the water vapor transport convergence in the boundary layer; and the total of 130 mm/month of water vapor are transferred from the boundary layer to the free atmosphere.

In July, the condensation and precipitation in this region is predominantly of the convective type, with relatively little large-scale upglide condensation and precipitation. Therefore, the transfer of water vapor upward from the surface is predominantly by small-scale turbulent mixing within the planetary boundary layer, with a handover to cumulus convection which carries the water vapor from the top of the boundary layer into the free atmosphere.

Of the 130 mm/month of moisture carried into the free atmosphere by the cumulus cloud towers, 36 mm/month (in the form of water vapor, liquid water droplets and ice crystals) are detrained from the clouds into the cloud environment (where the water droplets and ice crystals evaporate) and are removed from the region by the divergence of the water vapor transport in the free atmosphere. The remaining 94 mm/month return to the earth's surface as the convective precipitation. The excess of the evapotranspiration over the precipitation, 22 mm/month, is the moisture withdrawn from the soil.

According to this analysis, the convective precipitation draws all of its moisture from the water vapor in the planetary boundary layer; and the amount of water vapor supplied to the boundary layer by the surface evapotranspiration is an order of magnitude larger than the amount supplied by the water vapor transport convergence. This suggests that the surface evapotranspiration is the main determinant of the precipitation.

The winter season water budget over the central and eastern United States is very different from that shown in Fig. 6.1. In winter the water vapor transport convergence does not change sign with height, but is convergent at all levels and produces a net import of water vapor to the region (Rasmusson, 1968, Table 1 and Fig. 2). In winter, the condensation and precipitation is predominantly of the large-scale upglide condensation type (frontal cloud and precipitation) which draws from the water vapor at all levels in the troposphere. Moreover, in winter the net radiational heating of the ground is small (Budyko, 1963, Plate 15) and, consequently, over the unforested part of this region the evapotranspiration is small. In winter, therefore, the land-surface evapotranspiration cannot have much influence on the precipitation or other fields. It is only in the tropics and in the summer season extratropics, where evapotranspiration is large and where the precipitation is of the type that draws its water vapor from the planetary boundary layer, that the land-surface evapotranspiration can be of major importance.

With respect to the tropics and the summer season extratropics, two questions immediately come to mind:

(i) If the surface evapotranspiration, by some means, is greatly reduced, can the boundary layer water vapor transport

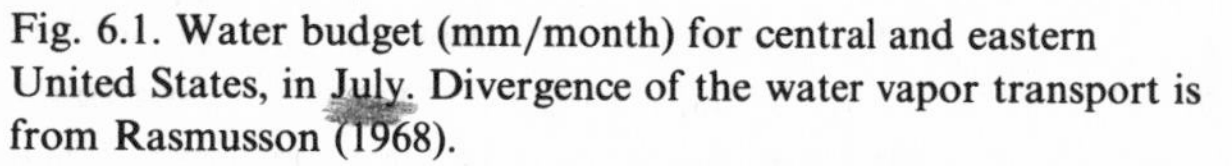

Fig. 6.1. Water budget (mm/month) for central and eastern United States, in July. Divergence of the water vapor transport is from Rasmusson (1968).

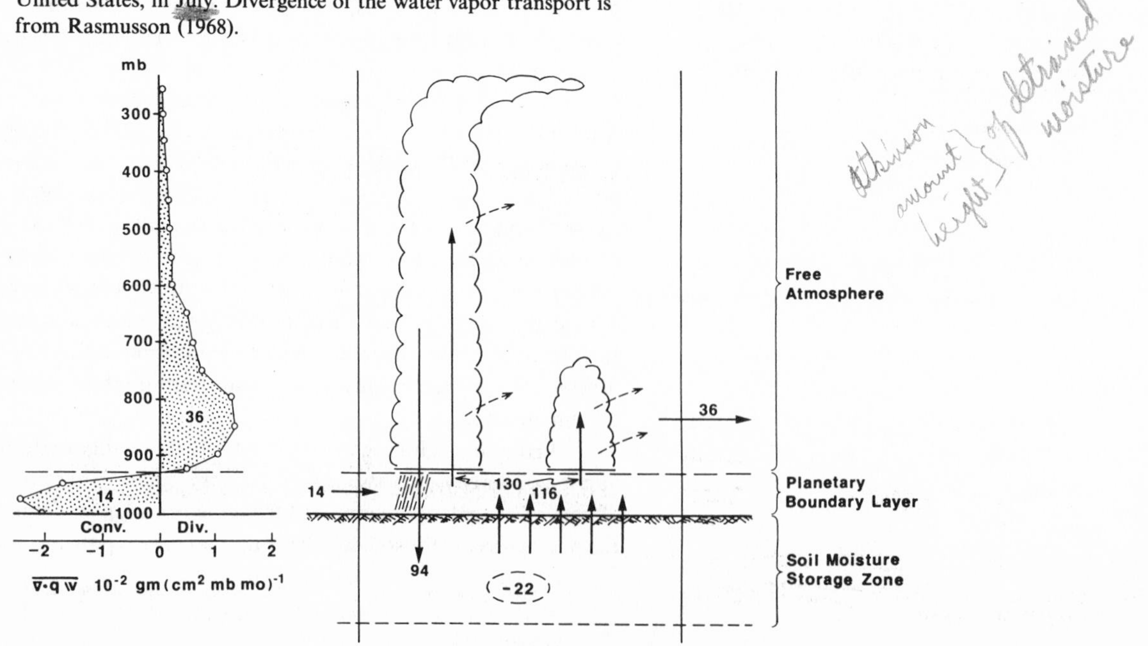

convergence increase by a corresponding amount and, in that way, maintain the precipitation?

(ii) If the surface evapotranspiration and the boundary layer water vapor transport convergence remain the same, can an increased detrainment and water vapor transport divergence in the free atmosphere stop the precipitation?

● To answer the first question, we write the water vapor transport convergence as $-\nabla \cdot q\mathbf{v} = -\mathbf{v} \cdot \nabla q - q\nabla \cdot \mathbf{v}$, where q is the water vapor mixing ratio and $\mathbf{v}$ is the horizontal velocity of the air.

The quantity $-\mathbf{v} \cdot \nabla q$ is positive when the air that leaves the region is drier than the air that enters. But if this drying is due to the removal of water vapor from the boundary layer by cumulus convection, then the convection will stop as soon as there is a small reduction in the boundary layer water vapor content. The observed rate of precipitation cannot be maintained by $-\mathbf{v} \cdot \nabla q$ over a distance which is greater than just a few cumulus convection cells, say a total distance of a few kilometers.

The other term, $-q\nabla \cdot \mathbf{v}$, can also maintain the observed rate of precipitation only over a restricted domain in the extratropics. The water vapor mixing ratio in the boundary layer of a maritime tropical air mass over the extratropical continents in summer, is of the order of ten parts per thousand. Therefore, a boundary layer that is 100 mb (1 km) deep must have a horizontal velocity convergence, $-\nabla \cdot \mathbf{v}$, of 0.37/day (0.43×10^{-5}/second) to produce a water vapor transport convergence of 3·7 mm/day. The characteristic velocity of the boundary layer air in the extratropics, in summer, is 2–3 m/second; and the angle between this vector velocity and the streamline of the non-divergent flow, integrated over the depth of the boundary layer, is about 10°. Thus, if we consider a circular region with radius r, we have $-\nabla \cdot \mathbf{v} = 0.43 \times 10^{-5}$ second $=$ (3 m/second sin 10°) $2\pi r/\pi r^2$; or $r = 240$ km, and 480 km is the limiting diameter of the region where water vapor transport convergence in the boundary layer can produce the observed rate of precipitation.

In the extratropics, therefore, there is a size limit, of the order of a few hundred kilometers, beyond which boundary layer water vapor transport convergence cannot compensate for diminished evapotranspiration. It is only near the equator, where the divergent component of the velocity field is larger and the planetary boundary layer is deeper, that there can be appreciable water vapor transport convergence over a much larger sized area.

●● The answer to the second question: 'Can an increased detrainment and water vapor transport divergence in the free atmosphere stop the precipitation?', depends on whether the free atmosphere is supplied with dry air into which the cumulus cloud towers can detrain. That will happen only if, in addition to the boundary layer mass (and water vapor) convergence, there is also a mass convergence in the uppermost troposphere. Then, the cumulus cloud towers can detrain all of the water into the subsiding and diverging dry air of the middle troposphere, and not produce any precipitation at all. The best known example of extensive fields of non-precipitating cumulus clouds of this kind are the Trade Wind cumuli over the tropical oceans, where the subsiding air in the middle troposphere has its origin in the high-level outflow above the intertropical convergence zone. We also see such fair-weather cumulus clouds removing water vapor from the boundary layer, without producing precipitation, west of the trough lines and east of the ridge lines of the fast-transient and slow-transient waves in the extratropical westerlies, where both the longitudinal and the latitudinal scale can be as large as a few thousand kilometers. We can say, therefore, that on a scale larger than a few hundred kilometers, in the extratropics, land-surface evapotranspiration is a necessary (but not sufficient) condition for convective precipitation. The upper tropospheric circulation must also be favorable for precipitation.

Because so many interactive thermodynamical and hydrodynamical processes are involved, the best way to determine the overall influence of the land-surface boundary conditions on the rainfall, temperature and circulation is through experiments with atmospheric general circulation models. Existing general circulation models have been fairly successful in simulating the observed climate of the earth, including the principal geographical and seasonal characteristics of the precipitation (WMO, 1979). By making pairs of time-integrations, with all of the initial conditions and boundary conditions the same except for those which can affect the land-surface evapotranspiration, and comparing the two solutions, we can ascertain what the land-surface influence is.

In the existing general circulation models, the two boundary conditions that can affect the land-surface evapotranspiration are the soil moisture and the surface albedo. The soil moisture determines how large the evapotranspiration will be relative to the model calculated potential evapotranspiration (the evapotranspiration when soil moisture is fully available): the albedo is a major factor in determining the potential evapotranspiration itself.

The experiments that are being reviewed are grouped as follows:

(I) experiments with non-interactive soil moisture;
(II) experiments with interactive soil moisture;
(III) hybrid experiments.

In the first group, either different soil-moisture availabilities or different albedos are prescribed, and both of these parameters are kept constant with time. Such experiments reveal the sensitivity of the atmosphere to the boundary conditions. (These experiments are analogous to sensitivity experiments in which different non-interactive ocean-surface temperatures are prescribed: the so-called sea-surface temperature anomaly experiments).

In the second group, the soil moisture (but not the albedo) is interactive and changes with time according to the model's governing equations for soil moisture. When the albedos are the same in a pair of comparison runs, but the initial soil moistures are different, the integrations will either produce time-series that remain separate (intransitive) or converge to a common solution; and, if transitive, they will show how long it takes for the two initially different states to converge to a common state. When the albedos are different, this will be another kind of sensitivity experiment.

In the third group, the hybrid experiments, calculations with non-interactive and interactive soil moistures are compared. To the extent that the calculation with interactive soil moisture simulates the observed rainfall, temperature and circulation of the earth's atmosphere, the comparison will show how the earth's climate may be affected by such imposed changes in the land-surface evapotranspiration as might be brought about by large-scale deforestation or afforestation, by soil erosion or reclamation, or by large-scale irrigation.

6.2

List of the experiments

(I) ***Experiments with non-interactive soil moisture***

(A) *Different soil moistures, with same albedo*

(1) Shukla & Mintz (1981)

(2) Suarez & Arakawa (personal communication)

(3) Miyakoda & Strickler (1981)

(B) *Different albedos, with same soil moisture*

(4) Charney, Quirk, Chow & Kornfield (1977)

(5) Carson & Sangster (1981)

(II) ***Experiments with interactive soil moisture***

(A) *Different initial soil moistures, with same albedo*

(6) Walker & Rowntree (1977)

(7) Rowntree & Bolton (1978)

(B) *Different albedos, with same initial soil moisture*

(8) Charney, Quirk, Chow & Kornfield (1977)

(9) Chervin (1979)

(III) ***Hybrid experiments***

Non-interactive v. interactive soil moistures

(10) Manabe (1975).

(11) Kurbatkin, Manabe & Hahn (1979)

6.3

(I) Experiments with non-interactive soil moisture

6.3.1

(A) *Different soil moistures, with same albedo*

6.3.1.1

Shukla and Mintz (1981)

The experiment of Shukla & Mintz (1981) used the general circulation model of the NASA Goddard Space Flight Center, Laboratory for Atmospheric Sciences. The properties of the GLAS model and its ability to simulate the regional and season characteristics of the observed climate of the earth have been described by Shukla *et al.* (1981). In the experiment, one climate simulation is made in which the land-surface evapotranspiration, E, is everywhere made equal to the model calculated potential evapotranspiration, E_p, which makes the evapotranspiration coefficient, $\beta = E/E_p = 1$. In the other case, no land-surface evapotranspiration is allowed to take place at all ($\beta \equiv 0$). The prescribed albedo is the same in both cases, and is a very slightly modified version of the one given by Posey & Clapp (1964). For convenience, the two calculations are called the 'wet-soil' case and the 'dry-soil' case. Both calculations were started from the same initial observed atmospheric state on 15 June. The results that are shown here are the averages for July.

In the wet-soil case, the calculated land-surface evapotranspiration is relatively constant (within about ± 1 mm/day) between latitudes 20°S and 60°N, with an average value of 4.3 mm/day, corresponding to an evaporative cooling of the surface of 125 watt/m², as shown in Fig. 6.2. Here, the sensible heat transfer to the atmosphere is 21 watt/m². In the dry-soil case, however, the land-surface evapotranspiration is zero and the sensible heat transfer is 169 watt/m².

The dry-soil case gives rise to much less cloudiness over the continents than the wet-soil case and, as a result, a larger amount of solar radiation reaches and is absorbed by the ground, 258 instead of 172 watt/m². The increased solar heating of the ground, as well as the elimination of the evaporative cooling, makes the ground warmer; and the higher ground temperature produces a greater long wave radiation emission from the ground, 550 instead of 419 watt/m². The atmosphere also becomes warmer in the dry-soil case, and there is an increase in the atmospheric long wave 'back radiation' to the ground; but because of the reduction in the cloudiness, the increase in the back radiation, from 393 to 461 watt/m², is only about half as large as the increase in the radiation emitted by the ground. The end result of all these large, but partially compensating, changes in the radiation transfers, is that there is only a relatively small change in the net (all-wavelength) radiational heating of the land surface: an increase of only 23 watt/m² from the wet-soil to the dry-soil case.

The top panel of Fig. 6.3 shows the global precipitation distribution in the wet-soil case. Over most of North America and most of Eurasia the precipitation is within about 1 mm/day of the local evapotranspiration. Only over southeast China does the precipitation exceed evapotranspiration by as much as 4 mm/day. Over South America there is heavy rain near the equator, which is about 2 mm/day greater than the land-surface evapotranspiration. Across Africa, at about 10°N, there is a band of rain which is about 4 mm/day greater than the local evapotranspiration. On the other hand, across Africa at about 25°N, and across Africa and South America at about 15°S, the precipitation is 2–3 mm/day smaller than the evapotranspiration. Thus, although in the wet-soil case there is a fairly uniform transfer of water vapor to the air by the land-surface evapotranspiration, within the tropics and subtropics there are convergences and divergences of the water vapor transports by the large-scale atmospheric circulation, which enhance or diminish the precipitation by substantial amounts.

The dry-soil case, shown in the bottom panel of Fig. 6.3,

Fig. 6.2. Surface energy transfers (watt/m²) averaged for the continents between 20°S and 60°N, in experiment of Shukla and Mintz (1981).

R_S solar raditional heating of the ground.

R_L long wave radiational cooling of the ground (difference between radiation emitted by ground and radiation absorbed by ground).

$R_N = (R_S - R_L)$ net (all-wavelength) radiational heating of the ground.

LE latent heat transfer from ground to atmosphere (evaporative cooling of the ground).

H conductive-convective heat transfer from ground to atmosphere.

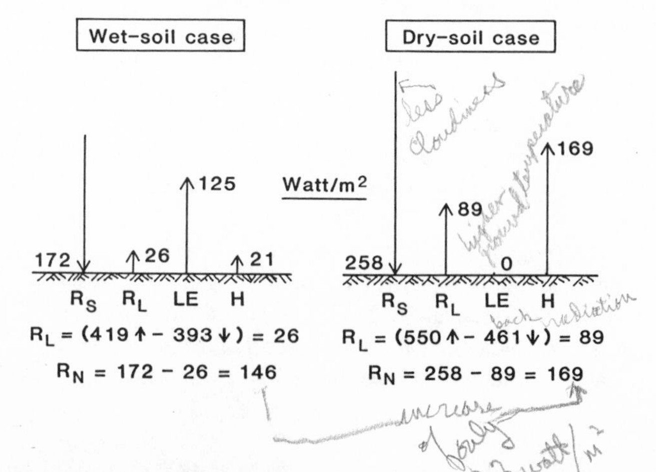

produces almost no precipitation at all over Europe and most of Asia; and over most of North America the precipitation is only about a quarter to a half of that of the wet-soil case. Over the equatorial part of South America, on the other hand, the rainfall in the dry-soil case is about the same as in the wet-soil case: i.e., about 6 mm/day; but, now, the water vapor which produces that precipitation comes only from the ocean.

Across north Africa, the rainband is about 400 km (one model grid interval) farther north in the dry-soil case than in the wet-soil case, and weaker by 3–4 mm/day. The precipitation in the dry-soil case is about the same as the amount by which the precipitation exceeded evapotranspiration in the wet-soil case: which is to say that the convergence of the water vapor transport by the atmospheric circulation is about the same in the two cases.

Perhaps the most surprising difference of all, when comparing the dry-soil case with the wet-soil case, is the southward and westward displacement of the region of maximum precipitation in southeast Asia. Over Bangladesh, the convergence in the water vapor transport from the ocean in the dry-soil case more than compensates for the absence of surface evapotranspiration. It is in the dry-soil case that the calculated precipitation most closely resembles the observed summer rainfall of southeast Asia.

Fig. 6.4 shows the ground-surface temperature. In the dry-soil case, in which there is no evaporative cooling of the ground and, because of the reduced cloudiness, more solar radiation is absorbed by the ground, the surface temperatures north of latitude 20°S are about 15°–30°C warmer than in the wet-soil case.

As shown in Fig. 6.2, the total non-radiational heat transfer to the atmosphere ($H+LE$) is not greatly different in the two cases (146 v. 169 watt/m²); but in the dry-soil case all of this is sensible heat transfer, which is confined to the planetary boundary layer. In the wet-soil case, by contrast, the larger part of the transfer is in the form of latent heat which warms the free atmosphere and not the boundary layer, whether immediately and locally realized by convective condensation and precipitation or realized at some later time and distant place. Thus, there is a different vertical distribution, and sometimes a different horizontal distribution, of the diabatic heating. This can produce significant differences in the thermally forced atmospheric circulation and, by the geostrophic adjustment process, corresponding differences in the horizontal pressure distribution.

Fig. 6.5 shows the surface pressure fields. The top and center

Fig. 6.3. Precipitation (mm/day) in wet-soil case (top) and dry-soil case (bottom), in experiment of Shukla & Mintz (1981). (Precipitation greater than 2 mm/day is shaded.)

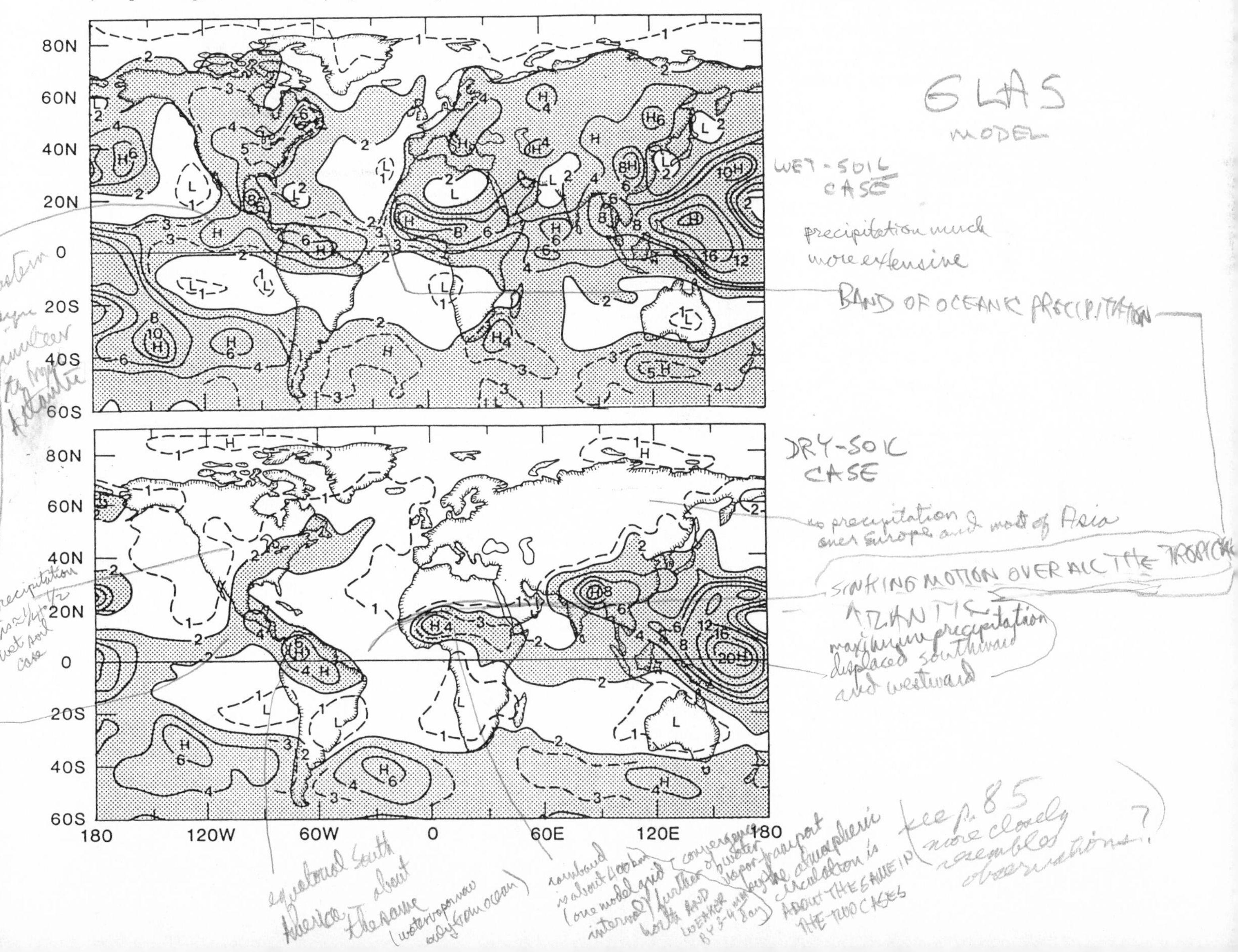

panels show the surface pressures reduced to sea level, in the two cases. The bottom panel shows the difference between the two surface pressures, without reduction to sea level. It is here that we see the change in the surface geostrophic wind. Over most of the land the surface pressures are about 5–15 mb lower in the dry-soil case, which means enhanced cyclonic circulations over the continents.

In the wet-soil case the trough of low pressure across Africa coincides with the intertropical rainband, as may be seen by comparing the top panels of Figs. 6.3 and 6.5. This is the same relationship that we see over the tropical oceans. But in the dry-soil case the trough of low pressure is about 400–800 km north of the rainband; which is about the same relationship that is found over north Africa, in nature.

When the surface pressure is lower over the continents, it must be higher over the oceans. Most of the increase is in the mid-latitudes of the central and western North and South Pacific Oceans. Examination of the vertical motion field (not reproduced here) shows that in these ocean regions there is an increased subsiding motion in the dry-soil case. As Figs. 6.3 and 6.5 show, not only does the increased sinking motion suppress the oceanic precipitation, but the accompanying low-level horizontal velocity divergence, by generating anticyclonic vorticity, increases the anticyclonic circulation in these regions; and geostrophic adjustment produces the corresponding rise in surface pressure.

Over the Atlantic Ocean, the vertical motion field in the wet-soil case shows a band of rising motion and low-level velocity convergence, which coincides with the band of oceanic precipitation just north of the equator (top panel of Fig. 6.3). But in the dry-soil case there is sinking motion over all of the tropical Atlantic; and there is no oceanic rainband near the equator at all. Over the eastern half of the tropical Pacific, the same kind of change takes place, but it is not as pronounced. Thus, the change in the land-surface boundary condition also produces large changes in the circulation and rainfall over the oceans.

Fig. 6.4. Ground-surface temperature (°C) in wet-soil case (top) and dry-soil case (bottom), in experiment of Shukla & Mintz (1981).

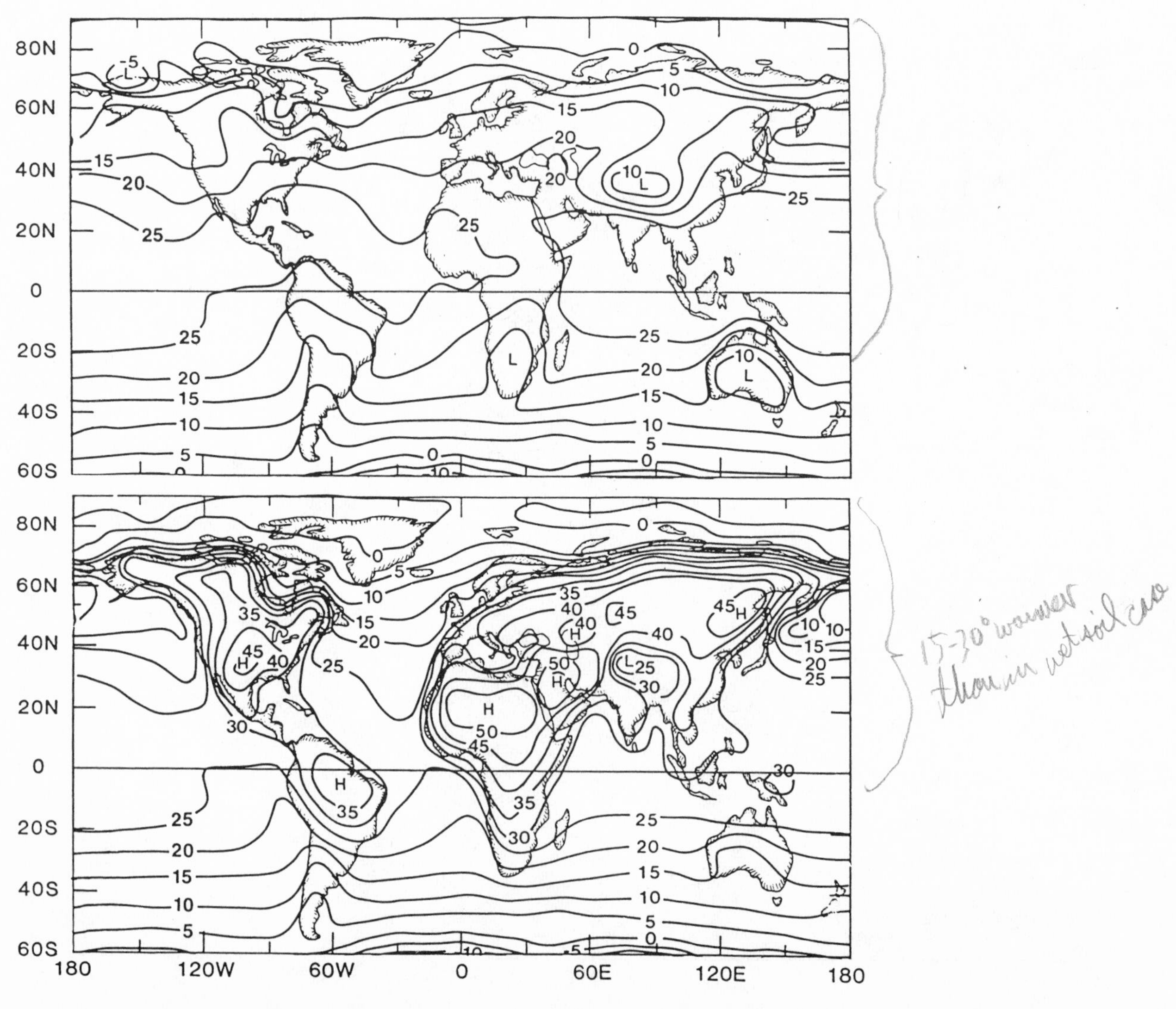

6.3.1.2

Suarez and Arakawa (*personal communication*)

At the reviewer's suggestion, the same wet-soil ($\beta = 1$) versus dry-soil ($\beta = 0$) sensitivity experiment was made with the University of California, Los Angeles (UCLA) general circulation model by M. Suarez & A. Arakawa (personal communication). (For a description of the model, see Arakawa & Lamb, 1977; Arakawa & Suarez, 1983; and Suarez *et al.* 1983.) There are substantial differences between the UCLA and the GLAS models, of which the most important, insofar as the present sensitivity experiment is concerned, may be the way in which the planetary boundary layer and cumulus convection are parameterized.

In the UCLA experiment the integrations for the two cases were started on the first day of July, with the initial state of the atmosphere taken from a previous general circulation simulation.

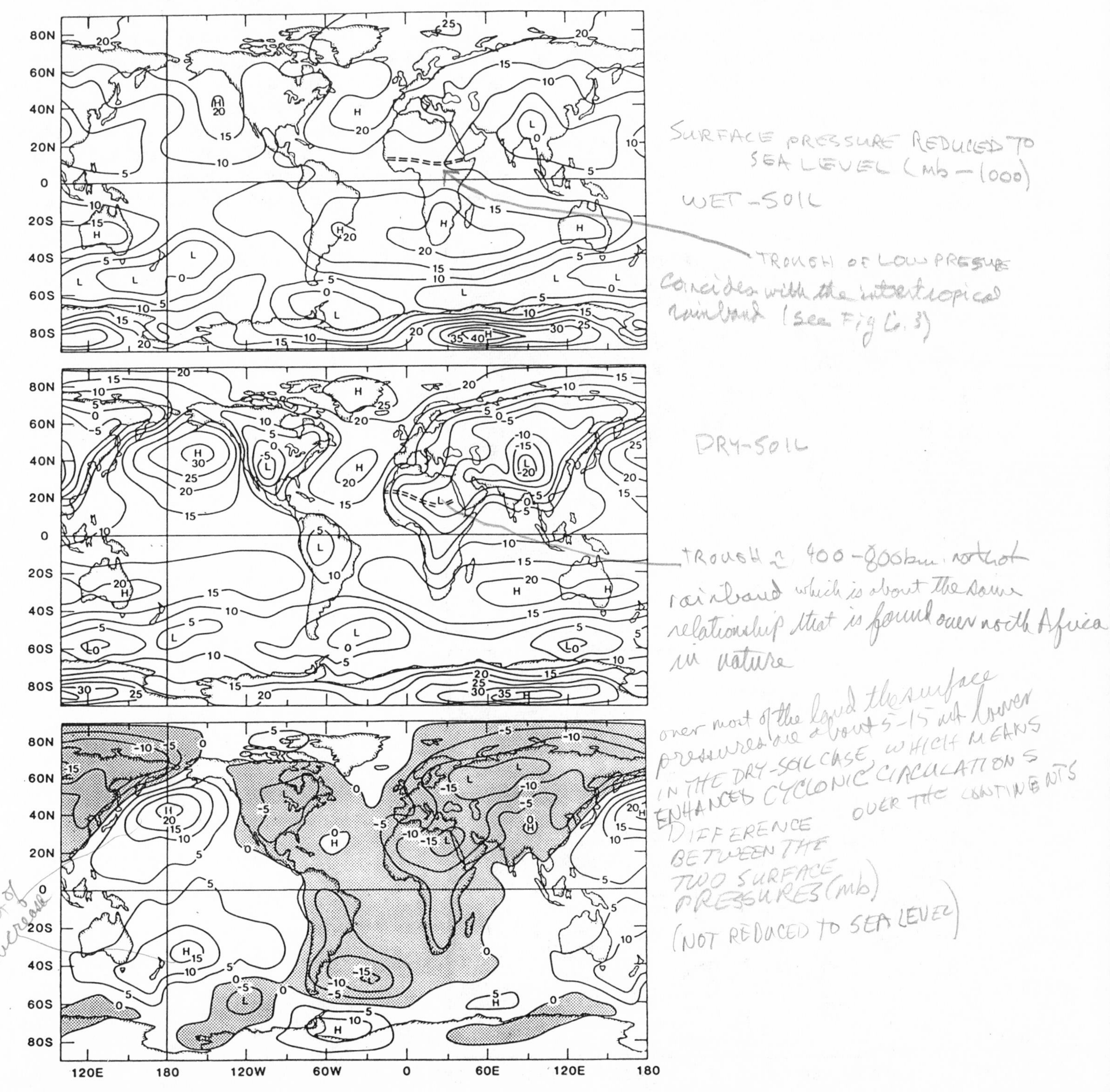

Fig. 6.5. Surface pressure reduced to sea level (mb minus 1000) in wet-soil case (top) and dry-soil case (center), in experiment of Shukla & Mintz (1981). Bottom map show the difference between the two surface pressures (mb).

The results that are shown here are for the 31-day period starting on 16 July. Again, the prescribed surface albedo follows Posey & Clapp (1964).

In the wet-soil case, the calculated land-surface evapotranspiration was relatively constant (within about ± 1 mm/day) between 20°S and 60°N, with an average value of about 6 mm/day. This is about 1.7 mm/day larger than in the wet-soil case of the GLAS experiment, and is probably a consequence of the fact that the UCLA model produces less cloud cover than does the scheme used in the GLAS model and, thereby, a greater net radiational heating of the ground.

Fig. 6.6 shows that the precipitation in the wet-soil case is about 6 ± 1 mm/day over almost all of extratropical North America and Eurasia and, therefore, does not differ from the local evapotranspiration by more than about 1 mm/day. Within the tropics, however, the precipitation exceeds the local evapotranspiration by about 10 mm/day over the Indochina peninsula, by about 3–6 mm/day over a few small land areas that are close to the sea (Guatemala, southern India, southeast China, Columbia, Venezuela and northeast Brazil), and by a few mm/day over a large area adjacent to the Somali coast of north Africa. These are regions, therefore, of substantial water vapor transport convergences.

In the dry-soil case, shown in the bottom panel of Fig. 6.6, there is almost no continental precipitation at all. Only in an east–west band across north Africa is there a significant amount of precipitation, 2–5 mm/day, produced by a convergence of the water vapor transported from the oceans. The axis of this rainband, at 10°N, is about 1000 km south of the axis of the low-pressure trough which, in the dry-soil case, is at about 20°N.

The change in the precipitation over the oceans is very large near some of the tropical and subtropical coastlines, and especially where there are embayments. In the wet-soil case there are pronounced minima over the Gulf of Mexico and the Bay of Bengal

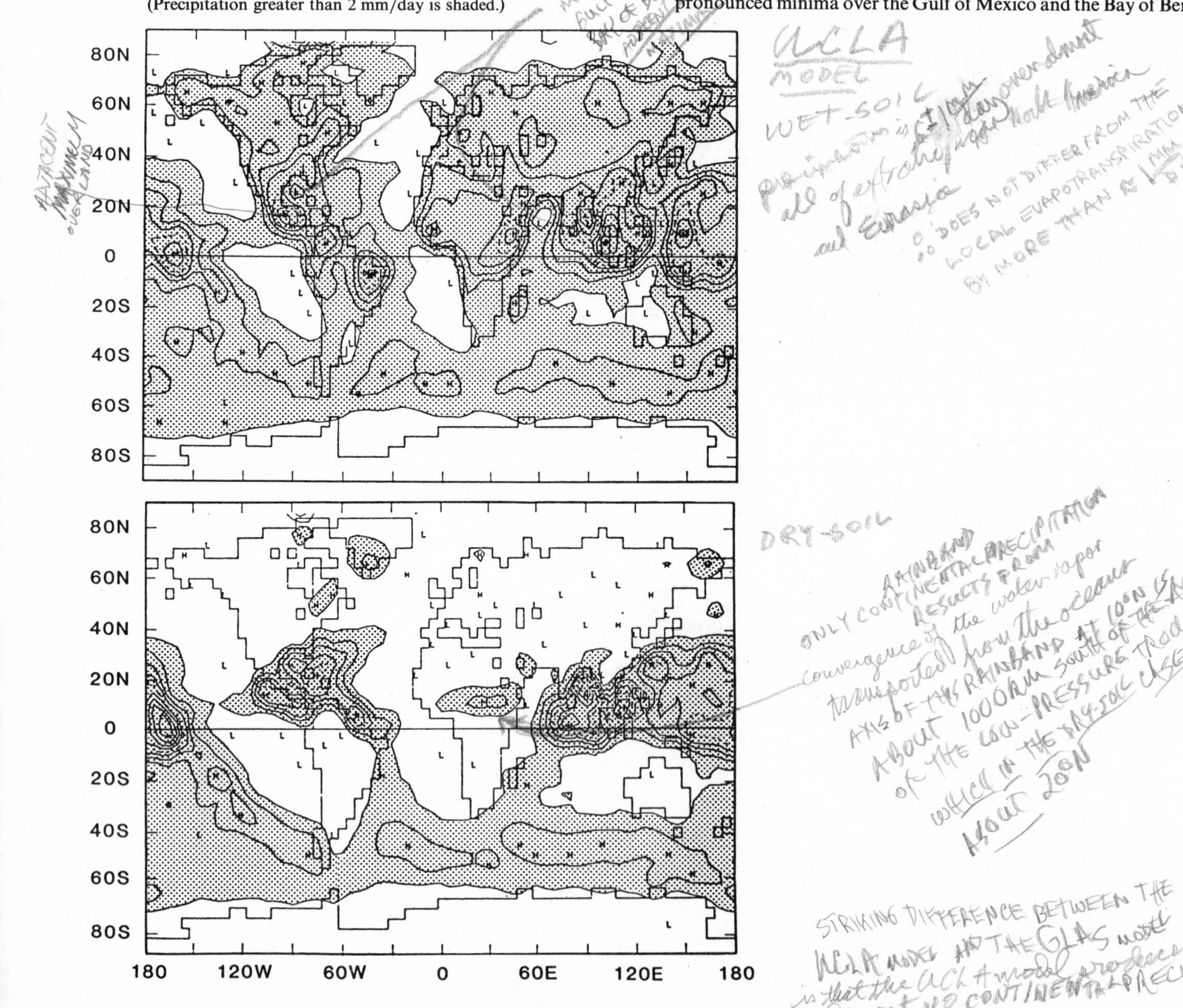

Fig. 6.6. Precipitation (mm/day) in wet-soil case (top) and dry-soil case (bottom), in the experiment of Suarez & Arakawa (personal communication). Contour interval is 2 mm/day. (Precipitation greater than 2 mm/day is shaded.)

(accompanied by pronounced maxima over the adjacent land areas). But in the dry-soil case, the minima are replaced by maxima over the ocean embayments. Similarly, along the coasts of central America and northeast Brazil the land precipitation decreases and the nearshore ocean precipitation increases in going from the wet-soil to the dry-soil case.

The striking difference between the experiments with the UCLA model and those with the GLAS model is that, except for the Sahel region of Africa, the UCLA model produces almost no continental precipitation in the dry-soil case.

Examination of the water vapor transport field by the investigators showed that there are regions, such as northeast Brazil, where within the planetary boundary layer there is a large convergence of the water vapor transported from the ocean, in the dry-soil case, but no rain. The interpretation they made (Suarez & Arakawa, personal communication) is that with dry soil there is a very large diurnal variation of the ground-surface temperature, which produces a very large diurnal variation in the depth of the model's planetary boundary layer, growing in thickness during the day and collapsing at sunset; and that it is this diurnal oscillation which, without producing clouds, transfers the water vapor from the boundary layer to the free atmosphere, where the transport is divergent. This transfer of water vapor from the boundary layer to the free atmosphere by diurnal 'boundary layer–free atmosphere mixing' is not unlike the transfer by detrainment from fair-weather, non-precipitating cumulus clouds, described in the introduction. The same condition of upper troposphere velocity convergence and middle troposphere subsidence must be satisfied. Here it is the result of upper-level outflow from the region of intense convective precipitation over the adjacent ocean.

6.3.1.3

Miyakoda and Strickler (1981)

Miyakoda & Strickler (1981) used an early version of the general circulation model of the NOAA–Princeton Geophysical Fluid Dynamics Laboratory (Smagorinsky *et al.*, 1969) to make and compare two different sets of 14-day numerical weather predictions for the northern hemisphere, in July, when different distributions of the soil moisture availability, β, were prescribed.

The surface albedo was fixed and followed Posey & Clapp (1964). The clouds were climatologically prescribed as a function of latitude and height. The convective-adjustment scheme was used, in which the moist convective heating of the atmosphere and the convective precipitation depend only on the relative humidity and on the temperature difference between adjacent levels in the vertical. Thus there is no penetrative convection and, consequently,

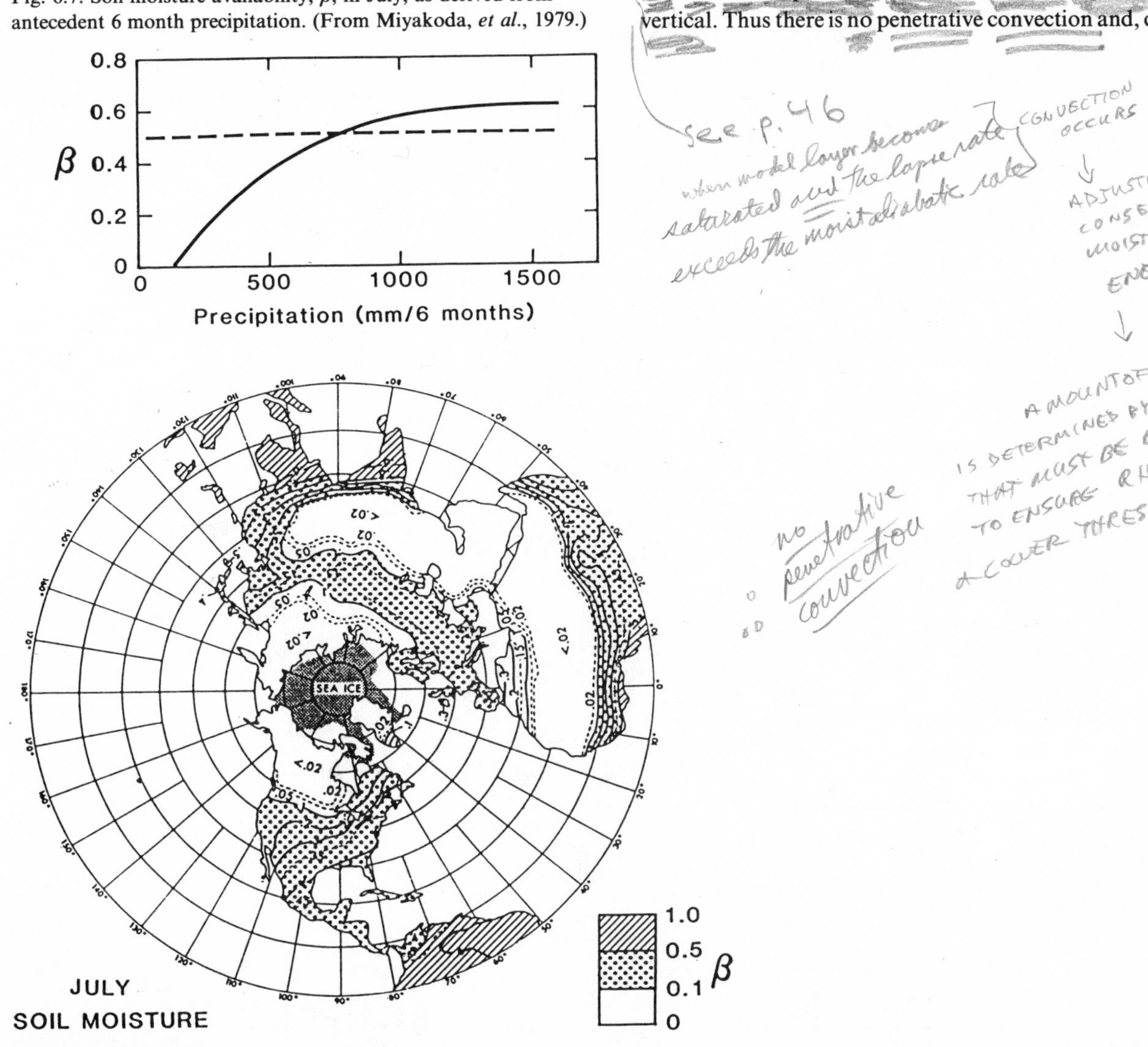

Fig. 6.7. Soil moisture availability, β, in July, as derived from antecedent 6 month precipitation. (From Miyakoda, *et al.*, 1979.)

Fig. 6.8. Evapotranspiration difference (top) and precipitation difference (bottom), when $\beta \equiv 0.5$ is replaced by $\beta(\lambda, \phi)$, in experiment of Miyakoda & Strickler (1981). (Contours for 0, ∓ 25, ∓ 100, ∓ 200, ∓ 500 mm/mo. Negative values are shaded.)

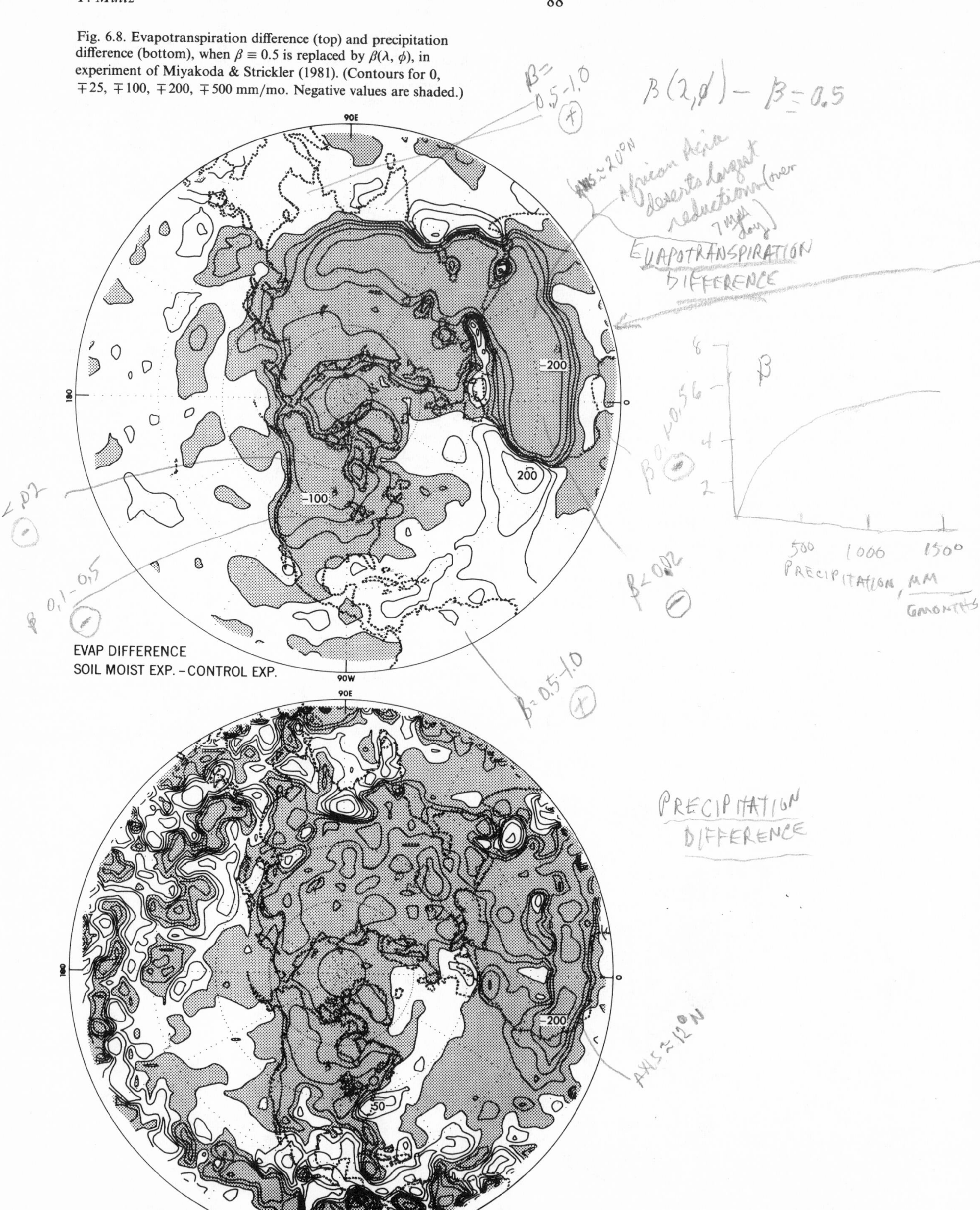

Fig. 6.9. Surface temperature difference (top) (contour interval 2.5 °C) and difference in height of the 1000 mb surface (bottom) (contour interval 10 m; negative values shaded), in experiment of Miyakoda & Strickler (1981).

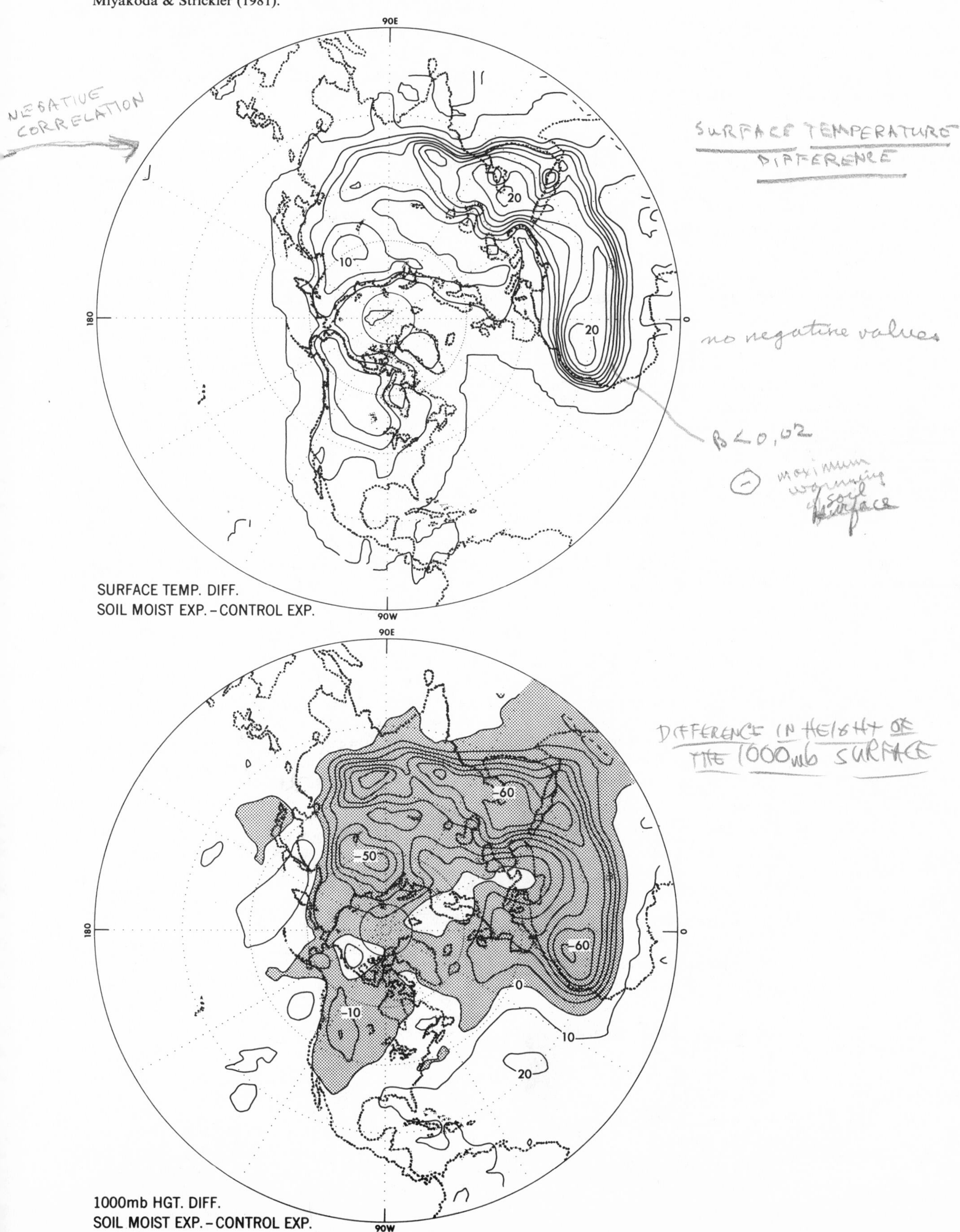

the sensitivity of the convective heating and precipitation to the amount of water vapor in the model's planetary boundary layer is not as great as in the cumulus convection parameterization schemes of the GLAS and the UCLA models.

In one case, β was everywhere set equal to 0.5. For the other case, the authors sought a more realistic field of β; and for this they took the observed normal distribution of precipitation for the antecedent six-month period, February through July, and relabeled the isohyets as lines of constant β, according to the arbitrary function shown in the top panel of Fig. 6.7 (Miyakoda *et al.*, 1979). No account was taken of the antecedent evapotranspiration. Consequently, as the bottom panel of the figure shows, over the northern forests and the wet tundra regions of Canada and Siberia, β was made as low as in the subtropical deserts (although, in nature, it is near the maximum value of 1.0).

Figures 6.8 and 6.9 show the differences in the evapotranspiration, precipitation, surface temperature and height of the 1000 mb surface, for the case of $\beta = \beta(\lambda, \phi)$ minus the case of $\beta \equiv 0.5$, when the ensemble average is taken of the last 12 days of three sets of 14-day forecasts.

We see, in Fig. 6.8, that where β is reduced there is, in general, a reduction in evapotranspiration and precipitation. The largest reduction in evapotranspiration, of more than 7 mm/day, is over the central part of the north African and Asian deserts, with the axis of the maximum evapotranspiration reduction at about latitude 20°N across Africa. But the axis of the largest reduction in precipitation is at about 12°N across Africa (the Sahel), where the precipitation decreases by about 12 mm/day. As shown in Miyakoda & Strickler (1981, Fig. 7), the rainband of the intertropical convergence zone across north Africa does not change its position, but its magnitude goes down from 20 to 8 mm/day when the land-surface moisture source in the Sahara is eliminated. There is, obviously, a large change that takes place in the water vapor transport.

In the case where the average land-surface evapotranspiration became smaller, the average ocean evaporation became larger. In spite of that, the average ocean precipitation decreased, in agreement with the experiments made with the GLAS and UCLA models. Presumably it is, again, an enhancement of the sinking motion over the oceans that suppresses the ocean precipitation.

If we compare the top panel of Fig. 6.9 with the top panel of Fig. 6.8, we see that there is a negative correlation between the change in the land-surface temperature and the change in evapotranspiration.

When we compare the upper and lower panels of Fig. 6.9, we see that there is a negative correlation between the change in surface temperature and the change in the height of the 1000 mb surface. This is true even where the land-surface is not at a high elevation.

This experiment shows two important things: (1) that it does not require an extreme change in the soil moisture availability (β), such as a change from 1 to 0, in order to produce large changes in the precipitation, temperature and motion fields; and (2) that the influence of the land-surface evapotranspiration on the atmosphere operates very quickly. Miyakoda (personal communication) reports that sizeable differences in the surface temperature and precipitation appeared within a few days.

(In an earlier study of the role of the surface transfers of sensible and latent heat in numerical weather prediction, Gadd & Keers (1970, Figs. 4, 5, 6) showed that even in a very short range (18-hour) prediction for north-western Europe and the British Isles, in August, the inclusion of evaporation and sensible heat transfer from the land and the sea surfaces made a noticeable improvement in the predicted rainfall over the land.)

6.3.2

(B) ***Different albedos, with same soil moisture***

6.3.2.1

Charney, Quirk, Chow and Kornfield (1977)

This experiment was made by Charney *et al.* (1977) with the NASA Goddard Institute for Space Studies general circulation model, the properties and performance of which have been described by Somerville *et al.* (1974) and Stone *et al.* (1977).

The three runs that are shown here use the prescribed field of non-interactive soil moisture availability, β, from Stone *et al.* (1977), who assumed that $\beta = 2 \times (RH - 15)/85$, $\beta_{\max} = 1$, where RH is the observed normal monthly mean relative humidity of the surface air. The observed relative humidities, for July, were taken from the tabulation by Schutz & Gates (1972). In the above formulation, $\beta = 1$ when the relative humidity is equal to or greater than 57.5%. Consequently, β was made equal to or close to 1, and the evapotranspiration therefore equal to or close to the potential evapotranspiration over most of the land surface of the earth. Only in a small region in the western United States and across the central Sahara was the prescribed July β smaller than 0.5.

In the run designated as case '2a', the ice-free and snow-free continents were assigned a surface albedo of 0.14; except that a higher albedo, 0.35, was assigned to the regions of the observed northern hemisphere deserts (see Fig. 6.19, below).

In a comparison run, '3a', the albedo was changed from 0.14 to 0.35 in three additional regions, the 'Sahel', 'Rajputana', and 'Western Great Plains', which are adjacent to deserts (Fig. 6.19). Otherwise everything was the same as in case 2a.

In a separate comparison run, case '4', everything was again the same as in case 2a, except that the change of albedo, from 0.14 to 0.35, was made in three regions that are within the observed rainy

Fig. 6.10. Evapotranspiration (top) and precipitation (bottom) (mm/day), in experiment with prescribed soil moisture availability (case 3a), of Charney *et al.* (1977).

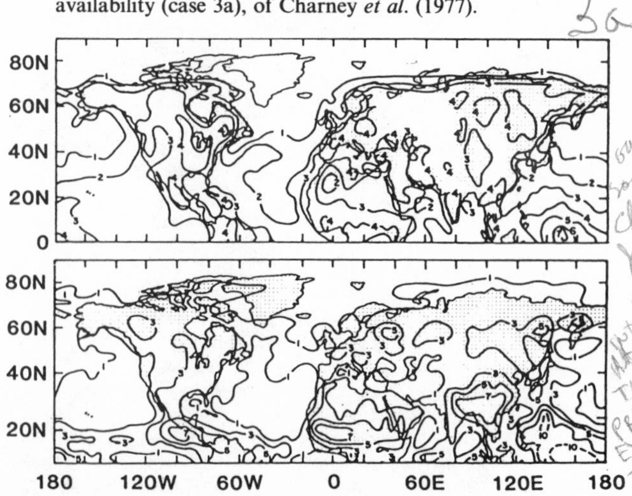

and vegetation covered areas of the earth. The locations of these regions, called 'Central Africa', 'Bangladesh', and 'Mississippi Valley' are given in the table.

The first numbered column of Table 6.1 shows the prescribed soil moisture availabilities for the six regions. In the Sahel region $\beta = 0.51$, and in the other regions β is 0.78 or more.

Fig. 6.10 shows the evapotranspiration and precipitation in the northern hemisphere, for case 3a. Over most of the continents the evapotranspiration and precipitation do not differ by more than about ± 1 mm/day. It is only over southeast China and Indochina, and where the intertropical rainband crosses Africa, that the precipitation exceeds the evapotranspiration by 3–4 mm/day. Near the Mediterranean coast of Africa and in the Middle East the precipitation is less than the evapotranspiration by about 2–4 mm/day.

Table 6.1 is a rearrangement of the data in Charney *et al.* (1977, Tables 4.1–4.4) and shows the components of the energy and water budgets at the earth's surface in the three desert-margin regions (case 2a v. 3a) and in the three humid regions (case 2a v. 4).

Table 6.1 *Components of the energy and water budgets, in experiment of Charney et al.* (1977)

Region	Case No.	(1) β	(2) $(1-\alpha)$	(3) R_S	(4) R_L	(5) R_N Energy balance	(6) LE	(7) H	(8) N	(9) T_A	(10) E Water balance	(11) $(-\overline{\nabla \cdot q\mathbf{v}})$	(12) P	(13) $\frac{\Delta(-\overline{\nabla \cdot q\mathbf{v}})}{\Delta E}$	(14) $\frac{\Delta P}{\Delta E}$
(1) Sahel	2a	0.51	0.86	169	58	111	107	4	0.70	26.0	3.7	3.7	7.4		
(16–20°N,	3a	0.51	0.65	177	84	93	81	12	0.46	25.7	2.8	1.2	4.0		
17.5°W–37.5°E)			−24%	+5%	+49%	−18	−26	+8	−34%	−0.3	−0.9	−2.5	−3.4	2.8	3.8
(2) Rajputana	2a	0.92	0.86	180	48	132	119	13	0.77	24.9	4.1	0.8	4.9		
(24–32°N,	3a	0.92	0.65	189	75	114	104	10	0.57	24.1	3.6	−1.3	2.3		
67.5–77.5°E)			−24%	+5%	+56%	−18	−15	−3	−26%	−0.8	−0.5	−2.1	−2.6	4.2	5.2
(3) Western Great	2a	0.78	0.86	186	62	124	122	2	0.67	21.1	4.2	−0.5	3.7		
Plains (32–48°N,	3a	0.78	0.65	185	79	106	93	13	0.52	19.0	3.2	−1.0	2.2		
107.5–97.5°W)			−24%	+1%	+27%	−18	−29	+11	−22%	−2.1	−1.0	−0.5	−1.5	0.5	1.5
(4) Central Africa	2a	0.94	0.86	170	56	114	125	−11	0.72	22.2	4.3	0.7	5.0		
(8–12°N,	4	0.94	0.65	171	69	102	104	−2	0.59	21.6	3.6	−1.7	1.9		
12.5°W–52.5°E)			−24%	+1%	+23%	−12	−21	+9	−18%	−0.6	−0.7	−2.4	−3.1	3.4	4.4
(5) Bangladesh	2a	1.00	0.86	149	38	111	113	−2	0.85	24.2	3.9	4.1	8.0		
(20–28°N,	4	1.00	0.65	140	44	96	107	−11	0.78	23.6	3.7	4.3	8.0		
77.5–87.5°E)			−24%	−6%	+16%	−15	−6	−9	−8%	−0.6	−0.2	0.20	0	−1.0	0.0
(6) Mississippi	2a	1.00	0.86	208	68	140	148	−8	0.57	22.1	5.1	−0.7	4.4		
Valley (32–48°N,	4		0.65	170	67	103	102	1	0.56	22.6	3.5	−0.2	3.3		
92.5–82.5°W)			−24%	−18%	−1%	−37	−4.6	+9	−2%	+0.5	−1.6	+0.5	−1.1	−0.3	0.7
Average for the		0.86	0.86	177	55	122	122	0	0.71	23.4	4.2	1.4	5.6		
six regions		0.86	0.65	172	70	102	98	4	0.58	22.8	3.4	0.2	3.6		
			−24%	−3%	+27%	−20	−24	+4	−18%	−0.6	−0.8	−1.2	−2.0	1.5	2.5

β average, for the region, of the prescribed soil moisture availability (ratio of evapotranspiration to potential evapotranspiration)
$(1-\alpha)$ fraction of the incident solar radiation that is absorbed by the ground (α = land-surface albedo)
R_S solar radiational heating of the ground (watt/m²)
R_L long wave (infrared) radiational cooling of the ground (difference between long wave radiation emitted by the ground and atmospheric 'back radiation' absorbed by the ground) (watt/m²)
$R_N = (R_S - R_L)$ net (all-wavelength) radiational heating of the ground (watt/m²)
LE latent heat transfer from ground to atmosphere (evaporative cooling of the ground)
H conductive–convective heat transfer from ground to atmosphere
N fraction of the sky covered by clouds of all types
T_A surface air temperature (°C)
E_T surface evapotranspiration (mm/day)
$(-\overline{\nabla \cdot q\mathbf{v}})$ vertically integrated convergence of the water vapor transport (mm/day)
P precipitation (mm/day)
$\Delta P/\Delta E$ ratio of precipitation change to evapotranspiration change
For each region the third line shows either the absolute change between the two cases or the percentage change, where the % sign indicates the latter.

We see, in column (10), that in the Western Great Plains (where β was assigned the value of 0.78) and in the Mississippi Valley (where β was made 1.0), the evapotranspirations with albedo of 0.14, are, respectively, 4.2 mm/day (130 mm/month) and 5.1 mm/day (158 mm/month). These are in fair agreement with the aerologically derived evapotranspiration over the central and eastern United States in July, of 3.7 mm/day (116 mm/month), shown in Fig. 6.1. More important, however, as an indication of the reliability of the model, is the fact that the vertical integral of the water vapor flux convergence ($-\overline{\nabla \cdot q\mathbf{v}}$), shown in column (11), is negative in the two regions, with values, respectively, of −0.5 mm/day (−16 mm/month) and −0.7 mm/day (−22 mm/month). This means that water vapor is being exported from these regions at about the same rate as the observed transport divergence, of 0.7 mm/day (22 mm/month), shown in Fig. 6.1.

In the other four regions, in the case with normal surface albedo, 2a, the vertical integrals of the water vapor transport convergence are positive, water vapor is being imported (so that precipitation is larger than evapotranspiration), which is what one would expect for these particular regions in the month of July.

Table 6.1 is replete with information about the performance of the model and its complex, non-linear response to the change in the surface albedo. But, for brevity, we will here examine only what happens in the Sahel, the region of greatest interest.

We see, in Table 6.1 (column 3), and in Fig. 6.11, that when the surface albedo is increased in the Sahel, from 0.14 to 0.35, the solar radiational heating of the surface does not become smaller: it becomes *larger*. This is because of the large decrease in the cloud cover, from 0.70 to 0.46 (column 8), which more than compensates for the increased albedo.

The cloud cover is less because (1) there is less evapotranspiration (a change from 3.7 to 2.8 mm/day); and (2) there is less convergence in the water vapor transport (a change from 3.7 to 1.2 mm/day).

The local evapotranspiration is reduced in the high-albedo case because there is more long wave radiational cooling of the ground (an increase from 58 to 84 watt/m²). Unfortunately, no record was kept of the ground-surface temperature, nor of the long wave emission by the ground; but it is most likely that it is the decrease in the downward long wave 'back radiation' from the atmosphere, $R_L\downarrow$, as a consequence of the decreased cloudiness, which increased the long wave cooling of the ground.

The sensible heat transfer from the ground to the atmosphere is small, in both cases.

By using arrows to indicate when a change in one parameter produces a change in another, we can describe what happens in the Sahel, in this experiment, as the coupling between a sequence of processes operating locally and a sequence that involves the large-scale atmospheric circulation. The local sequence is: albedo $\xrightarrow{\text{(changes)}}$ radiation $\xrightarrow{\text{(changes)}}$ evapotranspiration $\xrightarrow{\text{(changes)}}$ precipitation. The larger-scale sequence is: precipitation–condensation-heating $\xrightarrow{\text{(changes)}}$ large-scale circulation $\xrightarrow{\text{(changes)}}$ water vapor transport convergence $\xrightarrow{\text{(changes)}}$ precipitation.

The second of these two sequences is similar (but not identical) to the one in the Charney (1975) hypothesis on the dynamics of deserts. In that hypothesis the sequence is: albedo $\xrightarrow{\text{(changes)}}$ surface temperature $\xrightarrow{\text{(changes)}}$ large-scale circulation $\xrightarrow{\text{(changes)}}$ water vapor transport convergence $\xrightarrow{\text{(changes)}}$ precipitation. Here, the change in evapotranspiration plays no role. (A direct examination of the Charney hypothesis with a general circulation model would consist of a comparison of two runs, in both of which no evapotranspiration was allowed in the region of interest, but was allowed elsewhere, and the albedo in the region of interest was changed.) (See experiment in section 6.4.2.1.)

The relative importance of the two sequences of processes, when surface evapotranspiration does take place, may be seen in columns (13) and (14) of Table 6.1. We see, in column (13), that in the Sahel, Rajputana and Central Africa, the reduction in the water vapor flux convergence is between 2.8–4.2 times larger than the reduction in the evapotranspiration. But in the Western Great Plains, the reduction in the water vapor flux convergence is only half as large as the reduction in evapotranspiration. In Bangladesh and the Mississippi Valley, things go the other way: increasing the surface albedo again decreases the evapotranspiration; but it *increases* the water vapor flux convergence.

As indicated earlier, the response of the large-scale precipitation, temperature and motion fields to a change in the surface boundary conditions (whether soil moisture availability or albedo) will depend on many factors. Of particular importance is the horizontal scale and the latitude of the region in which the boundary condition is changed. Through the geostrophic adjustment process, the horizontal scale and the latitude determine whether the circulation change will be in the vertical plane (small-scale or low-latitude) or in the horizontal plane (large-scale and high-latitude). When the circulation change is in the vertical plane there is a positive feedback on the condensation heating, through water vapor transport convergence. But when the circulation change is in the horizontal plane there is a negative feedback on the condensation heating, because then the transport removes water

Fig. 6.11. The Sahel region energy budgets (watt/m²) (top) and water budgets (mm/day) (bottom), in experiment with prescribed soil moisture availability, of Charney *et al.* (1977). Case 2a is on the left, Case 3a is on the right. (For definitions of symbols, see notes to Table 6.1.)

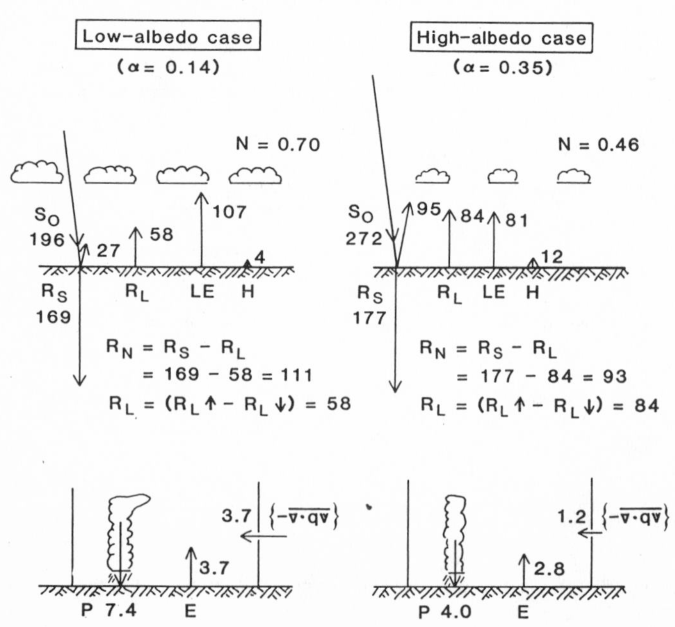

vapor, as well as sensible heat, from the region of the condensation heating.

6.3.2.2

Carson and Sangster (1981)

Another experiment in this category was made by Carson & Sangster (1981) with a low-resolution (N20) version of the British Meteorological Office five-layer general circulation model (Corby *et al.*, 1977). In both runs, evapotranspiration was made equal to the calculated potential evapotranspiration ($\beta \equiv 1$). In one case, the albedo of the snow-free land was everywhere set equal to 0.1. In the other case, it was everywhere set equal to 0.3. The remaining lower boundary conditions (sea-surface temperatures, sea ice, and

Fig. 6.12. Precipitation in the low-albedo case (top) and the high-albedo case (center), of experiment by Carson & Sangster (1981). The contours are for 1, 2, 5, 10 and 20 mm/day. (Light shading, precip. < 1 mm/day; heavy shading, precip. > 5 mm/day). The bottom panel is the difference in the precipitation: the low-albedo case minus the high albedo case (unshaded area is positive).

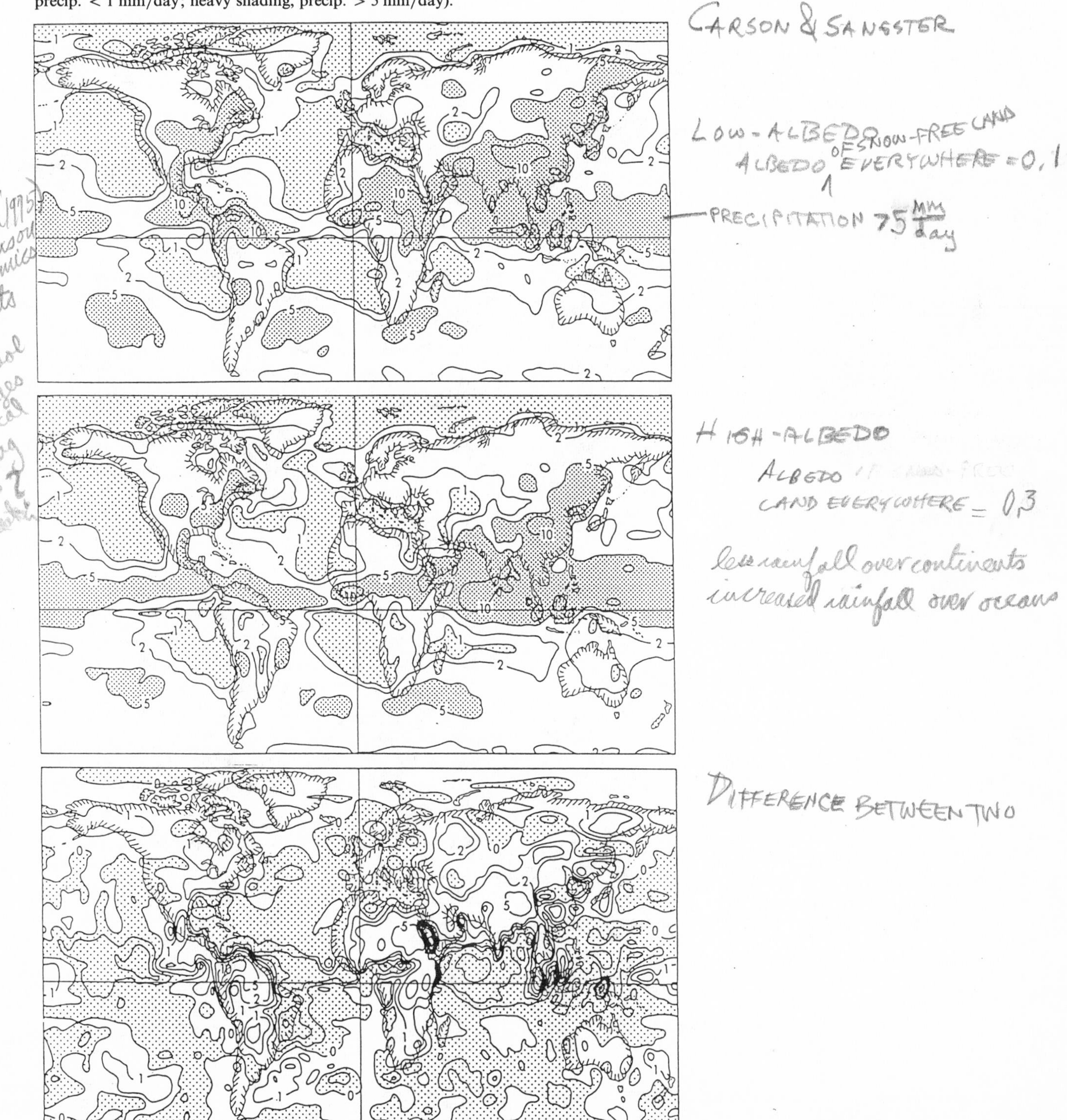

land snow cover), as well as the climatologically determined long wave radiational heating rates, were the observed July values.

Figure 6.12 shows the precipitation averaged over 90 days (days 21–110 of integration), where the top panel is the low-albedo case, the center panel is the high-albedo case, and the bottom panel is the difference between the two.

We see that the high-albedo case has less rainfall over most of the continental areas, but that over the oceans the rainfall is increased.

The averages of the land precipitation, and of other parameters, are shown in Table 6.2.

Like the experiment by Charney *et al.*, 6.3.2.1 above, this is an albedo change experiment with permanently wet soil. Here, too, the increase of albedo produces a decrease in evapotranspiration (−0.9 mm/day) and an even larger decrease in precipitation (−1.2 mm/day); but now the albedo is changed on the continental scale, whereas in 6.3.2.1 it was changed only over a few hundred kilometers in the widths of the various regions. It is not surprising, therefore, that in this experiment the contribution to the change in precipitation of the change in the water vapor transport convergence is only about a third as large as is the contribution by the change in the local evapotranspiration. Both experiments have about the same average water vapor transport convergence, 0.8 mm/day v. 0.85 mm/day. But in (4), where the albedo was changed from 0.14 to 0.35 over a number of small regions, the average change in the transport convergence in those regions was −1.2 mm/day. In (5), where the albedo is changed from 0.1 to 0.3 over all of the land, the change in the transport convergence is only −0.3 mm/day; showing, again, that the larger the horizontal scale the smaller is the role of the water vapor transport convergence in compensating for a decrease in the land-surface evapotranspiration.

6.4

(II) Experiments with interactive soil moisture

In all but one of the experiments that follow, the time-dependent soil moisture is governed by the equations:

$$\frac{\partial W}{\partial t} = P - E, \quad W_{\max} = W^*, \tag{6.1}$$

$$E = \beta E_p, \tag{6.2}$$

$$\beta = \frac{W}{kW^*}, \quad \beta_{\max} = 1, \tag{6.3}$$

where W is the available moisture stored in the soil, W^* is the available moisture storage capacity of the soil, P is the rate of precipitation, E is the rate of evapotranspiration, E_p is the rate of potential evapotranspiration, β is the soil moisture availability, and k is a prescribed coefficient (see Carson, 1981). In all of the models, E_p is evaporation calculated by an aerodynamic method, under the assumption that the vapor pressure at the surface is the saturation value for the calculated ground temperature.

6.4.1

(A) ***Different initial soil moistures, with same albedo***

6.4.1.1

Walker and Rowntree (1977)

Walker & Rowntree (1977) examined the interaction between time-dependent soil moisture and the calculated precipitation, temperature and circulation of the atmosphere, not in the global domain, but in a zonal channel between latitudes 16°S and 36°N, and extending over 32° of longitude with cyclic east–west boundary conditions. The land and sea distribution was made zonally symmetric, with land to the north and ocean to the south of 6°N latitude; this being an idealization of the western part of north Africa and the Gulf of Guinea.

The model was an 11-layer primitive equations model with 2° latitude–longitude resolution. The radiational part of the thermal forcing was taken as a constant radiational cooling of the atmosphere, of 1.2 K/day from the surface to the 200 mb level, with radiative equilibrium at higher levels (which means a constant radiational cooling of the atmosphere of 110 watt/m^2); and with a constant net radiational heating of the land surface, R_N, of 150 watt/m^2. Thus, over the land, there was a prescribed horizontally uniform radiational heating (of 40 watt/m^2) of the atmosphere– earth system; but over the ocean, the surface temperature, and not the surface radiation flux, was the prescribed boundary condition. The prescribed, zonally-symmetric ocean temperatures, from 16°S to 6°N, were the observed August normals at 0° longitude. The moist-convective adjustment scheme was used to obtain the convective precipitation and moist-convective heating of the air.

The available soil moisture and the land-surface evapotranspiration were calculated with the equations given at the beginning of this section; with W^* taken as 150 mm, and k taken as 0.333. Therefore, $\beta = 1$ when $W \geqslant 50$ mm.

Two integrations were made in which everything was the same, except that:

In case 1 (the initially dry-soil Sahara), W was initialized at zero in the latitude zone 14–32°N; and at 100 mm in the land zones 6–14°N and 32–36°N.

Table 6.2. *Albedo experiment of Carson & Sangster (1981) 90-day means (days 21–110), permanent July*

	Global Averages Over Land						
	(1)	(2)	(3)	(4)	(5)	(6)	(7)
Surface albedo	LE	H	E	$(-\overline{\nabla \cdot qv})$	P	$\frac{\Delta(-\overline{\nabla \cdot qv})}{\Delta E}$	$\frac{\Delta P}{\Delta E}$
0.1	104	35	3.6	1.0	4.6		
0.3	78	21	2.7	0.7	3.4		
Difference	−26	−14	−0.9	−0.3	−1.2	0.3	1.3

For definition of symbols, see Table 6.1.

Fig. 6.13. Variation with time of the zonally averaged precipitation (top) and soil moisture (bottom) in the case where, initially, the soil in the Sahara is completely dry. (Experiment of Walker & Rowntree, 1977.)

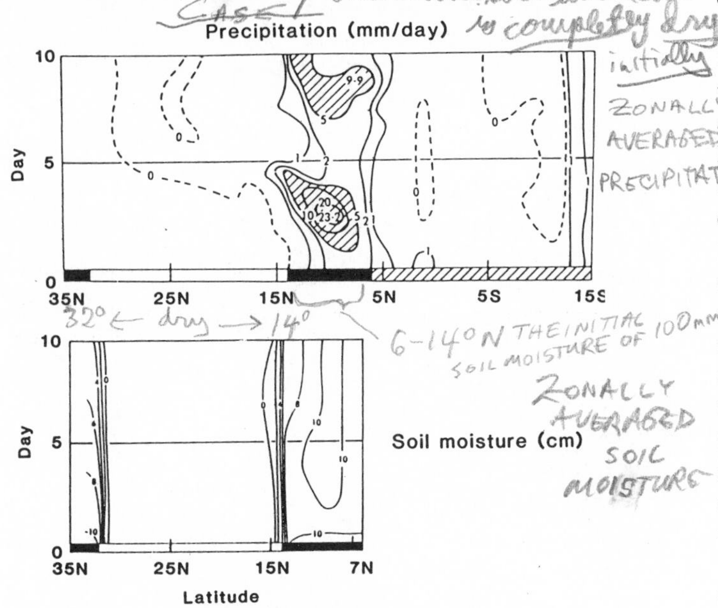

Fig. 6.14. Variation with time of the zonally averaged precipitation (top) and soil moisture (bottom) in the case where the initial soil moisture in the Sahara is 100 mm. (Experiment of Walker & Rowntree, 1977.)

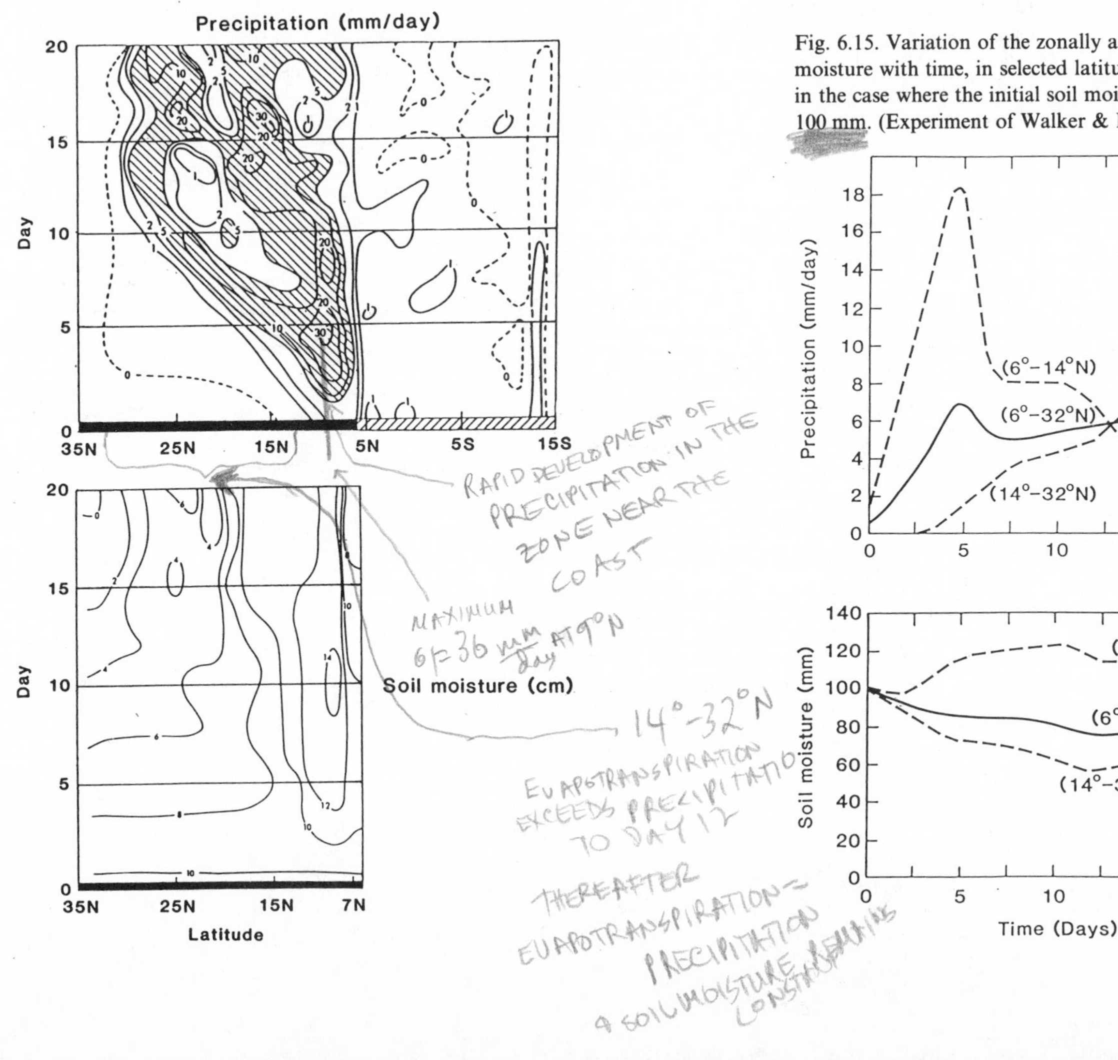

In case 9 (the initially moist-soil Sahara), W was initialized at 100 mm over all of the land region, 6–36°N.

Figures 6.13, 6.14 and 6.15 show the time-evolutions of the soil moisture and precipitation in the two cases.

In case 1 (Fig. 6.13), where the soil was initially dry between 14° and 32°N, it remains dry. There is almost no net water vapor transport into that region and, therefore, there is no precipitation and no water is added to the soil there. On the other hand, in the land region 6–14°N the initial soil moisture, of 100 mm, goes down to about 90 mm over the first seven days, showing an excess of evapotranspiration over precipitation which averages about 1.4 mm/day. The corresponding seven-day water vapor transport divergence, of about 1.4 mm/day, is the difference between a large northward transport of water vapor across the coastline by the mean meridional circulation (Walker & Rowntree, 1977, Fig. 4(*a*)) and an even larger equatorward eddy-transport of water vapor by the wave disturbance which developed and moved westward across the region. By the end of the integration period, this part of the system also appears to have reached a steady state, except for a short-period and small-amplitude variation produced by transient waves in the flow.

In case 2, the initially moist-soil Sahara, (Figs. 6.14 and 6.15), there is a rapid development of precipitation in the zone near the coast, which, after about two days, exceeds the evapotranspiration rate and the soil moisture starts to increase. The average precipitation in this coastal zone reaches 18 mm/day on day 5; with a maximum of 30 mm/day at 9°N. After that the precipitation rate

Fig. 6.15. Variation of the zonally averaged precipitation and soil moisture with time, in selected latitude zones across north Africa, in the case where the initial soil moisture in the Sahara is 100 mm. (Experiment of Walker & Rowntree, 1977.)

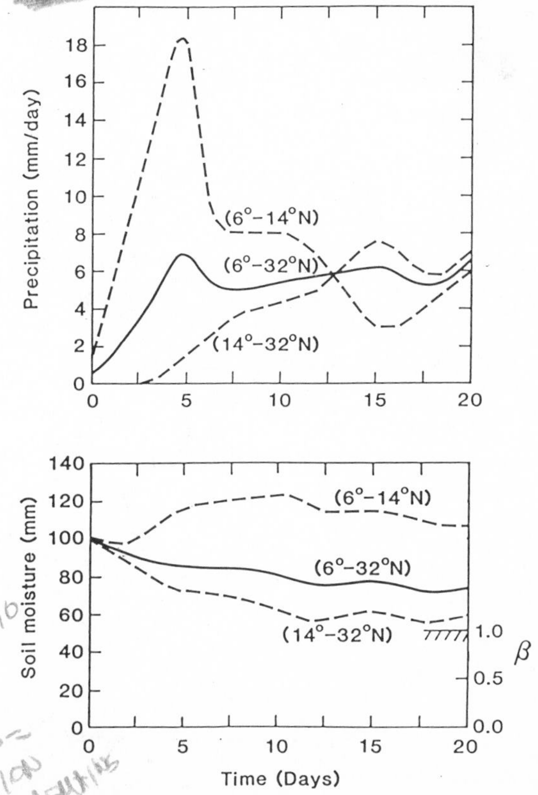

in this zone decreases rapidly and seems to be starting an oscillation about an average rate of around 6 mm/day. The prescribed $R_N = 150$ watt/m² would provide enough energy for $E = E_p = 5{\cdot}2$ mm/day; but the calculated evapotranspiration may be smaller or larger than this, depending on whether the sensible heat transfer at the surface is upward or downward.

Over the rest of the land region, 14–32°N, the evapotranspiration exceeds the precipitation until about day 12; and, thereafter, except for an oscillation produced by the transient wave disturbances, evapotranspiration equals precipitation and the soil moisture remains constant.

From what we see in these figures, it appears unlikely that the solutions for the initially dry-soil Sahara and the initially moist-soil Sahara will approach one another no matter how long the integrations were to continue. It seems safe to say that this highly simplified soil moisture–atmosphere system is intransitive.

6.4.1.2 *Rowntree and Bolton (1978)*

Rowntree & Bolton (1978) made an interactive soil moisture experiment with the five-layer, 500 km grid size, version of the British Meteorological Office general circulation model (described by Corby *et al.*, 1977).

For the calculation of the soil moisture and evapotranspiration, W^* was taken as 200 mm and k as 0.5; so that $\beta = 1$ when $W \geqslant 100$ mm.

Three 50-day integrations were made, all starting from the same initial atmospheric conditions on 27 May, but with different initial distributions of soil moisture.

In one run, designated C (for control), the initial soil moisture was set at 50 mm at all land points over the globe. Therefore, the initial soil moisture availability, β, was 0.5 everywhere.

In the run designated W (for wet-soil case) the initial soil moisture was set at 150 mm at all of the European land points that are within the region enclosed by the rectangle in Fig. 6.16; but with an initial value of 50 mm at all other land points over the globe. Thus, the initial β was 1 in the European region but, again, 0.5 at all other land points over the globe.

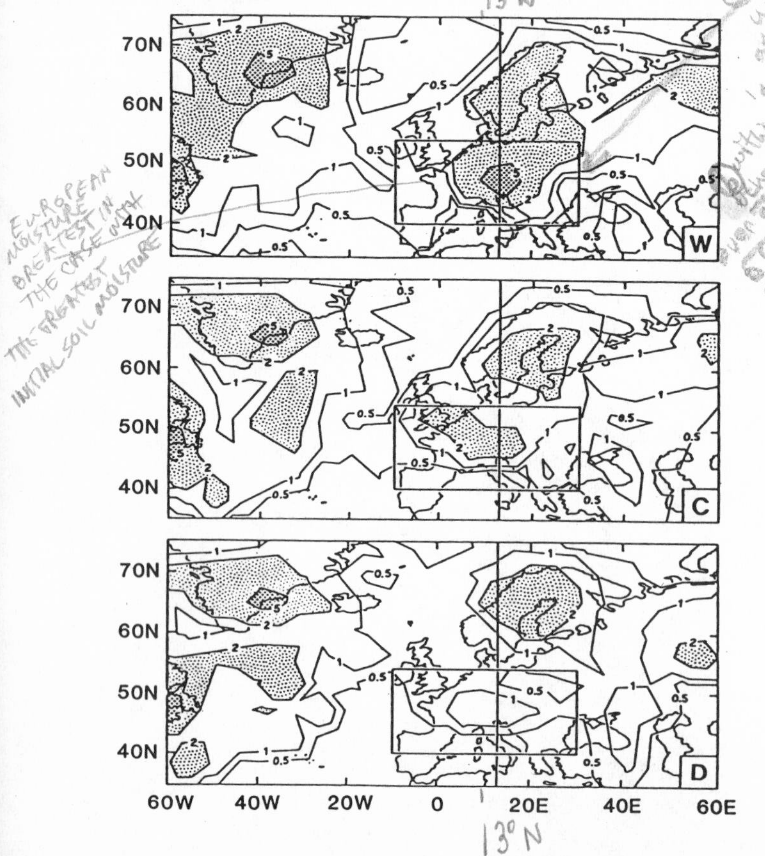

Fig. 6.16. Precipitation (mm/day) averaged for 15 June–15 July, in experiment of Rowntree & Bolton (1978). Center panel: control run, where the initial soil moisture, on 27 May, was 50 mm everywhere. Top panel: case where the European land points within the indicated rectangular region had an initial soil moisture of 150 mm. Bottom panel: case where the European land points within the indicated rectangular region had zero soil moisture initially

Fig. 6.17. Distribution of precipitation along the 13°E meridian, for the three fields shown in Fig. 6.16.

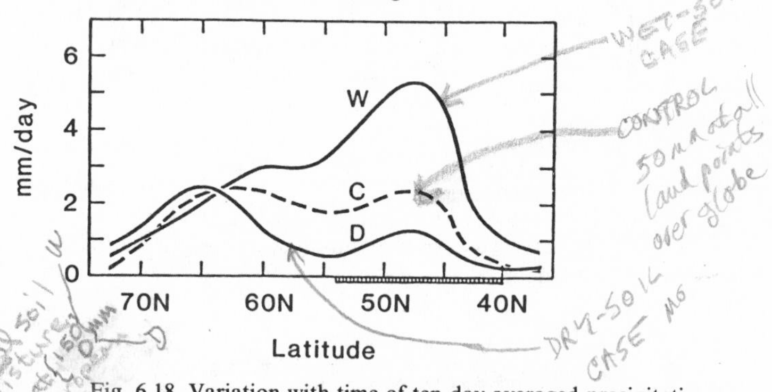

Fig. 6.18. Variation with time of ten-day averaged precipitation (top) and soil moisture (bottom), averaged for the European land points within the rectangular region shown in Fig. 6.16, for the three cases in the experiment of Rowntree & Bolton (1978).

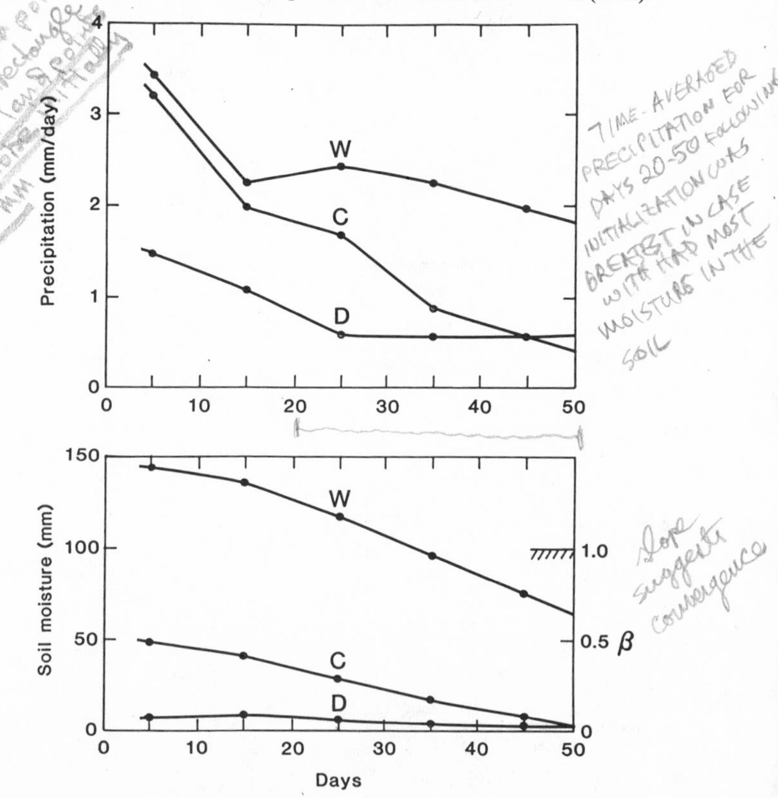

In the run D (the dry-soil case) the soil moisture in the European region was initialized at 0 mm (but, again, at 50 mm elsewhere). Now, the initial β, and hence the initial evapotranspiration, was zero in the European region.

Fig. 6.16 shows maps of the three rainfall distributions, averaged for the 30-day period, 15 June–15 July; and Fig. 6.17 shows meridional profiles of those time-averaged rainfalls along longitude 13°E, where the maximum rainfall occurs. In general the differences between the 30-day rainfalls are comparable to what are produced by the natural variability of a model atmosphere with fixed surface boundary conditions. But within the European region the time-averaged precipitation, for days 20–50 following initialization, was greatest in the case which initially had the most moisture in the soil, and smallest when the soil was initially devoid of moisture.

The changes in the precipitation and in the soil moisture with time are shown in Fig. 6.18, where the values are ten-day averages for the indicated European land region. We see that even after one-and-a-half months there are still large differences between the precipitation rates and between the soil moistures, when we compare the initially wet-soil case (W initially 100 mm) with the other two cases (W initially 50 mm and 0 mm). The slopes of the rainfall curve and soil moisture curve, for the initially wet-soil case, suggest that the system is transitive, but that the time required for convergence is several months.

6.4.2

(B) ***Different albedos, with same initial soil moisture.***

6.4.2.1

Charney, Quirk, Chow and Kornfield (1977)

Charney *et al.* (1977) performed an experiment in which they compared two runs which had different albedo distributions, but the same initial soil moistures which could interact with the atmosphere.

The distribution of the land-surface albedo in the two cases is shown in the bottom panel of Fig. 6.19. In their run designated '2b', the albedos of 'permanent desert' (the regions with dotted shading) were assigned the value of 0.35; and everywhere else over the ice-free and snow-free land surface of the globe the albedo was taken as 0.14. In the comparison run, designated '3b', three regions adjacent to the permanent deserts, the 'Western Great Plains', 'Rajputana' and 'Sahel' (shown by the cross-ruled shading) were also assigned an albedo of 0.35. In both runs, the initial soil moisture was taken to be zero everywhere.

The change in the time-dependent soil moisture was calculated, in half-hourly time steps, from Eqs. (6.1) and (6.2) given at the beginning of this section, and with the function $\beta = \beta(W, W^*, E_p)$ (Charney *et al.*, 1977, p. 1368) which is shown in Fig. 6.20. (Diagram by personal communication from Y. Sud.) Over the range of E_p between 1.4 and 6.4 mm/day, this formulation for β was taken from Denmead & Shaw (1962), who obtained it from measurements of daily (24 hr) evapotranspiration and potential evapotranspiration, together with measured soil moisture.

In the experiment, this formula for β was inadvertently applied to the calculation of half-hourly values of evapotranspiration, and this made the calculated daily evapotranspiration an order of magnitude too small (because during the mid-day hours, when E_p is about π times as large as its 24-hour average, the β obtained in this way is extremely small for almost all values of

Fig. 6.19. Bottom panel: the assigned albedos in experiments of Charney *et al.* (1977). Unshaded land areas have an albedo of 0.14 in all cases, and dot-shaded areas have an albedo of 0.35 in all cases. In the cross-ruled areas the albedo was changed from 0.14 (in cases 2a and 2b) to 0.35 (in cases 3a and 3b). Top panel: precipitation (mm/day) in case 3b.

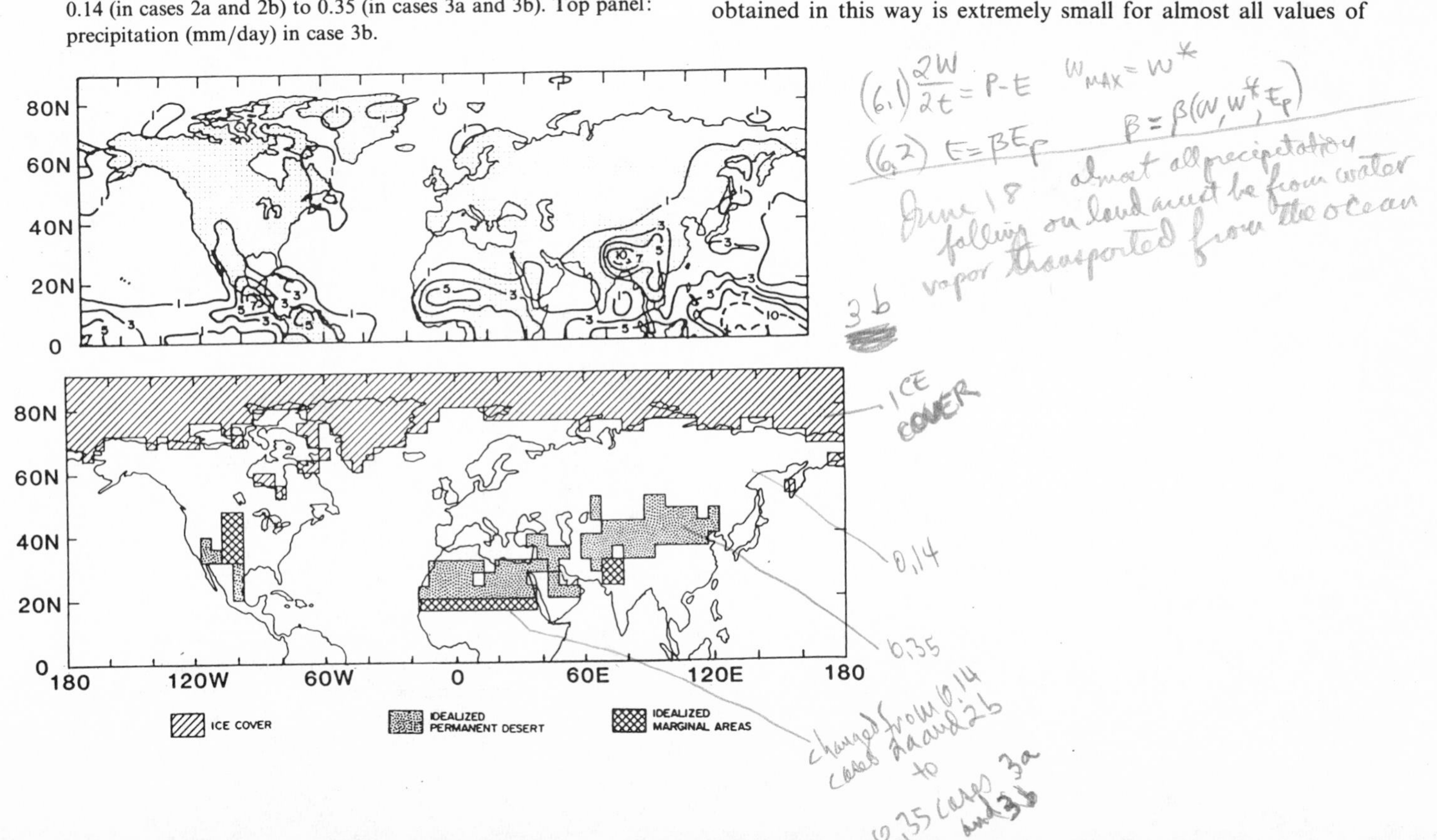

W/W^*; and during the night, when β can approach 1, E_p is negligible).

The top panel in Fig. 6.19 shows the calculated July precipitation in case 3b. At the beginning of the run, 18 June, almost all of the precipitation that falls on the land must be from water vapor transported from the ocean; and this situation must continue (because evapotranspiration is negligible) until the accumulation of the water in the soil brings W close to the soil storage capacity W^*, at which point β becomes equal to 1 for all values of E_p. For a precipitation of 5 mm/day, which is the mean July precipitation across north Africa in case 3b, it takes about 20 days, or until about 8 July, for the soil moisture, W, to approach or reach W^*, at which point β approaches or reaches 1. Elsewhere, β remains negligible, even until the end of July. It is therefore not surprising that in this experiment with interactive soil moisture, the precipitation, for days 14–44 following initialization with zero soil moisture, should resemble the persistently dry-soil case shown in the lower panel of Fig. 6.3.

Table 6.3 shows the energy and water balances of the three regions, in the two runs, for July (days 14–44 of the integration). We see, by comparing columns (11) and (12), that during this period (which, as indicated, is a transient stage for the soil moisture and evapotranspiration) the dominant term in the supply of water vapor for precipitation, in all cases, is the water vapor transport convergence. In the Sahel, the evapotranspirations for the month are 0.14 and 0.34 mm/day, which are only 4% and 12% of the monthly averaged precipitation rates. But almost all of this must be due to the evapotranspiration near the end of July, at which time, or shortly thereafter, the accumulated precipitation will make W approach, or equal, W^*.

With the small amounts of evapotranspiration, the cloud cover (column 8) is not very different in the low- and high-albedo cases. Consequently, unlike experiment 6.3.2.1 above, the change in the net radiational heating of the ground, R_N, is almost entirely due to the change in the solar radiational heating of the ground, R_S; and this produces the large change in the sensible heat transfer,

Fig. 6.20. The soil moisture availability function, $\beta = \beta$ (W, W*, E_p), used by Charney *et al.* (1977) in the albedo experiment with interactive soil moisture.

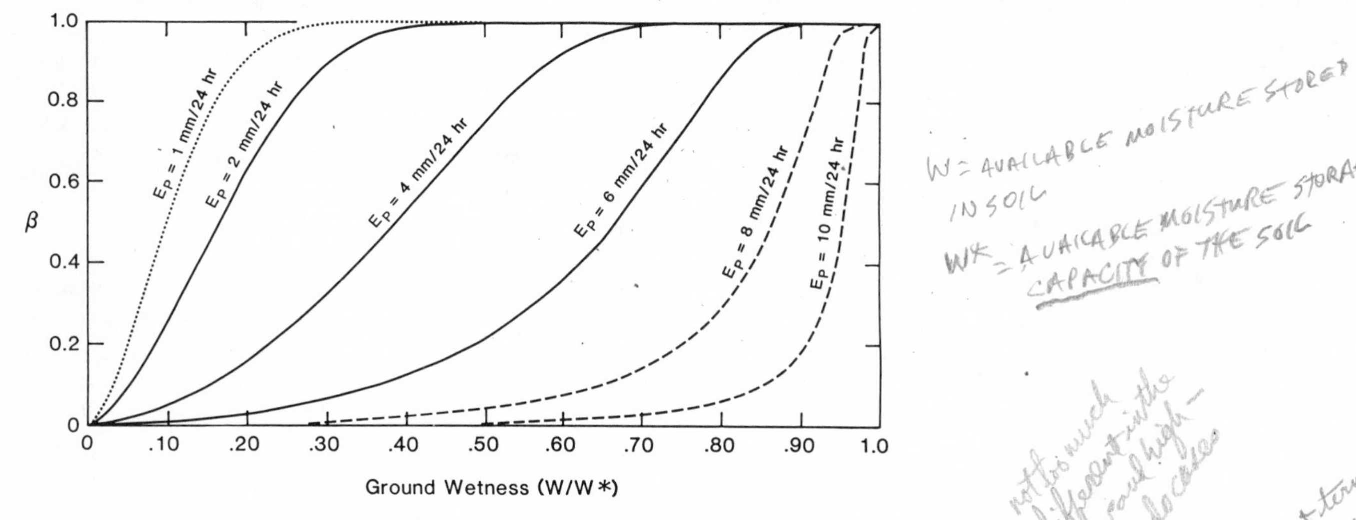

Table 6.3. *Components of the energy and water budgets, in experiment of Charney et al.* (1977)

Region	Case No.	(2) $(1-\alpha)$	(3) R_S	(4) R_L	Energy balance (5) R_N	(6) LE	(7) H	(8) N	Water balance (9) T_A	(10) E	(11) $-\nabla\cdot q\mathbf{v}$	(12) P
Sahel	2b	0.86	259	139	120	4	116	40	39.2	0.14	3.9	4.0
	3b	0.65	213	140	73	10	63	35	36.1	0.34	2.4	2.7
		−24%	−46	1	−47	6	−53	−12%	−3.1	0.20	−1.5	−1.3
Rajputana	2b	0.86	269	134	135	3	132	43	35.1	0.10	2.0	2.1
	3b	0.65	219	129	90	8	72	42	34.2	0.26	2.1	2.4
		−24%	−50	−5	−45	5	−60	−2%	−0.9	0.16	0.1	0.3
Western Great Plains	2b	0.86	302	172	130	0	130	21	37.5	0.00	0.8	0.8
	3b	0.65	245	161	84	3	81	20	31.4	0.10	0.3	0.4
		−24%	−57	11	−46	3	−49	−5%	−6.1	0.10	−0.5	−0.4

For definition of symbols, see Table 6.1

H. It is this change in the sensible heating of the planetary boundary layer which produces the change in the vertical velocity shown in the bottom panel of Fig. 6.21, the associated change in water vapor transport convergence, shown in column (11) of Table 6.3, and the change in precipitation shown in column (12) and in the top panel of Fig. 6.21. As the experiment stands, it does illustrate the mechanism of the Charney (1975*a*) 'dry-soil' desertification hypothesis: but not if the integrations were to continue; for then W will everywhere approach, or equal, W^* and the processes which depend on evapotranspiration will become important.

We note, furthermore, that this simple picture of the coupling of the water vapor transport convergence to the surface albedo does not hold for the Rajputana and Western Great Plains regions, even when the evapotranspiration is negligibly small. In both of these regions the changes in R_S, R_N, and H are about the same as the changes in the Sahel; but, unlike the Sahel, there is almost no change in the water vapor transport convergence. As already indicated, the response of the atmospheric circulation to a change in the boundary layer heating, H, will depend very much on the horizontal scale and latitude of the heating perturbation, as well as on the orientation of the region with respect to external moisture sources and the way in which the altered flow encounters the mountain barriers.

6.4.2.2

Chervin (1979)

Chervin (1979) used the NCAR general circulation model (described by Washington & Williamson, 1977) to examine the effect of a change in the land-surface albedo when the soil moisture is fully interactive. The change in soil moisture was calculated with the equations given at the beginning of this section, with $W^* = 150$ mm and $k = 0.75$.

The control was the average of a master run, which started from a state of rest and isothermalcy and was integrated for 120 days, plus four other runs, each of which started from day 30 of the master and ran until day 120. All of these were perpetual July integrations, in which the sun declination, the ocean-surface temperatures, and the snow-free land albedo (which followed Posey & Clapp, 1964) were held constant in time.

The run with a different albedo was also started from day 30 of the master and ran until day 120. The change in the albedo consisted of replacing the Posey & Clapp values by a constant albedo of 0.45 within two regions: (1) a large region over north Africa, extending from the Atlantic to the Red Sea and from latitude 7.5°N to the Mediterranean, and therefore covering the zone of the July intertropical convergence rain, as well as the Sahara Desert; and (2) a smaller region over the US High Plains (97.5–107.5°W, 27.5–52.5°N).

Over Africa, the control (Posey & Clapp) albedo varied from about 0.35 in the northern Sahara to about 0.08 near the southern boundary of the region where the albedo will be changed. Over the US High Plains, the control albedo was between 0.07 and 0.17.

Fig. 6.22 shows the change over Africa in the precipitation, soil moisture and ground temperature, and in the vertical velocity at 3 km elevation. The values shown are the averages of the last 60 days of the modified albedo case, minus the ensemble average of the last 60 days of the five control runs. The stippled areas in the diagram show the regions of $r = |\Delta_{60}|/\sigma_{60} \geqslant 3$, where Δ_{60} is the prescribed change response (i.e., the difference between the 60-day mean in the prescribed change case and the ensemble average of the 60-day means of the five control runs); and σ_{60} is the standard deviation of the 60-day means of the five control runs. According to Chervin & Schneider (1976), $r \geqslant 3$ implies an approximately 5% significance level in rejecting the hypothesis that the prescribed change response is the result of random fluctuations and not the result of the prescribed surface albedo change.

The maps in Fig. 6.22 show that the changes are greatest at and near the zone where the albedo is increased from 0.08 to 0.45. In this zone there is a decrease in the average upward motion of the air of about 2 mm/second (200 m/day), with $r > 3$; a decrease in the average precipitation of about 4 mm/day, with $r \geqslant 2$, [$\Delta_{60}\ P \approx (4-8)$ mm/day, $\sigma_{60}\ P = 2$ mm/day]; and a decrease in the average soil moisture storage of about 50 mm, with $r > 3$. There is a decrease of the ground-surface temperature, with $r > 3$, over almost all of the region of the albedo change, but not along its southern edge. There, the ground-surface temperature *increases*, by about 0.5°C, with $r > 3$.

The paradoxical rise in the ground-surface temperature, in the region of the largest increase of albedo (the change from 0.08 to 0.45), can be attributed to the fact that in that zone, 7.5–12.5°N, where there is the largest decrease in precipitation, the soil moisture, W, goes down from about 100 mm in the control to 50 mm in the high-albedo case. With $W^* = 150$ mm and $k = 0.75$, this reduces β from 0.9 to 0.45 and, consequently, there is a large reduction in the evapotranspiration and the evaporative cooling of the surface in that zone. There is also a contribution to the

Fig. 6.21. Zonally averaged precipitation (top) and vertical velocity in the middle troposphere (bottom), over Africa, when the albedo in the Sahel (16–20°N) is increased from 0.14 to 0.35. (Figure from Charney, 1975*a*.)

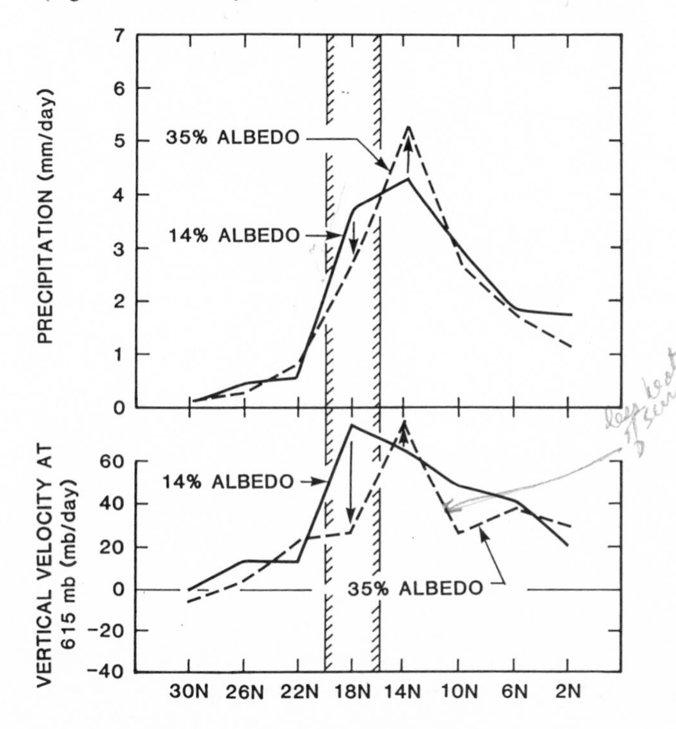

temperature rise from the accompanying reduction of the cloudiness in that zone (R. Chervin, personal communication).

Over the US Great Plain region there was almost no change in the vertical velocity at 3 km elevation; an average decrease of about 1 mm/day in the precipitation, with $r > 3$ over about half of the region; almost no change in the soil moisture storage; and a decrease, averaging about 2°C, with $r > 3$, in the ground-surface temperature.

6.5

(III) Hybrid experiments

6.5.1

Non-interactive v. interactive soil moisture

These are experiments in which a calculation with interactive soil moisture is compared with one in which the prescribed soil moisture is held fixed for the duration of the experiment. If we regard the interactive case as a simulation of nature, then the case with the prescribed, fixed soil moisture can be thought of as showing how the climate would be changed if the land-surface evapotranspiration were to be brought under man's control: as, for example, by large-scale irrigation or by a change or a complete removal of the vegetation cover.

Fig. 6.22. The change in vertical velocity at 3 km elevation (top left), precipitation (top right), ground temperature (bottom left), and soil moisture (bottom right), in the albedo change experiment of Chervin (1979).

6.5.1.1

Manabe (1975)

A massive irrigation simulation experiment was made by Manabe (1975) with one of the Geophysics Fluid Dynamics Laboratory (GFDL) general circulation models. The model used the moist-convective adjustment method for calculating the convective precipitation and the moist-convective heating of the air; solar and long wave radiation transfers calculated with a non-interactive cloud distribution, prescribed as a function of latitude and elevation; and an albedo for ice-free and snow-free land that follows Posey & Clapp (1964).

In the 'natural case' (the interactive soil moisture case) the soil moisture was governed by the equations given at the beginning of Section 6.4, with $W^* = 150$ mm, and $k = 0.75$. In the 'irrigation case' (the non-interactive soil moisture case) β was everywhere held equal to 1.

The natural case simulation produced a rainband across north Africa in which, averaged between 15 and 30°E, there was a rainfall maximum of about 6 mm/day at latitude 5°N. In the irrigation case, the maximum rainfall was about 12 mm/day at latitude 8°N (see Manabe, 1975, Fig. 3).

The solid line in Fig. 6.23 shows the change in the precipitation, the irrigation case minus the natural case. The dashed line shows the corresponding difference in the evapotranspiration. (The reduction in evaporation, between 31 and 37°N, is over the Mediterranean Sea, presumably because the air was more humid from the massive land irrigation.)

Between 18 and 30°N the increase in precipitation is somewhat

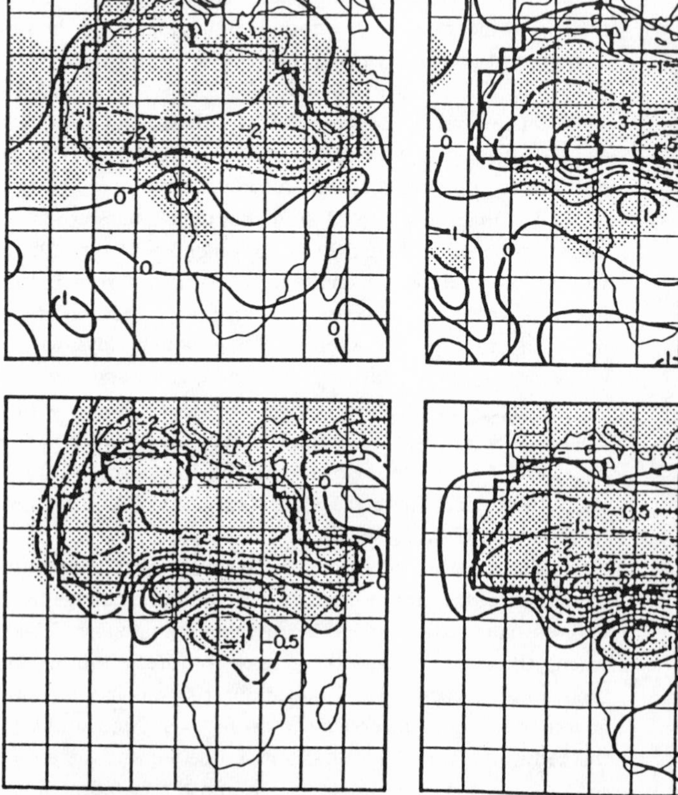

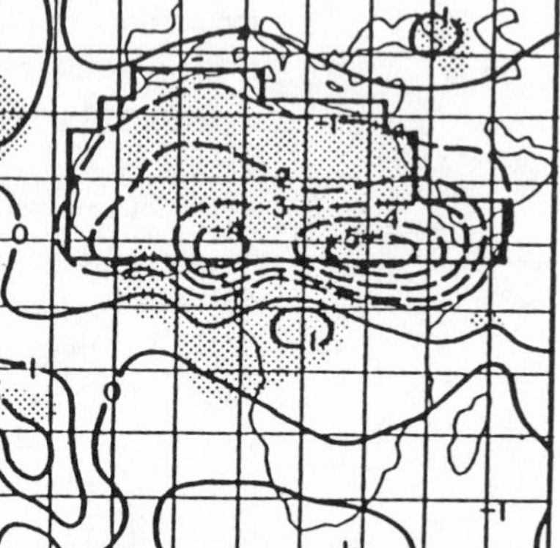

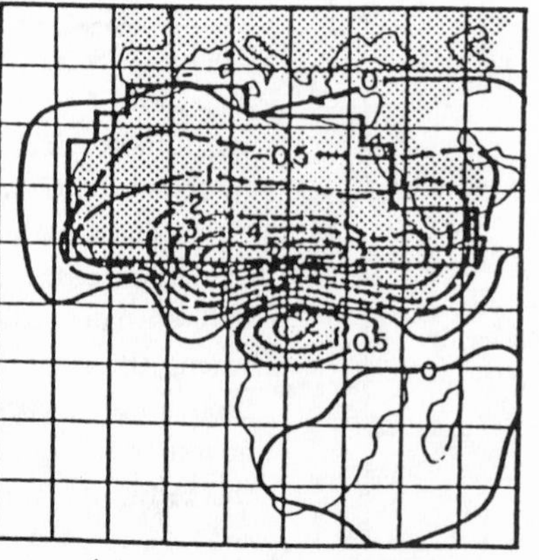

less than the increase in evapotranspiration, while between 12 and 15°N, where the evapotranspiration increases by only about 1.5 mm/day, the precipitation goes up by about 5.5 mm/day. On the other hand, at the equator, where there is an increase of evapotranspiration of 0.7 mm/day, the precipitation *decreases* by 3.5 mm/day. It is obvious, therefore, that there are large changes in the water vapor transport convergence.

Fig. 6.24 shows the circulation in the meridional plane, $(v\mathbf{j}+w\mathbf{k})$, averaged between 15 and 30°E (where v is the northward component of the horizontal velocity, w is the vertical velocity, $\mathbf{j}$ is unit horizontal vector directed northward, and $\mathbf{k}$ is unit vertical vector directed upward). Although the eastward component of the horizontal velocity does not appear here, its divergence, $\partial u/\partial x$, enters into the calculation of the vertical velocity, w.

We see that at the equator there is a change from upward motion to downward motion in the free atmosphere, which must be accompanied by a change from horizontal velocity convergence to horizontal velocity divergence in the boundary layer. It therefore is the decrease in the boundary layer water vapor transport convergence $(-\overline{\nabla \cdot q\mathbf{v}} \approx -\overline{q\nabla \cdot \mathbf{v}})$ which makes the precipitation decrease by 3.5 mm/day (from 5 to 1.5 mm/day).

Between 12 and 15°N, on the other hand, weak ascending motion changes to very strong ascending motion; and the accompanying large increase in the boundary layer water vapor transport convergence, added to the small increase in evapotranspiration, makes the precipitation increase by 5.5 mm/day (from 0.7 to 6.2 mm/day).

Fig. 6.23. Change in evapotranspiration (broken line) and in precipitation (solid line), averaged between 15 and 30°E, in the hybrid experiment of Manabe (1975).

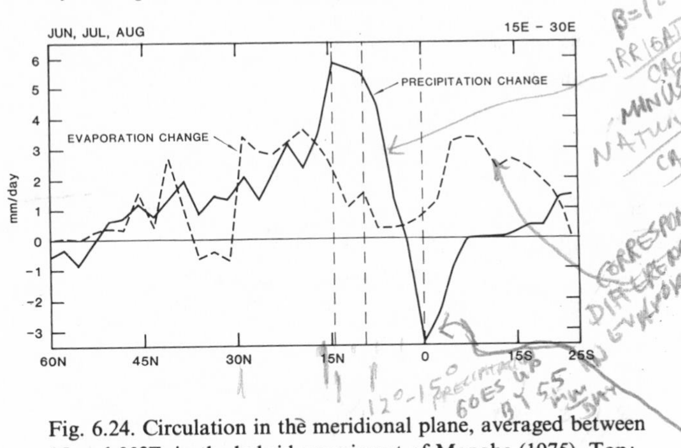

Fig. 6.24. Circulation in the meridional plane, averaged between 15 and 30°E, in the hybrid experiment of Manabe (1975). Top: interactive soil moisture case. Bottom: moist soil case. (Figure by personal communication.)

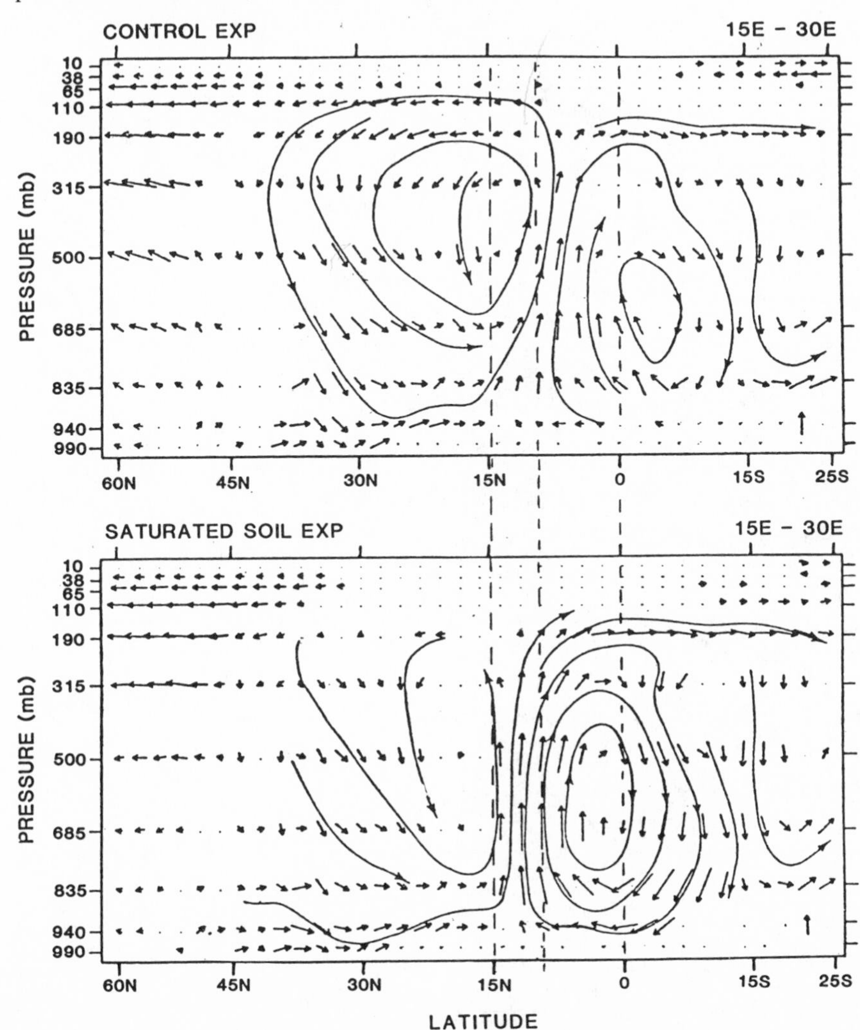

6.5.1.2

Kurbatkin, Manabe and Hahn (1979)

The irrigation experiment (6.5.1.1) (which makes $\beta = 1$ everywhere) is physically realizable in the real world. But it is not so easy to make $\beta = 0$ everywhere in the real world. If the rainfall were to be very constant in time, so that the surface of the earth was always wet, then it would be very difficult (although possible) to prevent evaporation. But where rainfall occurs in an intermittent way, and most of the water infiltrates to a depth of more than a few centimeters, then the removal of the vegetation would stop the transpiration and, thereby, would greatly reduce the transfer of water vapor to the air.

Kurbatkin, Manabe & Hahn (1979) made a hybrid experiment, in which a simulation with $\beta = 0$ everywhere was compared with a simulation with interactive soil moisture. For this experiment they used the M-21 version of the GFDL spectral model (Manabe, *et al.*, 1979), with moist-convective adjustment, prognostic clouds, and a prescribed albedo that follows Posey & Clapp (1964). In the interactive soil moisture case, the soil moisture and evapotranspiration were calculated with the equations given at the beginning of Section 6.4, with $W^* = 150$ mm, and $k = 0.75$.

The integration in the interactive case was over a period of two years and eight months. The results that are shown here are averages for the months of July and August at the end of the integration period. The non-interactive case, with $\beta = 0$, was initialized from the interactive case at the beginning of the last June and run until the end of August.

The top panel in Fig. 6.25 shows the simulated precipitation in the interactive soil moisture case, the center panel shows the precipitation in the no-evapotranspiration case, and the bottom panel shows the difference between the two (no-evapotranspiration case minus interactive case). With no land-surface evapotranspiration, there is less precipitation over the continents and also less precipitation over most of the oceans, with the largest decreases, of up to 5 mm/day, over India, the north Indian Ocean and over the western part of the tropical North Pacific. Only over the

Fig. 6.25. Precipitation (cm/day) averaged for July and August, in experiment of Kurbatkin *et al.* (1979). Top: interactive soil moisture case. Middle: no land-surface evapotranspiration case (contours 0.05, 0.1, 0.2, 0.5, 1.0, 5.0 cm/day. Dotted shading: precip. < 0.1 cm/day; ruled shading: precip. > 0.5 cm/day): Bottom: no-evapotranspiration case minus interactive case (shaded area is negative).

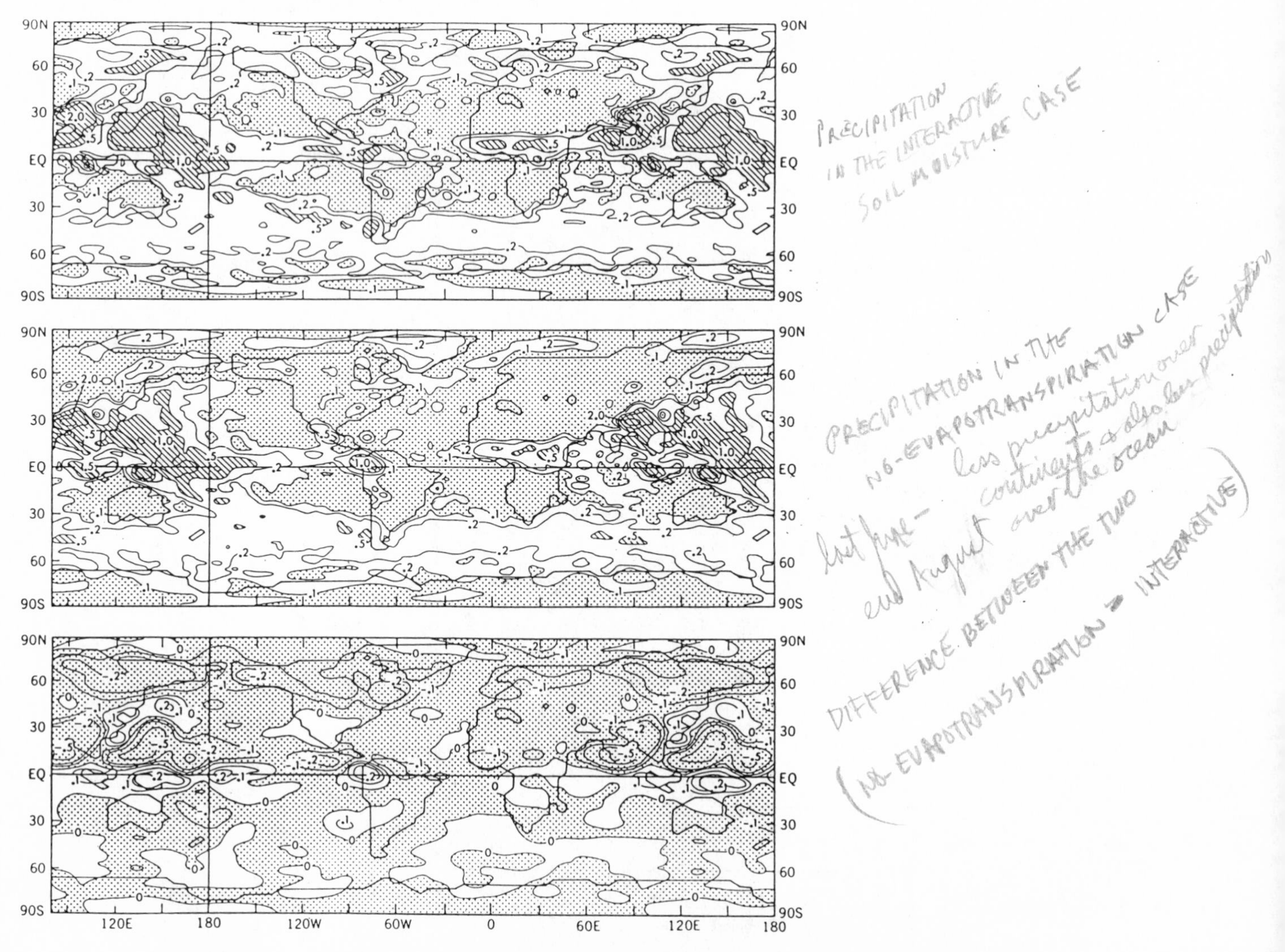

mid-latitude east coast of Asia and the adjacent Pacific, and in some longitudes along the equator (but not over the Atlantic Ocean sector), is the precipitation larger in the $\beta = 0$ case, and by as much as 2 mm/day.

Fig. 6.26 shows the sea-level pressure fields. We see that in the no-evapotranspiration case the pressure is lower over most of the continental areas. But over north-central Asia, over the northeast Atlantic, and especially over the extratropical central and western North and South Pacific Oceans, the pressures are higher in the no-evapotranspiration case. Although only the $\beta = 0$ case has the same evapotranspiration boundary condition that was used in the experiment with the GLAS model (in 6.3.1.1 above), there are many correspondences, as can be seen by comparing the bottom panels of Fig. 6.5 and Fig. 6.26, especially over the central and western North and South Pacific and over the South Atlantic.

The authors show (Kurbatkin *et al.*, 1979, Fig. 4) that in the $\beta = 0$ case there is, on the average, an increase in the large-scale ascending motion in the free atmosphere over the continents; and they remark that in spite of this increased ascending motion there is less cloudiness over the continents, and therefore a greater absorption of solar radiation by the ground and a higher surface temperature than in the interactive soil moisture case. The decrease in the precipitation over the oceans, they point out, is produced by the increase in descending motion (or weakening of ascending motion) over the oceans.

6.6 **Summary and conclusions**

All of the experiments show that the model simulated climates are sensitive to the land-surface boundary conditions which affect evapotranspiration. When soil moisture availability or surface albedo are changed regionally (or globally), changes in the precipitation, the temperature and the motion field of the atmosphere take place over the corresponding region (or over the globe), which are clearly above the level of the natural variability of the model simulated climates (those which are caused by the shear-flow instabilities of the atmosphere).

(Whether a change in the boundary condition of a given region has a significant effect on the climate of some distant region is not known. Such atmospheric 'teleconnections' are not self-evident in these experiments: and, like other kinds of forcing from

Fig. 6.26. Sea-level pressure (mb), averaged for July and August, in experiment of Kurbatkin *et al.* (1979). Top: interactive soil moisture case. Middle: no land-surface evapotranspiration case. Bottom: no-evapotranspiration case minus interactive case (shaded area is negative).

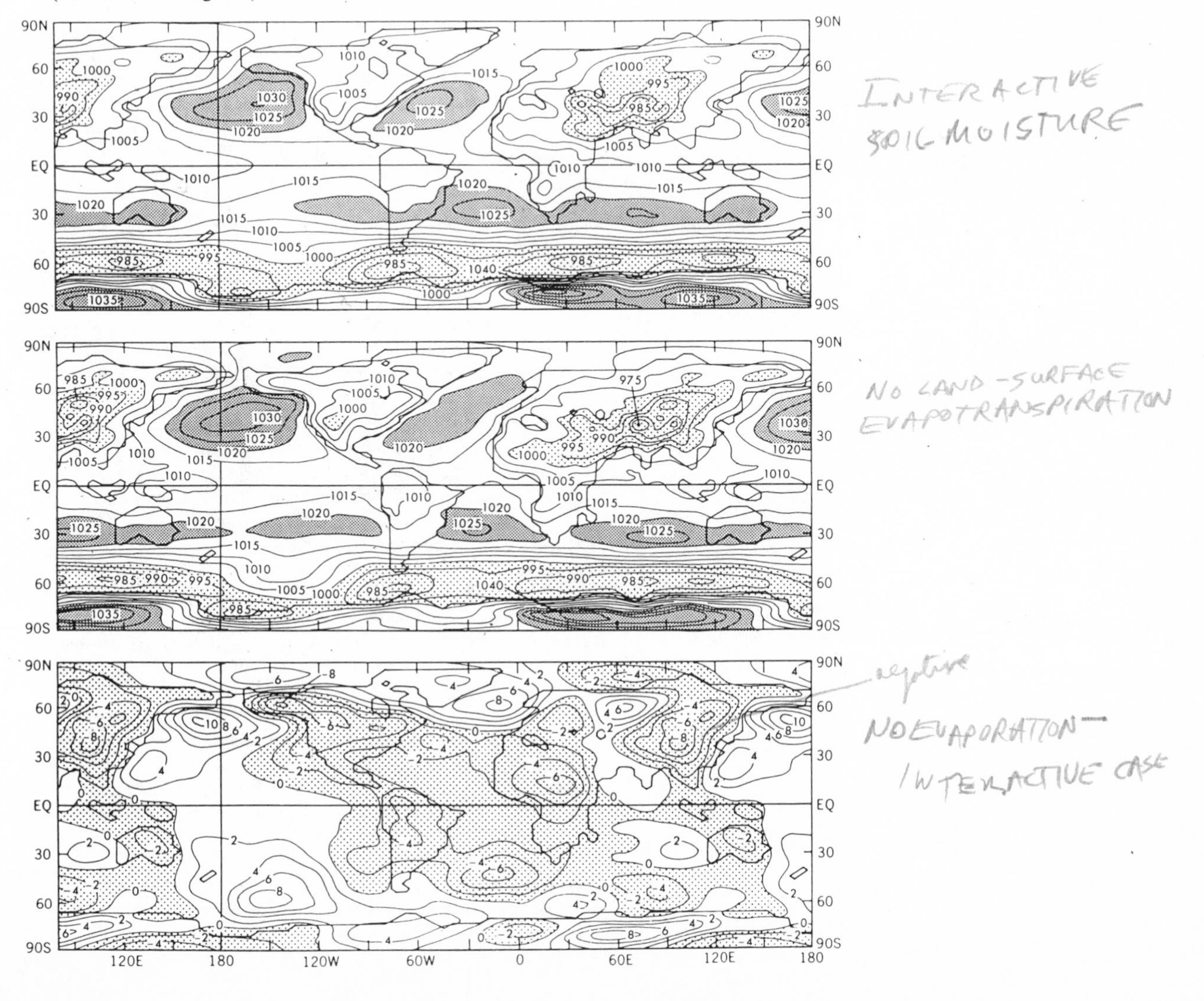

a distance, they require some kind of statistical analysis to separate the signal from the background noise.)

Not only are the regional (local) influences large and consistent from one experiment to another; they are also easily understood, in physical terms, after the analysis has untangled the non-linear interactions. Thus we find, for example, that under some circumstances an increase in the surface albedo reduces the cloudiness and, in that way, increases (not decreases) the ground temperature. Or, as another example, cutting off the evapotranspiration over Asia increases (not decreases) the Indian monsoon rainfall, because it changes the orientation, with respect to the Himalayan mountain barrier, of the moist boundary layer air stream from the ocean. It is this kind of behavior, easily understood after the fact, which is very difficult to anticipate beforehand.

The magnitudes of the changes in climate, which are produced by modifying the soil moisture availability or the surface albedo, as shown in these experiments, are about as large as the changes produced by the seasonal change in the declination of the sun (and, the author hazards the guess, larger than the changes produced by the observed seasonal changes in ocean-surface temperatures and extent of the sea-ice). It is very likely, therefore, that the land-surface evapotranspiration process, whose time scale depends on the magnitude of the soil moisture storage relative to the difference between the precipitation and evapotranspiration rates, is the most important boundary process that can produce anomalies in the time-averaged state of the atmosphere (changes in climate) on the monthly, seasonal and annual time scales.

A shortcoming of all existing general circulation models is that they calculate the potential evapotranspiration with a formulation which is appropriate for evaporation from an open water bucket, but not from a vegetated surface (and, especially, not from tall (forest) vegetation). (See, for example, Shuttleworth & Calder, 1979; and Sellers & Lockwood, 1981.) But, hopefully, this will be corrected and future models will take into account the vegetation influence on the water and energy transfers to the atmosphere in a realistic way.

Acknowledgement

The author thanks the several scientists who provided unpublished or supplementary information about their experiments; Dr J. Shukla and Dr Y. Sud who made helpful comments on the review; and Ms J. Reckley for her extraordinary patience and kindness in the typing, and retyping, of the manuscript.

Since this article was written, four additional papers have appeared, all of which show that the atmosphere is sensitive to the land-surface evapotranspiration.

Rind, D., 1982. 'The influence of ground moisture conditions in North America on summer climate as modeled in the GISS GCM'. *Mon. Wea. Rev.*, **110**, 1487–94.

Rowntree, P. R. & J. A. Bolton, 1983. 'Simulation of the atmospheric response to soil moisture anomalies over Europe'. *Quart. J. R. Met. Soc.*, **109**, 501–26.

Sud, Y. C. & M. Fennessy, 1982. 'A study of the influence of surface albedo on July circulation in semi-arid regions using the GLAS GCM'. *J. Climatology*, **2**, 105–25.

Yeh, T. C., R. T. Wetherald & S. Manabe, 1984. 'The effect of soil moisture on the short-term climate and hydrology change – a numerical experiment'. *Mon. Wea. Rev.* (in press).

References

Arakawa, A. & V. R. Lamb (1977). 'Computational design of the basic dynamical processes of the UCLA general circulation model'. *In: Methods in Computational Physics*, Vol. 17, pp. 173–265, Academic Press, N.Y.

Arakawa, A. & M. J. Suarez (1983). 'Vertical differencing of the primitive equations in sigma coordinates'. *Mon. Wea. Rev.*, **111**, 34–45.

Baumgartner, A. & E. Reichel (1975). *The World Water Balance: Mean Annual Global, Continental and Maritime Precipitation, Evaporation and Runoff.* Elsevier Publishing Co., Amsterdam, 179 pp. and plates.

Benton, G. S., M. A. Estoque & J. Dominitz (1953). *An evaluation of the Water Vapor Balance of the North American Continent.* The Johns Hopkins University Dept. Civil Engineering. Scientific Report No. 1, July 1953, 101 pp.

Budyko, M. I. (ed.) (1963). *Atlas of the Heat Balance of the Earth.* With N. A. Efimova: Maps of the radiation balance for the continents. Plates 14–26. Main Geophysical Observatory, Leningrad, 69 pl.

Carson, D. J. (1981). 'Current parameterization of land surface processes in atmospheric general circulation models. *In: Proceedings of the JSC Study Conference on Land-Surface Processes in Atmospheric General Circulation Models*, Greenbelt, USA, 5–10 January 1981. WMO, Geneva.

Carson, D. J. & A. B. Sangster (1981). 'The influence of land-surface albedo and soil moisture on general circulation model simulations'. GARP/WCRP: *Research Activities in Atmospheric and Oceanic Modelling*. (ed. I. D. Rutherford). *Numerical Experimentation Programme*, Report No. *2*, pp. 5.14–5.21.

Charney, J. G. (1975). 'Dynamics of deserts and drought in the Sahel'. *Quart. J. R. Met. Soc.*, **101**, 193–202.

Charney, J. G. (1975*a*). 'Drought, a biophysical feedback mechanism.' *Proceedings of Conference on Weather and Food, Endicott House, Mass. Inst. Tech.* Cambridge, Mass., 9–11 May, 1975, 10 pp.

Charney, J. G., W. J. Quirk, S. H. Chow & J. Kornfield (1977). 'A comparative study of the effects of albedo change on drought in semi-arid regions'. *J. Atmos. Sci.*, **34**, 1366–85.

Chervin, R. M. (1979) 'Response of the NCAR general circulation model to changed land surface albedo'. *Report of the JOC Study Conference on Climate Models: Performance, Intercomparison and Sensitivity Studies*, Washington, DC, 3–7 April, 1978. GARP Publ. Series, No. *22*, Vol. 1, pp. 563–81.

Chervin, R. M. & S. H. Schneider (1976). 'On determining the statistical significance of climate experiments with general circulation models'. *J. Atmos. Sci.*, **33**, 405–12.

Corby, G. A., A. Gilchrist & P. R. Rowntree (1977). 'The UK Meteorological Office 5-layer general circulation model. *In: Methods in Computational Physics*, Vol. 17, pp. 67–110.

Denmead, O. T. & R. H. Shaw (1962). 'Availability of soil water to plants as affected by soil moisture content and meteorological conditions'. *Agron. J.*, **54**, 385–439.

Gadd, A. J. & J. F. Keers (1970). 'Surface exchanges of sensible and latent heat in a 10-level model atmosphere. *Quart. J. R. Met. Soc.*, **96**, 297–308.

Korzun, V. I. (ed.) (1978). 'World water balance and water resources of Earth'. *Report of the USSR Committee for the International Hydrological Decade. Studies and Reports in Hydrology*, Vol. 25. Unesco Press, Paris, 663 pp. and plates.

Kurbatkin, G. P., S. Manabe & D. G. Hahn (1979). 'The moisture content of the continents and the intensity of summer monsoon circulation'. *Meteorologiya i Gidrologiya*, **11**, 5–11.

Manabe, S. (1975). 'A study of the interaction between the hydrological cycle and climate using a mathematical model of the atmosphere'. *Proceedings of Conference on Weather and Food, Endicott House, Mass. Inst. Tech.*, Cambridge, Mass., 9–11 May 1975, 10 pp. (and additional figure by personal communication).

Manabe, S., D. G. Hahn & J. L. Holloway, Jr (1979). 'Climate simulations with GFDL spectral models of the atmosphere: effect of spectral truncation'. *Report of the JOC Study Conference on Climate Models: Performance, Inter-comparison and Sensitivity Studies*, Washington, DC, 3–7 April 1978. GARP Publ. Ser., No. 22, Vol. 1, pp. 41–94.

Miyakoda, K., G. D. Hembree & R. F. Strickler (1979). 'Cumulative results of extended forecast experiments. II: Model performance for summer cases'. *Mon. Wea. Rev.*, **107**, 395–420.

Miyakoda, K., J. Smagorinsky, R. F. Strickler & G. D. Hebree (1969). 'Experimental extended predictions with a nine-level hemispheric model'. *Mon. Wea. Rev.*, **97**, 1–76.

Miyakoda, K. & R. F. Strikler (1981). 'Cumulative results of extended forecast experiment. Part III: Precipitation'. *Mon. Wea. Rev.*, **109**, 830–42.

Posey, J. W. & P. F. Clapp (1964). 'Global distribution of normal surface albedo'. *Geophisica Internacional*, **4** (1), 33–48.

Rasmusson, E. M. (1968). 'Atmospheric water vapor transport and the water balance of North America. Part II: Large-scale water balance investigations'. *Mon. Wea. Rev.*, **96**, 720–34.

Rowntree, P. R. & J. A. Bolton (1978). 'Experiments with soil moisture anomalies over Europe'. *The GARP Programme on Numerical Experimentation*: Research Activities in Atmospheric and Ocean Modelling (ed. R. Asselin). Report No. 18. WMO/ICSU, Geneva, August 1978, p. 63.

Rutter, A. J. (1975). 'The hydrological cycle in vegetation'. *In Vegetation and the Atmosphere, Vol. II, Case Studies* (ed. J. L. Monteith). Academic Press, New York, pp. 111–54.

Suarez, M. J., A. Arakawa & D. A. Randall (1983). 'The parameterization of the planetary boundary layer in the UCLA general circulation model: formulation and results'. *Mon. Wea. Rev.* (in press).

Schutz, C. & W. L. Gates (1972). *Global Climatic Data for Surface, 800 mb, 400 mb: July*. Rand Report. R-915-ARPA, The Rand Corporation, Santa Monica, Calif., 880 pp.

Sellers, P. & J. G. Lockwood (1981). 'A numerical simulation of the effects of changing vegetation type on surface hydroclimatology'. *Climatic Change*, 3, 121–36.

Shukla, J. & Y. Mintz (1981). 'Influence of land-surface evapotranspiration on the earth's climate'. (Presented at the *JSC Study Conference on Land-Surface Processes in Atmospheric General Circulation Models*, Greenbelt, USA, 5–10 January 1981.) Published in *Science*, **215**, 1498–501, 1982.

Shukla, J., D. Randall, D. Straus, Y. Sud & L. Marx (1981). 'Winter and Summer simulations with the GLAS climate model. *NASA Technical Memorandum 83866*, Goddard Space Flight Center, Greenbelt, Md, 282 pp.

Shuttleworth, W. J. & I. R. Calder (1979). 'Has the Priestley–Taylor equation any relevance to forest evaporation? *J. Appl. Meteor.*, **18**, 639–46.

Somerville, R. C. J., P. H. Stone, M. Halem, J. E. Hansen, J. S. Hogan, J. Tenenbaum (1974). 'The GISS model of the global atmosphere'. *J. Atmos. Sci.*, **31**, 84–117.

Stone, P. H., S. Chow & W. J. Quirk (1977). 'July climate and a comparison of the January and July climates simulated by the GISS general circulation model'. *Mon. Wea. Rev.*, **105**, 170–94.

Walker, J. M. & P. R. Rowntree, (1977). 'The effect of soil moisture on circulation and rainfall in a tropical model'. *Quart. J. R. Met. Soc.*, **103**, 29–46.

Washington, W. M. & D. L. Williamson (1977). 'A description of the NCAR global circulation models'. *Methods in Computational Physics*, **17**, 111–72.

WMO (1979). *Report of the JOC Study Conference on Climate Models: Performance, Intercomparison and Sensitivity Studies. Washington, D.C., 3–7 April 1978*. (ed. L. Gates) GARP Publication Series No. 22. WMO, Geneva.

S. I. Rasool
Laboratoire de Météorologie Dynamique
Ecole Normale Supérieure Paris 75231
and IBM Scientific Center
Paris 75116

On dynamics of deserts and climate

Abstract

A major part of the earth's surface is currently non-productive because it is dry and hot. However, the climate of many of these regions have changed both over the long-term (thousands of years) and short-term (decades) scales. The wet and dry episodes in the Sahel, the recent droughts in Australia and Brazil, the frequent breakdown of the Indian Monsoon are good examples of climatic fluctuations which have major human and social repercussions. The onset and persistence of these climatic events can have causes as varied as the anomalies in the global circulation, patterns of the sea surface temperature, changes in the heat balance of the surface, concentration of the atmospheric dust, etc. In this article, we examine the relative importance of these processes and suggest that, today, with the availability of the global measurements of these parameters from the satellites, a more coherent and focussed program of study can and should be formulated.

7.1
Introduction

Today, over one-quarter of the land surface is either desertic or semi-desertic, making a relatively large area of the earth unproductive for human needs (Fig. 7.1). However, during the period 5000 to 8000 years BP, when the global temperature of the earth was approximately 2°C higher, much of today's dry land, especially in Africa and in the Indus Valley, was quite vegetated. On the other hand, about 10000 years earlier, during the glaciation maximum, the climate of the earth was much more arid than today. Fig. 7.2 shows how the southern edge of the Sahara has moved in the last 20000 years. Although such large-scale changes in climate and in the extent of the deserts on earth appear to be a rule rather than an exception, the physics of the change is not well understood. These are very large changes and, as can be imagined, have played a very important role in the cultural evolution of north Africa and south Asia.

On a shorter time scale, the situation is even more intriguing. Fig.7.3 shows a composite of the levels of several lakes in Africa, the discharge of the Nile and the climate of Mauritania and Senegal. All curves indicate that, during the last 4000 years, there have been periods, lasting much longer than a century, when the boundary regions of the current deserts received above-normal rainfalls and therefore enjoyed a much wetter climate than today. On a still shorter time scale, over the last 50 years, the rainfall patterns in the Sahel, north west India, Brazil and Australia show periods of climate anomalies when droughts or monsoon breakdowns have occurred, persisting, in some cases, for as much as a decade (Fig. 7.4) (Hare, 1977; Nicholson, 1983; Moura & Shukla, 1981). Of these, the Sahel drought of 1972 seems to be still persisting (Lamb, 1982; Nicholson, 1983) and has been responsible for major economic and social upheavals in north Africa.

In this chapter, we will attempt to examine the currently formulated mechanisms for these changes in desert climate,

Fig. 7.1. Extent of arid and semi-arid regions on the earth. The arrows indicate the position of *cold* oceanic currents which contribute to the aridity downwind. (After Tannehill 1947.)

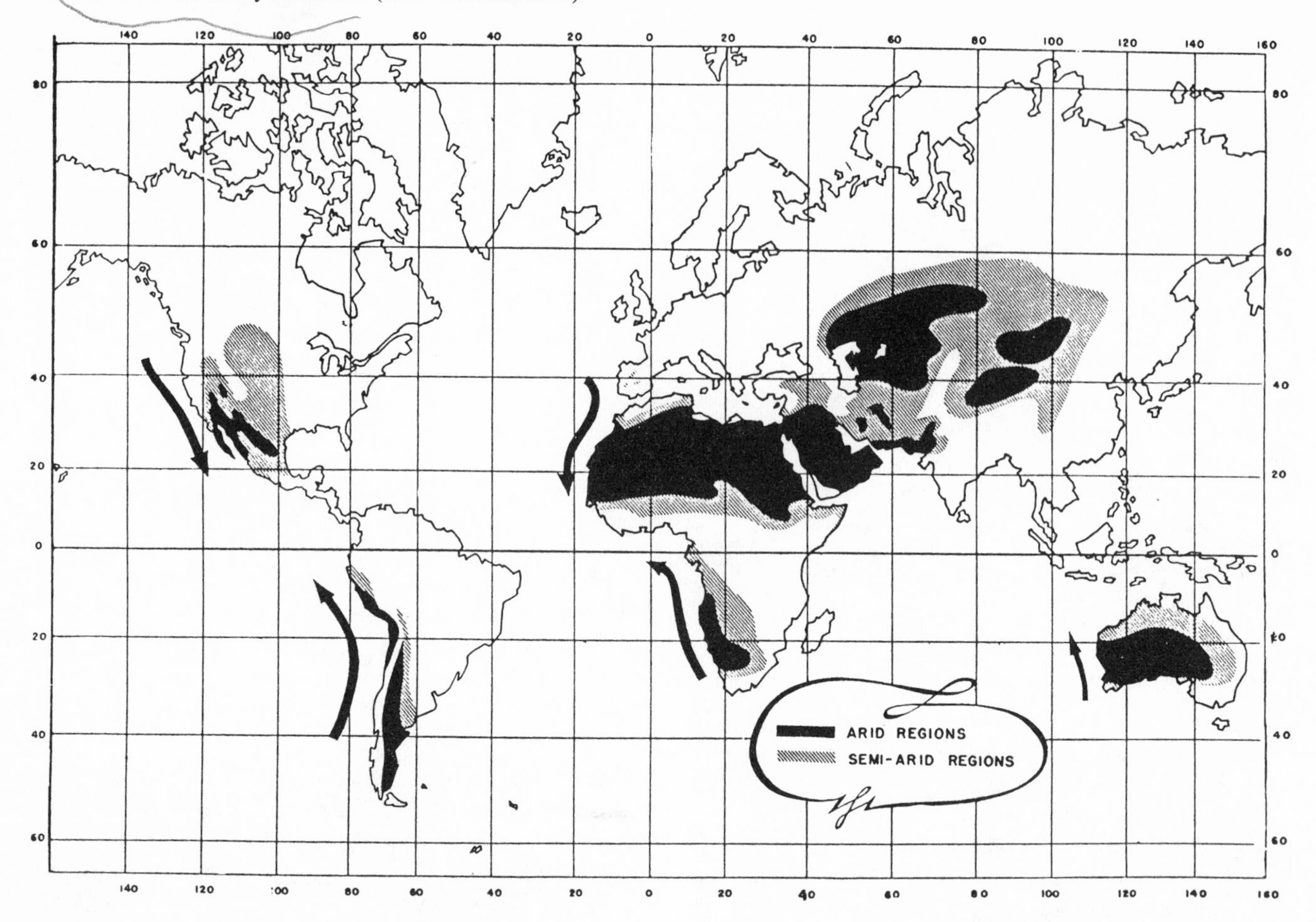

Fig. 7.2. Climate change in the Sahel. Isohytes (in mm/year) represent today's precipitation averages. Open triangles define the approximate southern limit of *active* modern dunes. Solid triangles show the southern limit of *fixed* Pleistocene dunes. Open squares mark today's Sahel–Sahara boundary on the basis of vegetation type. Solid circles mark the maximum Holocene extent of Lake Paleochad. The source areas of the major current and Holocene river systems are shaded. Also shown are the positions of Skylab and Landsat images used to study the dunes and water courses. (From Talbot 1980.)

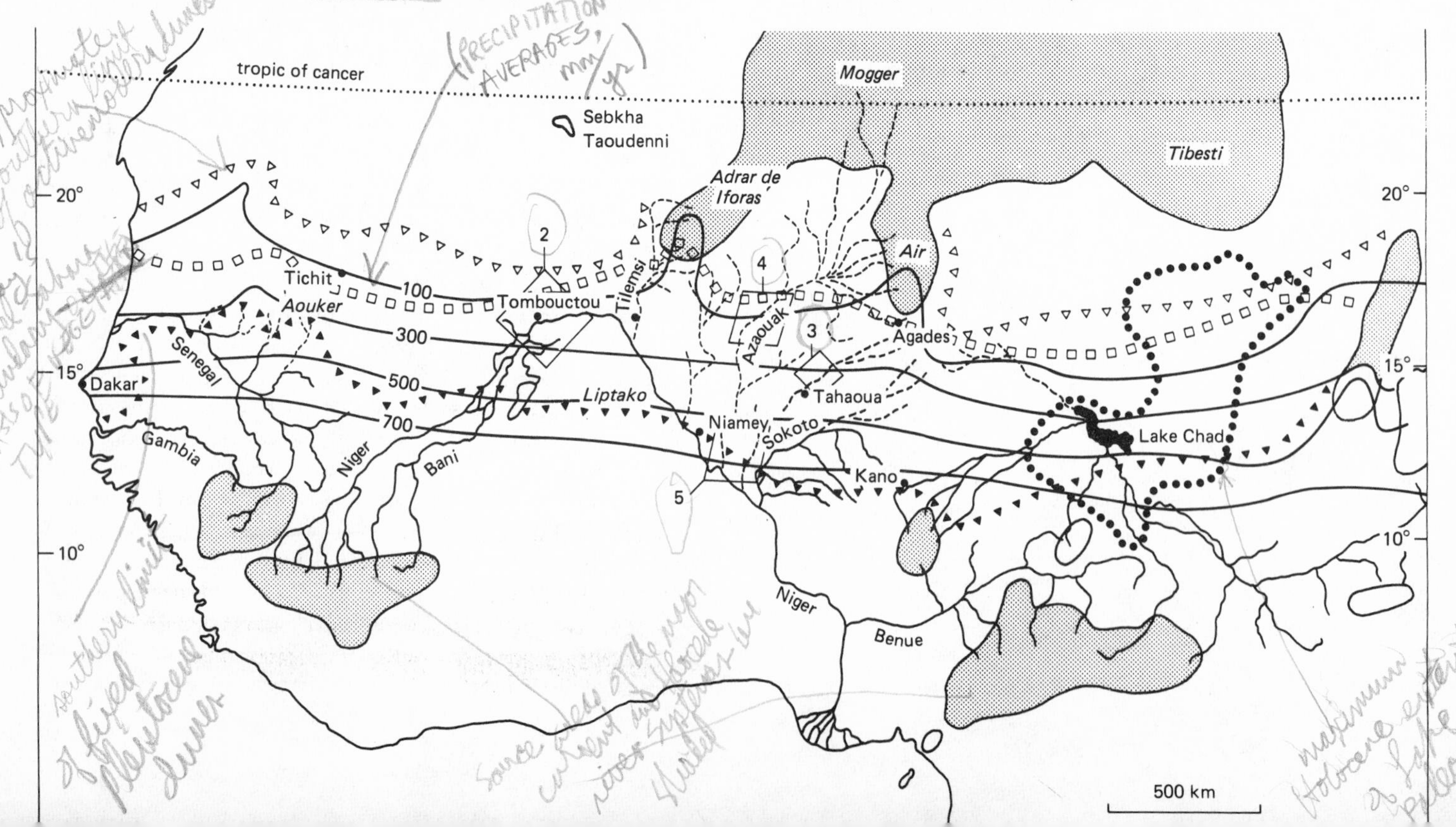

Fig. 7.3. Fluctuation of lake levels and climate in Africa during the last 4000 years. (From Nicholson 1980.)

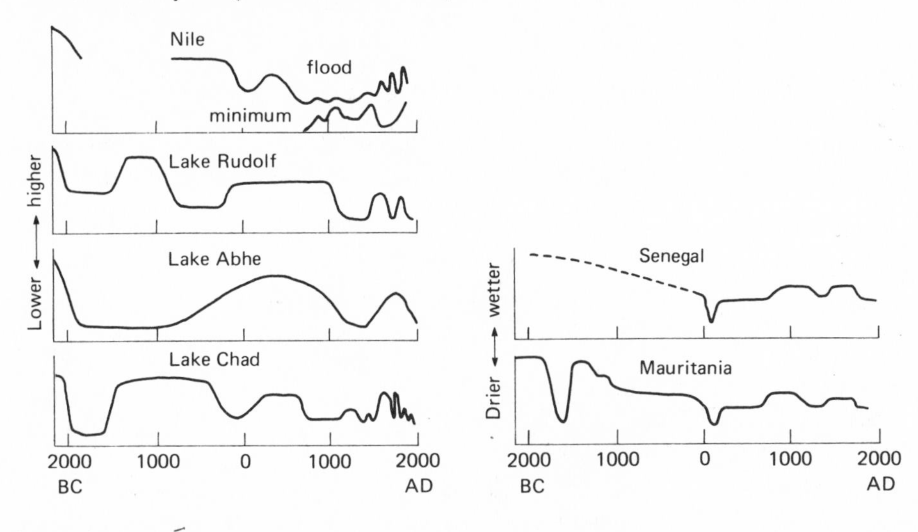

emphasize the discrepancies between theory and observation and conclude by summarizing what research activities are needed in order to improve our understanding of the dynamics of the desert climate so that, eventually, we will be able to predict such changes in advance.

Fig. 7.4. Precipitation history of the last 40 years at various arid stations around the world. (After Nicholson 1981; Hare 1977; Moura & Shukla 1981.)

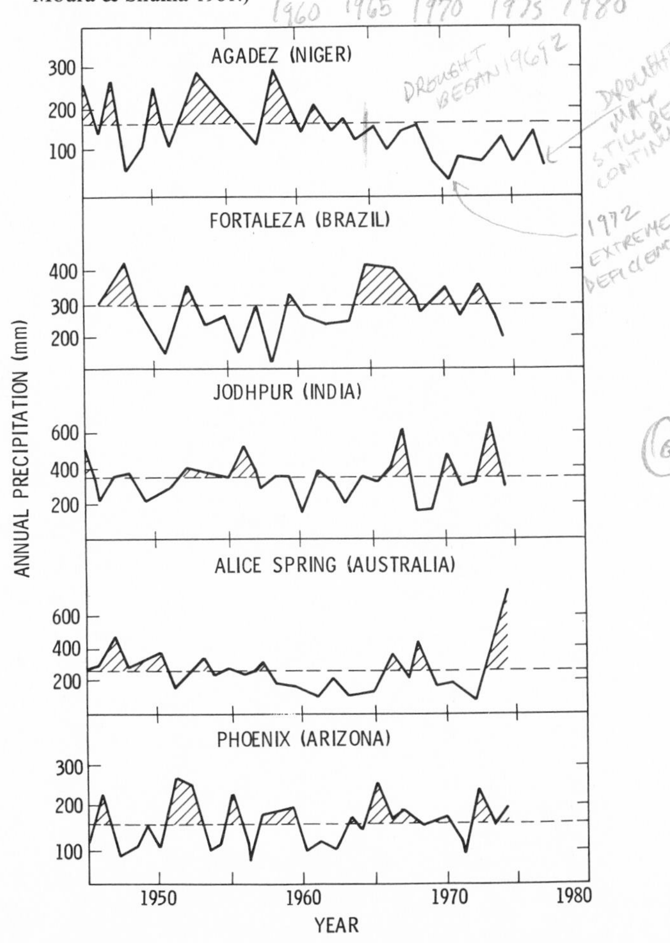

Deserts exist on earth for either one or a combination of the following reasons, as well summarized by Hare (1977):

Down leg of the Hadley cell. Fig. 7.5(*a*) shows the sub-tropical latitudes where widespread, persistent subsidence occurs because they correspond to the down leg of the Hadley cell. These subsidence zones move in latitude with the seasons. The latitudes with maximum variability are of course those which are semi-desertic, while those with minimum variability have subsidence throughout the years and are therefore perpetual deserts. Superposed on this meridional circulation is the zonal Walker circulation (Fig. 7.5(*b*)) which complicates the dynamics of the tropical and sub-tropical atmosphere but has to be taken into account in any discussion of the variability of climate over the sub-desertic regions of the world.

Lee of mountains. At several locations around the world, western North America, western southern Argentina and parts of inner Asia, dryness is caused by mountain barriers which make air descend on the lee side of the mountains (Fig. 7.1).

Absence of water vapor and/or of rain-inducing disturbances. Desertic conditions may also be caused either by the absence of water vapor in the air or by the absence of rain-inducing disturbances or both. The best examples of these circumstances are the down wind sides of cold oceanic currents (Fig. 7.1), inner continental regions which are remote from sources of water vapor, and central North America which is often humid but rainless because of the absence of cyclonic activity.

Persistent changes in the efficiency of these mechanisms can bring about large-scale variations in the rainfall patterns over the arid regions. In this chapter, we will examine the mechanisms of climatic changes over two different time scales.

(1) long term, approximately 20000 years;
(2) short term, approximately 100 years.

The first time scale addresses the slow background changes in the aridity of the earth, probably driven by changes in global energetics and the circulation of the atmosphere. Understanding the physics of these changes is imperative for a better understanding of the dynamics of the global climate. The second time scale addresses the habitability of the earth within a human life time. If we really understand the mechanism of these short-term fluctuations, it may give us a handle on the predictability of the climate.

7.2

Long-term changes of aridity

Data on the paleoclimates of the arid and semi-arid regions have been collected and analyzed by many authors. Extensive summaries exist (Fairbridge, 1976; Nicholson & Flohn, 1980; Williams & Faure, 1980; Sarnthein 1978). The sources of these data are varied, such as lake levels, stratigraphic and paleoecological studies, sub-fossil pollen legacy, extent of the fixed dune field and even the sediments in the neighbouring oceans and concentration of dust in the ice cores in the Antartic. From these studies a few salient points can be extracted:

The extent of aridity on the earth was high during the glacial of 18000 years BP. The monsoons which supply moisture to todays' semi-arid land of north Africa and India were weak. There is also some indirect evidence that the average wind velocities were higher (Sarnthein, 1981)

The situation reversed during the period between 10000 BP and 5000 BP, when monsoon rains were strong, aridity was less extensive and global wind patterns were calmer.

Today's aridity conditions appear to be more like those of the glacial period than those of the interglacial period of 5000 BP.

How can one explain these slow but important long-term changes? One of the most widely discussed mechanisms (first proposed by Milankovitch) for global climate change for these time scales is that of the changing orbital parameters of the earth, i.e. obliquity, perhilion and eccentricity, which combine to produce changes in the relative distribution of solar radiation as a function of latitude (Hays *et al.*, 1976). Many model calculations have attempted to simulate the climate of the glacial period, 18000 BP, with varying degree of success (Gates, 1976; Mason, 1976; Manabe & Hahn, 1977). To date, at least two mechanisms have been invoked to explain these extreme excursions in global aridity. First, the general circulation models (GCMs), when run with the boundary conditions specified from studies of geological records of 18000 BP, such as lower ocean surface temperatures, higher latitudinal temperature gradient, extended polar ice caps, suggest a very much reduced convective activity. Because most of the rainfall in the GCM, for latitudes between 30°N and 20°S, is of convective origin, the reduction in precipitation is maximum in the tropics and

Fig. 7.5. (*a*) Mean atmospheric motion in the meridional plane. (Adapted by Hare 1977.) Latitudes of average subsidence are indicated by heavy horizontal bars. Units are 10^{12} g s^{-1} mass transport. (*b*) Schematic of the Walker Circulation (From Newell 1979.)

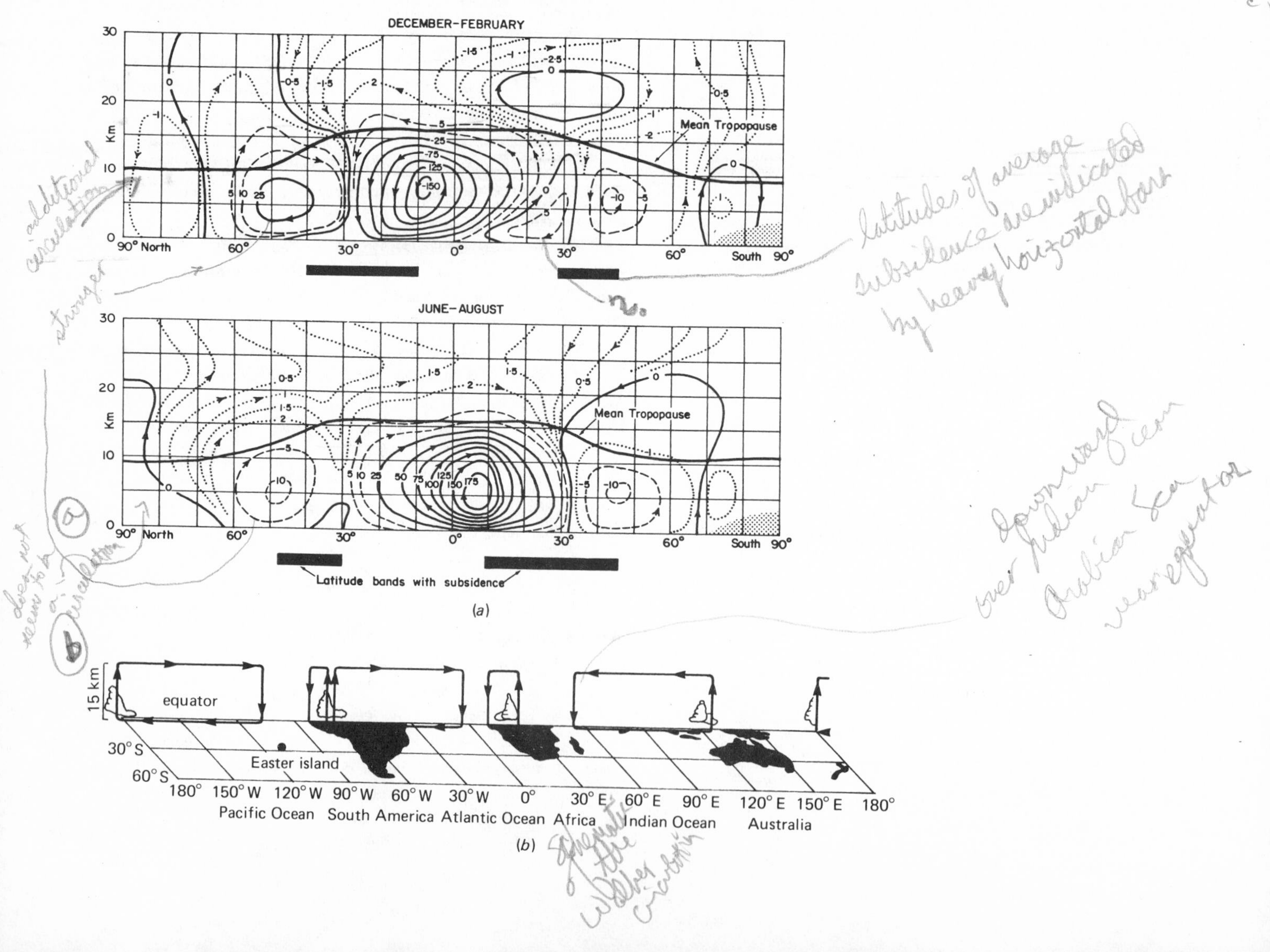

sub-tropics, explaining the increase in aridity during the glacial maximum. Another way of looking at the same problem, in rather simple terms, is through consideration of baroclinic stability of the atmosphere, in which the meridional extent of the tropical Hadley Cell and the latitude of the intertropical convergence zone (ITCZ) are determined principally by the meridional and vertical temperature gradients of the atmosphere. A simple empirical relationship, originally discussed by Smagorinsky (1963) and applied to the problem of climatic shifts by Korff & Flohn (1969), Bryson (1973) and recently by Greenhut (1981), implies that

$$\tan \phi = \frac{h}{a}\frac{\xi}{\eta}, \qquad (7.1)$$

where

ϕ is the latitude of the sub-tropical high,
h is the scale height of the atmosphere $\sim$ 8 km,
a is the radius of the earth, 6370 km,
ξ and η are related to vertical lapse rate and meridional temperature gradient.

In fact, Flohn (1980) has published data (Fig. 7.6) which shows an $\sim$ 10° change in latitude of the sub-tropical high for a 15 °C change in meridional temperature gradient. It has also been suggested that the movement of the latitude of the sub-tropical high is positively correlated with the movement of ITCZ with an amplification factor of $\sim$ 2.5 (Bryson, 1973; Beer *et al.*, 1977). It therefore follows that if the temperature gradient between the equator and the high latitudes increases (as during the period of the glacial) the sub-tropical high moves southwards but the ITCZ moves southwards even further. This therefore broadens the zones of aridity around the world.

A second approach is that of Kutzbach (1981) in which he compares the modern conditions of solar insolation with those of 9000 years BP, and obtains the following results:

- The increased solar radiation of the period June to August 9000 years BP (approximately 6%) caused an increased heating of land surface relative to surrounding oceans.
- Surface temperature was 5 K higher in July 9000 years BP than in the modern simulation over central Asia and 2 K higher for the mid-latitudinal average for land relative to oceans.
- The geopotential height at the 900 mb level decreased by close to 100 m in mid-latitudes, more over land than over oceans, producing a stronger flow of air from ocean to land.
- Stronger monsoon (surface winds increasing by as much as 50%) produced from 10 to 50% more rain over Africa and south Asia than falls today.
- Over the entire northern hemisphere land area the model produces evidence of 8% more precipitation but the increase over land was compensated by a decrease over the ocean and therefore the global average rate was the same as that of today.

This experiment was one of the first simulations of the desertic climates for changes in the distribution of solar flux over land and oceans. Although it is not immediately clear whether the postulated increase in the strength of monsoons can explain the observed enlargements of paleolakes in Africa and Asia (the waters in some of these lakes and rivers originate in distant tropics), it does provide an important insight into at least one mechanism of climate change. It is possible that this mechanism, along with the effects of other geophysical changes during the glacial and interglacial such as the changes in the extent of polar caps, latitudinal gradient of ocean temperatures and varying strength of the ocean currents, may together explain the large-scale changes in the extent of the global deserts from the very dry period of 18000 years BP, through the wet period of approximately 9000 BP, to the relatively dry conditions of today. As Kutzbach has pointed out himself, a number of very important problems still remain:

- Models have simulated either January or July conditions. It is possible that a full annual cycle simulation will give a drastically different result, especially because of phase lag in ocean–land temperature differences.
- Feedback of ocean–atmosphere–ice sheet has been ignored.
- Feedbacks involving land surface changes and possible changes in atmospheric CO_2 on these time scales have also to be considered.

In this context, the recent observations of ice cores from the Antartic are quite intriguing (Petit *et al.*, 1981). They indicate that during the last glacial period, the amount of dust in the atmosphere was greater by a factor of 20 than today (Fig. 7.7). Suggestions have

Fig. 7.6. Latitude of the subtropical anticyclone belt in all months plotted against the actual temperature difference (ΔT) between equator and poles, 300–700 mb. (From Flohn 1980.)

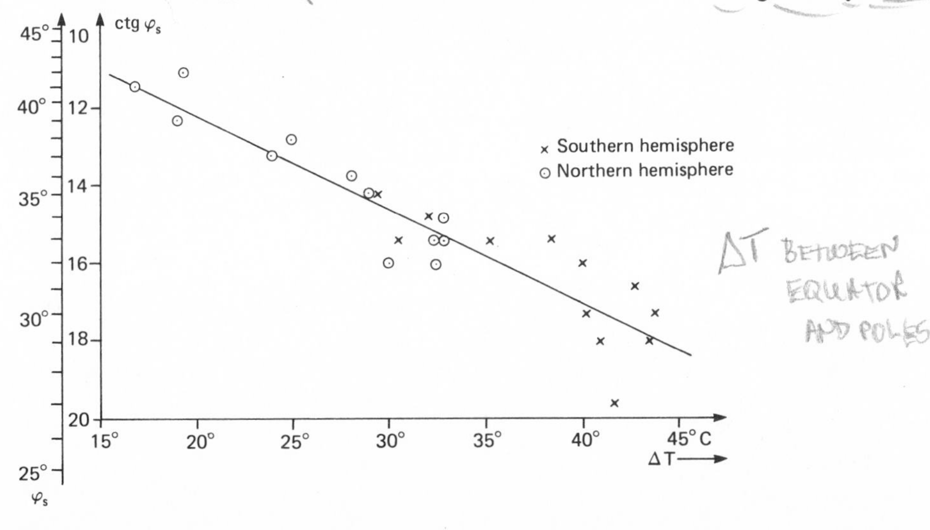

been made that the dust was carried to the pole by high winds prevalent at that time. Analysis of sediments in the central Atlantic also indicate that the Saharan dust was probably being carried further west 18000 years BP (Sarnthein, 1981). These measurements provide critical tests for models for glacial and interglacial climates relative to the nature of the general circulation of the atmosphere during periods of climatic extremes.

7.3 Short-term climatic fluctuations in arid lands

On the shorter time scale, ~ 50 years, Fig. 7.4 demonstrates the precipitation history of several typical arid stations around the world. (Hare, 1977; Nicholson, 1983). Although, at first glance no coherency of a global pattern is evident in this diagram, a number of salient features can be discerned:

Year-to-year rainfall is highly variable. The coefficient of variation (defined as $\sigma/\tilde{p}$ where σ is the standard deviation and $\tilde{p}$ is the mean annual precipitation) is > 25% in most dry lands and is as much as 40% in the deserts.

There is no regular pattern in the interannual variability except perhaps a very weak 2–3-year rhythm in some areas, while a 10- , 20- or 30-year recurrence interval occurs in the others.

In sub-Saharan Africa particularly, climatic anomalies seem to persist for several years and sometimes for as long as one full decade. It appears that the drought in the Sahel, which commenced in 1969, with extreme deficiency in rainfall recorded in 1972 and 1973, may still be continuing (Lamb, 1982; Nicholson, 1983).

Global connections in climatic anomalies have been noted. Breakdowns in the Indian monsoon have been positively correlated with the occurrences of El Nino in the eastern equatorial Pacific (Fig. 7.8 Rasmusson & Carpenter 1983) and the Southern Oscillation index seems to be a good measure of drought and rainfall index in Australia.

The drought in the Sahel zone appears to be continental in scale on either side of the equator (Nicholson, 1981), with some indication of positive correlation with the rainfall pattern in the Carribean and a negative correlation with the droughts in Brazil.

The rainfall patterns in central and western North America have been successfully correlated with large-scale temperature anomalies in the equatorial Pacific. Likewise, the central Atlantic temperatures seem to influence the precipitation over the Sahel, while the Arabian Sea temperatures and wind patterns control the intensity of the monsoon over India.

There are three different kinds of mechanisms which are invoked today for explaining the excessive fluctuations in the climate of dry lands.

7.3.1 *Global connections*

It now appears quite plausible that the global circulation of the atmosphere 'oscillates' in response to large-scale changes in oceanic conditions (Figs. 7.5(*b*) and 7.8). What was known to be the Southern Oscillation now appears to be a global phenomenon. Modeling of this phenomenon is difficult and is progressing very slowly because it involves the interaction of the atmosphere with the oceans and most models do not yet simulate this efficiently. On the other hand, the circumstantial evidence of the global connection of anomalous climatic patterns on a seasonal time scale is growing continuously. The year 1972 is a good example, when a number of major climatic anomalies occurred around the world. Fig. 7.9, published by Oguntoyinbo & Odingo (1979) summarizes the situation graphically. The droughts in the Sahel, Australia and China were concurrent with excessive rains north of the Sahara and Spain, failure of the wheat crop in the Soviet Union, a heavy ice pack in eastern Canada and a strong El Nino event in the eastern Pacific. To model the climate of 1972 and understand its anomalous behavior is a considerable challenge to climate research.

7.3.2 *Feedback mechanisms*

The other question that one would like to see answered is why, especially in sub-Saharan regions, the anomalous climate seems to persist for several years, sometimes as long as a decade. Here, the answers proposed so far all appear plausible but none has been really tested with actual data in the 'field'. All involve a positive feedback mechanism involving the energy balance at the surface and lower atmosphere, which causes the anomaly to persist.

Fig. 7.7. Analysis of Dome C (Antartic) ice core showing, on the left, ^{18}O isotope variation with core depth, which directly translates into temperature v. time B P; on the right, for the same core, the density of insoluble dust particles showing a factor of 20 increase in dust in the atmosphere during the glacial. (After Petit *et al.* 1981.)

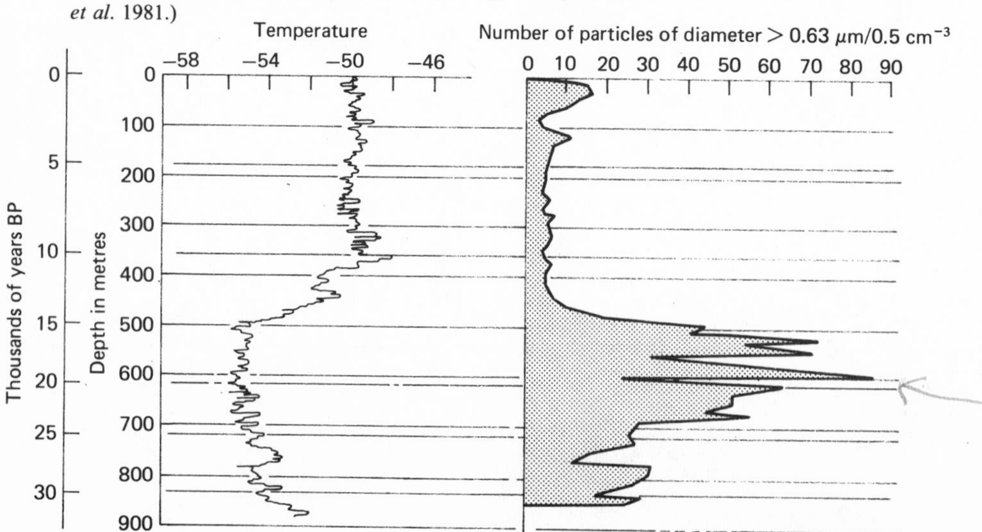

Fig. 7.8. Drought area index (from Bhalme & Mooney (1980)) and *below* – standardized summer monsoon precipitation anomalies (in terms of standard deviation SD) for India. Correlation with years of moderate or strong El Niño, indicated by shaded bars. (From Rasmusson & Carpenter 1983.)

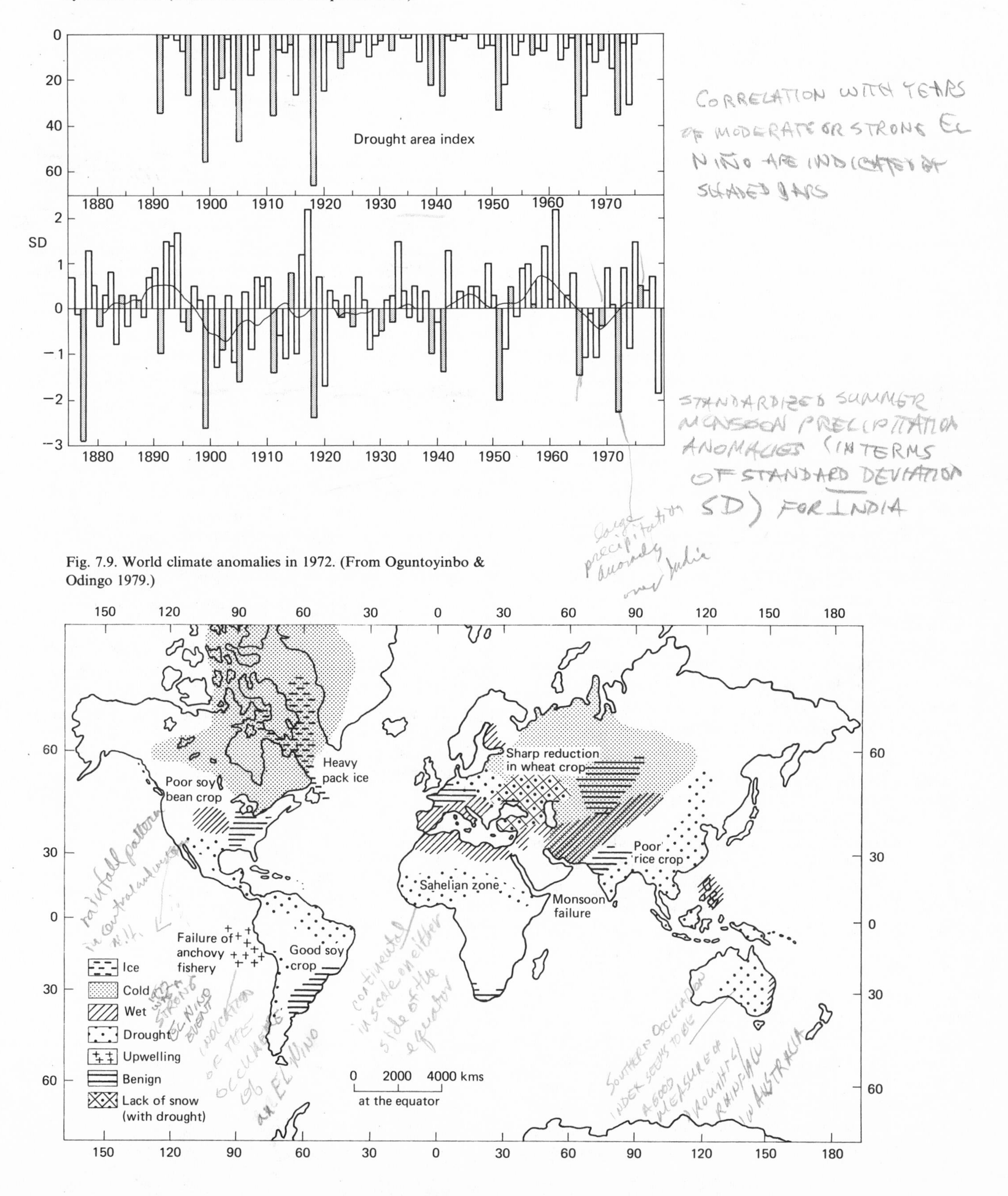

Fig. 7.9. World climate anomalies in 1972. (From Oguntoyinbo & Odingo 1979.)

The transfer of energy at the surface can be written as (see, for example, Bolle, 1982):

$$R = F_0(1-A) - R_T + R_A - G_0 = H + LE, \qquad (7.2)$$

where

F_0 is incoming solar flux, both direct and diffused;
A is the albedo of the surface;
R_T is the thermal flux emitted by the surface $= \sigma T_0^4$ (where T_0 is the ground temperature);
R_A is the thermal flux returned from the atmosphere to the surface;
G_0 is the energy conducted into the soil;
H is the energy removed from the surface by convection, also called sensible heat;
LE is the energy removed from the surface by evaporation of water vapor.

Over the arid regions, A is usually high (30–40%), but with some vegetation this could decrease to 20%; R_T is also high during the day but low at night; R_A is usually small but the values of H and LE fluctuate considerably between day and night, summer and winter and from year to year. For example, depending on whether the ground is wet or very dry, vegetated or bare, the term LE could be very high or very low. At the same time, depending on the gradient between the ground and the air, H could be high or low. In fact, H/LE, called the Bowen ratio, can vary in deserts from 0.1 to more than 20 and can therefore completely change the way the energy is exchanged at the surface. The feedback mechanisms for persistence of droughts in the semi-desertic regions discussed below invoke a change in one or more of the parameters in equation (7.2) at the beginning of the drought, leading to changes in other parameters and resulting in a lasting change in the energy balance at the surface.

7.3.2.1
Soil moisture and evapotranspiration

Evaporation of soil moisture impacts the climate in two ways: (1) it provides the water vapor to the atmosphere, which can eventually return as precipitation; and (2) it removes energy from the surface, thereby changing the energy balance and influencing the surface exchange processes, both of matter and energy.

In the tropics and sub-tropics, most of the rain is of convective origin rather than of large-scale motions as it is at higher latitudes. The source of water vapor in the atmosphere is therefore either the ground or the neighboring oceans. The feedback mechanism for prolonging a drought can therefore work in the following way, with evapotranspiration as the main driver. Once the drought starts, the vegetation begins to die and soil moisture diminishes. Both these circumstances combine to decrease the evapotranspiration considerably, resulting in a major decrease of convective rainfall. In a modeling experiment, Walker & Rowntree (1977) have shown that if the Sahara is initially assumed dry, no rainfall is developed, while if one assumes that the surface is wet, with 10 cm of moist soil, major depressions develop producing widespread rain. In another experiment with more extreme conditions, Shukla & Mintz (1981) ran one model with all land surface fully vegetated and wet and another with no vegetation and dry. Although the boundary conditions imposed were quite unrealistic, the models' results are very instructive (See Fig. 6.4). They found that in the wet case, Europe, Asia and North America receive high precipitation, while in the dry case they have almost no precipitation, with the exception of south east Asia, India, central Africa and the central Americas, where water vapor is transported from the oceans to produce heavy rains.

It is clear that what is needed are experiments using more-realistic changes in vegetation and soil moisture, which occur in desertic regions in the first years of drought and during the months preceding the summer rains. There are two reasons why no experiments of this kind have so far been performed: first, no reliable estimates of real changes in evapotranspiration are available for large areas like the Sahel and, second, the 'noise' in the current numerical models is such that the impact of anything less than a drastic change in the value of one parameter is not easily discernable. As discussed later in this paper, progress in both these areas is now being made. Soon it may be possible to obtain information on land surface changes from satellite observations and to use such information as input to coupled numerical models to study their impact on climate.

7.3.2.2
Surface albedo

One of the more widely discussed feedback hypotheses is that of Charney (1975) who suggested that excessive grazing by man in the sub-Sahara may increase the surface albedo by more than a factor of two, from 14% to 35%, thereby drastically changing the energy balance of the surface, and leading to a sinking motion resulting in drying and maintenance of desert conditions. Fig. 7.10 shows the results of the simulation experiment in which the amount of rainfall over the Sahel decreased by 50% because of a rather large assumed change in surface albedo. However, as pointed out by a number of authors since then, including Charney *et al.* (1976) and Mason (1979), the problem is much more complex than is implied by such a simple formulation; a more complex interactive calculation is necessary. When the ground is wet, evapotranspiration must be allowed to vary with the albedo and ground temperature; the resulting cloud amounts need also to be taken into account in the calculation of the new planetary albedo. Such an interactive

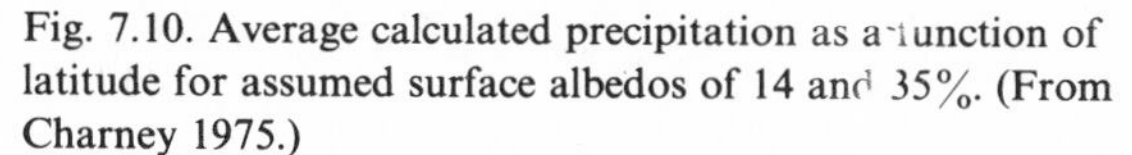

Fig. 7.10. Average calculated precipitation as a function of latitude for assumed surface albedos of 14 and 35%. (From Charney 1975.)

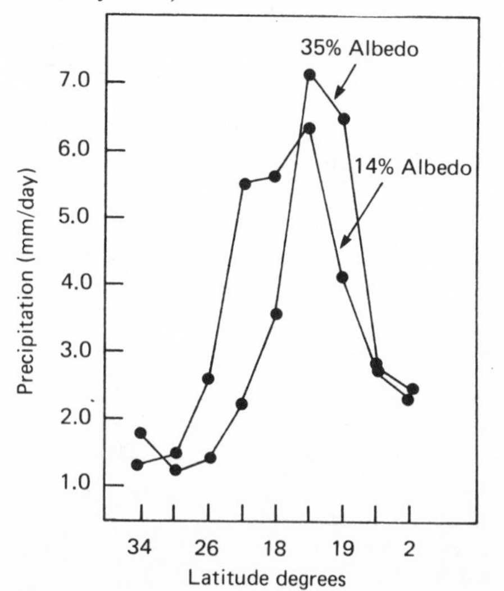

calculation is of course highly complex and consensus of opinion on the exact nature of the feedback mechanisms is yet to emerge.

Again, it is clear that before we proceed to evaluate these hypotheses with more sophisticated models, it is imperative that we determine the 'real' change in surface albedo that occurs during a long spell of drought. Is the change really as much as was assumed by Charney (14–35%) or are the changes less when measured over large areas such as the Sahel? In this context, a study performed by Courel & Habif (1982) and Rasool *et al.* (1982) is very relevant. Data on surface reflectivities from two different satellites, Landsat and Meteosat, were analyzed for the period between 1972 and 1980 and compared with earlier measurements made by Norton *et al.* (1979) by ATS over the Sahel. (Fig. 7.11)

Although the sampling of data was not as frequent and expansive as one would have liked, the results are quite different from what has been usually assumed for albedo values and their changes for the Sahara and sub-Saharan region. The results can be summarized as follows:

The surface albedo of Sahara as measured by Landsat and Meteosat and weighted for the entire solar spectrum is closer to 45% rather than 35% usually assumed by modelers (Rasool *et al.*, 1982).

The normal surface albedo of the Sahel is close to 20% instead of 14%.

Fig. 7.11. Time history of surface albedo over the Sahel between 1967 and 1979 as measured by ATS, Landsat and Meteosat: open triangles, ATS 3 (Norton *et al.* 1979); open squares, Landsat; open diamonds, Meteosat; filled symbols are for the wet season. (From Rasool *et al.* 1982.)

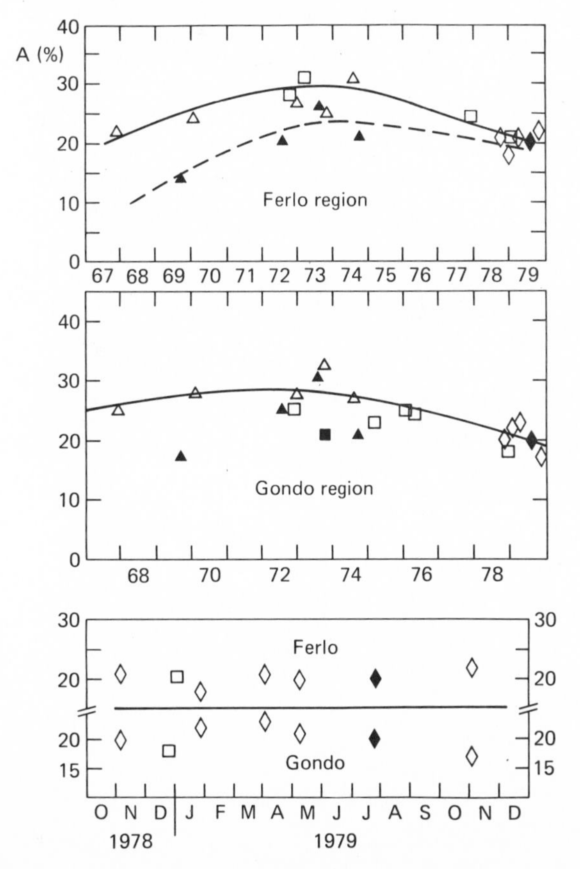

The changes related to drought are less than assumed by modelers (20–28%) rather than (14–35%).

In many parts of the Sahel zone, the surface albedo has decreased since 1973, implying that a runaway process of perpetuating the drought through an increase in albedo *did not occur*.

These results, if confirmed with more data and over other desertic regions, will have an important bearing on the modeling experiments. With higher albedo values over the land, the energy gradient between ocean and land will be different and, also, with relatively smaller changes in albedo, as observed during the drought, it is not clear that a change in albedo, on its own, can explain the long-term persistence of droughts. It is possible that a change in evapotranspiration aided by the decrease in vegetation cover, together with a modest increase in albedo may impact the rainfall pattern. Again, the data so far have been very sparse and models too coarse to run a realistic case for checking such a hypothesis.

7.3.2.3
Meridional temperature gradients, Saharan dust and the latitude of the ITCZ

Earlier in this paper, we discussed the possibility that the latitudinal extent of the deserts may be determined by the gradient of temperature between the equator and the higher latitudes, thereby explaining in part the high aridity of the earth during the glacial period of 18000 BP. Following Korff & Flohn, Greenhut (1981) has recently invoked the possibility that small, but persistent, changes in the meridional and vertical gradient of temperature may have resulted in the deficiency of rainfall in the Sahel during the early 1970s. Analyzing data on atmospheric and surface temperatures published by Angell & Korshover (1978), Greenhut showed that the interannual changes in the value of ξ and η (see equation (7.1)) can be directly correlated with the latitudinal shift in rainfall patterns over the Sahel.

However, in these calculations significant (95%) correlation is achieved only when the eastern hemisphere data are used to calculate the shift in the latitude of ITCZ. This is an intriguing result and probably reflects the fact that the intensity of the Tropical Easterly Jet plays a major role in determining the climate of the Sahel.

The Sahara is a major source of dust in the troposphere. Between 100 and 400 million tons of dust is blown over the Atlantic every year and reaches Barbados and Miami regularly (Prospero *et al.* 1981) (Fig. 7.12). Carlson (1979) has evaluated the climatic impact of the Saharan dust layer over the Atlantic and has found that aerosol concentrations, of optical depth $\tau = 0.7$, which are frequent over the eastern Atlantic, could enhance the heating rates in the 1000–500 mb layer by as much as 1 °C/day. What it does to the actual temperature and therefore to the meridional temperature gradient is a question which needs to be explored. It may provide a way of explaining the persistence in the drought because, as the drought begins, more dust blows over to the Atlantic and heats the atmospheric layers, thereby increasing the meridional temperature gradient, moving the ITCZ southward and thus aiding the continuation of the drought in the Sahel. In order to test this suggestion, we need to perform a similar analysis, as was done by Greenhut, but using only the temperatures over the Atlantic and then comparing the results with those obtained for the same period, using the data from the east of the Sahel. Such analysis may also begin to quantify the relative influence of easterly v. westerly circulation on the Sahelian rainfall, a problem of longstanding controversy (Newell, 1979).

Fig. 7.12. An analysis, of satellite derived turbidity over the Atlantic for (*a*) July 9 and (*b*) June 27, 1974. Note optical thickness (τ_d) as high 3.6 near the coast of Africa on 9 July. (From Carlson 1979.)

7.4
Conclusions

In this chapter, we have attempted to deal with the changing climate of the arid lands in general, but with special emphasis on the droughts in the Sahel, particularly on the problem of the various feedback mechanisms which cause the droughts to persist for as long as a decade or more. From the discussion above, a few salient points emerge.

The excessive aridity of the earth during the glacial (18000 BP) and the relatively humid climate of 9000 BP can both be explained, at least to the first order, by the varying *equator* to *pole*

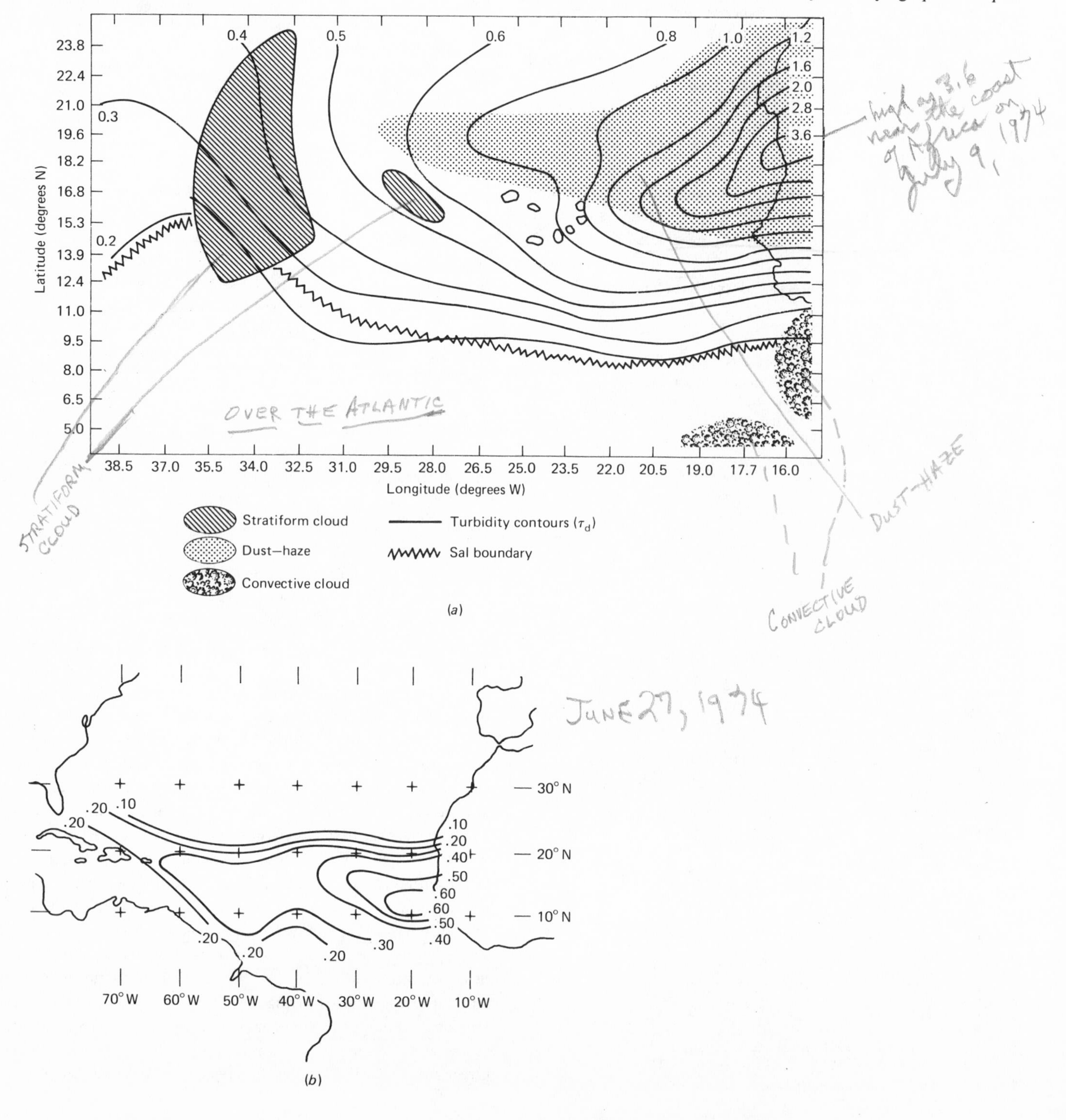

and *ocean* to *land* energy gradients. However, whether it was the atmospheric dust, changing CO_2 concentration or a different ocean circulation which, through a positive feedback mechanism, accentuated these paleoclimates is far from clear.

The major fluctuations in climate which have been recorded during this century in sub-Saharan Africa, Brazil, western North America, north western India appear to be all of natural origin although they may be accentuated by some local feedback mechanism.

A consistent global pattern in these climatic fluctuations is not yet discernable although long-distance tele-connections between phenomena such as El-Nino and Indian monsoon precipitation anomalies, and between eastern hemisphere meridional temperature gradients and the Sahelian rainfall are beginning to be discussed.

Sahelian droughts appear to be peculiar in that they last for as long as a decade. The drought which started in the early 1970s continued at least until 1980.

A number of feedback mechanisms for accentuating and perpetuating the Sahelian droughts proposed earlier in the decade can now be tested using actual data and relatively sophisticated climate models. Some studies show that the drought related surface albedo changes in the Sahel were not as large as was invoked by theoreticians. Also, the albedo seems to have decreased since 1973,

Fig. 7.13. A Meteosat image of West Africa clearly showing the boundary between the Sahara and the Sahel.

although the drought has continued. Another study demonstrates that the removal of vegetation and consequent changes in the evapotranspiration may be a more significant driver for the droughts than just the changes in the surface albedo. And the role of Saharan dust over the Atlantic modifying the global climate is beginning to be seriously discussed.

With operational meteorological satellites now having the capability to observe, very frequently, both the *land* and the *sea surface* (Fig. 7.13), it is possible to mount a program of observation and research for understanding the variability of the dry climates. These data, once calibrated and combined with those from Landsat, other research satellites and ground observations, can now be used to provide a 10–20-year climatology of parameters such as surface albedo, heat balance, surface temperature, dust, cloud cover and perhaps even soil moisture. Analyses of these parameters for periods before, during and after a major climatic change, especially in the arid lands, can begin to give us a better feel for the mechanisms of the onset and persistence of the change. Two examples are in order.

Recently, Chahine & Susskind published a global map of *surface temperature* of the earth as derived from NOAA satellites, corrected for clouds and atmospheric effects. Fig. 7.14 is that for January 1979. Data like these are presumably going to be available, on a month-by-month basis, for the 1980s. This diagram is of course quite instructive for a variety of reasons but it is particularly relevant to the large meridional temperature gradients in the northern hemisphere over the continents. These excessive gradients are probably one of the more-important controlling factors in the seasonal dynamics of the ITCZ as roughly drawn on the map for January (Garnier, 1976). Many months and years of data such as these, together with data for the atmosphere, including analyses of the positions of the ITCZ are required to test the original hypothesis of Korff & Flohn. But, already, the potential seems to be there for understanding the cause and effect relationship between the climate of the higher latitudes, the position of the ITCZ and the amount of rainfall in the sub-tropics.

The other example is that of changes on the land surface, which can now be studied on a large scale using data from satellites. Fig. 7.15 is from Bolle (private communication), in which he has derived a change in the vegetation cover which occurred in Tunisia between August 1980 and March 1982 as observed by an AVHRR sensor of the NOAA satellite.

These two examples show how the already available satellite data can be applied to the problems of climatic change over arid lands. The data are there in plenty. The problem seems to be in authenticating their validity and getting them to the right people asking the right questions.

Fig. 7.14. A global map of mean 'skin' temperature of the earth's surface for January 1979. The data were derived from NOAA satellites using the 3.7 μm window channels in combination with additional microwave and infra-red data from the two sounders. (After Chahine & Susskind 1982.)

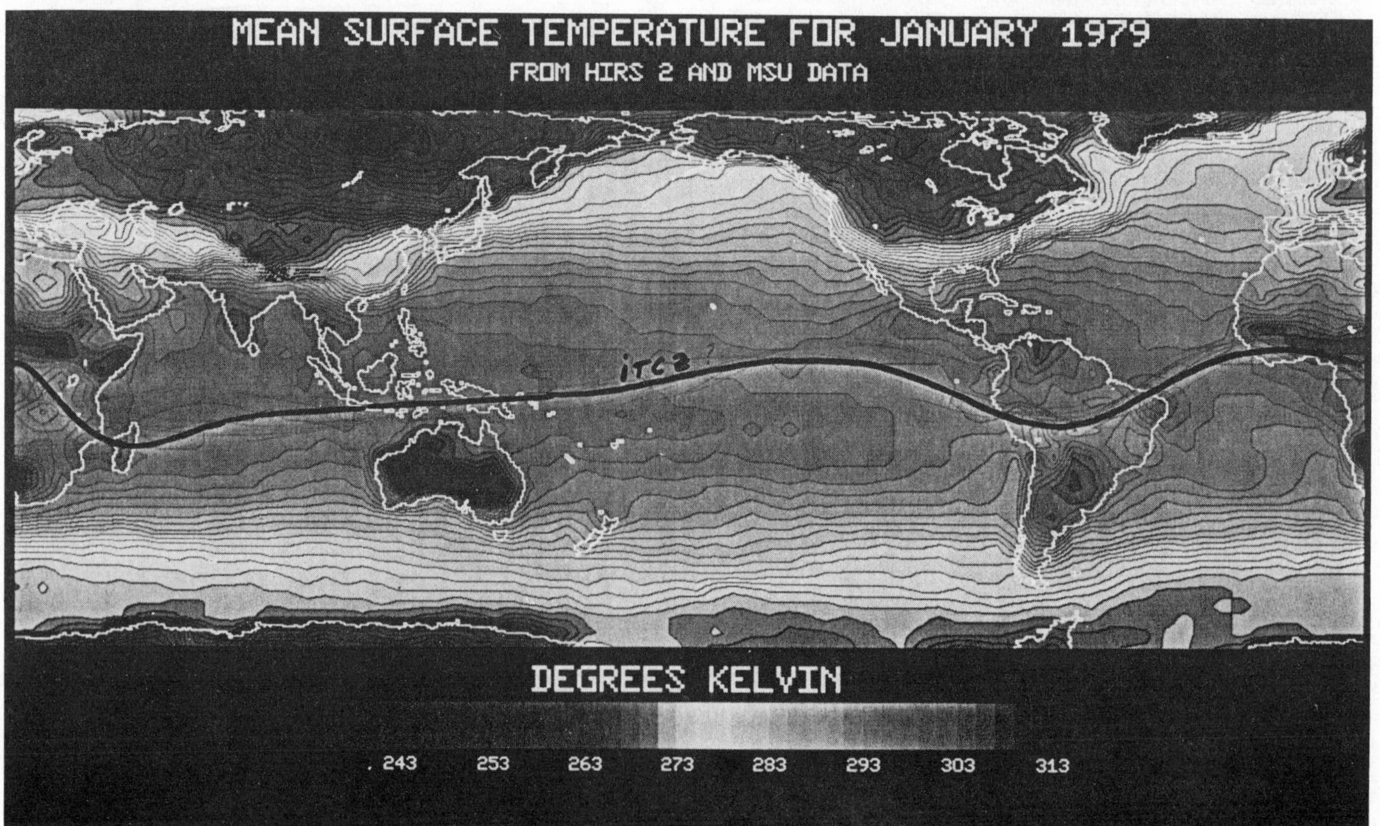

References

Angell, J. K. & Korshover, J. (1978). *Monthly Weather Review*, **106**, 755–70.

Beer, T., Greenhut G. K. & Tandoh S. E. (1977). *Monthly Weather Review*, **105**, 849–55.

Bhalme, H. N. & Mooney, D. A. (1980). *Monthly Weather Review*, **108**, 1197–1211.

Fig. 7.15. Seasonal change of vegetation cover in Tunisia as monitored by AVHRR imagery on board NOAA satellite (normalized differences of channels 2 and 1). Top: 11 August 1980. Bottom: 4 March 1982. Bright: maximum vegetation. Dark: dry and minimum vegetation. Original in color with 24 steps. (From H. Bolle 1982.)

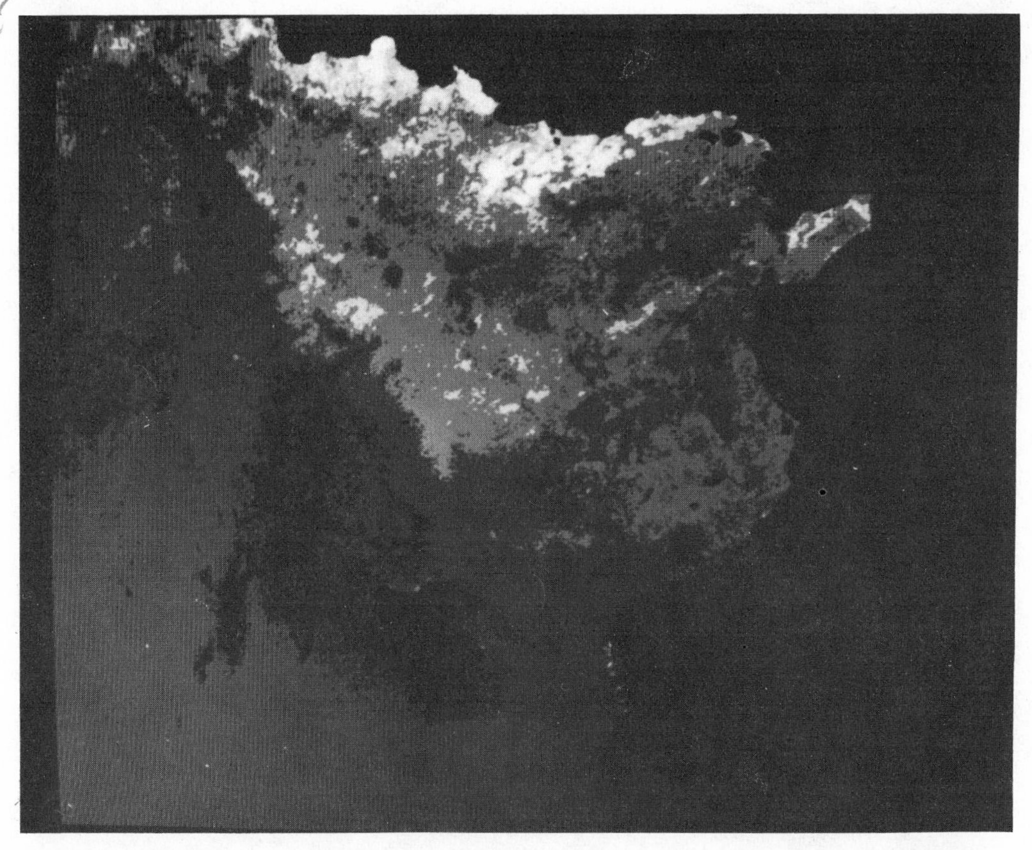

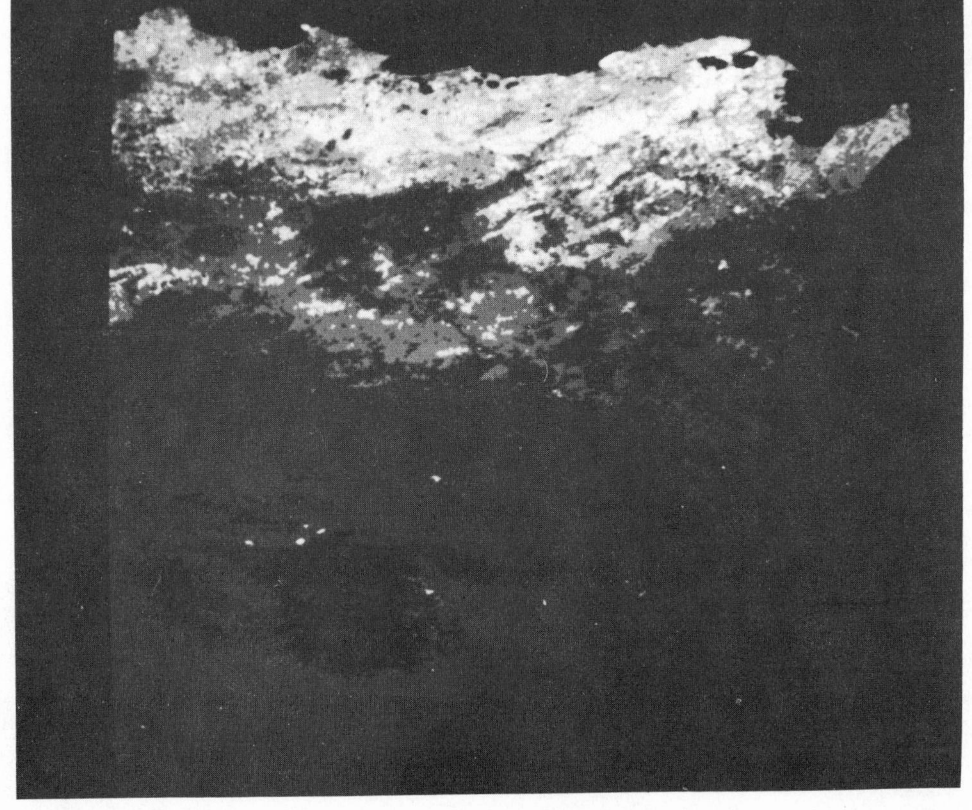

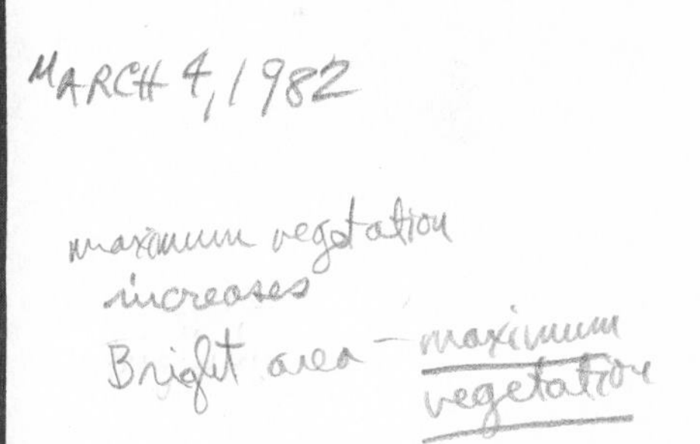

Bolle, H. J. (1982). *The Handbook of Environmental Chemistry*, Vol. *1*, B, Spring Verlag, Berlin.
Bolle, H. J. (1982). *Private Communication.*
Bryson, R. A. (1973). *Ecologist*, **3**, 366–71.
Carlson, T. N. (1979). *Monthly Weather Review*, **107**, 322–5.
Carlson, T. N. & Benjamin, S. G. (1980). *Journal of the Atmospheric Sciences*, **37**, 193–213.
Chahine, M. & Susskind, J. (1982). Private Communication.
Charney, J. G. (1975). *Quarterly Journal of the Royal Meteorological Society*, **101**, 193.
Charney, J. G., Stone, P. H. & Quirk, W. J. (1976). *Science*, **191**, 100–2.
Courel, M. F. & Habif, M. (1982). *24th General Assembly COSPAR*, Ottawa.
Fairbridge, R. W. (1976). *Quaternary Research*, **6**, 529.
Flohn, H. (1980). *World Climate Conference*, WMO 537.
Garnier, R. (1976). *Asecna*, **1**, 29.
Gates, W. L. (1976). *Journal of the Atmospheric Sciences*, **33**, 1844.
Greenhut, G. K. (1981). *Monthly Weather Review*, **109**, 137–47.
Hare, F. K. (1977). 'Climate and desertification', *In: Desertification: Its Causes and Consequences*, Pergamon Press, Oxford.
Hays, J. D., Imbrie, J. & Shackleton, N. J. (1976). *Science*, **191**, 1121.
Korff, H. C. & Flohn, H. (1969). *Ann. der Meteor.*, NF4, **163**.
Kutzbach, J. E. (1981). *Science*, **214**, 59.
Lamb, P. (1982). *Nature*, **299**, 46.
Louis, J. F. (1974). *The Natural Stratosphere of 1974*, CIAP Monograph 1, pp. 6–29.
Manabe, S. & Hahn D. G. (1977). *Journal of Geophysical Research*, **82**, 3889.
Mason, B. J. (1976). *Quarterly Journal of the Royal Meteorological Society*, **102**, 473–98.
Mason, B. J. (1979). *World Climate Conference*, WMO 537.
Moura, A. D. & Shukla, J. (1981). *Journal of the Atmospheric Sciences*, **38**, 2653.
Newell, R. E. (1979). *American Scientist*, **67**, 405.
Nicholson, S. (1980). *The Sahara and the Nile*, A. A. Balkema, Rotterdam.
Nicholson, S. (1981). *Monthly Weather Review*, **109**, 2191.
Nicholson, S. (1983). *Monthly Weather Review* (in press).
Nicholson, S. & Flohn, H. (1980). *Climate Change*, **2**, 313.
Norton, C. C., Mosher, F. R., Hinton, B., Martin, D. W., Santek, D. & Hohlow, W. (1979) *Journal of Applied Meteorology*, **19**, 633–44.
Oguntoyinbo, J. A. & Odingo, R. S. (1979). *World Climate Conference*, WMO 537.
Petit, J. R., Briat, M. & Royer, A. (1981). *Nature*, **293**, 391.
Prospero, J. M., Glacuum, R. A. & Nees, R. T. (1981). *Nature*, **289**, 570.
Rasmusson, E. M. & Carpenter, T. H. (1983). *Monthly Weather Review*, **111**, 517–28.
Rasool, S. I., Courel, M. F. & Kandel, R. (1982). *24th General Assembly COSPAR*, Ottawa.
Sarnthein, M. (1978). *Nature*, **272**, 43.
Sarnthein, M. (1981). *Nature*, **293**, 193.
Shukla, J. & Mintz, Y. (1981). *Science*, **215**, 1498–1500.
Smagorinsky, J. (1963). *Monthly Weather Review*, **91**, 99–164.
Talbot, M. R. (1980). *In: The Sahara and the Nile*, A. A. Balkema, Rotterdam.
Tannehill, J. (1947). *Drought: its causes and effects*, Princeton Univ. Press.
Walker, J. & Rowntree, P. R. (1977). *Quarterly Journal of the Royal Meteorological Society*, **103**, 29–46.
Williams, M. A. J. & Faure, H. (eds.) (1980). *The Sahara and the Nile*, A. A. Balkema, Rotterdam.

The cryosphere†

N. Untersteiner
University of Washington, Seattle, Washington, USA

Abstract

From the viewpoint of climate and human activities, the significance of the five elements of the terrestrial cryosphere may be characterized as follows:

Seasonal snow cover responds rapidly to atmospheric dynamics on time scales of days and longer. In a global context the seasonal heat storage in snow is small. The primary and large effect is exerted by the high albedo of snow-covered surfaces. In the northern hemisphere, at its winter maximum, snow covers 50% of the land surface and 10% of the ocean surface.

Sea ice plays a complex role in the climate system on time scales of seasons and longer. Its seasonal cycle of extent has a similar, though smaller, effect on the surface heat balance than that of snow on land. But sea ice also acts as a barrier to the exchange of moisture and momentum between the atmosphere and the ocean. In some regions sea ice is related to the formation of deep water masses and may play a role in long-term climatic change. In applied terms, sea ice is an impediment to marine commerce along the Eurasian north coast and virtually blocks the short shipping route between the industrial centers of the North Atlantic and North Pacific regions.

The ice sheets of Antarctica and Greenland are quasi-permanent topographic features (on the shorter climatic time scales) and, because of their high albedo, act as elevated cooling surfaces for the atmospheric heat balance. They contain 80% of the existing fresh water on earth. Changes in their volume could cause large changes in mean sea level.

Mountain glaciers are a small part of the cryosphere in volume and surface area, but many exist in densely populated areas where they affect human activities in many ways. In addition, their rapid movement and high rates of accumulation and ablation (and therefore rapid response to climatic change) may entail a significant contribution to eustatic sea level changes.

Permafrost is a manifestation of past and present climate, changing significantly on time scales of centuries and longer. It affects surface ecosystems and river discharge into the ocean that, especially along the estuaries and vast shelf areas of the Eurasian continent, and influences the convective regime of the ocean. In the northern hemisphere, permafrost underlies regions with considerable populations and natural resources, and poses engineering problems.

8.1 Introduction

The low latitudes of the earth occupy the largest fraction of its surface and receive by far the largest amount of solar energy to drive the thermodynamic engine whose average and fluctuating state we experience as climate and climate change. The low-latitude heat source is roughly compensated for by a high-latitude heat sink that is less well defined because hemispheric asymmetries, the land–sea distribution, oceanic heat transports, and continent-sized topographic features complicate the picture. Considering that the earth–atmosphere radiation balance is negative at latitudes greater than 30° the concept of a 'polar heat sink' becomes difficult to define. It seems timely to recall that a century ago the First International Polar Year (Weyprecht, 1875) was conceived with the notion that fundamental insight into the behavior of the atmosphere could be gained from observations in the polar regions. Recently discovered mechanisms have placed the source of climatic unrest primarily in tropical and subtropical regions. Neither the First (1882–83) nor the Second (1932–33) International Polar Years have done much directly toward understanding the mechanisms of

† The wide range of ice-related subjects in climatology makes it infeasible to offer a comprehensive list of primary references. The citations given in the text are a selection emphasizing examples of recent work and review papers, monographs and topical conference proceedings containing large bibliographies. The most exhaustive collection of literature on cryospheric subjects is the *Bibliography on Cold Regions Science and Technology*, published by the US Army Cold Regions Research and Engineering Laboratory, Hanover, New Hampshire 03755.

climate, but they have made a large and lasting contribution toward earth science in general by demonstrating both the need and the utility of planned and coordinated projects among many countries in the service of a global science.

Current research generally does not favor the polar regions as a cause or source of climate unrest, at least on the time scales of centuries and less. However, there is both direct observational and proxy evidence that the most active *scene* of climate unrest is the polar regions. This ranges from large year-to-year- anomalies of the seasonal snow cover to the enormous expansions of ice during the Quaternary glaciations on land and sea.

It was recently shown by North *et al.* (1982) that certain zonally averaged climate parameters can have large variances near the poles, for purely geometric reasons. This introduces a useful note of caution in the analysis and documentation of climatic trends as a function of latitude. On the other hand, time series at fixed high-latitude stations, like snow–no-snow or sea ice–no-sea ice, do not lend themselves to the type of analysis used by North *et al.*, and they do indicate great climatic sensitivity at high latitudes (and a high signal-to-noise ratio) simulated by general circulation models (e.g., Manabe & Wetherald, 1980).

A characteristic of the cryospheric data base is its uneven coverage in time. Best documented are, of course, the past few decades. Then, for a few thousand years there is practically no record. Glaciers and ice sheets changed little in that time, and there are no methods (proxy data) to reconstruct seasonal snow cover. Owing to the massive effort of one particular project (CLIMAP 1981), information about the cryosphere 18000 years ago is much more detailed than for the intervening millennia.

8.2 Ice inventory

In terms of its geologic history, the earth carries today a larger than 'normal' cryosphere (e.g. Tarling, 1978). The total amount of water in all earthly forms is estimated to be 1348×10^6 km³. Of this, 97.4% is sea water; 0.0009% is atmospheric vapor; 0.5% is ground water, mostly at great depths; 0.1% is contained in rivers and lakes; and 2.0% is frozen. This last figure means that nearly 80% of the fresh water on earth exists in the form of ice and snow. Transitions between the solid and liquid phases of water occur at a temperature lying close to the center of the range of temperature commonly found in the terrestrial environment. Therefore, the cryosphere is the most changeable feature, on all time scales, of the physical constitution of our planet.

Today, perennial ice covers 11% of the earth's land surface and an average of 7% of the world ocean. The ratio of thickness to diameter for mountain glaciers and continental ice sheets lies between $1:10^2$ and $1:10^3$. Their 'metabolic rate', or the residence time of solid precipitation in the ice mass, ranges from 10^2–10^3 years in a fast-moving mountain glacier to 10^4–10^5 in the antarctic ice sheet. In contrast, the ratio of thickness to diameter of sea ice is of the order of $1:10^6$, and the residence time of a particle of frozen sea water in the southern and Arctic Oceans is 1–10 years.

For the present discussion, it is convenient to subdivide the terrestrial cryosphere according to the time scale of the dominant physical processes by which it interacts with the other components of the climate system:

Ice sheets and permafrost	10^3–10^5 years
Temperate glaciers	10–10^3 years
Sea ice and snow	10^{-2}–10 years

Sources and sinks for snow and ice are their interfaces with the atmosphere and (in the case of sea ice) with the ocean. A measure of the 'metabolic rate', related to the cryospheric response to climate changes, may be seen in the ratio of the annual mass balance (occurring at the surfaces) to the total volume. This ratio is for

Snow and sea ice	10^0
Mountain glaciers	10^{-2}
Ice sheets	$10^{-4 \text{ to } -5}$

Not considered in this comparison are dramatic events of dynamic instability, such as the different types of surges known to

Table 8.1. *Estimated global inventory of land and sea ice*

			Area (km²)	Volume (km³)
Land ice	Antarctic ice sheet		14×10^6	28×10^6
	Greenland ice sheet		1.8×10^6	2.7×10^6
	Mountain glaciers		0.35×10^6	0.24×10^6
	Permafrost	cont.	8×10^6	
		discont.	17×10^6	(ice content) $0.2–0.5 \times 10^6$
	Seasonal snow (avg. max)	Eurasia	30×10^6	$2–3 \times 10^3$
		America	17×10^6	
Sea ice	Southern Ocean	max	20×10^6	3×10^4
		min	2.5×10^6	5×10^3
	Arctic	max	15×10^6	5×10^4
		min	8×10^6	2×10^4

Not included in this table is the volume of water in the ground that annually freezes and thaws at the surface of permafrost ('active layer') and in regions without permafrost but with subfreezing winter temperatures.

occur in mountain glaciers and deemed theoretically possible in ice sheets.

In most applications of the general laws of mechanics, thermodynamics, and radiation to problems in geophysical fluid dynamics, it can be stipulated that there is no mass divergence. The mathematical treatment of some cryospheric problems is made difficult by large mass divergences that are not easily observed.

In keeping with the scientific emphasis of the World Climate Research Programme and its view toward practically beneficial applications, the following discussion will emphasize phenomena occurring on time scales of seasons to centuries.

Fig. 8.1. (*a*) Snow and ice in the northern hemisphere at their minimum and maximum extent. (*b*) Snow and ice in the southern hemisphere at their minimum and maximum extent (from Meier, 1983). It should be noted that the summer sea ice extent in Fig. 8.1 appears larger than the corresponding value shown in Table 8.1. The reason is that sea ice 'edge' or 'boundary' are definable only in the context of the method by which the data were obtained or by the purpose for which the data are to be used. The best set of data for Antarctic sea ice was derived from passive microwave images (ESMR on NIMBUS 5, beginning in December 1972). For reasons of accuracy and unambiguous interpretation of these data, the ice limit was taken to be at concentrations of $> 15\%$, which tends to make the total ice extent appear large. Ships reporting the encounter with ice are likely to place the 'ice margin' at a far higher concentration.

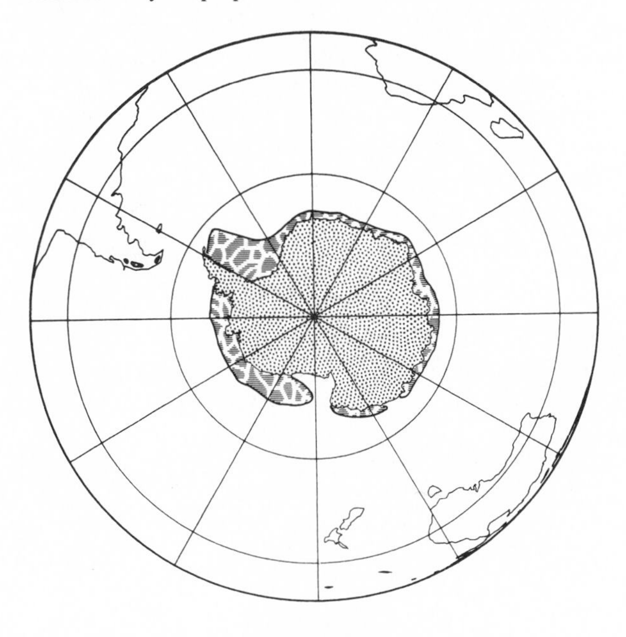

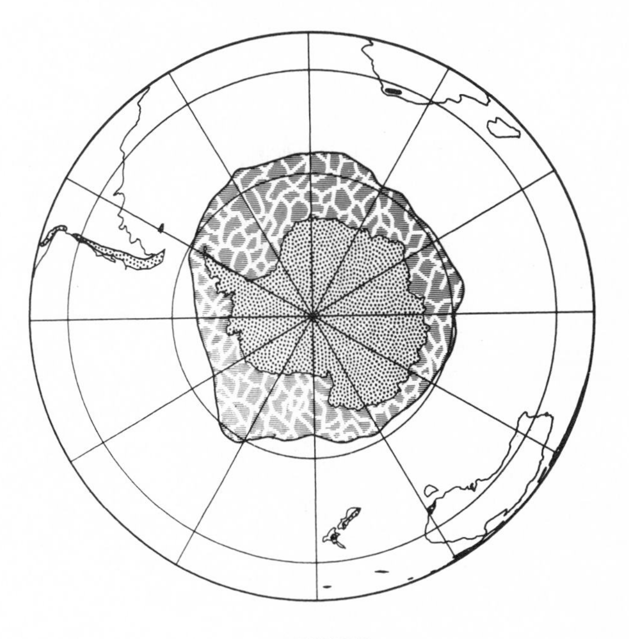

(*a*)

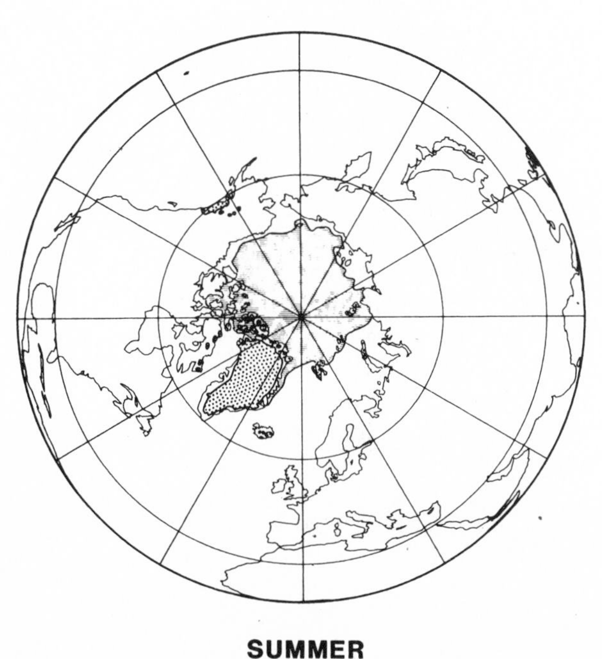

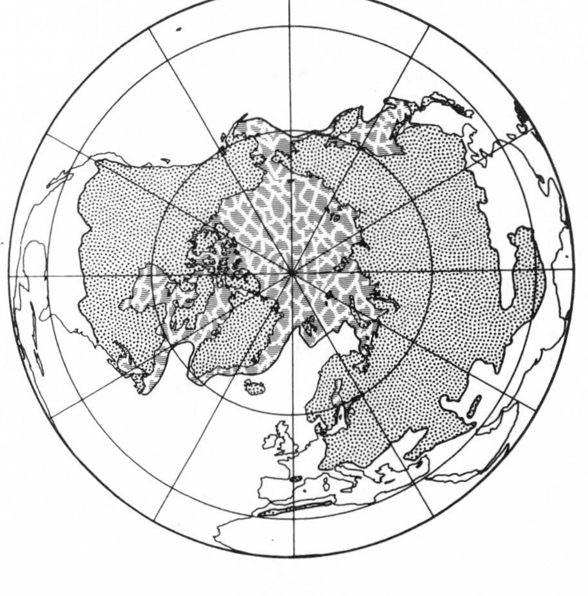

(*b*)

Table 8.1 is a compilation (from numerous sources) of the surface area and volume of all constituents of the cryosphere. Variable amounts of data, different methods of data acquisition and interpretation and (in some cases) genuine variations make it impossible to assign meaningful errors to individual numbers.

Singling out that part of the cryosphere that affects the seasonal change of surface radiation balance in Table 8.2, it is evident that, at the seasonal maximum, snow and ice cover twice as much area in the northern hemisphere as in the southern hemisphere, with the seasonal amplitude being three times larger in the north than in the south (see also Fig. 8.1).

8.3 Seasonal snow

The self-enhancing effect of a snow cover (positive albedo feedback) has been recognized in a qualitative way for a long time (e.g., Croll, 1867. For a summary of early work see Brooks, 1949).

Table 8.2 *Seasonal variation of snow and sea ice in the northern and southern hemispheres at the times of maximum extent (in 10^6 km²).*

Globe = 510 × 10⁶ km²				
	Northern hemisphere		Southern hemisphere	
	Land	Ocean	Land	Ocean
	100	154	49	206
min	2	8	14	2.5
max	49	15	14+	20.0
	Land and ocean		Land and ocean	
min	10		16.5	
max	64		34.0	
	% of hemisphere			
min	4		6.5	
max	25		13.5	

The numbers for the northern hemisphere contain a small inaccuracy because the maximum snow cover occurs about one month earlier (February) than the maximum sea ice extent. (Minimum snow cover is the snow surface of ice sheets.)

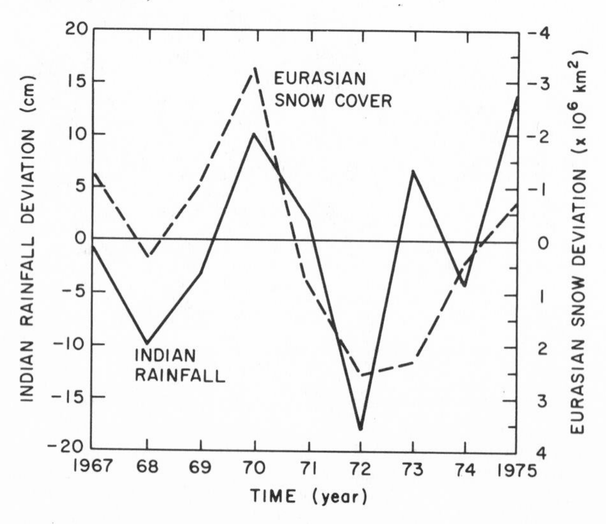

Fig. 8.2. Year-to-year variation of winter snow-cover deviation over Eurasia south of 52°N, and the corresponding variation of summertime area mean rainfall departure for India. (Hahn & Shukla, 1976.)

In several studies, Namias (1975, 1981) sought to establish relationships between snow cover and atmospheric circulation, mainly for the purpose of long-range forecasting. Hahn & Shukla (1976) described a strong apparent relationship between the extent of snow in central Asia and the Indian monsoon (Fig. 8.2). On the same subject, Fu & Fletcher (1983) relate the amount of rainfall in India to soil temperature in Tibet and sea surface temperature in eastern Pacific. The large 'signal' of snow, and its large anomalies, have been an intriguing subject for forecasters and climatologists for many decades (for a selection of references, see Kukla, 1981). While a great deal of phenomenological information has been collected, definitive results and insights into physical mechanisms applicable to practical problems have been sparse.

An encouraging aspect of studying the relationships of climate and snow cover is the growing amount of accurate snow-cover data being assembled from satellite observations (Robok, 1980; Kukla *et al.*, 1981). Fig. 8.3 shows the 14-year average snow cover on the northern hemisphere (Dewey & Heim, 1981). The maximum occurs on about 1 February. The latest maxima occurred during week 8 in the winters of 1966–67 and 1970–71, and the earliest in week 1 in the winters of 1968–69 and 1969–70. The same data also show that the snow-cover maximum occurs four weeks later in Eurasia than in North America. The standard deviation (in per cent of the total snow area on each continent) of the same months during the 14-year time span is 50% higher in Eurasia than in North America. A generally increasing trend of total snow cover (Fig. 8.4) is more pronounced in Eurasia than in North America.

A large impetus toward studying snow and ice by means of models were the energy balance models (EBM) first formulated by Budyko (1969) and Sellers (1969). The models employed greatly simplified physics, and they predicted unstable expansions of snow and ice in response to small perturbations of the solar constant (positive albedo feedback). These dramatic predictions stimulated a rapid development of EBMs with more sophisticated physics (for a recent review see North *et al.*, 1981) striving for an increasingly realistic simulation of the present climate, as well as the response to external and internal perturbations in search for explanations of the climatic record, especially on the long time scales.

At the same time, general circulations models (GCMs), with their much longer history of development, and aided by powerful computers, became increasingly amenable to the inclusion of snow and ice effects. At present, the albedo feedbacks of clouds and snow on the ground are considered the most problematic aspect of GCM modelling (Schneider & Dickinson, 1975).

Owing to the complexity of modern GCMs and EBMs, much emphasis was placed on recommending sensitivity studies (e.g., GARP, 1975) in which only one parameter should be perturbed in order to allow an unambiguous interpretation of the model calculations. It appears now less clear that single-parameter sensitivity studies with any kind of model are necessarily meaningful because the assumption that nature performs the perturbation of a single parameter without affecting others is not realistic. This does not, of course, apply to parameters external to the climate system, such as continental drift, mountain building, volcanism, or solar variation.

A third category of model, studying the stochastic behavior

of the climate system as a collection of interactive subsystems with different time constants (Hasselmann, 1976; North & Cahalan, 1981) emphasized the possibility that climatic variations represent internal fluctuations of a complicated nonlinear system, which can occur even under fixed external conditions.

The ultimate success of any model calculation is the degree with which the model can duplicate observations. An excellent illustration of this was provided by Hahn (1981), partially reproduced in Fig. 8.5. Whether or not the agreement between simulation and reality should be declared satisfactory is a question that cannot be asked outside a specific context or purpose. But the example is convincing evidence for the crucial need for more data.

Snow-cover observations at ground stations have been taken for many decades. Like precipitation data, these observations are susceptible to small-scale effects, and the accuracy of regional extrapolations encompassing many types of terrain have been of variable and uncertain accuracy. The collection of reliable and detailed data from satellites is becoming large enough to lend itself to analyses of the kind shown in Fig. 8.5.

Perhaps the most enigmatic aspect of seasonal snow is its relation to the onset of glaciations. Major glaciations during the Quaternary started on similar northern latitudes and on low continental terrain. The build-ups were gradual and the decay rapid. The configuration of land and sea was about the same as it is today, and the Eurasian center of glaciation was near the ocean while the North American was almost midway between two oceans. Astronomical theory has come a long way toward reconstructing global, hemispheric, and seasonal variation of radiation input on the long time scales. Hardly a beginning has been made toward using that information to explain the enormous events that occurred on the surface of the earth during the past million years. Even if a model of these glaciations consistent with astronomical theory, evidence of volcanic activity, sea level, eustatic movements, ice-sheet

Fig. 8.3. Extent of northern hemisphere snow cover (from Dewey & Heim, 1981). Solid lines are the 14-year (1966–1980) average in 10^6 km². Crosses mark the snow cover during a given winter. The time of maximum extent varies by 2 months (latest: (*a*) and (*b*) in week 8; earliest: (*c*) and (*d*) in week 1).

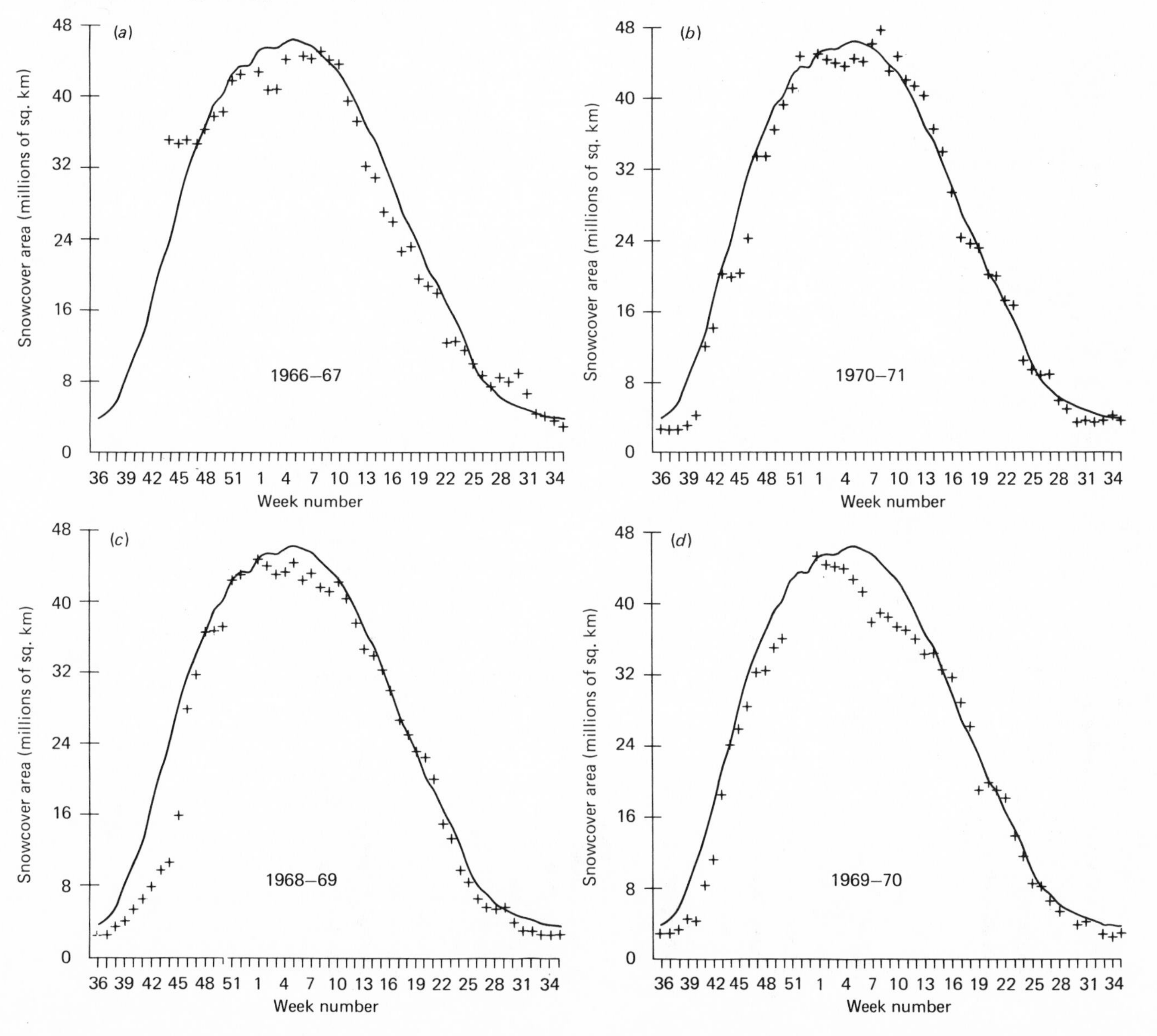

mechanics, paleovegetations, and the large features of atmospheric and oceanic dynamics could be constructed, it is not *a priori* clear that such a model would help in the practically important predictions of seasonal or annual climate anomalies, or the effects of another century of adding carbon dioxide to the atmosphere by fossil-fuel combustion.

8.4 Sea ice

The large horizontal extent and small thickness of sea ice make it a prime candidate for a sensitive response to climatic change and albedo feedback (Manabe & Stouffer, 1980). Contemporary average sea ice boundaries are shown in Fig. 8.1.

Unlike fresh water, sea water of normal salinity has its density maximum well below its freezing point. In a homogeneous ocean this would imply that during autumn the entire water column to the sea floor would have to be cooled to the freezing point before ice could form at the surface. In actuality, this type of convection is limited by a pycnocline at a depth of 30–60 m in the Arctic and 200–300 m in the Antarctic Oceans. The strength of this pycnocline is similar to that found in lower latitudes except that it is caused by a salinity rather than a temperature gradient. The origin and maintenance of this boundary for upper ocean mixing in the polar oceans is believed to be the result of ice formation and melting: salt rejection during freezing induces instability and an efficient down mixing of salt, while ice melting stabilizes the boundary layer, tending to lower the salinity of the upper ocean. In the central Arctic, which is a nearly mediterranean ocean, this effect is greatly enhanced by the addition of continental fresh-water runoff that amounts to about 0.1 Sv or, if evenly spread, a layer of 30 cm per year of fresh water over the entire basin. The presence and the dynamics of this surface layer are extremely important as they have a large influence on the onset of freezing, the heat storage, and the heat exchange with the deeper layers (Aagaard *et al.*, 1981). The basic sea ice problem is one of understanding its formation, motion, and decay as it interacts with the atmosphere and the ocean. As a geophysical material it is unlike any other, consisting of an

Fig. 8.4. 52-week running means of snow-cover areas for North America, Eurasia, and the northern hemisphere (from Dewey & Heim, 1981), in 10^6 km².

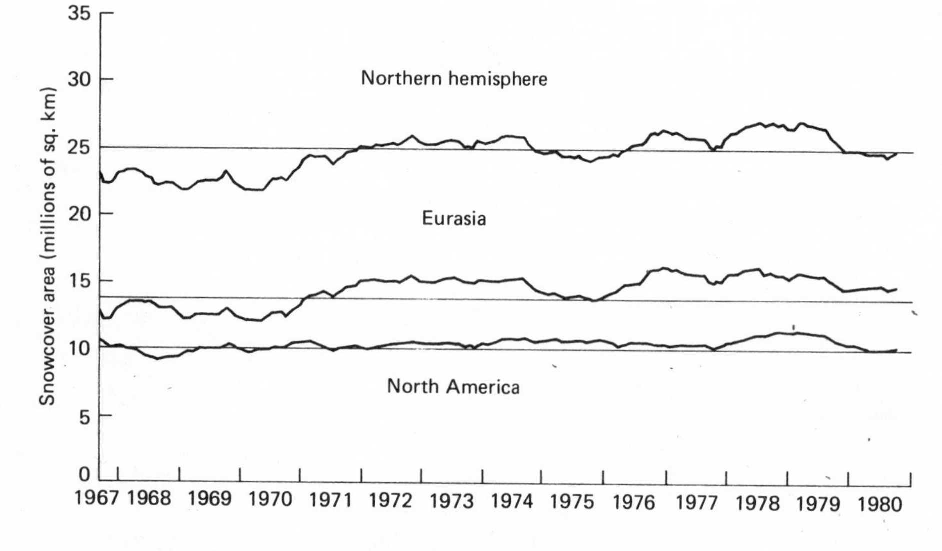

Fig. 8.5. Observed mean monthly snow cover and its standard deviation for the northern hemisphere, compared with values computed by a general circulation model. (From Hahn, 1981.)

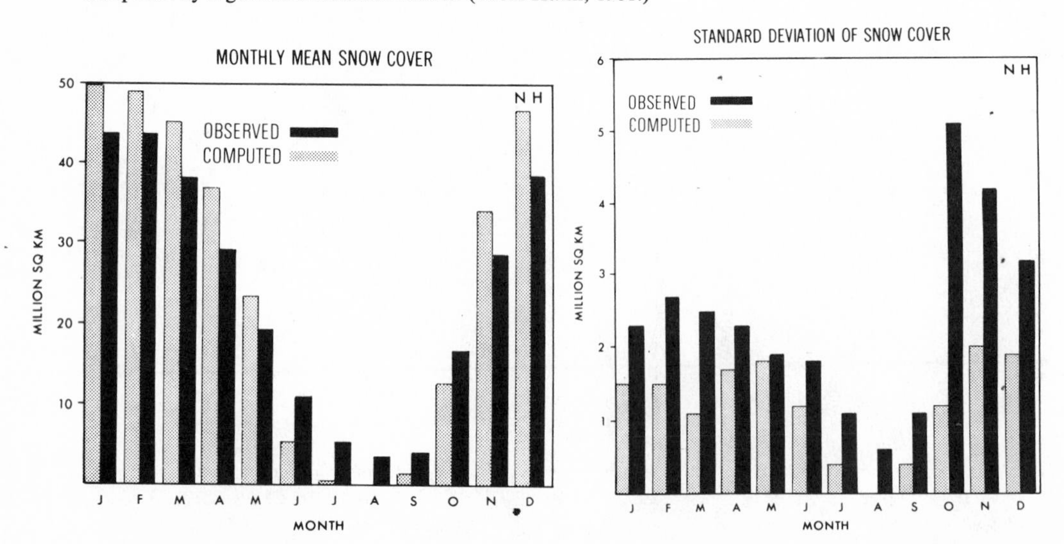

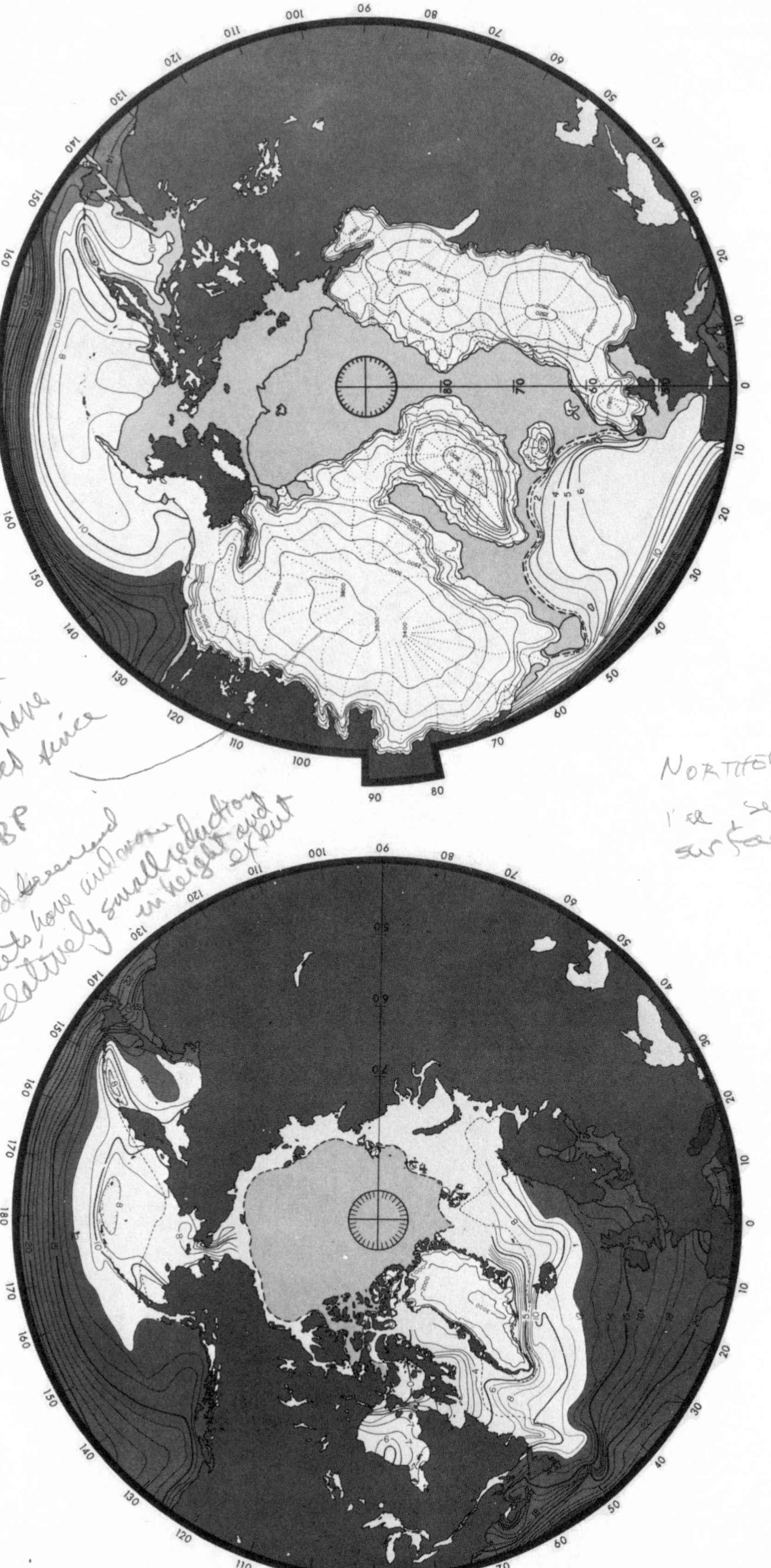

Fig. 8.6. (*a*) Northern hemisphere continental ice, sea ice, and sea surface temperatures in August at present and 18000 BP. (From CLIMAP, 1981.) (*b*) (over page) Southern hemisphere continental ice, sea ice, and sea surface temperatures in August at present and 18000 BP. (From CLIMAP, 1981.)

(*a*)

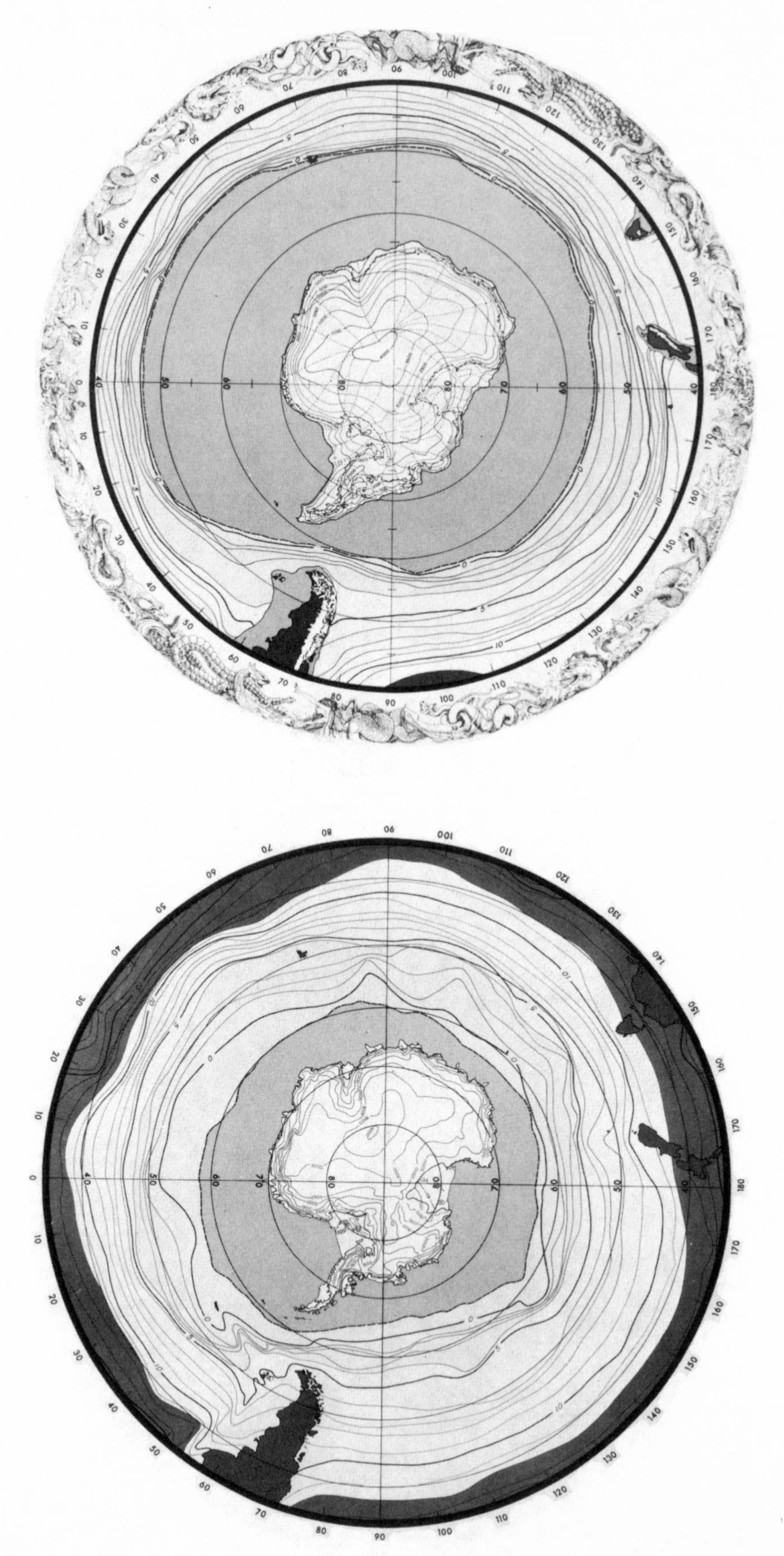

(b)

ensemble of fragments ranging in size from kilometers to millimeters, spread out to cover only a fraction of the sea surface or compressed to build ridges many meters high, and moving by the forces of wind and ocean currents.

The ultimate goal of representing this physical system is to formulate a mathematical model describing the ice itself, its response to atmospheric and oceanic forcing, and its feedback to these forcings. No such model exists at present, but steady progress is being made toward collecting the necessary elements.

A condition of continuity is easily formulated stating that the change in volume of ice in a unit area must be the result of advection, divergence, and melting–freezing. A momentum equation describes the relationship between ice acceleration (usually negligible) and surface stresses, Coriolis force, a gravity component due to sea surface tilt, and a term representing a degree of stiffness within the ice and resulting in the propagation of internal stresses. A third equation is needed to formulate the change in the above-mentioned ensemble of pieces of ice as a result of thermodynamic processes (thickening and thinning) and the mechanical transformation of the overall thickness distribution by mechanical processes (Thorndike *et al.*, 1975). And lastly, the resistance to deformation (internal ice stress in the momentum equation) must be stated as a constitutive equation relating strain rate to stress (Hibler, 1979).

The constitutive law of sea ice has been the subject of many studies and much debate, which appear to converge on the pragmatic notion that no generally applicable law exists and that the relation of strain rate to stress has to be chosen to fit the circumstances (scales in space and time, coastal boundaries, open ocean boundaries).

A large number of studies has been conducted of all elements of the system outlined above, described in reviews by Rothrock (1975), Hibler (1980), Pritchard, ed. (1980) and, recently, in comprehensive texts (Doronin & Kheisin, 1975; Untersteiner, ed., 1983*b*).

Fig. 8.7. Schematic illustration of processes at an unconfined sea ice boundary.

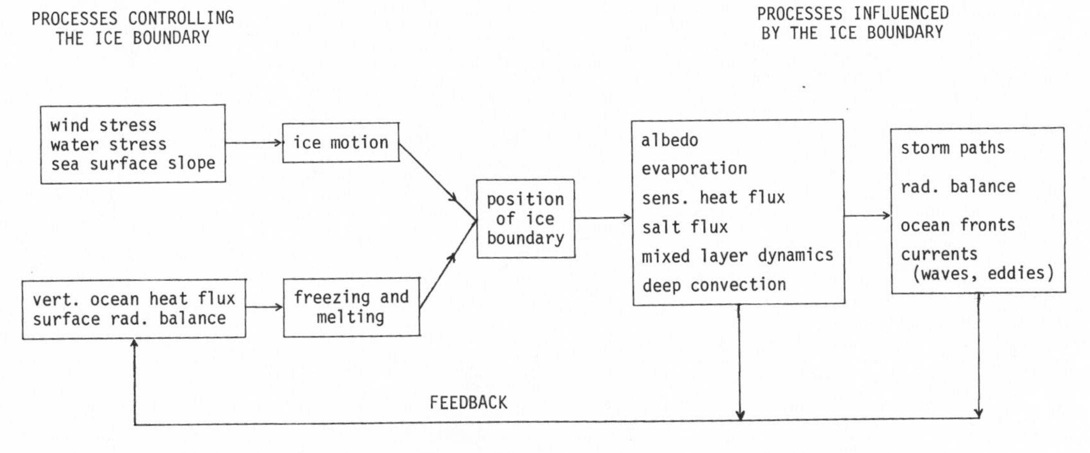

Fig. 8.8. Landsat image of Greenland Sea (180 × 180 km).

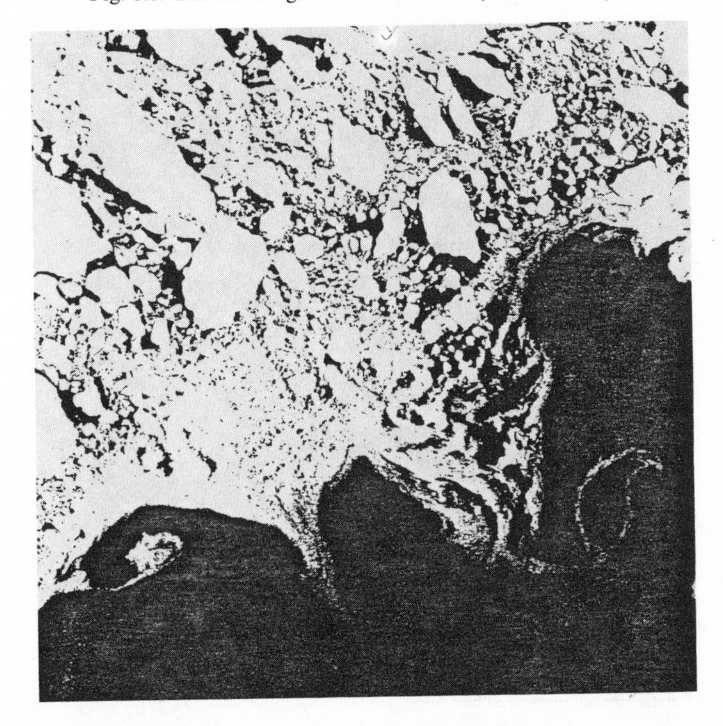

It has been speculated, notably by Fletcher (1965), Donn & Shaw (1966), and Budyko (1972), that one summer's heat storage in an ice-free Arctic Ocean would suffice to minimize ice formation during the subsequent winter so that the ice-free condition would become self-perpetuating. According to recent calculations by Foster (1975), it appears that the heat loss from an open Arctic Ocean during summer would be so rapid that only a small part of the heat stored in the upper ocean would be available to delay freezing in autumn. The sedimentary record from the central Arctic seems to indicate that the Arctic Ocean has been ice-covered not only during the entire Holocene but also during at least the last major glaciations of the Wisconsin (Würm) and Illinoian (Riss) periods (Hunkins & Kutschale, 1965; Ku & Broecker, 1965).

Fig. 8.6 shows a comparison between modern sea ice extent and that reconstructed for the last glacial maximum in both hemispheres (CLIMAP, 1981). A striking and unexplained difference in the southward expansion of sea ice 18000 years ago is evident between the Atlantic and the Pacific sector.

Since the early studies by Wiese (1924) and Walker (1947), special significance has been ascribed to the marginal ice zone as the locale for the strongest interactions between sea ice, the atmosphere, and the ocean. More recently (NAS, 1974, 1975; GARP, 1978; The Polar Group, 1980) it was emphasized that the marginal ice zone probably holds the key for understanding the relationship between sea ice and climate. An excellent view of the role of sea ice in climate variability has recently been prepared by Walsh (1983). An effort to simulate the seasonal cycle of sea ice extent in both hemispheres by means of a model that is essentially an EBM (Parkinson & Washington, 1979; similarly, Hibler & Walsh, 1982) has been successful, but it was forced by the existing meteorological data (which intrinsically contain the sea ice effect) and was 'tuned' by an observationally unknown vertical heat flux in the ocean, thus yielding little insight into the crucial interactive nature of the problem.

Fig. 8.9. Drift paths of 45 air-dropped data buoys between January 1979 and December 1981. The buoys are positioned by ARGOS several times per day to an accuracy better than 1 km. They contain quartz-oscillator pressure sensors whose readings are accurate to ±1 mb, with a drift of less than 0.2 mb per year (adapted from Untersteiner & Thorndike, 1982). The stippled arrows indicate an early (Gordienko, 1958) estimate of the mean circulation of the upper several hundred meters of water in the Arctic Basin.

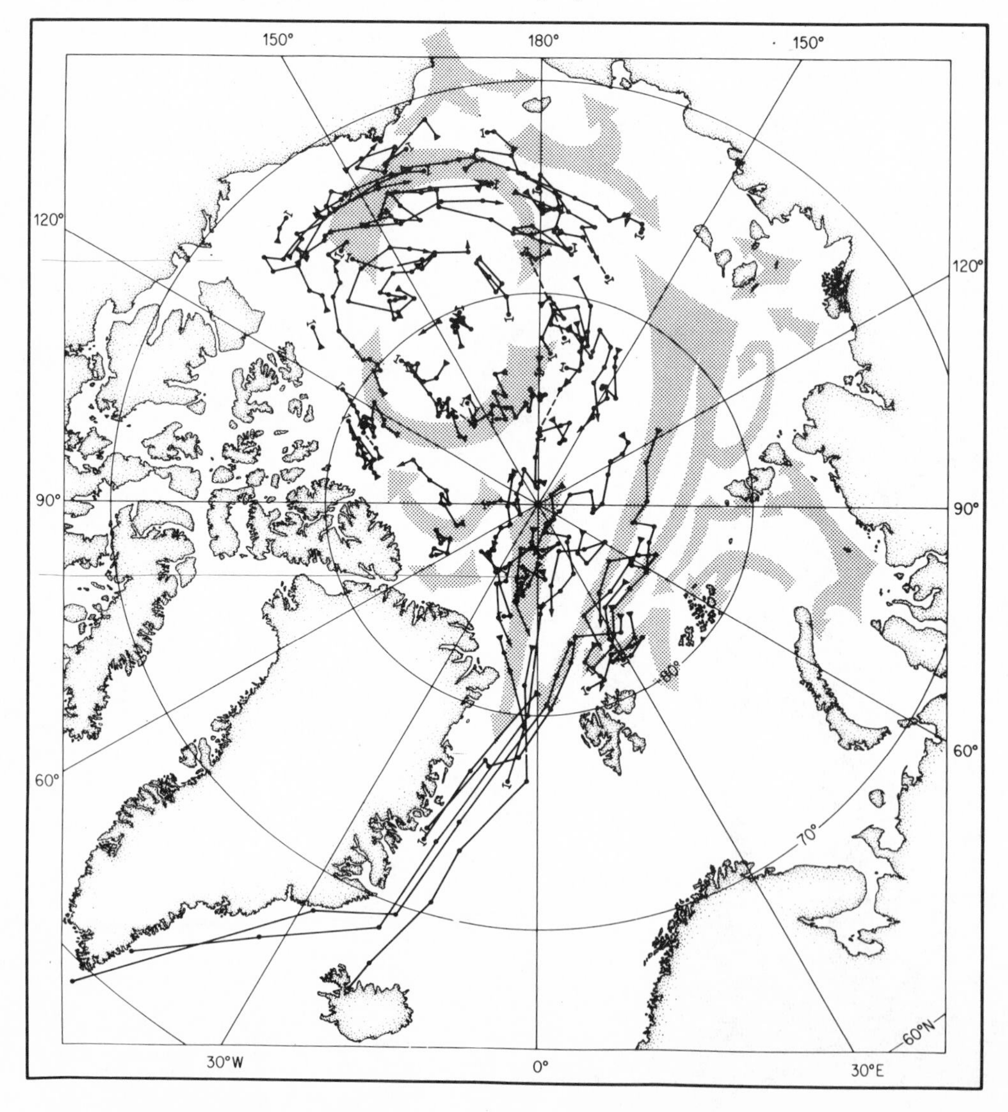

The physical processes controlling the position of the sea ice margin were the subject of essays (e.g., Wadhams, 1981) and workshops (e.g., Andersen *et al.*, eds., 1980). Only recently, and with the support of several institutions and countries, a concerted attack on the problem, the Marginal Ice Zone Experiment (MIZEX) is being mounted (Johanessen *et al.*, 1982; Untersteiner, 1983*a*), with field observations in summer 1983. A schematic representation of the processes involved is shown in Fig. 8.7. The Landsat image of the marginal ice zone, shown in Fig. 8.8 conveys an impression of the complexity of the problem.

As for snow (Section 8.3), satellite remote sensing of sea ice has been of enormous advantage. The ice–no-ice-signal is very strong in virtually the entire electromagnetic spectrum and has made the monitoring of sea ice extent a routine matter. More-sophisticated measurements – the monitoring of the fractional coverage by open water, first-year (thin) and multi-year ice–could form a basis for evaluating the ice-mass balance of large regions and provide clues about the energy exchange between atmosphere and ocean and its climatic impact. Owing to the complex physics of the microwave emissivity of sea ice, the interpretation of data is still experimental but more ground-truth experiments should make the method operational in the near future (e.g. Carsey & Zwally, 1983).

Another observational breakthrough was the development of inexpensive data buoys that are dropped on the ice by parachute and report location, atmospheric pressure, and temperature (Untersteiner & Thorndike, 1982). These buoys have, for the first time, provided a synoptic view of a large portion of ice drift in the Arctic Basin (Fig. 8.9). Daily maps of the geostrophic wind, converted to surface air stress by a simple model (Brown 1981) yield an air stress field of sufficient accuracy and resolution for both driving and testing models of large-scale sea ice dynamics. In addition, it has been shown by Thorndike & Colony (1982) that in the absence of large internal ice stresses (at some distance from shore) the vector difference between the purely wind-driven and the actual ice velocity can be used to monitor, by inference, the circulation of the ocean beneath the ice (Fig. 8.10). There is reason to anticipate that buoy technology will advance to include automatic measurements below the ice, for instance to observe the seasonal evolution of the arctic mixed layer and pycnocline (McPhee, 1980) with its controlling effect on ice-bottom melting and freezing.

8.5 Ice sheets

The two large ice sheets in existence today cover Greenland and the Antarctic continent. With topographic surface elevations of over 2000 m and a maximum height of over 3000 m in Greenland and over 4000 m in Antarctica, both are among the highest land masses on earth. The bedrock beneath the Greenland ice is dish-shaped, with its central portion well below sea level. The continental plate of Antarctica is more structured. The west Antarctic ice from opposite South America to about the meridian opposite New Zealand is underlain by rugged terrain that would form a group of islands if the ice were removed (Fig. 8.11). The west Antarctic ice sheet is grounded in water as deep as 2500 m. The east Antarctic ice sheet contains 80% of the ice volume and has little bottom topography, with a deepest point 500 m below sea level. Because of their extreme cold (mean January and July temperatures in central east Antarctica are −32 and −72 °C), annual precipitation is small and diminishes from the coast inland. For a comprehensive description of the climates of Antarctic see, for instance, GUGK (1966) and for the Arctic, Orvig (1970).

An important difference between the two ice sheets is that practically no surface melting occurs in Antarctica and the surface accumulation is balanced (or nearly so) by the discharge of icebergs

Fig. 8.10. Vector difference between the observed drift of data buoys and ice velocity computed from geostrophic wind and surface stress alone (neglecting internal ice stress). The vector field represents the circulation of the upper ocean in the Arctic Basin. (Arrows without rings are current observations by *Fram* in 1883–84 and *Sedov* in 1937–40). (From Thorndike & Colony, 1982.)

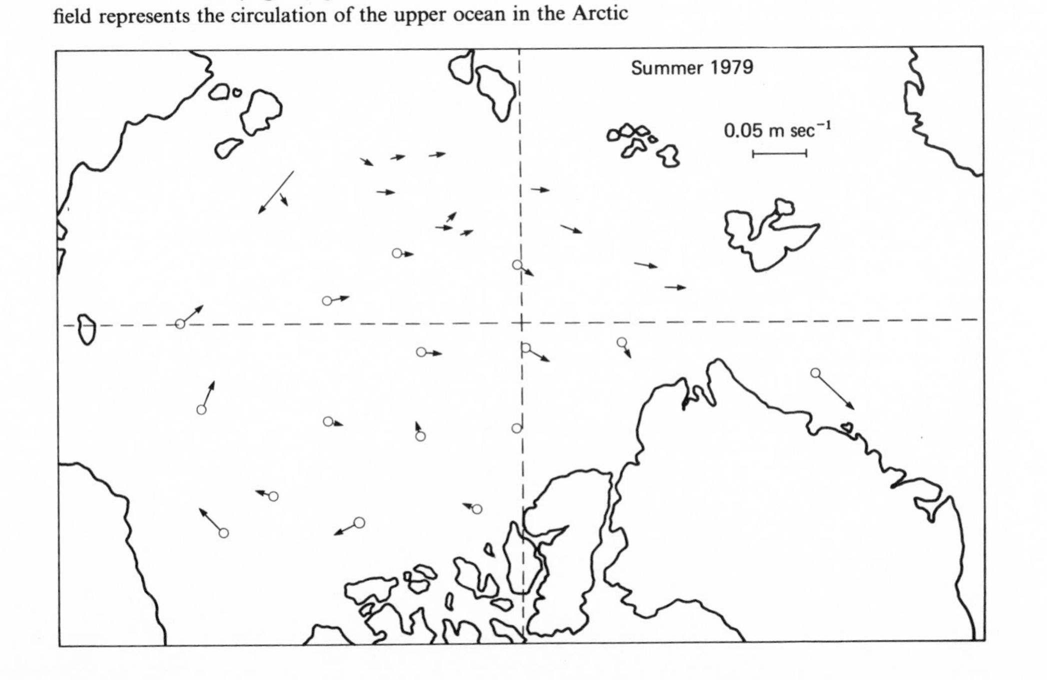

into the sea and by ablation under the large shelves of ice extending into the ocean (e.g. Thomas, 1979). In contrast, the Greenland ice sheet has significant surface melting in its southern portion and an equilibrium line with zero mass balance like a mountain glacier. Greenland has no major ice shelves. Ice not lost by ablation is discharged into the ocean by icebergs calving from glaciers.

Fig. 8.11. The west antarctic ice sheet, containing the two largest ice shelves of the continent, and a large area of ice grounded below sea level (to a maximum depth of 2500 m). (After Mercer, 1978.)

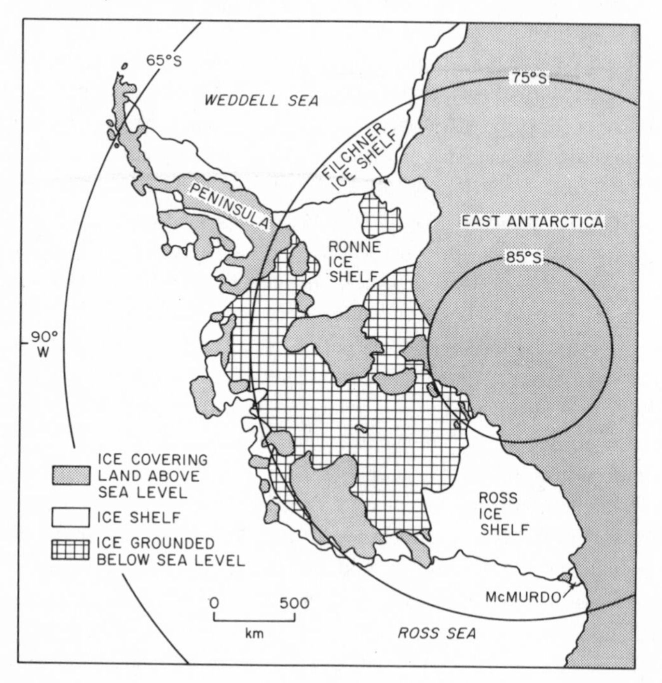

Neglecting crustal adjustment, the melting of the Greenland ice sheet would raise mean sea level by 8 m, of the west Antarctic ice sheet by 5–6 m, and of the east Antarctic ice sheet by 55 m. This raises the obvious question: what is the present mass balance of these ice sheets?

Mainly since the International Geophysical Year (1957–58), a data base has been built by means of seismic, gravity, and radio echo soundings of ice thickness, ice-flow velocity measurements, iceberg discharge, and snow stratigraphy. Mass-balance estimates derived from such observations are shown in Table 8.3 (quoted from Meier, 1983). Relative to each other the discrepancies are large. For instance, estimates of the Antarctic iceberg discharge vary between 300 and 1500 km³ per year. This is between equal to and five times larger than the observed mean annual sea level rise of about 1 mm per year (Gornitz *et al.*, 1982). Relative to the total ice volume of the ice sheets, the spread in mass-balance estimates is small, indicating that the ice sheets cannot be far from equilibrium.

As seen in Fig. 8.6, two continent-sized ice sheets have disappeared since 18000 BP while, by comparison, the Antarctic and Greenland ice sheets underwent a relatively small reduction in height and extent.

Much has been written about a possible global event associated with a rapid change in the west Antarctic ice sheet caused by a CO_2-induced warming of the atmosphere (for a summary, see e.g. Bentley, 1982 and several topical studies in Allison, ed., 1981). Opinions range widely, from predicting an impending disintegration of the west Antarctic ice sheet merely because of its present configuration, to the assertion of a continued stability of the ice sheets on time scales of millenia. The lack of data and the incompleteness of models make none of the arguments entirely convincing. But the role of the Antarctic ice sheet in the global

Table 8.3. *Mass balance determinations of the ice sheets, in km³ of water per year (from Meier, 1983)*

Author	Input (precipi-tation)	Evapo-ration	Blowing snow	Balance, accumula-tion area	area	Surface balance, ablation area	Ablation under ice shelves	Iceberg discharge	Net
Antarctic ice sheet (Antarctic peninsula generally excluded)									
Kosack	(1956)	2850	—	−1200	< 1650	−2050	0	−550	−950
Wexler	(1958)	1620	—	−280	1340	0	−1300	−40	0
Mellor	(1959)	2000	−63	−190	1700	−74	−48	−570	1000
Lister	(1959)	1920	—	−14	1906	−135	−675	−270	826
Loewe	(1959)	1200–2000	—	−20–−500	1200	0	0	−140–−300	1100–1200
Wexler	(1961)	945	—	−65	880	0	−220	−660	0
Kotliakov	(1961)	> 2650	low	low	2650	~0	−120	−1210	1320
Rubin	(1962)	1459–2236	0–−63	−20–−500	1439–1673	0–−74	−163–−323	−1276	0
Loewe	(1967)	1900	—	—	~1900	−10	−200	−1450	240
Greenland ice sheet									
Loewe	(1935–36)				425	−295		−150	−20
Bauer	(1955)				446	−315		−215	−84
Benson	(1961)				500	−272		−215	+13
Bader	(1961)				630	−120–−270		−215	+145–+295
Bauer	(1967)				500	−330		−280	−110
Weidick	(1978)				500	−300		−200	0
Radok *et al.*	(1982)				576–466	−69–−139		−327 to −507	0

climate system does provide an opportunity to at least enumerate, the complexities:

- The most important surface parameters, temperature and snow deposition, are not modeled well by present GCMs. The surface radiation balance (strong inversions) and the shape of the continent create a near-surface katabatic wind regime that affects surface temperature and transports snow radially outward (of unknown quantity, see Table 8.3). This surface regime is beyond the resolution of present GCMs. Also, the moisture transport into the continent is affected by the sea ice-controlled evaporation around the continent. In turn, the sea ice extent is controlled by the atmosphere and the ocean in a way that could be represented only by a fully interactive model.
- A large body of knowledge exists of the dynamics of ice sheets (for a recent review see Paterson, 1980) in response to mass balance, temperature, and bedrock configuration. While many useful results have been obtained with such models, confidence in their predictions is limited (a) by our lack of understanding of the bedrock sliding mechanism and the condition of the ice at its base, and (b) by uncertainties about the flow law (temperature, pressure, crystal fabric) at great depths.
- Much of the ice discharge of Antarctica occurs through its two major ice shelves. These are plates of ice, partially resting on bedrock, partially extending out into shallow water, and buttressed along the sides by a bay-like coastline. The mass balance of these shelves, especially ablation or accretion at the underside, is not known. Any change in ocean temperature or sea level could affect the mass balance at the underside as well as the grounding line, thus completely upsetting the dynamic regime not only of the ice shelves themselves but also of the drainage basins that feed them.

A global warming induced by increased fuel combustion, whose effect on the cryosphere might be quite straightforward in some respects (recession of snow and permafrost boundaries) would have competing consequences on the Antarctic ice sheet: more melting by a warmer ocean and more precipitation by a warmer atmosphere. Despite the ingenious work underway to solve these problems (e.g., Källen *et al.*, 1979; Budd, 1981; Bindschadler

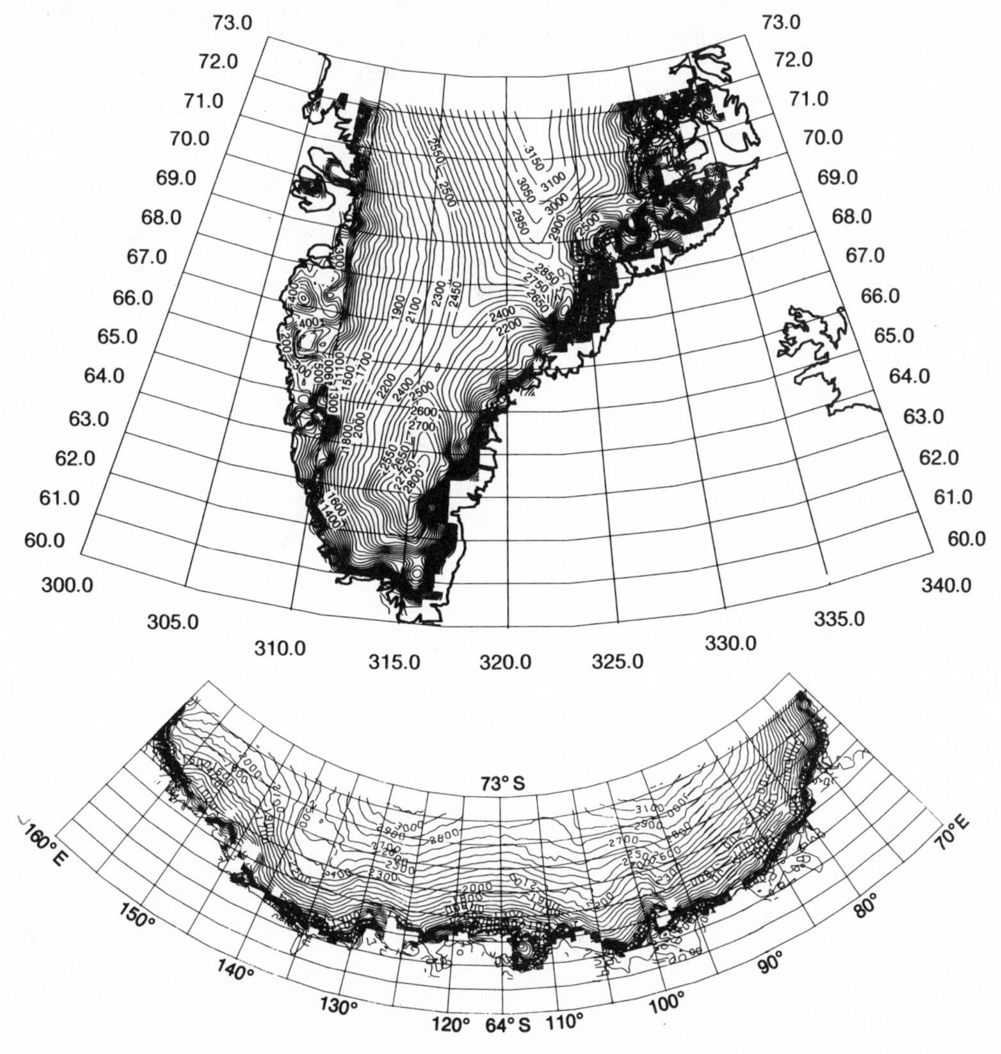

Fig. 8.12. Preliminary results obtained by radar altimeter (Seasat) of the absolute topography of portions of the Greenland and Antarctic ice sheets. (From Zwally *et al.* 1983). The maps are truncated because of the orbital inclination of Seasat. If a map of such accuracy and resolution could be obtained of the west antarctic ice sheet, the precursors of major dynamic events could be detected by periodic (e.g., decadal) re-mapping.

& Gore, 1982), it seems safe to anticipate that their solution lies decades in the future.

The great thickness of ice sheets represents many years of snow accumulation, which makes them unique repositories of paleoclimatic information. Like ocean sediments, ice contains horizons of volcanic particulate matter that can serve as datum lines, and the record in ice cores is different from those in ocean sediments in several useful ways. At locations with sufficiently large accumulation rates, strata can be identified with individual years for thousands of years into the past. Unlike in sediment cores, there are few remnants of biota in ice cores, but they contain bubbles of fossil gas that provide clues to the composition of air and height at the time of deposition in the past. With the increasing sophistication of microanalytic techniques, the study of ice cores has become an established tool for paleoclimatic reconstructions (Raynaud & Lebel, 1979; Thompson & Moseley-Thompson, 1981; Petit *et al.*, 1981). The extraction of deep cores from ice sheets serves the additional purpose of measuring ice temperature and crystal fabric at depth, which are data needed for ice-sheet mechanics (e.g. Lorius et al., 1979; Robin, 1981).

Airborne radio echo sounding of ice thickness has been in progress for some years, but complete maps of the existing ice sheets will not be available in the near future, owing to the size of the area to be mapped.

Battery-operated transmitters on the ice surface in conjunction with Doppler shift measurements by satellite are sufficiently accurate to measure ice-surface velocities at many points.

Radar altimetric measurements tested with observations from Seasat (Zwally *et al.*, 1983, see Fig. 8.12) yield absolute surface topographies. A new airborne laser altimeter promises efficient, if spatially limited, measurements of surface elevation changes. The Seasat orbit limited mapping to 72° north and south latitude, but with future satellites in sufficiently high orbit and carrying radar or laser altimeters, the great ice masses of the world could be mapped at suitable intervals (NASA–ICEX, 1979) to monitor not only their volume change and its relation to global sea level, but also to detect topographic changes that might precede dramatic dynamic events in the ice sheet.

The 80-m resolution of Landsat images is adequate to monitor the iceberg discharge from Antarctic ice shelves and glaciers. A sufficient effort of that kind is not in progress but could make a large contribution toward understanding the Antarctic mass balance.

Comprehensive plans and recommendations for acquiring new data and testing models are being advanced by international organizations (SCAR, 1982).

As indicated in Table 8.3, an estimated 100–300 km^3 of meltwater drain annually from southern Greenland into the ocean. The morphology of outlet glaciers and valleys naturally favors the construction of dams for the utilization of these large quantities of water for hydroelectric power generation. This idea, mainly promoted by Stauber (1967), has been studied in some technical and economic detail (Partl, 1977). The potential output of 12–15 selected sites could be 60–120 GW, approaching the total consumption of the 1974 western European grid system. Transportation of the energy to consumer areas by cables, liquefied hydrogen from the electrolysis of sea water, or ammonia, have been considered but appear to pose (along with the enormous capital investments) problems not likely to be surmounted in the near future.

An amount of water several times as large as that from Greenland is discharged by the Antarctic ice sheet in the form of large tabular icebergs. The possibility of towing these icebergs to arid regions in Australia, South America, or the Arabian peninsula, has been the subject of both serious studies and outlandish propositions (Husseiny, 1978). It appears that, even under extreme assumptions about towing power and ice insulation, it is not feasible to transport significant amounts of ice across the equator. Towing icebergs to the arid region of Western Australia is considered deserving of further investigation.

8.6 Mountain glaciers

The accessibility of glaciers in populated regions, their tractability in terms of size for surveys and other observations, and their intriguing beauty have attracted scientific minds for two centuries. It was probably Walcher (1773) who stated for the first time the notion that glacier variations are related to those of the weather. In his classic work Tyndall (1861) described the transformation of snow into ice and the manifestations of viscous as well as plastic behavior of a glacier flowing downhill under the influence of gravity. He also stated the concept of 'mass balance' as a function of elevation, including the diminution of the rate of accumulation at great heights where the atmosphere is too cold to yield much precipitation (Fig. 8.13).

Some of the earliest indirect evidence for climatic change came from the study of certain geomorphic features and the

Fig. 8.13. Mass balance of a typical glacier in the European Alps (Austria). Upper curve: Accumulation and ablation in meters of water per unit area and year. The equilibrium line lies at about 2950 m altitude. Accumulation has a maximum at about 3400 m. Lower curve: Width of the glacier along a flow line. (From Smith & Budd, 1981.)

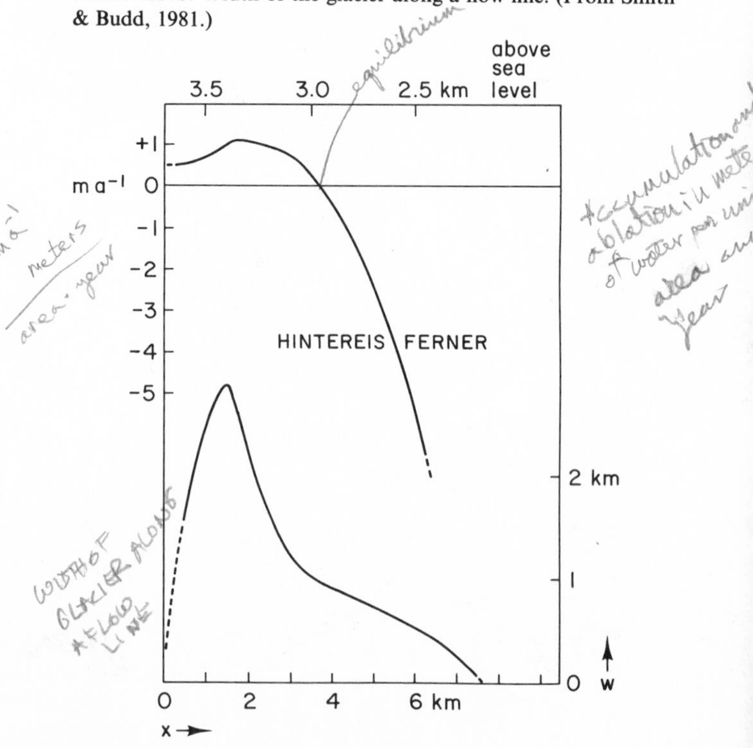

realization that they must have been caused by glacial variations in the past. Since the late nineteenth century, there have been literally hundreds of attempts to relate observed climatic trends and variations of the termini of glaciers in the Alps and other mountain ranges. The phrase that 'mountain glaciers are sensitive indicators of climatic change' has been in use for decades and has lost little of its original charm. Today we recognize that we are far from being able to elaborate on the above statement (for a review see Hoinkes, 1968).

The theory of glacier dynamics (for a comprehensive review see Raymond, 1980) contains the same elements as that of ice sheets: a statement of continuity (representing accumulation and ablation), a flow law relating strain rate to stress, as well as initial and boundary conditions. These require knowledge of the three-dimensional shape of the glacier, and an understanding of the physical mechanisms of the sliding of ice on bedrock. Since most mountain glaciers are wholly or in their lower reaches 'temperate', i.e., at the melting point, the choice of a realistic flow law has been narrowed by numerous experimental studies. In contrast, basal sliding may range from a nearly no-slip to a nearly all-slip condition, depending on temperature, roughness and, most of all, the presence of water either originating at the surface or generated by the sliding process itself. Although progress has been made during the past decade, basal sliding is still considered the least understood aspect of glacier dynamics.

After a general recession of glaciers in temperate regions for a century, an increasing number have begun to advance since the mid-twentieth century (Fig. 8.14). Only a minute fraction of the thousands of existing mountain glaciers are being monitored, and even among those few, individual variations are great. Individual net mass balances (m of water per m² per year) and movements of termini vary widely because mountain glaciers are essentially controlled by mesoscale effects. While the dynamic response of glaciers to a given change in their mass-balance regime is relatively well understood (see above) the mass-balance regime itself (snow accumulation, surface heat balance, and melting) and its relationship to large-scale climate is exceedingly difficult to unravel. Local orographic effects on precipitation patterns, wind regime, and the re-deposition of wind-blown snow can locally obscure large-scale climatic trends.

Attempts to use the sensitivity of mountain glaciers to climate change for reconstructing past climate has had some success for time scales of 10^3 years and longer (e.g., Porter, 1981). Studies (e.g., Smith & Budd 1981) of the possibility of reconstructing past climate from observed glacier changes over a time span of several hundred years generally use independent data such as instrument records, lake levels, and various paleobotanical evidence for comparison. Thus they may be seen as a proving ground for theories of glacier change rather than as actual reconstructions of past climate (Kruss, 1981).

Forecasting the behavior of a glacier under certain assumptions about the climate (mass balance) is an exercise of potentially great practical benefit. An early quantitative forecast of such kind, based on Nye's theories of glacier mechanics (see Raymond, 1980), predicted that a glacier in southern Alaska would not overrun the entrance to a copper mine near the terminus for the following 20 years (Untersteiner & Nye, 1968). While being only a best estimate based on scant data, this forecast has held true for 15 of its 20 years. Among the number of similar predictions made in the following years, one pertains to the very large Columbia Glacier (Alaska) and the potential disintegration of its lower portion which, within a matter of a few years, could scatter thousands of icebergs across the vital shipping lane for tankers from the port of Valdez to refineries in the Pacific region (Rasmussen & Meier, 1982).

Dramatic events associated with the mechanisms of basal sliding are glacier 'surges', during which a glacier can increase its speed by as much as two orders of magnitude. Evidence has been accumulated to indicate that as many as 5% of all mountain glaciers are capable of surging at periods between ten and 100 years. The destructive effects of surging glaciers are limited to local

Fig. 8.14. Percentage of advancing and retreating glaciers in Switzerland, 1891–1975, showing the increase of advancing glaciers since the mid-twentieth century. (From Kasser & Aellen, 1981.)

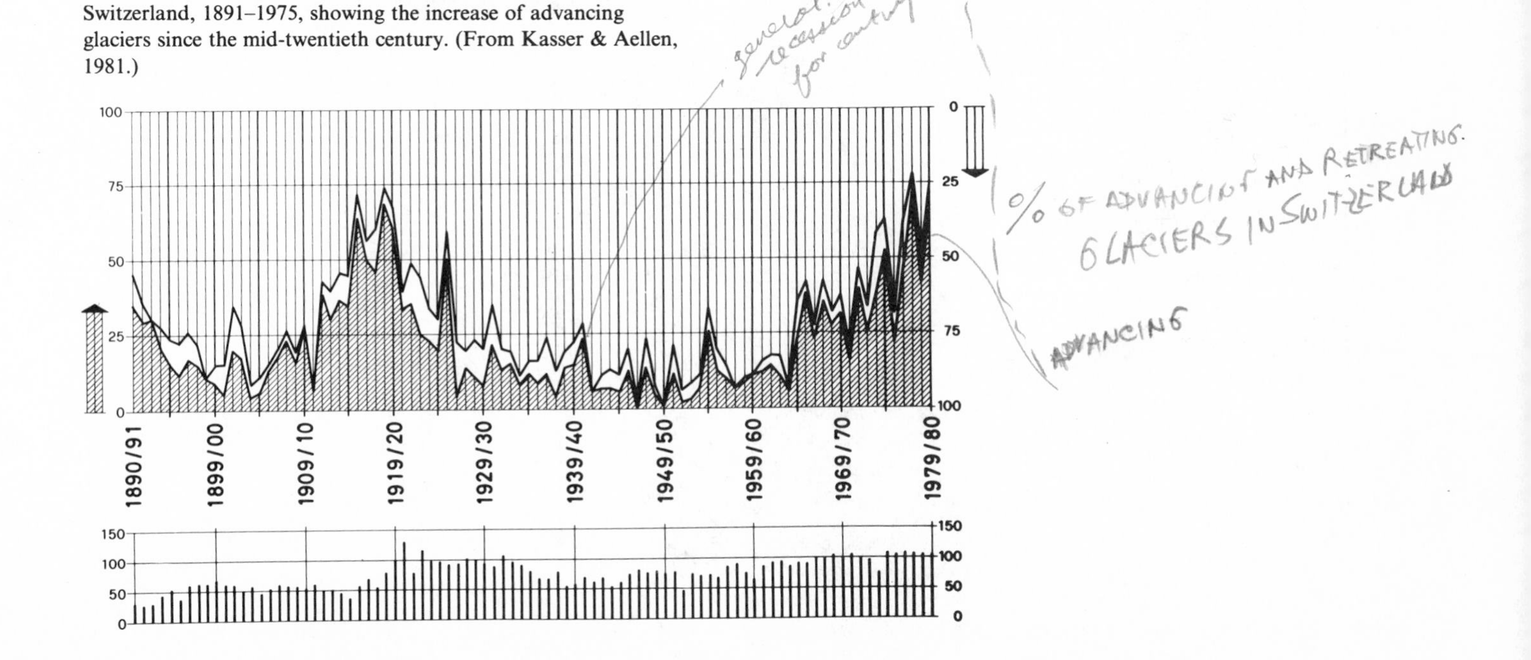

flooding and damage to transportation and other facilities. A surge of an ice stream of Antarctic size could be an event of global consequences (see Section 8.5).

A climatological aspect of mountain glaciers not to be neglected relates to the global average sea level. It is an established observational fact that the mean sea level has been rising for the past century at a rate of about 1 mm per year. While the importance of that phenomenon *per se* may be in question, its cause could signify important climatological events (see also Section 8.5). Based on data reported by the Permanent Service for the Fluctuations of Glaciers (IASH–ICSI, UNESO) and other sources, Meier (1983) concludes that the general recession of mountain glaciers since the mid-nineteenth century, and within the error limits of available data, that recession may have contributed as much as 300 km³ of melt water per year to the world ocean. Spread over 360×10^6 km² of ocean, this amounts to 0.8 mm per year, or a large fraction of the observed sea level rise.

At present the total area of all mountain glaciers on earth is so small that it seems safe to assume that their effect on the surface heat balance on global climate is not appreciable. This may be true even on the time scale of the entire Quaternary Age: the great ice sheets of that era did not descend from high glaciated mountains but formed on the flat terrain of northern North America and Eurasia (see Fig. 8.6).

This should not detract from the need for further study of mountain glaciers. They affect snow accumulation, retention, and runoff in a way desirable for irrigation and the generation of hydroelectric power; and they affect and occasionally threaten human activities in the populated mountainous regions of the world. While it is unlikely that the behavior of mountain glaciers will help to 'explain' global climate, we must learn to understand their response to mesoscale atmospheric phenomena that accompany climatic change.

Furthermore, they are natural-scale laboratories with which the processes affecting larger ice sheets, past and present, can be studied in a practical way. Because of the accessibility and variability of mountain glaciers, their study is likely to yield progress in understanding the sliding mechanism (e.g., Engelhardt *et al.*, 1979).

Fig. 8.15. Permafrost in the northern hemisphere. (From Washburn, 1980.)

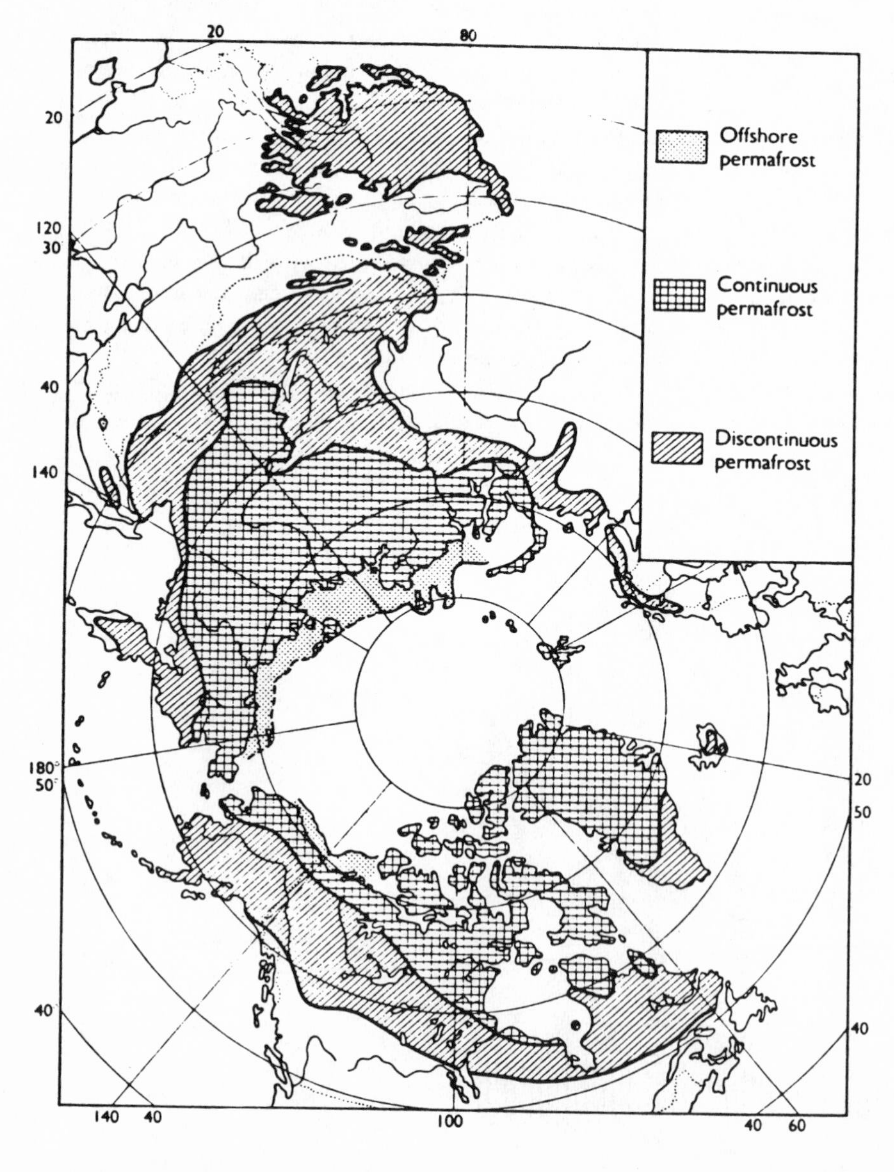

8.7 Permafrost

The definition of permafrost is surface material whose water content is perennially frozen. In many regions the uppermost, so-called active layer, thaws annually to a depth of one to a few meters.

Permafrost underlies about 20% of the earth's land surface. The greatest recorded permafrost depths are 1500 m in Siberia (Markha River) and 600 m in North America (Prudhoe Bay). The present distribution of permafrost in the northern hemisphere is shown in Fig. 8.15. Most of the ground underlying the Greenland and Antarctic ice sheets is assumed to be at temperatures below freezing.

A recent discovery owed to the offshore search for oil is the existence of permafrost under the shallow shelf waters of the Arctic Ocean.

Permafrost results from a delicate equilibrium between surface heat balance and geothermal heat flux, and is affected by the water content and thermal properties of the ground. In theory, vertical profiles of ground temperature should faithfully reflect the accumulated (if damped) trends of surface temperature over long spans of time. The evolution of vegetation, soil structure and composition, and interaction with the hydrosphere, tend to obscure this temperature record. Where these effects are weak, for instance in the cold and arid region of the Alaskan north coast, bore hole temperatures can be interpreted with precision. According to Lachenbruch & Marshall (1969) and Lachenbruch *et al.* (1982), temperature gradients beneath Barrow, Cape Thompson, and Simpson, to a depth of 100 m indicate a warming of 4 °C since the mid-nineteenth century and a cooling of 1 °C after about 1950. Relatively rapid changes occur in the regions of thin, discontinuous permafrost (Brown & Andrews, 1982).

The present-day distribution of continuous and discontinuous permafrost bears little semblance to the present-day distribution of any other physiographic, geologic, botanic, or climatic characteristics. There is no precise relationship between permafrost and modern climatological means. For instance, in Alaska the southern limit of continuous permafrost lies north of the −6–−8 °C mean annual isotherm, in Canada north of −8.5 °C, and in the USSR north of −7 °C. According to some reports, permafrost can be absent in regions with mean annual temperatures well below zero (for an extensive bibliography on permafrost, see Washburn, 1980).

There are indications that part of the existing permafrost is of Pleistocene origin, as evidenced by preserved tissue from Pleistocene animals, temperatures decreasing with depth, maximum permafrost thickness in areas not glaciated during the Pleistocene, and other observations. Even so, present permafrost boundaries in Eurasia and North America bear no semblance to the ice limits of

Fig. 8.16. Mean monthly discharge of four major rivers.

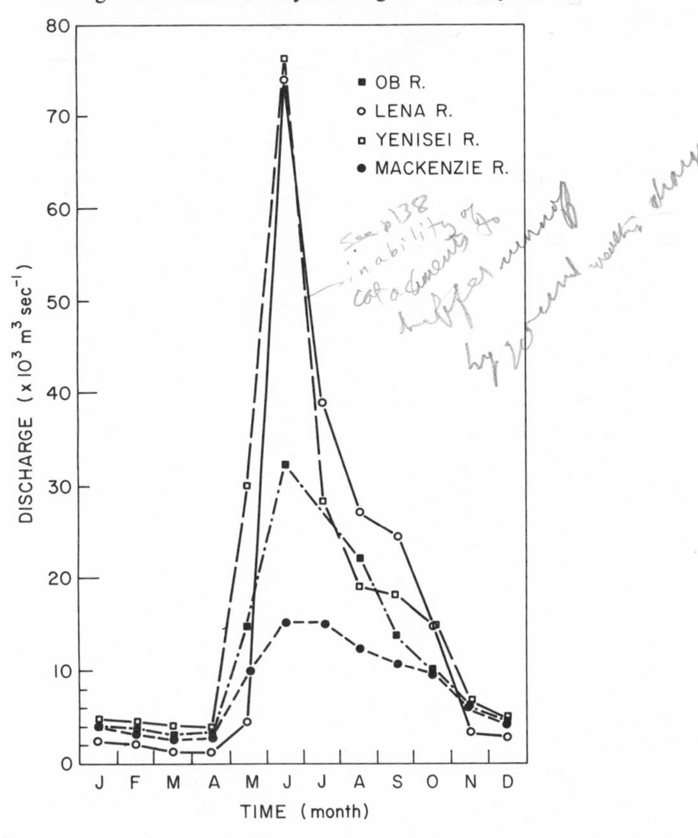

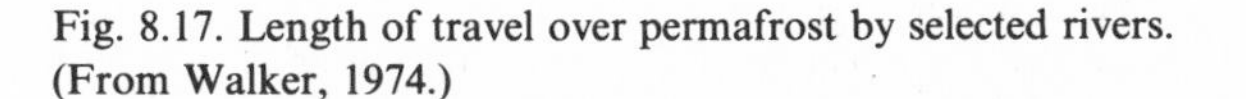

Fig. 8.17. Length of travel over permafrost by selected rivers. (From Walker, 1974.)

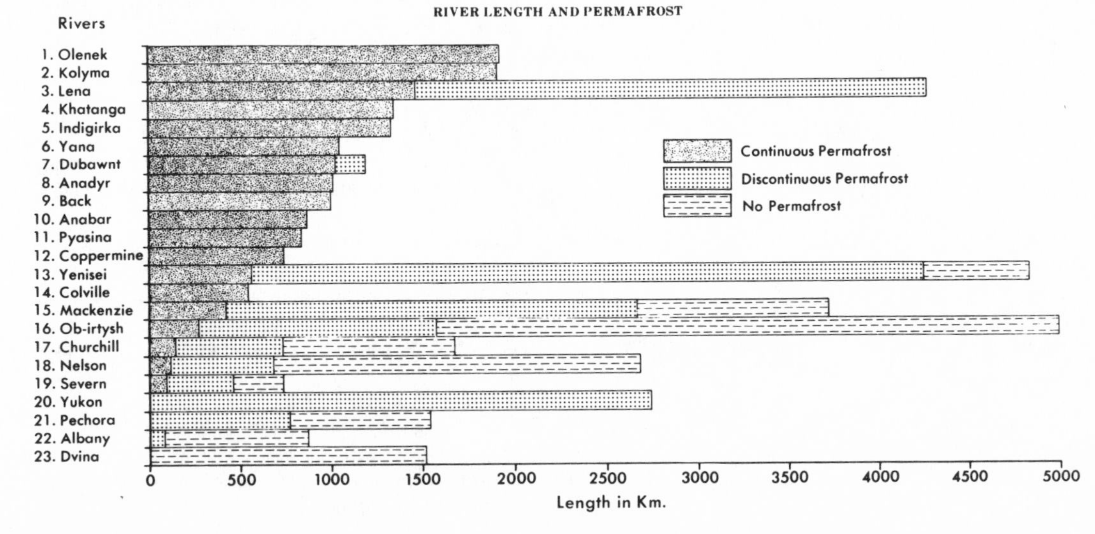

the last major glaciation. Conversely, some of the contemporary permafrost in formerly glaciated regions is of post-glacial origin (Washburn, 1980).

Permafrost is not only constantly and subtly responding to climatic variations, it is also highly susceptible to alteration by human activities, witness the environmental problems encountered in planning the Trans-Alaska pipeline (Lachenbruch, 1970). Perhaps the most important link between permafrost and climate lies in the fact that permafrost inhibits ground-water recharge and movement, restricts plant growth, and enhances runoff.

Like glaciers, permafrost responds to, and integrates, climatic change in a complex fashion that is difficult to unravel. However, the existence and changes of permafrost *per se* are of great economic importance.

Besides its role as a manifestation of long-term past climates, permafrost may have a more direct, short-term influence of climatic phenomena. Fig. 8.16 shows the seasonal discharge of water (hydrograph) of four rivers draining into the Arctic Ocean. All four have catchment areas lying in regions of seasonal snow cover. Only a small part of the course of the McKenzie and Ob rivers are underlain by continuous permafrost, while Yenisei and Lena flow over continuous and discontinuous permafrost most of their way (Fig. 8.17). The strikingly large amplitude of the hydrographs of Yenisei and Lena may, at least in part, be explained by the inability of their catchment to buffer runoff by ground-water charge. The consequence is a spike-shaped addition of fresh water to the shelf waters of the Kara and Laptev Seas and the attendant effects on their convective regime, ice formation, and regional climate.

In this connection, the argument has been advanced by Aagaard & Coachman (1981) that southward diversions of some Siberian river water for irrigation purposes would destabilize Siberian shelf waters and tend to intensify mixing of upper Arctic waters with the warmer Atlantic water masses below, thus possibly retarding or diminishing sea ice formation. Although an intriguing concept, its chance in reality seems small. Fig. 8.18 shows the interannual variability of the rivers Ob and Yenisei (cited from SCOR-WG 58, 1979). It appears that the projected diversions would be but a small fraction (Golubev & Vasiliev, 1978; Golubev *et al.*, 1978) of the natural variability of the river discharge that,

Fig. 8.18. Annual discharge of Yenisei and Ob. (From SCOR–MG 18, 1979.)

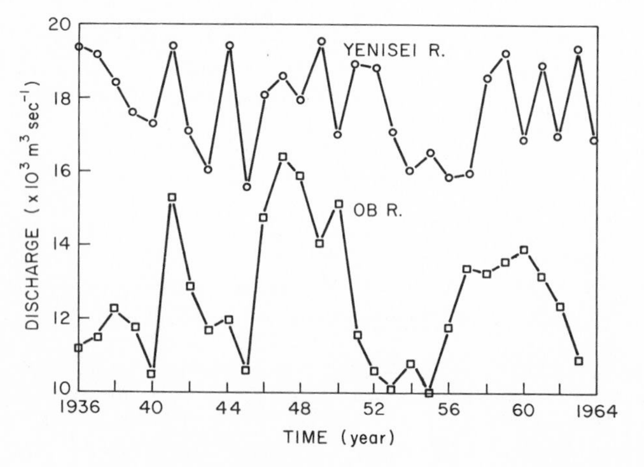

along with the natural variability of the seasonal weather patterns over the Siberian shelf, would make the effects of a man-made interference disappear into climatic noise. Furthermore, recent model calculations by Semtner (1983) of the effect of a reduction of discharge of Ob, Yenisei, Dvina, and Pechora are three times larger than that actually envisioned yields the result that almost no effect is seen in the halocline of the central Arctic. Even in the marginal seas the stability is not eroded to the point of convective instability because the modeled change in surface currents tends to confine the water with increased salinity to the shelf region, and the ice extent in the Barents and Kara Seas remains nearly unchanged.

Nevertheless, the Siberian Shelf is the largest in the world, and it receives an amount of fresh water about equal to one-half of the Amazon River, making it a region where the land hydrologic regime affects a larger area of ocean surface than anywhere else.

Acknowledgements

I am grateful to K. Brian, J. Brown, J. Fletcher, G. Herman, G. Kukla, P. Lemke, M. Meier, G. North, S. Porter, U. Radok, J. Walker, J. Walsh, and L. Washburn for information and advice received during the preparation of the review. R. Moritz, C. Raymond, and A. Thorndike reviewed the manuscript and made many improvements.

References

(*indicates publications with extensive bibliographies)

Aagaard, K. & L. K. Coachman (1981). 'Toward an ice-free Arctic Ocean'. *EOS, Trans. Amer. Geoph. Union*, **56**, 484–6.

Aagaard, K., L. K. Coachman & E. Carmack (1981). 'On the halocline of the Arctic Ocean'. *Deep Sea Res.*, **28A**, 529–45.

Allison, I., ed. (1981). *Sea Level, Ice, and Climatic Change*. Proc. Symp. Gen. Assembly IUGG, IAHS Publ. No. 131, 471 pp.

*Andersen, B. G., W. F. Weeks & J. L. Newton (1980). 'The seasonal ice zone, Proceedings of an International Workshop'. *Cold Regions Science and Technology*, **2**, 357 pp.

Bentley, C. (1982). 'Response of the west Antarctic ice sheet to CO_2-induced climatic warming. *In: Carbon Dioxide Effects Research and Assessment Program*, Vol. II, Part 1, US Dept of Energy, Washington, DC, 32 pp.

Bindschadler, R. A. & R. Gore (1982). 'A time-dependent ice sheet model: preliminary results'. *J. Geophys Res.*, **87 (C12)**, 9675–85.

*Brooks, C. E. P. (1949). *Climate through the Ages*, 2nd edn. Ernst Benn, London.

Brown, J. & J. T. Andrews (1982). 'Influence of short-term climate fluctuations on permafrost terrain'. *Carbon Dioxide Effects Research and Assessment Program*, Dept. of Energy, Washington, DC, Vol. II, Part 3, 30 pp.

Brown, R. A. (1981). 'Modelling the geostrophic drag coefficient for AIDJEX'. *J. Geophys. Res.*, **86**, 1989–94.

*Budd, W. F. (1981). 'The importance of ice sheets in long term changes of climate and sea level'. *In: Sea Level, Ice, and Climatic Change*. I. Allison, ed., Proc., Symp. Gen. Assembly IUGG, IAHS Publ. No. 131, pp. 441–71.

Budyko, M. I. (1969). 'The effect of solar radiation variations on the climate of the earth, **21**, 611–19.

Budyko, M. I. (1972). 'The future climate'. *EOS, Trans. Amer. Geoph. Union*, **53**, 868–74.

Carsey, F. D. & H. J. Zwally (1983). 'Remote sensing as a research tool'. *In: The Geophysics of Sea Ice*. N. Untersteiner, ed., Plenum Press, New York, (in preparation).

CLIMAP (Climatic Long-range Interpretation, Mapping, and Prediction) (1981) '*Seasonal reconstruction of the earth's surface at the last glacial maximum*'. *Geol Soc. of America. Map and Chart Series*, MC-36, 18 pp.

Croll, J. (1867). 'On the change in the obliquity of the ecliptic, its influence on the climate of the polar regions, and the level of the sea'. *Phil. Mag.*, **33**, 426–45.

Dewey, K. F. & R. Heim (1981). 'Satellite observations of variations in northern hemisphere seasonal snow cover'. *NOAA Technical Report NESS 87*, Washington, DC, 83 pp.

Donn, W. L. & D. M. Shaw (1966). 'The stability of an ice-free Arctic Ocean'. *J. Geophys. Res.*, **71**, 1087–95.

*Doronin, Yu. P. & D. E. Kheisin (1975). *Sea Ice*. Published in English translation by Amerind, New Delhi, for National Science Foundation, Office of Polar Programs, 1977, 323 pp.

Engelhardt, H. F., W. D. Harrison & B. Kamb (1979). 'Basal sliding and condition at the glacier bed as revealed by borehole photography'. *J. Glaciology*, **20**, 469–508.

*Fletcher, J. O. (1965). 'The heat budget of the Arctic Basin and its relation to climate.' *Rep. RM-444-PR, Rand Corporation*, Santa Monica, Calif., 179 pp.

Foster, T. D. (1975). 'Heat exchange in the upper Arctic Ocean'. *AIDJEX Bulletin*, No. 28, Univ. Washington, Seattle, Wash., pp. 151–66.

Fu Congbin & J. O. Fletcher (1983). 'The role of the surface heat source over the Tibetan Plateau in interannual variability of the India summer monsoon'. *J. Clim. and Appl. Met.*, submitted.

*GARP (ICSU–WMO) (1975). *The Physical Basis of Climate and Climate Modeling*. GARP Publ. Ser. No. 16, Geneva, 265 pp.

*GARP (ICSU–WMO) (1978). *The Polar Sub-programme*. GARP Publ. Ser. No. 19, Geneva, 47 pp.

Golubev, G. & O. Vasiliev (1978). 'Interregional water transfers as an interdisciplinary problem.' *Water Supply and Management*, **2**, 67–77.

*Golubev, G., G. Grin, G. Melnikova & G. Whetstone (1978). 'Bibliography in interregional water transfers'. *Water Supply and Management*, **2**, 187–211.

Gordienko, P. (1958). 'Arctic ice drift'. *In: Arctic Sea Ice*. Natl. Acad. Sci. Publ. No. 589, Washington, DC, 210–22.

Gornitz, V., S. Lebedeff & J. Hansen (1982). 'Global sea level trend in the past century'. *Science*, **215**, 1611–14.

GUGK (1966). *Atlas Antarktiki*. Glavnoe Upravlenii Geogezii i Kartografii, Moscow (in Russian).

Hahn, D. G. & J. Shukla (1976). 'An apparent relationship between Eurasian snow cover and Indian monsoon rainfall'. *J. Atm. Sci.*, **33**, 2461–2.

Hahn, D. G. (1981). 'Summary requirements of GCMs for observed snow and ice cover data'. *In: Snow Watch 1980*. G. Kukla, A. Hecht & D. Wiesnet, eds., World Data Center A–Glaciology, Univ. of Colorado, Boulder, Colo., pp. 45–53.

Hasselmann, K. (1976). 'Stochastic climate models'. *Tellus*, **28**, 473–84.

Hibler, W. D. (1979). 'A dynamic thermodynamic sea ice model'. *J. Phys. Oceanog.*, **9**, 815–46.

*Hibler, W. D. (1980). 'Sea ice growth, drift, and decay'. *In: Dynamics of Snow and Ice Masses*. S. C. Colbeck, ed., Academic Press, New York, pp. 141–209.

Hibler, W. D. & J. E. Walsh (1982). 'On modelling seasonal and interannual fluctuations of arctic sea ice.' *J. Phys. Oceanogr.*, **12**, 1514–23.

*Hoinkes, H. C. (1968). 'Glacier variations and weather'. *J. Glaciology*, **7**, 3–19.

Hunkins, K. L. & H. Kutschale (1965). 'Quaternary sedimentation in the Arctic Ocean'. *Proc. 7th INQUA Congress*.

*Husseiny, A. A., ed. (1978). *Iceberg Utilization*. Pergamon Press, New York, 760 pp.

Johanessen, O. M., W. D. Hibler, P. Wadhams, W. J. Campbell, K. Hasselmann & I. Dyer (1982) 'MIZEX, a program for mesoscale air-ice-ocean interaction experiments in arctic marginal ice zones'. Draft manuscript, 41 pp.

Källén, E., C. Crafoord & M. Ghil (1979). 'Free oscillations in a climate model with ice-sheet dynamics'. *J. Atmos. Sci.* **36**, 2292–303.

Kasser, P. & M. Aellen (1981). *Switzerland and Her Glaciers*. Swiss National Tourist Office, Kümmerly and Fry, Bern, 191 pp.

*Kruss, P. D. (1981). 'Numerical modeling of climatic change from the terminus record of Lewis Glacier, Mount Kenya'. Ph.D. Dissertation, Univ. of Wisconsin, Madison, Wisc., 128 pp.

Ku, T. L. & W. S. Broecker (1965). 'Rates of sedimentation in the Arctic Ocean'. *Proc. 7th INQUA Congress*.

Kukla, G. (1981). 'Snow covers and climate'. *In: Snow Watch 1980*. G. Kukla, A. Hecht & D. Wiesnet, eds., World Data Center A–Glaciology, Univ. of Colarado, Boulder, Colo., pp. 27–39.

*Kukla, G., A. Hecht & D. Wiesnet, eds. (1981). *Snow Watch 1980*. Report GD-11, World Data Center A–Glaciology, Univ. of Colorado, Boulder, Colo., 147 pp.

Lachenbruch, A. H. & B. V. Marshall (1969). 'Heat flow in the Arctic'. *Arctic*, **22**, 300–11.

Lachenbruch, A. H. (1970). 'Some estimates of thermal effects of a heated pipeline in permafrost'. *US Geological Survey, Circular No. 632*, 23 pp.

Lachenbruch, A. H., J. H. Sass, B. V. Marshall & T. H. Moses (1982). 'Permafrost, heat flow, and the geothermal regime at Prudhoe Bay, Alaska'. *J. Geophys. Res.*, **87, B11**, 9301–16.

Lorius, C., L. Merlivat, J. Jouzel & M. Pourchet (1979). 'A 30000-year isotope climatic record from Antarctic ice.' *Nature*, **280**, 644–8.

Manabe, S. & R. J. Stouffer (1980). 'Sensitivity of a global climate model to an increase of CO_2 concentration in the atmosphere'. *J. Geophys. Res.*, **85**, 5529–54.

Manabe, S. & R. T. Wetherald (1980). 'On the distribution of climate change resulting from an increase in CO_2 content of the atmosphere'. *J. Atmos. Sci.*, **37**, 99–118.

McPhee, M. G. (1980). 'A study of oceanic boundary layer characteristics including inertial oscillations at three drifting stations in the Arctic Ocean'. *J. Phys. Ocean*, **10**, 870–84.

*Meier, M. F. (1983). 'Snow and ice in a changing hydrological world'. *Hydrol. Sci. Jour.*, **28** (1), 3–22.

Mercer, J. H. (1978). 'West Antarctic ice sheet and CO_2 greenhouse effect: a threat of disaster'. *Nature*, **271**, 321–5.

*Namias, J. (1975). *Short Period Climate Variations: Collected Works of J. Namais, 1934–1975*. Univ. of California, San Diego, 905 pp.

Namias, J. (1981). 'Snow covers in climate and long range forecasting'. *In: Snow Watch 1980*. G. Kukla, A. Hecht & D. Wiesnet, eds., World Data Center A–Glaciology, Univ. of Colorado, Boulder, Colo., pp. 13–26.

National Academy of Sciences (1974). *US Contribution to the Polar Experiment (POLEX–GARP–North)*. Washington, DC, 119 pp.

National Academy of Sciences (1975). *US Contribution to the Polar Experiment (POLEX–GARP–South)*. Washington, DC, 37 pp.

North, G. R. & R. F. Cahalan (1981). 'Predictability in a solvable stochastic climate model'. *J. Atmos Sci.*, **38**, 504–13.

*North, G. R., R. F. Cahalan & J. A. Coakley (1981). 'Energy balance climate models'. *Rev. Geoph. and Space Phys.*, **19**, 91–121.

North, G. R., F. J. Moeng, T. L. Bell & R. F., Cahalan (1982). 'The latitude dependence of the variance of zonally averaged quantities'. *Mon. Wea. Rev.*, **110**, 319–26.

*Orvig, S. (1970). 'Climates of the polar regions'. *In: World Survey of Climatology*. H. Landsberg, ed., Vol. 14, 370 pp.

Parkinson, C. L. & W. M. Washington (1979). 'A large scale numerical model of sea ice'. *J. Geophys. Res.*, **84**, 311–37.

Partl, R. (1977). 'Power from the glacier: the hydropower potential of Greenland's glacial waters'. *Int. Inst. for Appl. Syst. Analysis*, Laxenburg, Austria, RR-77-20, 54 PP.

*Paterson, W. S. B. (1980). 'Ice sheets and ice shelves'. *In: Dynamics of Snow and Ice Masses*. S. Colbeck, ed., Academic Press, pp. 1–78.

Petit, J.-R., M. Briat & A. Royer (1981). 'Ice age aerosol content from east Antarctic ice core samples and past wind strength'. *Nature*, **293**, 391–4.

*Polar Group (1980). 'Polar atmosphere-ice-ocean processes: A review of polar problems in climate research'. *Rev. Geoph. and Space Phys.*, **18**, 525–43.

*Porter, S. (1981). 'Glacial evidence of Holocene climatic change'. *In:*

Climate and History. Wigley, P. M. L., M. G. Ingram & G. Farmer, eds., Cambridge Univ. Press, 82–110.

*Pritchard, R. S., ed. (1980). *Sea Ice Processes and Models*. Univ. of Washington Press, Seattle, Wash., 474 pp.

*Radok, U. (1978). 'Climatic roles of ice: A contribution to the International Hydrological Programme (IHP)'. *Hydrol. Sci. Bull.*, **23**, 333–54.

Rasmussen, L. A. & M. F. Meier (1982). 'Continuity equation model of the predicted drastic retreat of Columbia Glacier, Alaska'. *Geol. Survey Professional Paper* 1258–A, Washington, DC, 23 pp.

*Raymond, C. F. (1980) 'Temperate valley glaciers'. *In: Dynamics of Snow and Ice Masses*. S. C. Colbeck, ed., Academic Press, pp. 80–139.

Raynaud, D. & B. Lebel (1979). 'Total gas content and surface elevation of polar ice sheets'. *Nature*, **281**, 289–91.

Robin, G. de Q. (1981). 'Climate into ice: the isotopic record in polar ice sheets'. *In: Sea Level, Ice, and Climatic Change*. I. Allison, ed., Proc. Symp. Gen. Assembly IUGG, IAHS Publ. No. 131, pp. 207–15.

Robok, A. (1980). 'The seasonal cycle of snow cover, sea ice, and surface albedo'. *Mon. Weath. Rev.*, **108**, 267–85.

*Rothrock, D. A. (1975). 'The mechanical behavior of pack ice'. *Ann. Rev. Earth and Planetary Sci.*, **3**, 317–42.

Scientific Committee for Antarctic Research (SCAR) (1981). *Report of the SCAR Group of Specialists on Antarctic Climate Research.* Manuscript, 63 pp.

*Schneider, S. & R. Dickinson (1975). 'Climate modeling'. *Rev. Geophys. and Space Phys.*, **12**, 447–93.

*SCOR (1979). *The Arctic Ocean Heat Budget*. SCOR Working Group 58. Published by Geophys. Inst., Univ. of Bergen, 98 pp.

Sellers, W. D. A. (1969). 'A global climatic model based on the energy balance of the earth–atmosphere system'. *J. Appl. Met.*, **8**, 392–400.

Semtner, A. J. (1983). 'The climatic response of the Arctic Ocean to Soviet river diversions'. Unpublished manuscript.

Smith, I. N. & W. F. Budd (1981). 'The derivation of past climate from observed changes of glaciers'. *In: Sea Level, Ice and Climatic Change*. I. Allison, ed., Proc. Symp. Gen. Assembly IGG, IAHS Publ. No. 131, pp. 31–52.

Stauber, H. (1967). Anlage von Schmelzwasserkraftwerken im ewigen Eis. Austrian Patent No. 252 131 on February 1967. (Similar patents awarded in Norway, Switzerland, Canada, Iceland, USA, FRG, and Denmark.)

Tarling, D. H. (1978). 'The geological–geophysical framework of ice ages'. *In: Climatic Change*. J. Gribbin, ed., Cambridge Univ. Press, pp. 3–24.

Thomas, R. H. (1979). 'The dynamics of marine ice sheets'. *J. Glac.*, **24** (90), 167–78.

Thompson, L. G. & E. Mosley-Thompson (1981). 'Microparticle concentration variations linked with climatic change: evidence from polar ice cores'. *Science*, **212**, 812–15.

Thorndike, A. S., D. A. Rothrock, G. A. Maykut & R. Colony (1975). 'The thickness distribution of sea ice'. *J. Geophys. Res.*, **80**, 4501–13.

Thorndike, A. S. & R. Colony (1982). 'Sea ice motion in response to geostrophic winds'. *J. Geophys. Res.*, **87 (C8)**, 5845–52.

Tyndall, J. (1861). *The Glaciers of the Alps*. Cambridge Univ. Press, 446 pp.

Untersteiner, N. & J. F. Nye (1968). 'Computations of the possible future behavior of Berendon Glacier, Canada'. *J. Glaciology*, **7**, 205–13.

Untersteiner, N. & A. S. Thorndike (1982). 'Arctic data buoy program'. *Polar Record*, **21**, 127–35.

Untersteiner, N. (1983*a*) 'Ocean climate studies in the arctic seasonal ice zone.' *CCCO, Geneva*, 24 pp., in press.

*Untersteiner, N., ed. (1983*b*) *The Geophysics of Sea Ice*. NATO Advanced Study Institute of Air–Sea–Ice Interaction, Plenum Press, New York, in preparation.

*Wadhams, P. (1981). 'The ice cover in the Greenland and Norwegian Seas'. *Rev. Geoph. and Space Phys.*, **19**, 345–93.

Walcher, J. (1773). *Nachrichten von den Eisbergen in Tyrol*. Vienna, Josef Kurzböck.

Walker, G. (1947). 'Arctic conditions and world weather'. *Quart. J. Roy. Met. Soc.*, **73**, 226–56.

Walker, H. J. (1974). 'The Colville River and the Beaufort Sea: some interactions'. *In: The Coast and Shelf of the Beaufort Sea*. J. C. Reed & J. E. Sater, eds., the Arctic Institute of North America, Washington, DC, 513–39.

*Walsh, J. E. (1983). 'The role of sea ice in climate variability'. *Atmosphere–Ocean*, in press.

*Washburn, A. L. (1980). *Geocryology, A Survey of Periglacial Processes and Environments*. John Wiley, New York, 406 pp.

Weyprecht, K. (1875). 'Die Erforschung der Polarregionen'. *K.U.K. Geogr. Ges., Mitt.* (Vienna), **18**, 357–66.

Wiese, W. (1924). 'Polareis und atmospherische Schwankungen'. *Geogr. Ann.*, **6**, 273–99.

Zwally, H. J., R. A. Bindschadler, A. C. Brenner, T. V. Martin & R. H. Thomas (1983). 'Surface elevation contours of Greenland and the Antarctic ice sheets. *J. Geophys. Res.*, **88(C3)**, 1589–96.

The upper ocean and air–sea interaction in global climate

J. D. Woods
Insitut für Meereskunde an der Universität Kiel,
FR Germany

Abstract

The subject is presented in three parts. The first is concerned with establishing the role of the ocean in the planetary climate system, concentrating on the following themes: the global energy cycle, the global hydrological cycle, seasonal variation, interannual variation, the ocean as a sink of pollution, and palaeoclimatological changes. The aim of the first part is to establish a balanced picture of the ocean in climate from a planetary perspective, before embarking on detailed consideration of particular research problems.

The second part focusses on the task of the World Climate Research Programme (WCRP), to establish to what extent climate can be predicted, and the extent of man's influence on climate. The aim here is to clarify what one needs to know about air–sea interaction in order to achieve the WCRP objectives. This is done in three sections corresponding to the three streams of the WCRP strategy: (1) extended-range weather forecasting (order months), (2) short-term climate prediction (order years), and (3) longer-term climate prediction (order decades). In all cases the problem is to determine how the GPFO (the global patterns of fluxes of energy, water and gases from the ocean to the atmosphere) evolve during the period of the forecast, and how to measure the seasonal and interannual variations of the GPFO occurring now, with sufficient accuracy to permit tests of climate models in which the GPFO are predicted. An essential ingredient of such GPFO prediction is a model of the upper ocean boundary layer (OBL) that works accurately everywhere around the globe at all seasons. The emphasis in such OBL models depends on the time scale of climate prediction, i.e. on which stream of the WCRP is being considered. Detailed discussion of the OBL modelling problem for each stream leads to specifications for research in the context of the WCRP objectives

The third section reviews the present state of knowledge about the GPFO and OBL. In general it is concluded that existing knowledge and existing techniques for monitoring the surface fluxes and predicting changes in them, and in the OBL, do not meet the specifications established in part two. New methods, satellite sensing in particular, offer to improve our ability to monitor the GPFO, but even so the expected uncertainties are likely to remain unacceptably large in many cases. Contemporary methods of modelling the upper boundary layer may well be improved to meet the requirements of extended-range forecasting and short-term climate prediction, but new methods will be needed to meet the requirements of longer-term climate prediction, exemplified by the effect of CO_2 pollution over the next few decades. The role of geostrophic advection in determining OBL structure and therefore the rate of water mass conversion is discussed; as also is the effect of the OBL on ocean circulation dynamics, and inclusion of the OBL in global climate models

It is concluded that the present state of knowledge about air–sea interaction and the upper ocean is inadequate for the objectives of the WCRP, and that new methods are urgently needed to improve our ability to monitor the GPFO, and to model the structure of the OBL. Nevertheless, the upper ocean holds the key to climate prediction.

9.1 Introduction

The central role of the oceans in climate is acknowledged in every book on the subject, but the justification varies considerably depending on the author's viewpoint. In this chapter I will introduce the subject from a global perspective. Putting aside for a moment the practical applications of climate research, we will consider climate as a problem in planetary science. To an external observer the climate of a planet is its (atmospheric) response to solar and geothermal heating. The relative importance of the two heat sources depends on the planet's size and proximity to the sun, and on the time scale of interest. For the inner planets, including earth, direct geothermal heating of the atmosphere is negligible compared with solar heating; but geothermally powered plate tectonics can influence the climate, and does so dramatically on earth because there is a world ocean. Taking the broad perspective offered by comparative planetary climatology, the effect of the ocean on earth's climate can be summarized under three headings:

(1) It influenced the evolution of the planet; leading to the present form and composition of the land masses, the composition of the atmosphere, and the existence of a terrestrial biosphere.
(2) It responds to solar heating by evaporation, supplying water for a vigorous hydrological cycle on the continents, including the glaciers that form on continents drifting by the poles, a necessary condition for ice ages.
(3) It is one of the major elements of the present-day climate system, acting as an equal partner with the atmosphere in the global circulations that balance the planetary energy budget of the earth.

Here we are concerned with the role of the ocean in the present climate of the earth, and in particular the role of the upper ocean and air–sea interaction. (Oceanic circulation is considered by Wunsch in Chapter 10.) Successive sections of the present chapter progress from broad issues to specific research activities; establishing first the scientific problem and specifications for research, then reviewing progress so far and assessing the results.

The first section considers five aspects of the ocean's role in the planetary climate system; the mean energy and hydrological cycles, seasonal and interannual variations and the oceanic sink of pollution. The second section focusses on the influence of surface fluxes (of energy, water, gases and momentum) on climate prediction, paying special attention to the role of the upper boundary layer of the ocean. The next two sections contain reviews of contemporary research on the surface fluxes and the OBL respectively, and a discussion of their inclusion in models of global climate. The chapter closes with an assessment of the present state of knowledge in these two subjects in the context of climatology, and suggests priorities for future research. Prospects seem good for significant advances by the end of the century.

9.2
The role of the ocean in the planetary climate system,

9.2.1
The global energy cycle

9.2.1.1
Oceanic heating

Approximately 80% of the solar energy intercepted by our planet (173 petawatt) enters the atmosphere over the oceans (Fig. 9.1). About 50% of this energy flux reaches the bottom of the atmosphere after 25% has been reflected by, and 19% absorbed in, the atmosphere. Neglecting atmospheric bias between continental and oceanic regions, the ocean receives 40% and the continents 10% of the intercepted energy. After allowing for the oceanic albedo (6%) we conclude that the direct solar heating of the ocean has an annual mean of about 65 PW, nearly twice that for the atmosphere (about 33 PW). The ocean is the principal initial recipient of energy entering the planetary climate system. It also absorbs about two-thirds of the downward thermal radiation from the atmosphere (108 PW to the ocean). So the global and annual average heating rate of the ocean is about 175 PW (all values from Fortak 1979).

Almost all of this radiative flux into the ocean is absorbed in the top 100 m. The molecular absorption of radiation varies strongly with wavelength, so that the infra-red (IR) component of the solar input and all the atmospheric input are absorbed in the top centimetre, while the blue–green sunlight can sometimes reach deeper than 100 m. The vertical distribution of oceanic heating therefore depends on the spectrum of the surface irradiance. It also depends significantly on the turbidity of the sea water (Jerlov 1976), due mainly to the presence of plankton, tiny plants and animals whose concentration and depth distribution exhibit dramatic seasonal variations (Falkowski 1980). The rate of solar heating decreases quasi-exponentially with depth. If there were no heat transport, that would generate a temperature profile also of quasi-exponential form. The high specific heat of sea water prevents the amplitude of the temperature from growing too rapidly each day; it rarely attains 1 mk/day at a depth of 100 m even at the summer solstice (Fig. 9.2). However, the coefficient of thermal expansion of sea water is large too, so that the modest temperature gradient generated by solar heating in a single day can produce a density gradient sufficient to quench turbulence in the boundary layer. The result is that vertical turbulent transport of heat and other scalars, and momentum, is significantly affected by solar heating of the upper ocean.

9.2.1.2
Vertical heat flow in the ocean

It is normally assumed that the ocean is broadly in thermal equilibrium with the planetary climate system. That is because the ventilation time of the ocean is short compared with the interval

Fig. 9.1. The zonally averaged distribution of energy incident on the top of the atmosphere in one year, showing the partition between the continents (black) and the oceans (shaded) in each 5° band. More than half of the energy entering the earth's climate system is first absorbed inside the ocean.

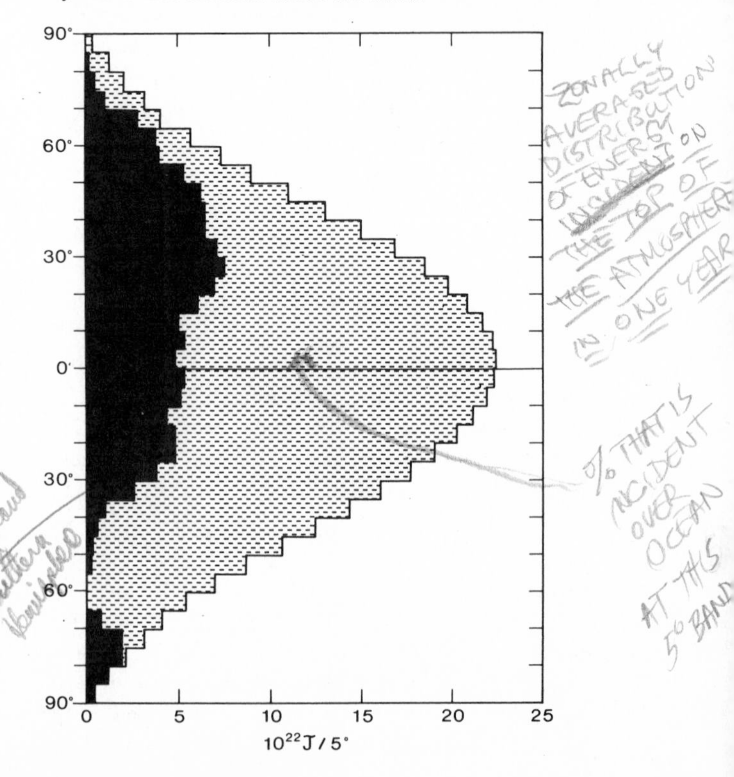

Fig. 9.2. The meridional variation of the maximum depth of solar heating of the ocean, and the annual maximum and minimum depths of the mixed layer along 30° W (from Woods, Barkmann & Horch 1984). The maximum depth of solar heating is here defined as the depth below which the annual accumulation of heat is 10MJ/m² y for Jerlov-type 1A water and no cloud. The depth at which it is 100 MJ/m² y is shown for comparison; that is equal to the annual heat accumulation resulting from doubled CO_2 according to Ramanathan (1981). Water warmed by the sun is never mixed in region A, is mixed during the winter in region B, and is mixed during all months of the year in region C. In region D, winter mixing extends below the maximum depth of solar heating.

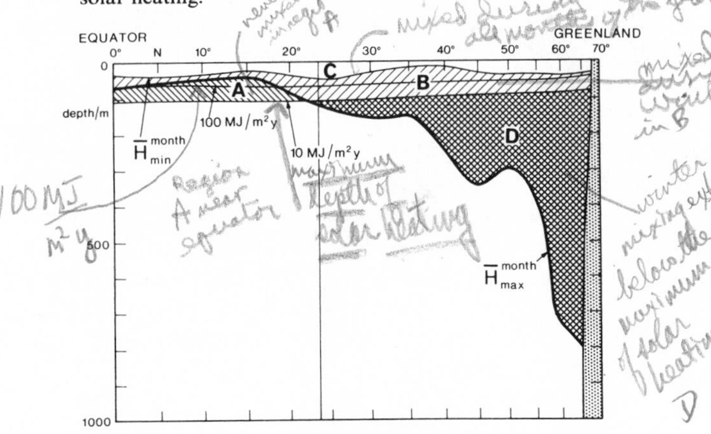

since the recovery from the last glaciation. So on average the ocean is losing energy at the same rate as it is received from the sun and atmosphere. But because the energy is absorbed at depth it must be transported upwards to the surface before it can be lost to the atmosphere. This upward heat flux takes time. The delay before the energy emerges from the ocean depends on the vertical distance to be travelled and the transport rate. Four transport processes are important: molecular conduction, up- and downwelling, convection, and turbulence.

Molecular conduction is present everywhere all the time, but is so slow that it contributes significantly only in the surface skin where the flow is laminar (apart from occasional white-caps) and the vertical distance short (about 1 cm). In calm weather the deeper flow may also be laminar, but the depth is too great for molecular conduction to effect significant transport. The dominant vertical transport mechanism is then upwelling (the oceanic equivalent of subsidence in the atmospheric boundary layer), which can raise water heated by the sun towards the surface at speeds of order 0·1 m/day. But downwelling at a similar rate (due to Ekman pumping) is more prevalent; it carries heat downwards against the climatological trend in a broad swath across the ocean (Fig. 9.3). The third process is buoyant convection driven by the sea water density increase accompanying the surface heat loss to the atmosphere. Except in calm, sunny weather, when the sun heats the skin of the ocean fast enough for molecular conduction to carry the

Fig. 9.3. The annual mean and seasonal mean distributions of the wind stress curl (10^{-8} kg/(m² s)) on the world ocean. (After Hantel 1972.) These maps are based on analysis of ship observations; greatly improved accuracy and systematic coverage will be obtained from satellite scatterometer measurements starting in the late 1980s.

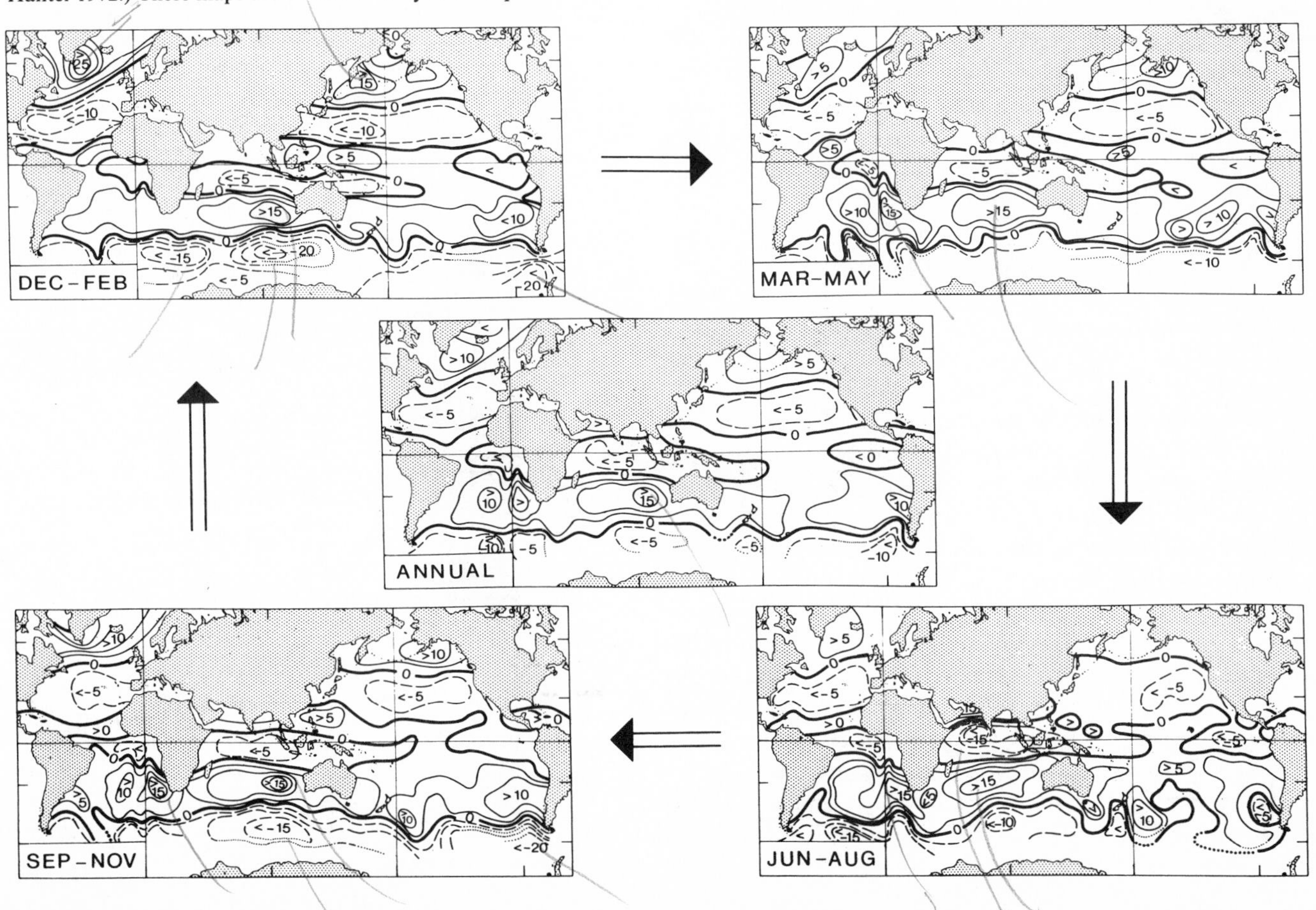

Fig. 9.4. (*a*). The daily maximum and minimum depths of the mixed layer at 43°N, according to Hofmann's (1982) model, at a site where the annual surface heat flux is zero. (*b*). The regional variation of the annual maximum depth of the convection of the convection layer in the north east Atlantic, deduced from bathythermograph data by Robinson, Bauer & Schroeder (1979). A somewhat different map has been published by Levitus (1982) on the basis of hydrographic data and a density criterion. Both show that the NE Atlantic is a region in which the upward heat flow in winter extends over many hundreds of metres. It has not yet proved possible to simulate this regional variation of annual maximum depth of the mixed layer. (*c*). The locations of the few sites at which is known that convection can extend to the ocean bed, according to Killworth (1983). Convection at these sites ventilates the deepest levels of the ocean at a rate that gives water mass residence times of many centuries. Deep polar convection created the cold water sphere in the last 30 million years.

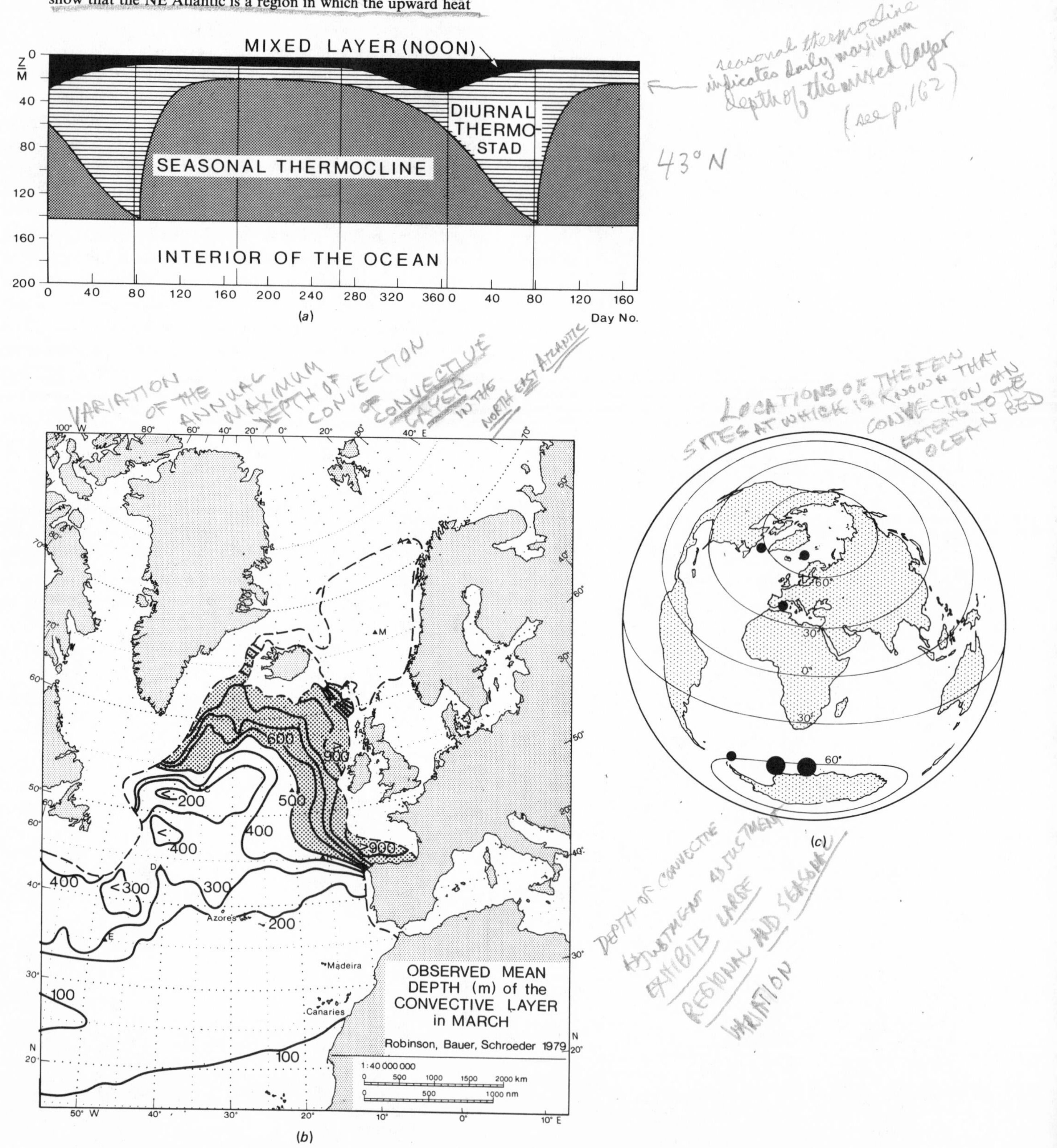

upward heat flux, the water becomes statically unstable and convection extends down to the depth of convective adjustment. In this *convectively mixed layer*, heat flows efficiently up towards the surface; the temperature decreases upwards, but so weakly that the mixed layer is normally thought of as being homogenous. The depth of convective adjustment exhibits large regional and seasonal variation (Fig. 9.4). It is one of the major oceanic variables in climate.

The fourth process is conduction by turbulence, which is always present in the mixed layer. The sources of turbulent kinetic energy (TKE) in the mixed layer are buoyant convection (loss of potential energy), wind–wave breaking and the Reynolds stress. Turbulence is also present, though less energetically and intermittently, below the mixed layer, where the sources of TKE are downward (turbulent) diffusion from the mixed layer and local generation by the Reynolds stress, internal wave breaking and double-diffusive convective. Models for calculating the vertical profiles of TKE and eddy diffusivity, $K = K(TKE, N^2)$, will be reviewed later. The vertical flux of a scalar depends on the eddy diffusivity and the scalar concentration gradient. Turbulent heat transport is something of a mixed blessing for climate: it is directed upwards in the convection layer, but downwards below. The mixed layer is cooled by this downward heat flux, reducing the rate of heat loss to the atmosphere. (Other scalars vary: oxygen is mixed down, nutrients are mixed up.) The intensity of TKE in the mixed layer is of little consequence for climate because convective overturning ensures that vertical scalar gradients are weak: below the mixed layer, where scalar gradients are strong, turbulent diffusion can be vital, especially in the tropics (Niiler 1982).

9.2.1.3

Energy release to the atmosphere

Energy is released from the ocean to the atmosphere in three ways:

≈ 80% by thermal radiation in the infra-red;
≈ 16% by evaporation, as latent heat;
≈ 4% by conduction, as sensible heat.

The energy is drawn from the oceanic mixed layer, whose temperature changes only slowly because the specific heat of sea water is high (4 MJ/(m³ K)). The three types of energy flowing into the atmosphere heat it in different ways. Some of the radiant energy is absorbed in the boundary layer, the rest passes through it to be absorbed in the free atmosphere above, or to escape directly to space. The sensible heat flux warms the atmospheric boundary layer. The latent heat does not heat the atmosphere until the water vapour carrying it condenses in clouds, and the air is then only warmed irreversibly when the water is precipitated back to the ocean (directly or after continental run-off). Some of the latent heat conversion occurs in clouds in the boundary layer, but most occurs higher in the atmosphere, and far away from the site of oceanic evaporation. The thermal capacity of air is so small that its temperature seldom deviates by more than a degree or so from that of the underlying oceanic mixed layer (Roll 1965, p. 326). The interface has an intermediate temperature, reflecting the sharp temperature gradient in the laminar flow skin where the vertical heat flux is carried by molecular conduction.

The magnitudes of the surface fluxes depend on the properties of the two mixed layers. The *thermal radiation* depends on the temperature of the surface skin, which may be a few tenths of a degree colder than the ocean mixed layer temperature. It is unaffected by the properties of the overlying air, and is therefore remarkably steady. Often, however, it is convenient to consider not the rate of radiative cooling of the ocean (80% of the total), but its excess over the heating of the ocean by thermal radiation from the atmosphere (which is less than the latent heat flux). The excess, known as the net longwave radiation at the sea surface, varies both with the ocean skin temperature and the atmospheric temperature, humidity and cloudiness. The *sensible heat flux* depends on the temperature difference between the two mixed layers, on the wind speed and on the spectrum of waves on the interface. The *latent heat flux* depends on the saturated vapour pressure of the ocean (a function of temperature and salinity) and the surface humidity pressure in the mixed layer of the atmosphere. The greatest surface

Fig. 9.5. The net annual energy flow through the sea surface, according to Budyko (1974). The zero line divides the world ocean into two regions. In the tropics more energy is received each year from the sun than is released to the atmosphere. The surplus is stored and transported by ocean currents to the second region, where it is released to the atmosphere. No such storage and transport occurs on the continents, where the annual energy budget is balanced everywhere. The simulation of that continental energy balance provides a useful test for climate models.

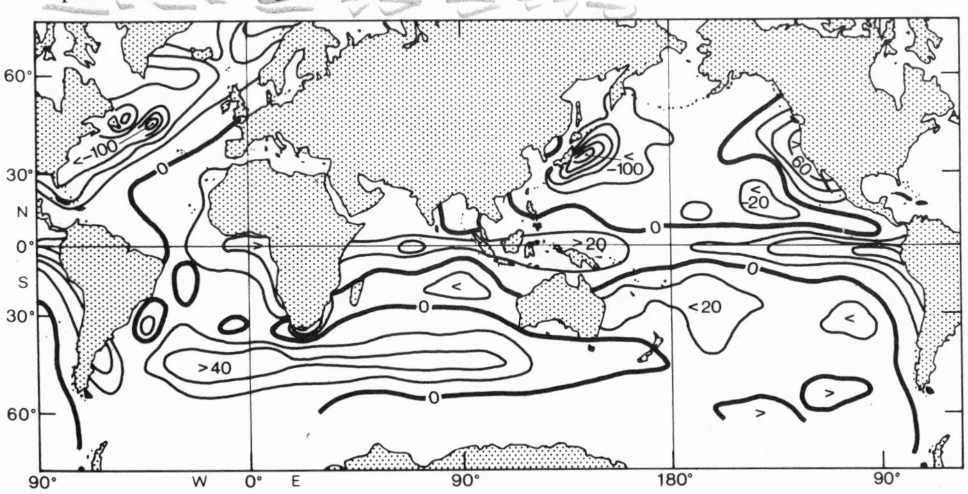

heat flux occurs where cold, dry air flows onto the ocean from wintery continents, the mixed layer rapidly becoming warm and humid. Evaporation into the warm dry air flowing off the African deserts rapidly moistens the Trade Winds, increasing the thermal radiation into the sea (i.e. reducing the net longwave radiation).

The global pattern of energy release from the ocean (Fig. 9.5) depends on the sum of the three components described above. To a first approximation, they are independent. But the rapid response of the atmospheric mixed layer to surface fluxes (humidity, cloudiness) influences the solar and atmospheric radiative heating of the ocean. This feedback is reflected in maps of total net radiation (insolation minus net longwave radiation), e.g. Hastenrath & Lamb (1978).

9.2.1.4

Heat storage in the ocean

We now consider two processes, heat storage and transport, that sharply distinguish the oceans from the continents in the global energy budget. We start with heat storage. Undue emphasis has been given in the past to the specific heat of sea water (4 MJ/(m³ K)), which is admittedly very large compared with that for air (1 KJ/(m³ K) at STP), but similar to that for land (sand = 1.3 MJ/(m³ K)). The ocean stores heat longer than the land for a combination of reasons, some of which involve positive feedback. Firstly, sea water is transparent, allowing energy to be absorbed at depths for which the escape time by molecular conduction is climatologically long. A glass lake would have the same property. Secondly, vertical fluid motions (convection, turbulence, upwelling), which on average accelerate the escape of heat from depth, often become feeble, or carry heat downwards. Convection only reaches as deep as the surface buoyancy flux requires and the pycnocline permits; Ekman pumping forces heat down; turbulence does too, but weakens when heat storage gives a stable thermocline. The motions controlling vertical heat transport are sensitive to the wind, which tends to be lighter in summer when storage is greatest. So is the rate of heat transfer through the interface by conduction and evaporation. The ocean has an unlimited supply of water for evaporation, in contrast to those continental regions where available soil moisture becomes exhausted, leaving conduction and thermal radiation to carry the heat flux into the air. Oceanic cooling is less efficient because it always

Fig. 9.6. The circulations of heat (10^{13} W) and fresh water (Kt/s) around the world ocean, as deduced by Stommel (1979) and Baumgartner & Reichel (1975) from the global patterns of energy and water fluxes from the ocean to the atmosphere (plus continental run-off). The experimental errors in the data base are comparable with the standard deviation of the megametre-scale regional variation, so such maps must be treated with caution.

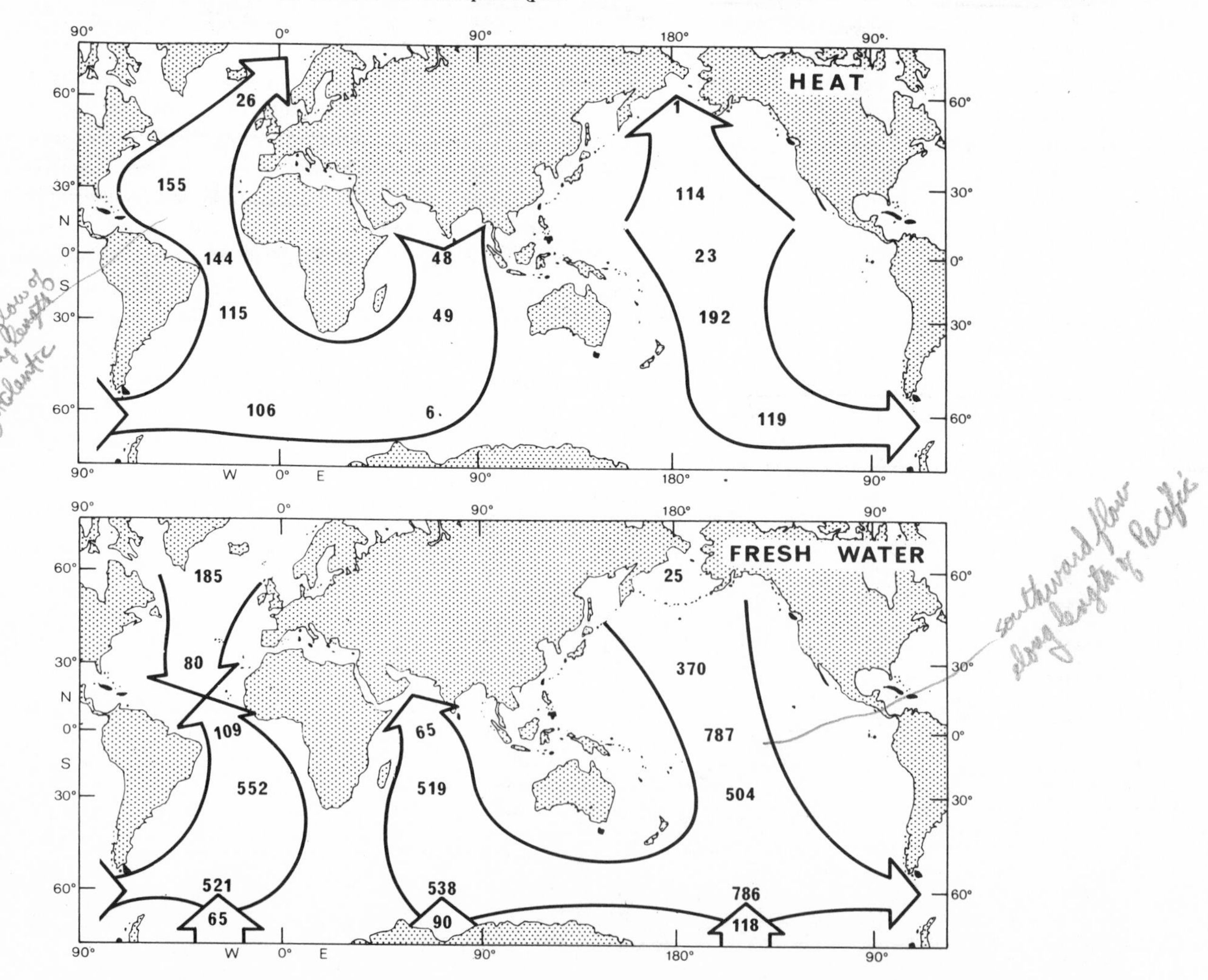

involves evaporation which moistens the air increasing the thermal radiation into the sea (GATE 1982).

9.2.1.5
Heat transport in the ocean

While heat is stored in the ocean it is displaced by currents. The distance travelled is not climatologically significant for most of the energy throughput because its residence time is less than one day. But some of the heat is stored for up to a year (a typical annual storage is 3 GJ/m²), giving time for currents to displace it, by a megametre or more in the case of major currents such as the Gulf Stream. Heat stored for many years may be released to the atmosphere in a different ocean basin. This large-scale redistribution of heat in the oceans was first inferred from the global patterns of annual mean heat flux through the bottom of the atmosphere (Fig. 9.5). The flux is zero over the continents, positive (downwards) in the tropical ocean, where heat income exceeds loss to the atmosphere, and negative at high latitudes where the loss exceeds income, the balance being achieved by divergence–convergence of the heat flux carried below the surface by the ocean currents.

The meridional mean ocean heat flux can be deduced from maps such as Fig. 9.5. The result of a recent analysis by Stommel (1980) is shown in Fig. 9.6. The fascinating feature of the map is that heat flows consistently northward along the entire length of the Atlantic Ocean, so that some of the heat released to the Atlantic Westerlies must have been first absorbed in the tropical Pacific Ocean. As will be shown later, such maps must be treated with caution because the surface fluxes from which they are constructed may be uncertain to 50 W/m², which is the annual mean heat flux averaged over the North Atlantic (area 20 Mm²) due to the 1 PW northward heat flow. Nevertheless, direct estimates of the oceanic heat transport give results that are not completely different (Hall & Bryden 1982; Bryden 1982) (Fig. 9.7). It is said that Wüst deduced the northward heat flux in the South Atlantic from analysis of the 'Meteor' expedition data, but did not publish his result because it seemed unreasonable. Wunsch (1983) compared the heat flux in the Atlantic calculated from the 'Meteor' expedition data and the IGY data 30 years apart, finding little change in the total heat flux integrated from coast to coast but considerable change in detailed distribution. Today we recognize that a mean heat flux in the range 0.1–1 PW between ocean basins is a characteristic feature of the global energy cycle. We note this is about 1% of the solar energy flux into the world ocean. The meridional component of the mean oceanic energy flux, zonally averaged, is comparable with that carried by the atmospheric circulation; larger in the tropics, smaller at higher latitudes (Vonder Haar & Oort 1973; Trenbarth 1980). It is less clear how much oceanic transport contributes to the seasonal cycle; external estimates (Oort & Vonder Haar 1976) suggest it may be significant, but internal evidence is less clear (Bryan 1982). Before leaving the subject of oceanic heat transport it is worth reemphasizing the point made earlier: that ocean currents can only effect significant transport when heat resides in the ocean for significant times. It does so because the ocean is transparent, because the processes responsible for the upward heat flux are lazy during the summer at all latitudes and year round in the tropics, and because Ekman pumping wins against them over a large area of the world ocean.

Fig. 9.7. The circulation of heat (10^{14} W) in the world ocean deduced from oceanographic measurements at a few transocean sections (Bryden 1982). Compare the values with Fig. 9.6. The two estimates are broadly similar but differ importantly in detail.

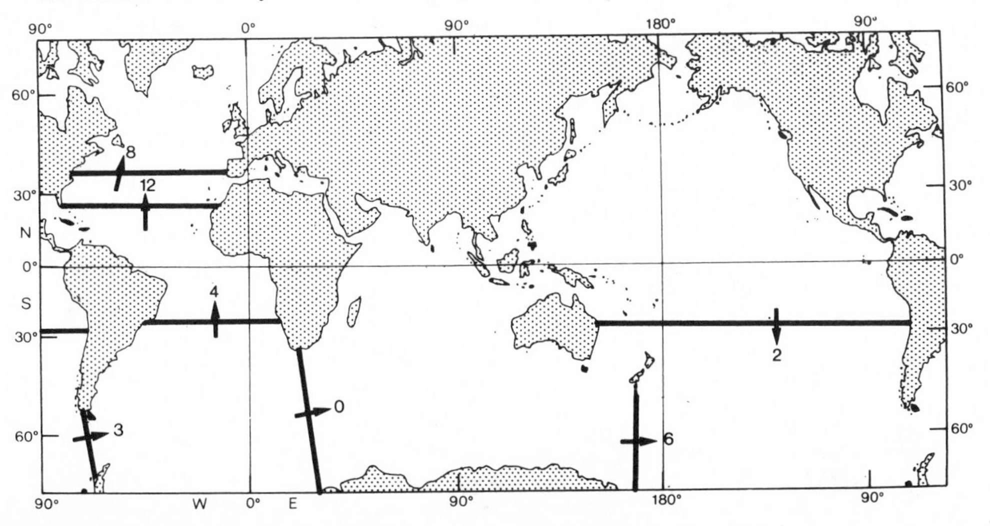

9.2.2
The global hydrological cycle

The global hydrological cycle is the quintessence of climate on earth. The impact of changes in the hydrological cycle are among the most worrying aspects of climate variation. In this section we consider the role of the ocean in the mean hydrological cycle. Baumgartner & Reichel (1975) have reviewed the world water balance, publishing maps and statistics of evaporation, precipitation and the difference between them. As other authors have done before, they draw attention to the poor data base and summarize the considerable differences between estimates published by different authors. Budyko (1974) discussed the seasonal variation of evaporation.

In the last section we saw that the escape of heat absorbed in the ocean is effected in part by evaporation. The global mean evaporation from the ocean is 1.2 m/y, a volume of 440 000 km³. This rate is equivalent to a mean residence time of about 3000 years, compared with about one year for soil moisture and ten days for the atmosphere. The meridional mean variation of annual evaporation exhibits maxima exceeding 1 m/y throughout the

Fig. 9.8 (*a*) The global water balance shown schematically on the basis of data in Baumgartner & Reichel (1975). The residence time of water in the ocean is 3000 years, in the soil one year, and in the atmosphere ten days. (*b*) The global pattern of annual mean rainfall over the ocean based on ship observations (After Dorman 1981). (*c*) The flow of freshwater from the continents into the world ocean. The areas of the segments are proportional to the annual volume flow from each river assuming a depth of one metre. The iceberg flow from Antarctica is shown schematically as two disks, scaled similarly. Note that most of the river flow goes into the Arctic–Atlantic 'estuary'. The freshwater supply to the Arctic Ocean helps maintain a near-surface pycnocline that is important for the maintenance of year-round sea ice.

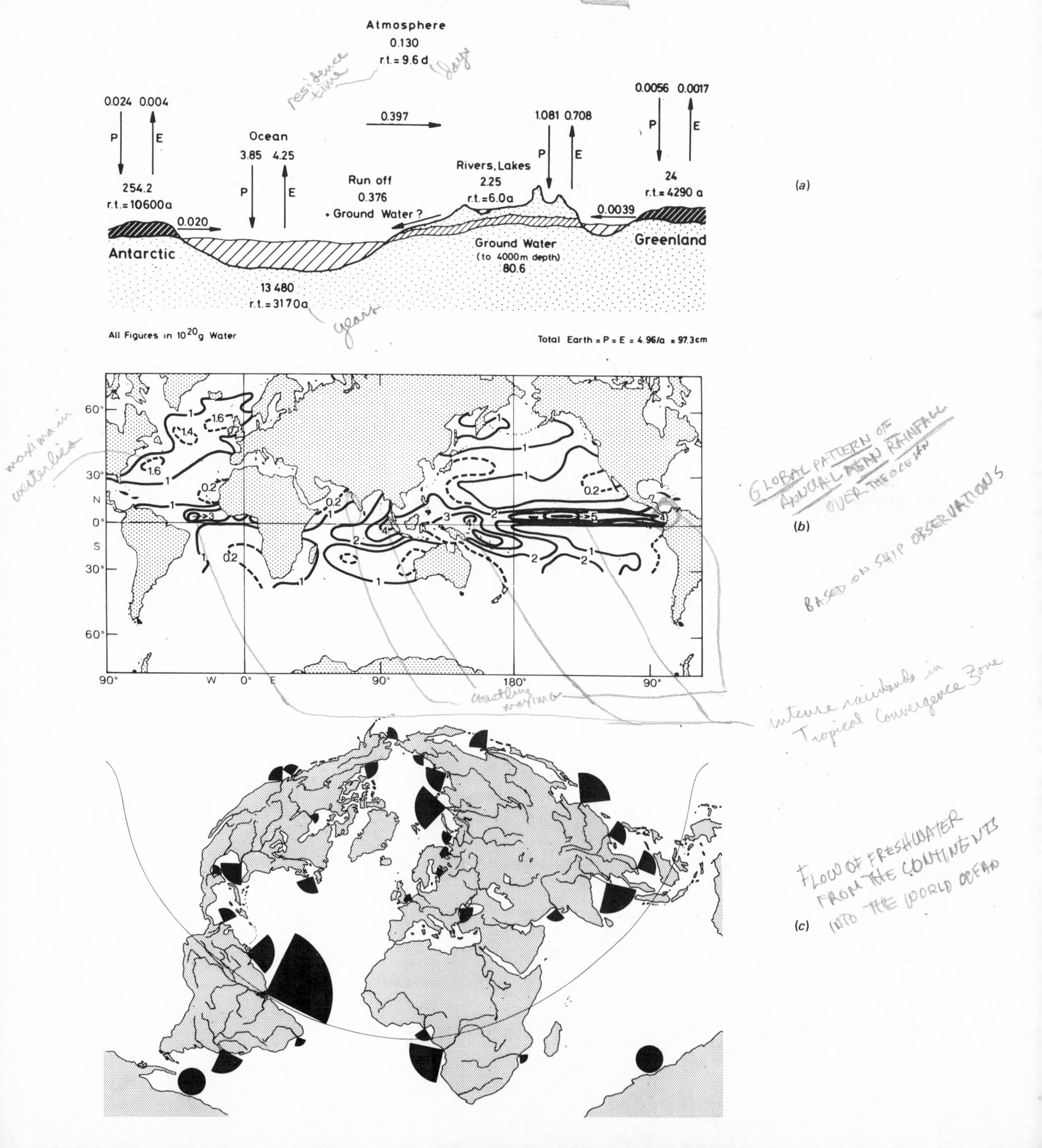

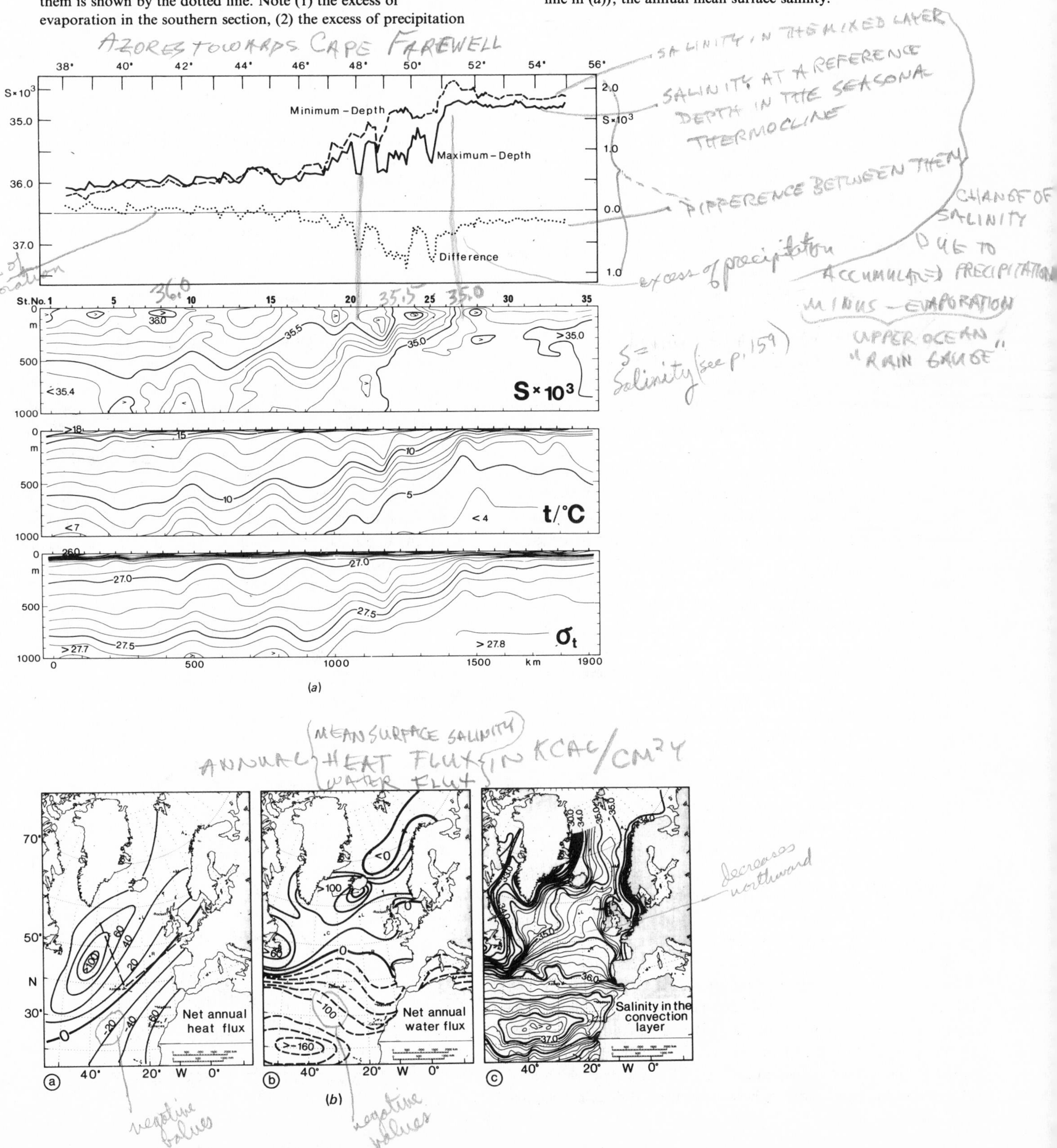

Fig. 9.9. (*a*) The upper ocean 'rain gauge'; the change of salinity due to accumulated precipitation-minus-evaporation and advection since the start of the heating season (approximately the vernal equinox). The diagram shows a section from the Azores towards Cape Farewell (Greenland). The top panel is derived from a batfish section: the salinity in the mixed layer is shown by the dashed line; the salinity at a reference depth in the seasonal thermocline is shown by the solid line; and the difference between them is shown by the dotted line. Note (1) the excess of evaporation in the southern section, (2) the excess of precipitation in the northern section, and (3) the effect of baroclinic geostrophic advection centred on 50°N, the polar front, marking the northern limit of the warm water sphere. The lower panels, based on a hydrographic section (kindly provided by Dr. J. Meincke), show the deeper structure. (*b*). The context for the section in Fig. 9.9. (*a*). The net annual heat flux in kcal/(cm² y) (note the ship track for the batfish section); the net annual water flux in cm (note that zero line coincides with that of the dotted line in (*a*)); the annual mean surface salinity.

tropics and decreasing monotonically towards high latitudes, becoming less than 0·2 m/y at 65°N and 50°S. The regional variation is characterized by a minimum near the equator, maxima of about 2 m/y in the tropics, and extratropical maximum of 2.5 m/y east of North America and 2 m/y east of China. Evaporation is strongest in winter, especially in the extratropical maxima.

The water vapour evaporated from the ocean rises and circulates around the atmosphere, eventually condensing to form clouds, and returning as precipitation either directly onto the ocean, or first onto the continents and then (after further evaporation–precipitation cycles) via rivers and glaciers back to the ocean. The cycle is summarized in Fig. 9.8(*a*) (from Dietrich *et al.* 1980). The Atlantic Ocean receives most of the continental run-off: taking account of the 33 great rivers with annual mean flow exceeding 2 kt/s (58% of the total run-off), the Atlantic receives 403, the Pacific 97 and the Indian Ocean 40 kt/s (Marcinek 1964). The run-off from the surface of continents is monitored routinely; ground water run-off is largely unknown.

The distribution of precipitation over the ocean is one of the most uncertain features of the hydrological cycle. Nevertheless, the principal features are well known (Fig. 9.8(*b*)). They include the intense rainbands at the Tropical Convergence Zone with maxima of 5 m/y in the Pacific and 3 m/y in the Atlantic and Pacific Oceans; mid-latitude maxima of over 1 m/y in the Westerlies; and intense maxima along coastlines (Indonesia, west Africa, central America, India and Chile). These maxima exhibit significant seasonal variation, with the meridional migration of the Tropical Convergence Zones and Westerlies, and the monsoon cycle.

At high latitudes, the sea freezes to give a layer of sea ice a few metres thick. The thickness is limited by the rate at which heat conduction through the ice can remove latent heat released at the lower surface during freezing. The seasonal variation of ice extent is a major feature of the earth's climate (Walsh & Johnson 1979; Walsh & Slater 1981). The mean annual ranges of sea ice area are 3–20 $(\mathrm{Mm})^2$ in the southern hemisphere and 8–15 $(\mathrm{Mm})^2$ in the northern. The residence time of water stored in sea ice is rarely more than two years around the Antarctic, but can reach six years in the geographically constrained Arctic. For a recent review of sea ice see Untersteiner (1985).

The water balance at the ocean surface (precipitation minus evaporation; melting minus freezing; continental run-off, including rivers and icebergs) causes the salinity of the upper ocean to deviate from its global mean value (34.72 kg/t). The observed distribution of salinity in the ocean represents a balance between the net surface water flux and the circulation of fresh water (i.e. salinity deficit) by ocean currents (Fig. 9.9). The vertically integrated fresh water circulation can be inferred from the distribution of net surface water flux; a recent example is shown in Fig. 9.6(*b*). Note the different character of the global circulations of fresh water and heat, in particular the northward heat flow along the whole length of the Atlantic and the southward water flow along the length of the Pacific. These maps depend only on the net surface fluxes, which are still rather uncertain. An alternative approach is to deduce the circulations from the measured three-dimensional distribution of temperature and salinity (see Wunsch 1983, Chapter 10 of this volume). Thus the oceanic circulation of fresh water, which is of little direct interest to atmospheric modellers (it does not significantly influence the rate of evaporation), is of indirect interest because the salinity anomalies provide information that can contribute to the calculation of the oceanic circulation of heat. Changes of salinity can also effect the surface fluxes of energy and water to the atmosphere through their influence on sea ice (Goody 1980) and deep convection (Killworth 1982; Woods 1982), with possible implications for long-term climate change (Rooth 1982).

9.2.3

Seasonal variation

The regular astronomical cycles of day length and noon solar elevation provoke changes on the continents, in the oceans and in the atmosphere that interact to produce the regular seasonal changes in the climate. In this section we briefly review the nature of the separate response of those three components of the planetary climate system to the annual cycle of insolation, and then consider how they interact to give the seasons.

9.2.3.1

Direct response of the ocean and continents to the astronomical cycle

The direct response of the oceans to the seasonal cycle of insolation is most noticeable outside the tropics, where heat is stored in the spring and summer, and sea ice forms in the autumn and winter. The annual temperature range of the open (ice-free) ocean rarely exceeds 10 K because the ocean's transparency and downward mixing permit storage to be distributed through the top 100 m. The annual range of heat storage and surface temperature increases towards the poles, following the meridional variation of the annual range of day length. Solar heating also affects the depth of convection. The (daily maximum) mixed layer depth has an annual minimum at the summer solstice and a maximum near the vernal equinox (i.e. at the end of winter). The magnitude of the minimum depth is rather uniform, but that of the maximum increases with latitude, and has large regional variation. The summer minimum and the meridional variation in winter can be explained in terms of direct solar forcing: the regional variation in winter is a consequence of the wind and currents. The autumn growth of sea ice is due to the decrease in day length, with some help from the strengthening Westerlies in the northern hemisphere where continental forcing is also a factor. The maximum sea ice extent is reached towards the end of winter, the minimum in July

Fig. 9.10. The annual cycle of the areal extent of sea ice in the Southern Ocean during 1973, 1974 and 1975. The variation is equal to the area of the Antarctic ice cap.

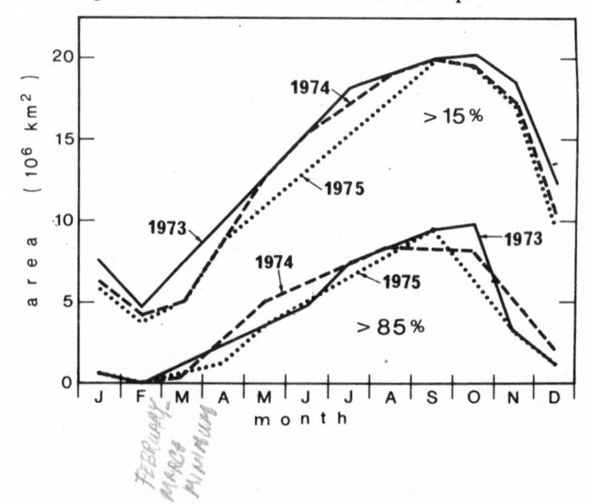

(northern hemisphere) and March (southern hemisphere, Gordon 1981). The variation in recent years is shown in Fig. 9.10.

The consequences of differences between land and sea are revealed in the January surface temperature distribution (Fig. 9.11). In summer the land gets hotter because energy received from the sun cannot be stored below the surface, as it is in the sea; also, in places such as the Australian desert (see Fig. 9.11) where soil moisture is exhausted, the heat cannot escape at low temperature by evaporation, only at high temperature by radiation and conduction. In the winter hemisphere, the surface temperature difference between land and sea depends largely on snow and ice cover. Snow cover virtually eliminates the daily heating by the sun. In the interior of the continents (e.g. Siberia in Fig. 9.11), the snow cover persists and the temperature falls very low. But warm moist air soon melts snow lying on continents exposed to west winds coming off an ocean that is releasing stored heat; the sunshine then further warms the land: in such regions the winter surface temperatures on land and sea are similar (e.g. western Europe in Fig. 9.11). When the ocean freezes it loses the power to warm downwind continents. Snow then settles on the sea ice, reflecting the sunshine away, while the ice acts as a lid shutting off heat transfer by evaporation and reducing conduction. Except in leads and polynyas the ocean then behaves like the land. These differences between land and sea are neatly summarized by Monin's (1975) famous map (Fig. 9.12) in which contours of the annual range of surface temperature delineate the extra-tropical continents.

9.2.3.2

Seasonal response of the atmospheric circulation

The seasonal cycles of ocean surface temperature and ice cover can be simulated to a first approximation with a model in which the atmospheric forcing (cloudiness, surface cooling, wind stress) is kept constant (Kraus & Turner 1967; and, for ice, Untersteiner 1983*b*). The error is much smaller than the annual range of land–sea difference, and is therefore adequate, as a first approximation, for exploring the annual cycle of atmospheric circulation. The atmospheric response is, of course, very complicated; but its essence can be captured by considering three features: the Westerlies, the Trades and the Monsoons. Schlesinger & Gates (1981) and the review by Held (1982) summarize the results of recent numerical investigations.

The Westerlies occur in their purest form in the southern hemisphere, where the annual cycle comprises modulation with the

Fig. 9.11. The surface temperature of the world in January 1979 based on satellite observations, showing the extremely cold winter regimes (Canada, Siberia), the extremely hot summer regimes (Australia), and the mild winter regime of continents exposed to maritime air masses (notably western Europe). (Based on a figure in Goody 1982.)

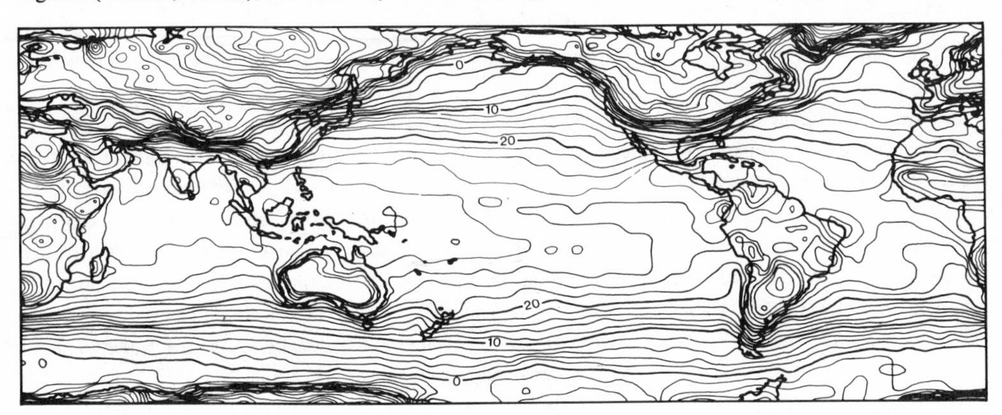

Fig. 9.12. The world map of annual range of surface temperature. (After Monin 1975.)

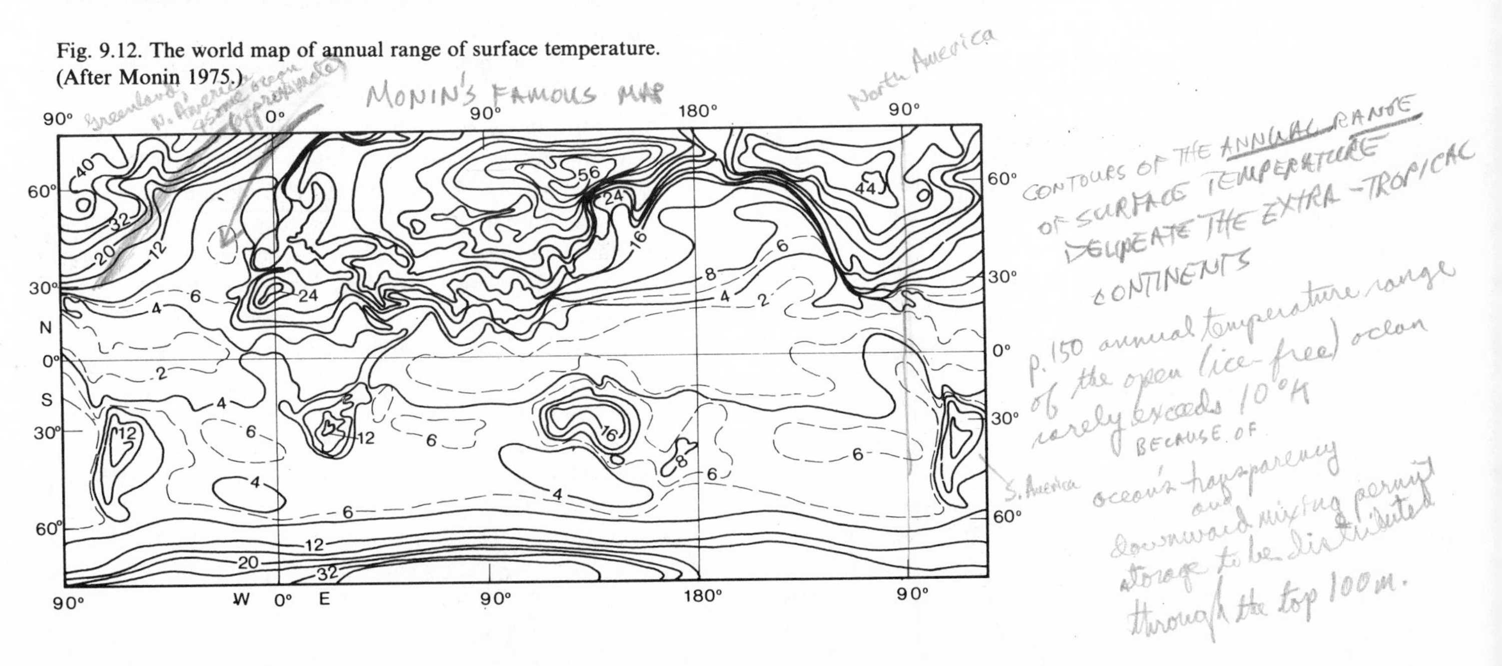

meridional temperature gradient and a meridional migration forced by the waxing and waning of sea ice around the Antarctic continent (±50% modulation of the effective area). The annual cycle of the Westerlies in the northern hemisphere is influenced less by the sea ice changes than by the winter cooling of the continents; the penetration of the warm, ice-free Atlantic Ocean to high latitude provides a thermal (and, by contrast with Greenland, an orographic) stimulus and phase key for planetary waves in contrast to the southern hemisphere where most is sea. The difference between the distribution of land and sea in the two hemispheres is reflected in the strengths of the respective Westerlies. The Trade Winds parasitic on the Westerlies reflect that imbalance, and converge north of the equator. The location of the convergence migrates annually as the Trades respond to the Westerlies as they strengthen and advance equatorward in the winter hemisphere, weaken and retreat poleward in the summer hemisphere. The meridional migration of the Tropical Convergence Zone, with its important rainband, is most clearly observed in the Pacific Ocean (Horel 1982). In the Indian and, to a lesser extent, in the Atlantic Oceans, the migration is modified by the annual development of the monsoon circulation. The monsoons develop as a result of the different annual temperature range on the continents and the ocean. The monsoon circulation is most pronounced over the Indian Ocean where the annual cycle of forcing comes from the Asian continent to the north, and the oceanic Westerlies to the south.

9.2.3.3

Oceanic response to the seasonal cycle in the atmosphere

The seasonal cycle in the atmospheric circulation produces regular changes in the ocean, which were ignored above. The most important changes occur in the tropics where the annual range of insolation is relatively small. The meridional migration of the Trade Winds modifies the near-equator pattern of Ekman convergence. The tropical thermocline is tilted meridionally in summer creating the north equatorial counter currents. Along the equator the thermocline is seasonally depressed to the west of the Pacific and Atlantic Oceans by Ekman convergence (Meyers 1979). In the Pacific (Niiler 1981) and Atlantic (Merle 1980), but not in the Indian Ocean (McPhaden 1982), the oceanic heat content varies locally, more as the result of these disturbances than by changes in the surface heat fluxes. The annual reversal of the monsoon circulation over the Indian Ocean produces a corresponding reversal of the Somali current (Lighthill 1969). The meridional migration of the Tropical Convergence Zone modulates the solar heating (cloud cover), the surface water balance (precipitation, evaporation in the doldrums) and the depth of the mixed layer (surface buoyancy flux, wind stress); see Stommel (1979*a*).

Outside the tropics the annual retreat of the Westerlies, especially in the northern hemisphere, leaves clear skies and light winds (the Azores high) which increase heat storage and decrease the mixed layer depth. The results of running a seasonal thermocline model with and without the annual variation of atmospheric forcing are shown in Fig. 9.13. The most important consequences of atmospheric circulation in polar seas is the equatorward dispersal of sea ice by Ekman flow due to the Westerlies. Gill (1973) has shown that the total amount of freezing is increased by the removal of ice in this way before it reaches the limiting thickness set by thermal conduction. This enhancement, and the separation of freezing and melting, leads to accumulation of salt in the Weddell Sea, producing dense Antarctic bottom water, the deepest component of the meridional circulation of the world ocean. The vigour of this circulation (the oceanic equivalent of the Hadley circulation driven by deep convection in the tropical atmosphere) is sensitive to the seasonally varying strength of the Westerlies.

Fig. 9.13. The relative importance of the seasonal cycles of insolation and weather in determining the seasonal cycle of ocean mixed layer temperature. A mixed layer model (à la Kraus & Turner 1967) has been integrated over a year for a site where the net annual surface heat flux is zero. The astronomical cycle is the same in all four cases. The surface weather is described by the Bunker climatology in one run (dotted); in each of the others one meteorological variable is kept constant.

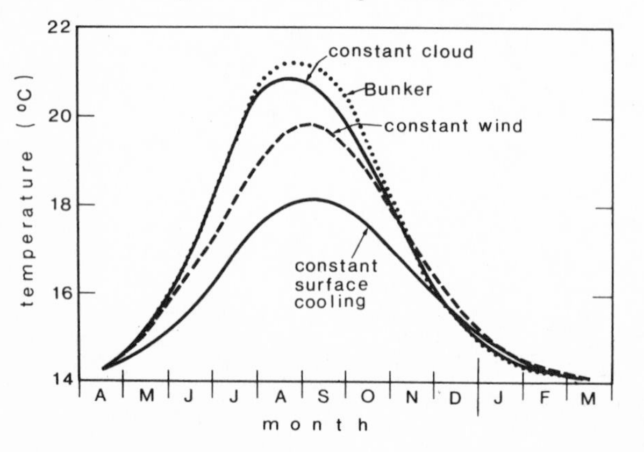

9.2.4

Interannual variation

The ocean is the central element of the climate system in any discussion of interannual variation. Deviations from the regular seasonal cycle of forcing by the sun or atmosphere produce a response in the ocean. That, in turn, leads to a modification of the GPFO, provoking a change in the climate. This section summarizes the response of the ocean to weather fluctuations and to more-widespread changes in the general circulation, the effect on surface fluxes and the climate consequences.

9.2.4.1

Weather forcing

Tests with one-dimensional models of the upper ocean show that the boundary layer structure is sensitive to random fluctuations in clouds, wind stress and surface cooling (Fig. 9.14), especially in early spring. These random fluctuations simulate the passage of weather systems with (Eulerian) time scales much shorter than the time taken by the upper ocean to recover from the heat content anomalies they provoke. The onset of spring heating in the ocean, when the mixed layer depth decreases sharply, can fluctuate by three weeks depending on the weather at the vernal equinox (Hofmann 1982). Random fluctuations of the oceanic heat input can integrate to significant anomalies which persist because they are distributed over a depth range of order 100 m, allowing only a small change in mixed layer temperature, and therefore only a small surface heat flux anomaly. It has been shown (Frankignoul & Hasselmann 1977; Reynolds 1977; Hasselmann 1981) that the spectrum of sea surface temperature (SST) fluctuations in the North Pacific are consistent with a simple statistical model based on the above idea.

9.2.4.2

General circulation forcing

We now consider the consequences of interannual variation in the major features of the atmospheric circulation: the Westerlies, Trades and monsoons. Although the basic character of the Westerlies is similar every year, there are inevitable variations in the planetary waves, and in the sequences of weather systems that grow on them. Slight changes can be accentuated by the continents: by winter snow cover or summer drought. The direct consequence in the ocean is fluctuating distributions of mid-latitude anomalies of heat content and surface temperature (Barnett 1981*c*). The climatological data base has been analysed into empirical orthogonal functions (EOF) by Barnett & Davis (1975) to reveal recurrent patterns and their variation during the last few decades (Fig. 9.15).

In the tropics the forcing is by the Trade Winds and monsoon, which also exhibit significant interannual variation. These anomalies modulate the Ekman advection of warm surface water, changing the local heat content of the upper advection of warm surface water, deepening the thermocline and increasing the local heat content. The magnitude of the heat anomalies is illustrated in Fig. 9.16, derived from island tide gauge data (Wyrtki 1979). An equatorial Kelvin wave then propagates the thermocline depth anomaly rapidly eastwards and then polewards along to the American coasts (Wyrtki 1975; Busalacchi & O'Brien, Kindle 1981), and Rossby waves spread it slowly back across the tropics from the east. Boundary layer processes then create an extensive sea surface temperature anomaly (SSTA). The tropical SSTAs have been subjected to EOF analysis by Navato *et al.* (1981) (Fig. 9.17). Particular attention has been paid to the El Niño years when anomalously warm surface water forms off Peru, with disastrous consequences for the world's major fishery, causing a downstream sequence of economic consequences that make the El Niño one of the major problems of climate impact. The sequence of SSTAs for

Fig. 9.14. The depth of the mixed layer is sensitive to cloud cover. Random changes in the range $5/10 \pm 5/10$ produce large fluctuations in the daily minimum mixed layer depth around the winter solstice, and in the daily maximum depth around the vernal equinox. (Based on Hofmann 1982.)

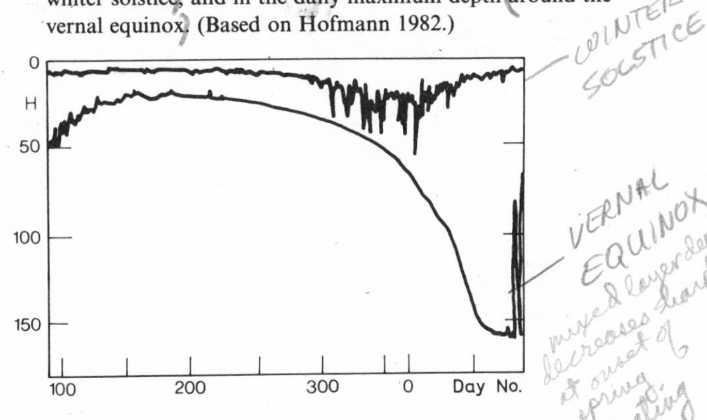

Fig. 9.15. The first six empirical orthogonal functions for sea surface temperature anomalies in the north Pacific. (Based on Barnett & Davis 1975.)

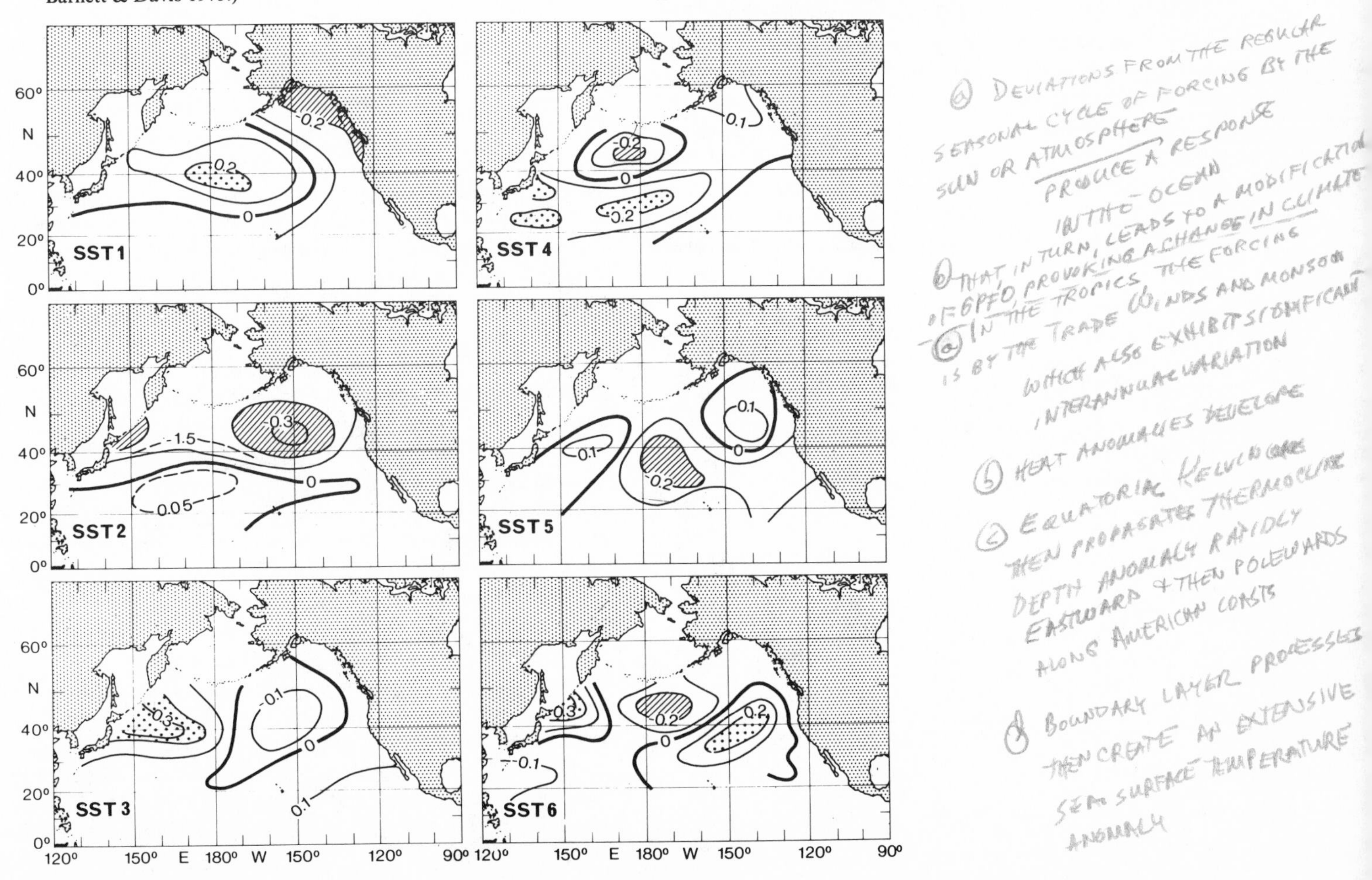

Fig. 9.16. An example of the change in sea level (cm) associated with interannual variaton of upper ocean heat content in the western tropical Pacific. (Based on a figure in Wyrtki 1979.)

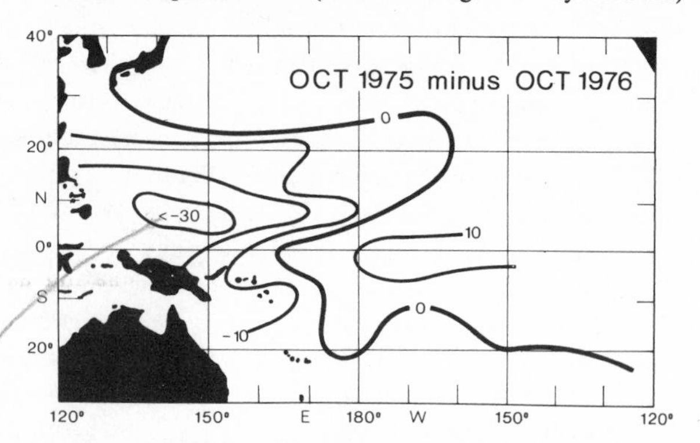

Fig. 9.17. The first empirical orthogonal function for tropical sea surface temperature anomalies. A composite picture based on separate analyses for the Atlantic, Indian and Pacific Oceans by Navato, Newell, Hsiung, Billing & Weare (1981).

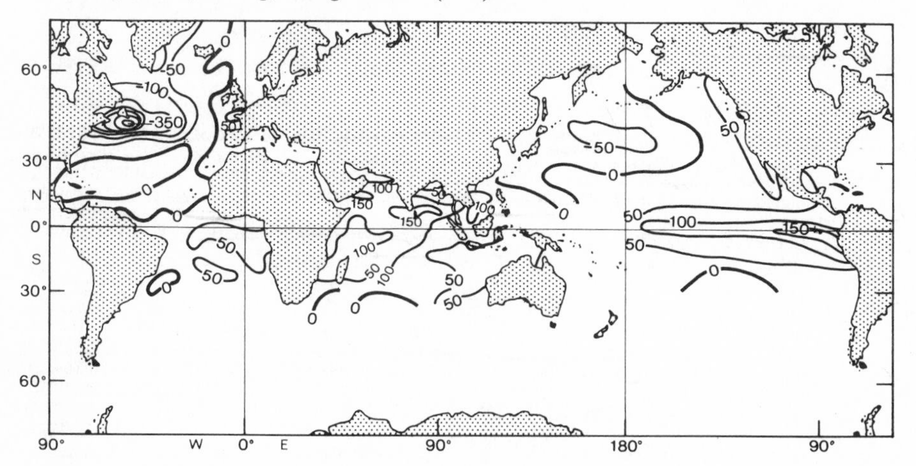

an El Niño year, based on a composite of size El Niño events (1951, 1953, 1957, 1965, 1968 & 1972), is shown in Fig. 9.18 (from Rasmussen & Carpenter 1982). McCreary (1982*a*,*b*) and Lau (1982) have developed models relating the Southern Oscillation to the El Niño.

9.2.4.3

Atmospheric response to the global pattern of fluxes of energy from the ocean

The GPFO influences the climate over the continents in two ways: by determining the temperature and humidity of maritime air masses advected over land, and by influencing the atmospheric circulation. The first of these is more obviously important; the latter more intriguing.

The changes of temperature and humidity of air masses being transformed by flow over the ocean was studied in the GARP AMTEX (Air Mass Transformation Experiment). The flow of maritime air over the land affects the climate of islands and continents to a distance inland that can be expressed as a mixing

Fig. 9.18. The sequence of sea surface temperature anomalies during an El Niño event, based on a composite of several events by Rasmussen & Carpenter (1982). This analysis has been widely used as the boundary condition for atmospheric general circulation models in studies of climate response to the El Niño. (isotherm interval 0·2 K; broken lines for cold anomaly).

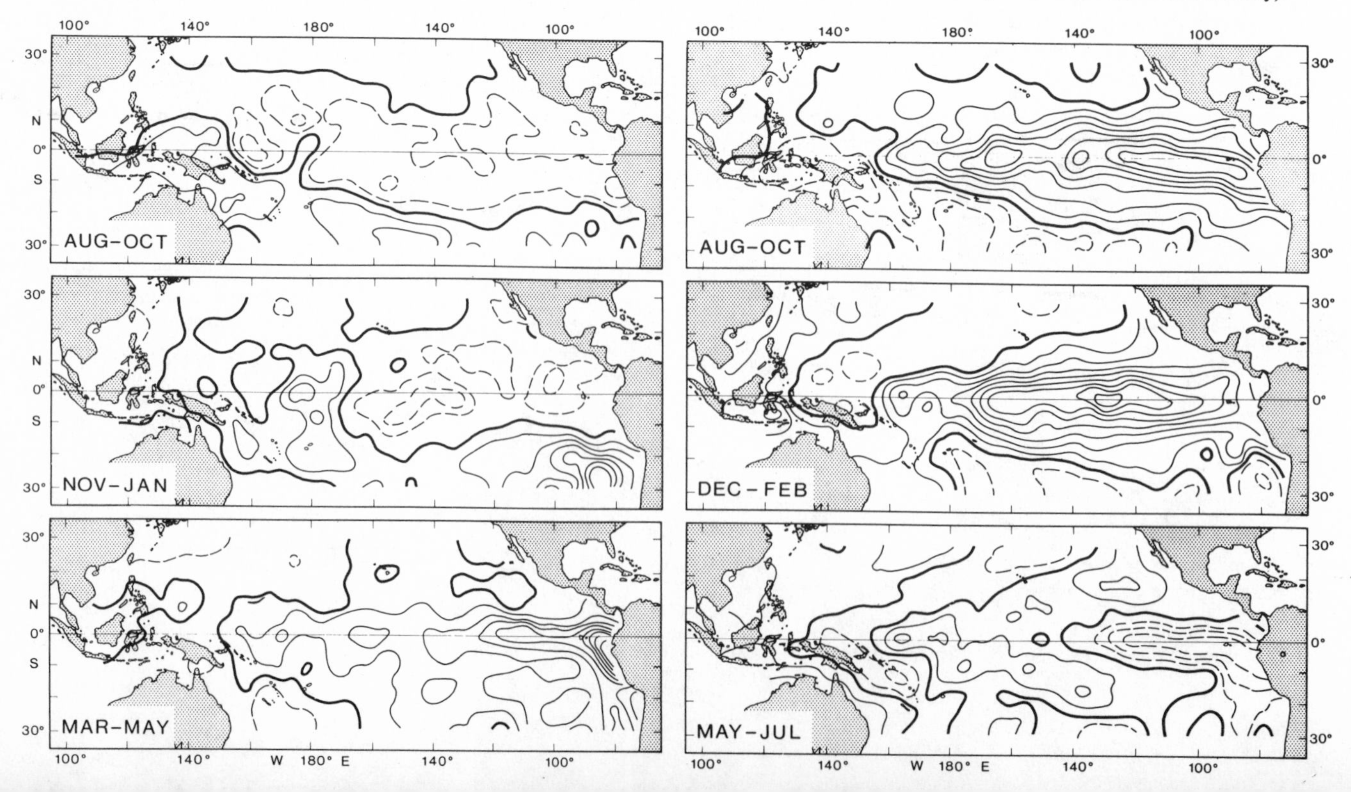

length controlled by the synoptic-scale air trajectories and by the rates of exchange of energy and moisture with the land surface and the upper layers of the troposphere. The mixing length ranges from a few hundred kilometres along the subtropical desert coasts of Australia and Africa, to a few thousand kilometres in western Europe. The climate in such regions is extremely variable, depending on the wind direction, which may bring mild maritime air one day and, depending on the season, oppressively hot or bitterly cold continental air the next. Sea ice cover in winter decreases the source regions of maritime air masses, an important factor for the habitability of Canada, Russia and Scandinavia. The classical textbooks, atlases and monographs on climatology (e.g. Lamb 1972) provide extensive documentation of the role of maritime air masses in island and continental climate. The air mass temperature fairly rapidly reaches equilibrium with the SST; its humidity depends on the moisture flux to the upper troposphere, by vertical motion in fronts and clouds. In regions of subsidence the inversion that acts as a lid on the boundary layer is sharpened and clouds form in the boundary layer. Regional anomalies in the SST are locally important. The best-known example is the exceptionally warm surface water in the northeast Atlantic which not only determines air mass temperature but also hinders sea ice formation in winter, so keeping open the sources of maritime air formation for western Europe. On a smaller scale, the cold currents flowing along the coast of California, Peru, northwest and southwest Africa moderate the desert climate of the hinterland.

The second way in which the GPFO influences continental climate is through its effect on the large-scale circulation of the atmosphere. In earlier sections we discussed the importance of the ocean as the source of most of the energy for atmospheric motion and the importance of the different annual ranges of temperature on the continents and ocean. Here we consider the more subtle effects arising in atmospheric circulation due to the GPFO. To a first approximation, with one exception, it might reasonably be said that GPFO forcing is not very important, since it emerges from the data only after considerable statistical effort; it contributes only a small fraction of the natural variability over the land (Barnett 1981*a,b*). Experiments with atmospheric general circulation models (GCMs) (reviewed by Haney 1980, Chervin *et al.* 1980) show that, to a first approximation, the GPFO is shaped by, more than the shaper of, the major wind systems; the atmospheric circulation gets its energy where it can, and is not too fussy about the detailed distribution. The exception is sea ice cover, which blocks the surface fluxes. The seasonal waxing and waning of sea ice cover around Antarctica causes significant meridional migration of the Westerlies and their vassal, the southeast Trades (Webster 1981).

There are, however, some second-order effects that can be important locally, and perhaps even globally in a limited number of cases. The energy flow from the surface is a critical factor for tropical cyclones, which form less frequently if the SST is low (Henin & Donguy 1980; Anthes 1982). The response of extra-tropical cyclones to variations in the pattern of surface fluxes is weaker (Webster 1981). Nevertheless, Namias (1965, 1969) 'hypothesized that anomalous large-scale oceanic thermal patterns might strongly influence the atmosphere' (Namias & Cayan 1981). Statistical studies by Davis (1976, 1978) and Namias (1976) showed that in general the SSTA follow atmospheric sea level pressure (SLP) by a month or so, but that the autumn sea level pressure anomaly (SLPA) in the north–east Pacific is correlated significantly with the summer SSTA (Fig. 9.19). Nevertheless, to quote Namias & Cayan (1981) again, 'while substantial progress has been made in studying large-scale air–sea interactions, our understanding is woefully inadequate to achieve reliability in prediction'. The subject has become more difficult since 1846 when Sabine confidently concluded that: '(In November 1776 and winter 1821–22) the warm water of the Gulf Stream spread itself beyond its usual bounds...to the coast of Europe, instead of terminating as it usually does about the meridian of the Azores.' That conclusion, based on Rennel's (1832) maps of North Atlantic SST change from 1780–1820, was the start of a great folk myth that became etched into the memory by the cold winters of the little ice age. Modern analysis does not support it, but the man on the Clapham omnibus still believes that fluctuation of the Gulf Stream controls the winter climate of Europe.

The Sabine–Namias hypothesis is concerned with air–sea interaction at mid-latitude. Now we consider the response of atmospheric circulation to the tropical component of the GPFO, remembering that the surface area of the tropical ocean is half of the total. Recent statistical studies (Horel & Wallace 1981; Wallace & Gutzler 1981; Keshavamurty 1982; Shukla & Wallace 1983) have revealed significant correlations between SLPA in North America and SSTA on the Pacific equator. Dynamical studies by Opsteegh & van den Dool (1980), Hoskins & Karoly (1981), Webster (1982), Simmons (1982), Hoskins (1983) have shown that planetary waves propagating through the atmosphere along great circles are capable of carrying the signal from the tropics to other locations around the globe under certain conditions. In particular, the Westerlies have to overlap the oceanic energy source because the planetary waves cannot propagate through the meridional shear of the Westerlies–Trades system. The conjuncture happens during the northern winter when the Westerlies are closest to the equator, and when the oceanic Kelvin–Rossby wave process (described above) has created surface energy flux anomalies of up to 100 W/m² across a broad range of tropical latitudes. Thus planetary wave teleconnections, first in the ocean and then in the atmosphere,

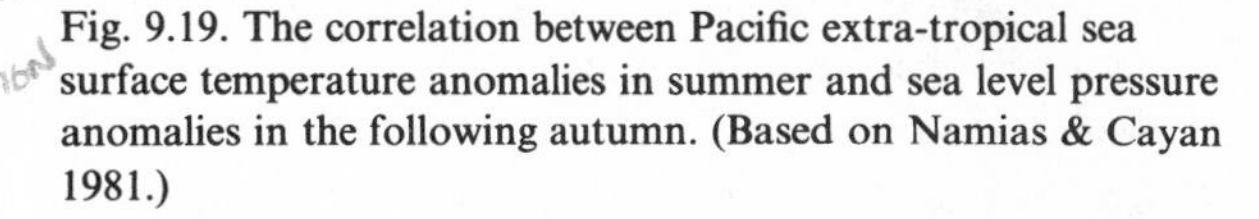

Fig. 9.19. The correlation between Pacific extra-tropical sea surface temperature anomalies in summer and sea level pressure anomalies in the following autumn. (Based on Namias & Cayan 1981.)

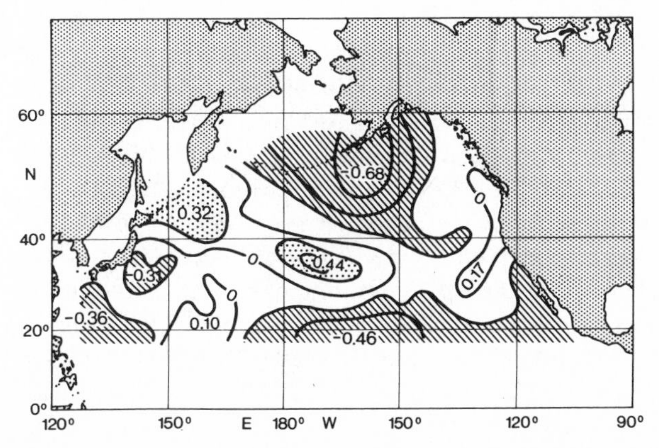

combine at one season of the year to radiate climatic disturbances around the world.

The empirical evidence of global correlations between SSTA, SLPA and precipitation was discovered by Walker (1923), who called the phenomenon 'The Southern Oscillation', and Bjerknes (1969) who adopted the term 'teleconnections' from the earlier work of Angström (1935); see, for example, Julian & Chervin (1978). Recent studies have established links between the El Niño and droughts in the east monsoon season (Quinn, Zopf, Short & Kuo Yang 1978), and rainfall in the Australian region (Nicholls & Woodcock 1981). The existence of a significant response in the atmospheric circulation, which affects climate on distant continents, has focussed attention onto the potential of the El Niño oceanic wave guide delay for short-term climate prediction (Ramage 1981; McCreary 1982*a*, *b*; Lau 1982). The Southern Oscillation is only part of the global pattern of variability of the ocean and atmosphere, and many other examples exist. For instance, Moura & Shukla (1981) have studied the SST patterns associated with droughts in northeast Brazil and have carried out numerical experiments that clarify the role of the ocean. Merle (1980) has discussed the relationship between warm water in the Gulf of Guinea and drought in the Sahel. Masuzawa & Nagasaka (1982) referred to the relationship between surface temperature along the 137° section and the timing of the end of the wet season in Japan. Shukla (1975) and Shukla & Misra (1977) have studied the relationship between temperature anomalies in the Arabian Sea and rainfall over India. Pan & Oort (1983) have mapped global anomalies synchronized with SSTA in the Equatorial Pacific (120°W).

9.2.5

The ocean as a sink of pollution

Gaseous, particulate and thermal pollution of the atmosphere is beginning to have a significant impact on climate and is expected to increase in the future (Bach, Pankrath & Kellogg 1979). The ocean moderates the anthropogenic climate change by acting as a sink of pollution, following injection by solution, precipitation or run-off. Attention has been focussed primarily on carbon dioxide, which is being injected into the atmosphere at a rate that promises to double the concentration in the next century (Bolin, Degens, Kempe & Ketner 1979). The natural carbon cycle in the ocean and its ability to accommodate some of the increasing supply depends on a complex combination of physical, chemical, biological and geological processes that are still far from understood (Bolin 1981). The physical processes include the mechanism of gaseous transfer through the sea surface (Liss 1973; Hasse & Liss 1980; Favre & Hasselmann 1977); vertical mixing in the boundary layer (Kraus 1977) and into the ocean's interior (Siegenthaler 1983); circulation and mixing in the interior (Warren 1981, Reid 1981, Wunsch 1983). Chemical processes allow the mixed layer slowly to accommodate far more CO_2 than can be taken up by physical solution (Revelle & Suess 1957). Further potential for accommodating CO_2 comes from the incorporation of carbon into planktonic plants and the zooplankton that feed on them and subsequently sink through the ocean carrying carbon (in the form of calcium carbonate) down through the ocean. The supersaturation of sea water relative to $CaCO_3$ decreases downwards and in the deepest levels it is undersaturated (Li, Takashi & Broecker 1969). So the shells reach and accumulate on the bottom if it is not too deep, but redissolve before reaching the greatest depths. The accumulation is particularly rapid on the continental shelves, where biological activity is more vigorous; the flux of sediments from the shelf regions into the deep oceans may constitute a significant mechanism by which the ocean accommodates excess carbon (Walsh 1980 Broecker 1982).

The accommodation of carbon is only half of the story; the ocean also contributes to ameliorating the climate change that occurs as the result of the fossil-fuel CO_2 that remains in the atmosphere. It does so by reducing the thermal pollution created by the effect of atmospheric CO_2 on longwave radiation. It has been calculated that when, in the next century, the CO_2 concentration of the atmosphere has doubled, the thermal radiation from the atmosphere into the ocean will have increased by a few percent (Ramanathan 1981). The depth of the mixed layer is sensitive to the energy balance close to the surface (Woods 1980). Assuming, for simplicity, that the cloud cover (i.e. solar heating of the ocean) remains unchanged, the increased longwave heat flux into the ocean will cause the mixed layer to become shallower, delaying the escape to the atmosphere of solar energy absorbed at depth. The ocean temperature will rise until the heat can escape, when the convection gets deeper in winter. So the immediate thermal consequence of higher atmospheric CO_2 is a rise in the *summer* SST. The decreased mixed layer in winter will allow more of the absorbed solar energy to reside in the ocean for more than a year. The resulting secular rise in ocean temperature will lead to higher annual energy loss to the atmosphere, driving the convection deeper at high latitudes. Calculation of the equilibrium SST for doubled CO_2 depends on detailed modelling of these OBL processes, including the downward heat flow into and circulation around the warm water sphere.

9.2.6

Palaeoclimatological changes

It has been assumed so far that the boundary conditions of the planetary climate system remain constant. Here we consider the role of the ocean in climate changes arising from variation in the energy output of the sun, the orbit of the earth, and continental drift. These become important on time scales of many thousands, or millions, of years. The existence of fossil evidence of the climate changes on such time scales (Shackleton 1982) makes it attractive to test climate models by seeking to simulate the conditions pertaining under boundary conditions quite different from today's; the changes in the ocean and the cryosphere are particularly important in such simulations (Berger 1981). Revelle (1981) has written a stimulating introduction to contemporary preoccupations in palaeoclimatology.

9.2.6.1

The solar constant

The empirical evidence suggests that changes in the solar constant have not exceeded 1% in recent years, but it is generally agreed that solar luminosity is likely to have increased by about one-quarter in the last 4 Gy during the development of the terrestrial biosphere on earth (Lovelock 1979). McCrae (1975) has suggested that the solar constant may be modulated by the passage of the solar system through the varying interstellar bands of dust and gas as it circulates around the galaxy (period, 250 My). The effects of a change in the solar constant have been studied by Manabe & Wetherald (1980), Wetherald & Manabe (1975), Bryan & Manaabe (1981).

9.2.6.2

Orbital change

The annual cycle of solar heating is not exactly the same for the two hemispheres. The earth is a little closer to the sun in the Austral summer, giving a 6% increase in the total rate of energy interception compared with the Boreal summer. The magnitude and the phase of the difference changes with the precession of the equinoxes, the tilt of the earth's rotation axis to the ecliptic, and the eccentricity of the earth's orbit around the sun. The varying pattern of the annual cycle of insolation due to these changes has been calculated most accurately by Berger (1977, 1981). It was suggested by Croll (1875) and Milankovich (1941) that the waxing and waning of the North American and European glaciers in the last few tens of thousands of years was the result of these changes in annual heating (see the review by Imbrie & Imbrie 1979). Recently, the theory has attracted considerable attention (e.g. Mason 1976) following the production of evidence from fossil plankton that the volume of the oceans (and, therefore, of the glaciers) and the oceanic temperature distribution have varied with the calculated insolation changes (Emiliani 1954, 1981; Shackleton & Opdyke 1977; Hays, Imbrie & Shackleton 1976; Duplessy 1981). The global pattern of SST 18000 years ago, when the glacier volume was at a maximum, has been mapped by the CLIMAP project members (1976) (Fig. 9.20). The southward extent of sea ice and of the polar front in the North Atlantic are particularly well documented (Ruddiman & McIntyre 1973). Kraus (1973) considered the oceanic circulation at that time. Broecker (1982) has proposed that the 40% reduction in atmospheric CO_2 during the last glaciation (Delmas *et al.* 1980; Berner *et al.* 1980; Fig. 9.21) was due to an ocean chemical response to the lowered sea level (see also Southam & Hay 1981). Berger (1981) and Rooth (1982) have discussed the salinity changes that occurred in the ocean as the glaciers melted (15000–9000 years ago).

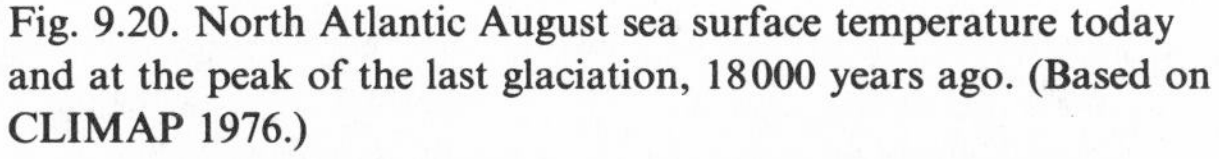

Fig. 9.20. North Atlantic August sea surface temperature today and at the peak of the last glaciation, 18000 years ago. (Based on CLIMAP 1976.)

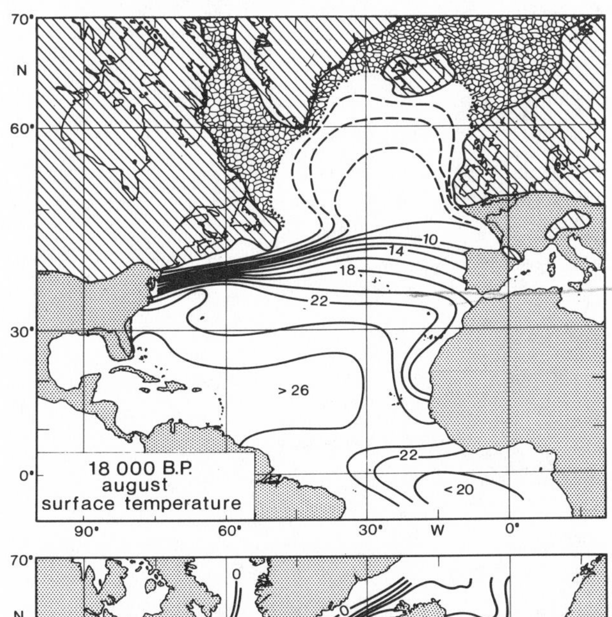

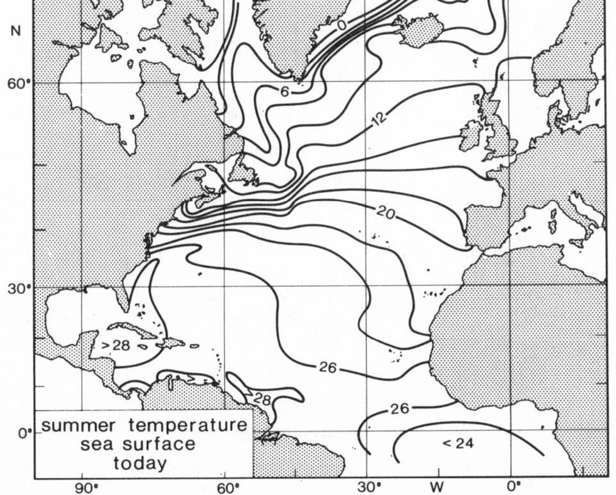

9.2.6.3

Continental drift

There are two ways in which continental drift interacts with the oceans to change the global climate. Firstly, it opens and closes the narrow sills between the water masses of differing thermohaline properties in the major basins. There is evidence from fossils in ocean sediments that links changes in the pattern of ocean temperature with such events (Berger 1981). Examples include the opening of the Drake Passage, the closing of the tropical flow between Pacific and Atlantic when Meso-America drifted into position as a land bridge between North and South America, the

Fig. 9.21. Evidence from an Antarctic ice core that the atmospheric CO_2 concentration (dots) was half the present value at the peak of the last glaciation 18000 years ago. The change of ocean volume as the ice melted (15000–10000 years ago) is shown by the change in $^{18}O:^{16}O$ ratio (continuous curve). (Based on Delmas, Ascencio & Legrand 1980.)

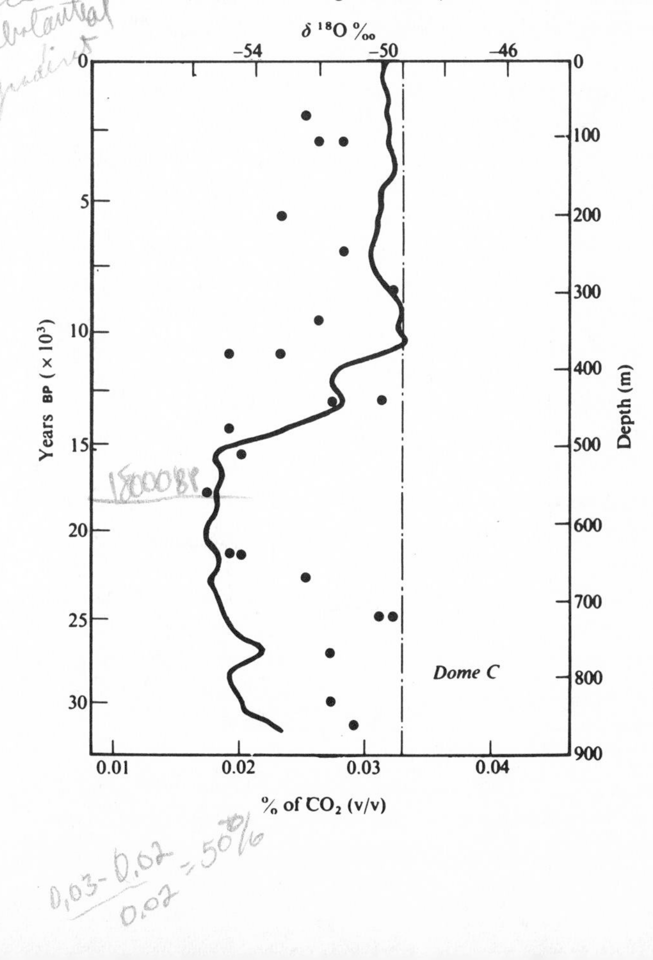

Fig. 9.22 (*a*). The development of the cold water sphere of the ocean during the present ice age (last 50 million years) as shown by the $^{18}O:^{16}O$ ratio in benthic foraminifera. The upper curve, for planktonic foraminifera, shows that sea surface temperature followed the deep temperature until about 30 million years ago. (From W. Berger 1981.) (*b*). The dramatic increase in temperature difference (K) between the surface and bottom water during the past 30 million years, as deep polar convection generated the cold water sphere. (From W. Berger 1981.)

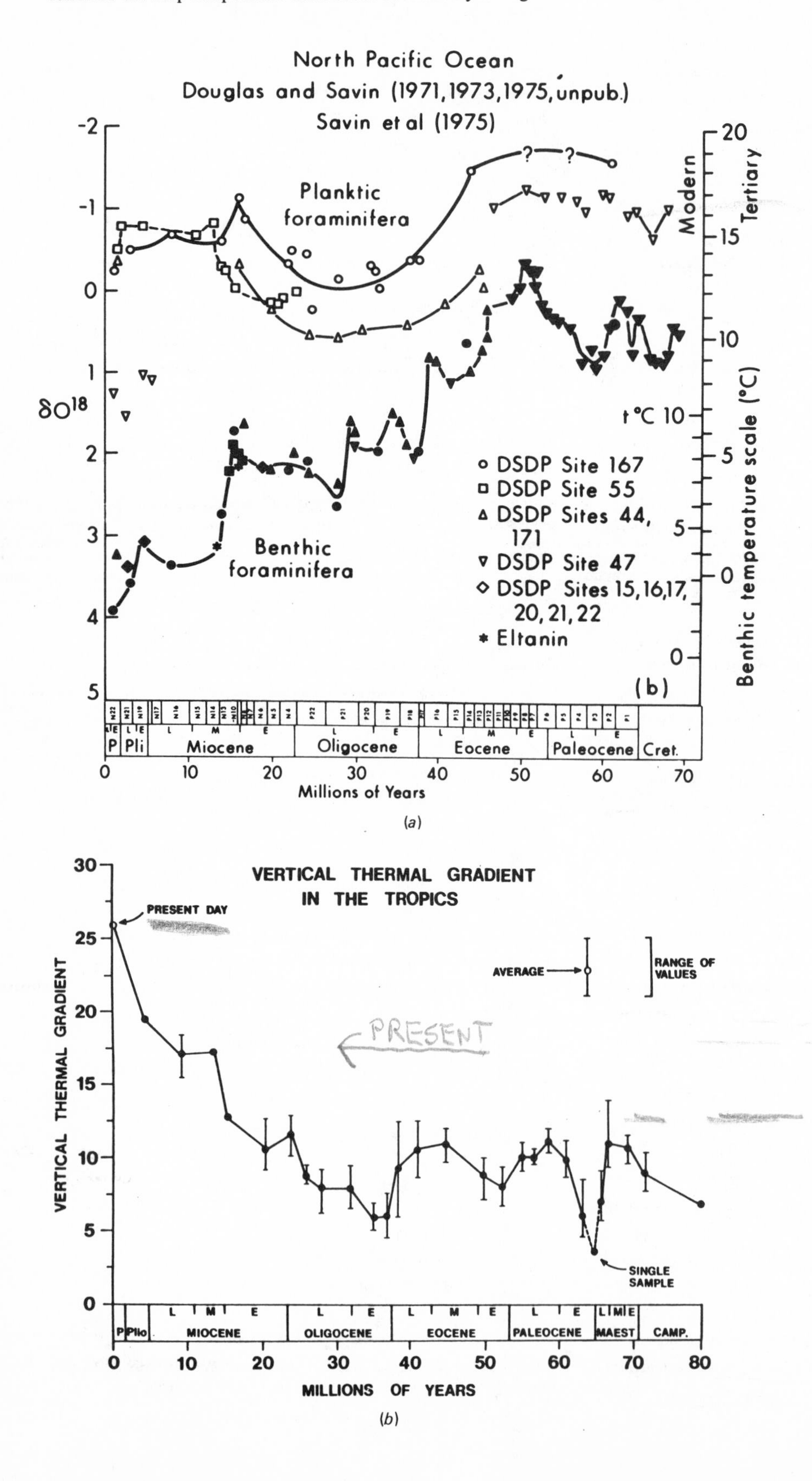

opening of the important sill between the Norwegian–Greenland Sea and the North Atlantic; and, more recently, the temporary (half-a-million-year) closing of the straits of Gibraltar, leaving the Mediterranean to dry out and become a burning desert 3 km below the level of the adjacent Atlantic (Hsü 1972; Drooger 1973). The effect of sudden intrusions of relatively fresh Arctic water 60 My ago (Gartner & Keany 1978), and saline Mediterranean water 5·5 My ago may, like the fresh water supply from the melting glaciers only 10000 years ago, have significantly changed the OBL, the global carbon cycle, sea ice, and the global climate (Berger 1981). Discussing the role of the deep ocean circulation in climate, Stommel (1958) vividly illustrated the vulnerability of narrow sills to continental drift when he noted that man now constructs dams with more than 10% of the material needed to reclose the straits of Gibraltar.

Secondly, when a continent's drift takes it to high latitude it provides a platform on which snow accumulates to make glaciers. Such polar ice caps are present today in Antarctica and Greenland; we are living in an ice age, though not in a period of maximum glaciation. Ice ages seem to have occurred at intervals of order 250 My. The present one began to affect SST some 37 My ago. The change of volume of the world ocean as water was lost to the waxing glaciers is recorded in the $^{18}O:^{16}O$ ratio in fossil foraminifera in ocean sediments (Shackleton & Opdyke 1977; Vincent & Berger 1981). Douglas & Woodruff (1981) have summarized evidence from fossils of benthic foraminifera showing that the temperature of the deep ocean was 10 K warmer before the Antarctic ice cap formed (Fig. 9.22(*a*)). Worthington's (1981) census of the water masses of the world ocean shows that water colder than 6 K now fills three-quarters of the world ocean. Presumably it is in equilibrium with the present rate of deep water formation (see discussion of the Pacific by Worthington 1981). The 'cold water sphere' of the ocean, discovered in 1751 by Ellis (see Warren 1981), is a characteristic feature of an ice age produced by continental drift.

How much does the existence of the cold water sphere influence the SST and therefore the atmospheric climate? Analyses of fossils of planktonic foramifera (which lived in the boundary layer of the upper ocean) show that the temperature of the upper ocean was fairly closely coupled to the abyssal temperature before the ice caps formed 40 My ago, but that the surface water then warmed as the cold water from polar seas cascaded down to fill the deep ocean (Fig. 9.22(*a*)). The changing temperature difference across the oceanic thermocline is shown in Fig. 9.22(*b*). This evidence of a doubling of the thermocline stratification over the last 40 My is a gauntlet thrown down by the palaeoceanographers; we wait for climate modellers to respond to the challenge it represents. The deep ocean is now being ventilated more vigorously than ever before, yet the upper ocean, which interacts strongly with the atmosphere, seems oblivious of the fact. It seems free to establish a contract with the other elements of the planetary climate system to produce a milder climate than the presence of polar ice caps would lead us to expect. Our failure so far to explain this observation puts at risk any prediction of climate change involving ventilation of the deep ocean, and encourages us to treat the upper ocean – the warm water sphere – as though it were decoupled thermally from the underlying cold water sphere, when we study decadal climate variation.

9.3 Air–sea interaction and climate prediction

In the previous section we viewed the ocean–climate problem from a planetary perspective. Now we adopt the position of a meteorologist concerned with predicting climate change, and ask what he needs to know about the fluxes of energy, water and chemicals through the ocean surface. It will be assumed that he has an atmospheric GCM suitable for climate research. What information is needed about the surface fluxes? Are climatological mean values of the seasonal cycle adequate? Is it necessary to monitor the fluxes? Or must they be predicted, and if so how? Does the atmospheric GCM have sufficient information for predicting the surface fluxes, or must one take account of the ocean circulation too? Is a coupled model necessary?

The answers to these questions depend on the kind of climate change that is being studied. We shall consider three kinds:

(1) extended-range weather forecasting (up to a month or so);
(2) short-term climate prediction (up to a decade or so);
(3) climate change due to pollution (up to a century or so).

These divisions correspond to streams 1, 2 and 3 in the WRCP (Houghton & Morel 1983, Chapter 1 of this volume). The last two correspond to Lorenz's (1975) climate prediction of the first and second kind because (2) is dominated by spontaneous internal changes in the planetary climate system, while (3) is dominated by changes in the boundary conditions of the planetary climate system. The specification for the surface fluxes will be considered for each in turn. It is concluded that for all three kinds of climate change, one needs to have quite a detailed understanding of the processes occurring in the OBL, where solar energy is absorbed and transported vertically by convection, turbulence and upwelling–downwelling, as described in earlier sections.

9.3.1 *Extended-range weather forecasting*

The term 'extended-range' is used here to distinguish stream-one prediction from 'short-term' weather forecasting, as performed operationally today. We start with a few remarks about the latter, by way of introduction to extended-range weather forecasting.

For the purpose of *short-term* weather forecasting, up to a week ahead, it is sufficient to assume that the temperature of the oceanic mixed layer remains constant during the period of the forecast: there is no need to calculate how it changes. Variation in the fluxes of energy and water vapour pressure (q_a) and wind speed (U_{10}, conventionally at a height of 10 m) in the atmospheric boundary layer, and in the cloud cover (C), according to the bulk formulae:

$$\begin{aligned} &\text{Sensible heat:} && Q_S = C_H\cdot(T_s - T_a)\cdot U,\\ &\text{Latent heat:} && Q_L = C_E\cdot(q_s - q_a)\cdot U,\\ &\text{Net IR radiation:} && Q_R = C_R\cdot T_S^4,\end{aligned}$$

where the saturated water vapour pressure (q_s) is a known function of ocean mixed layer temperature (T_s) and salinity (S_s), and the transfer constants C_H, C_E and the transfer function C_R (Clouds, q_a) are determined empirically.

In practice it is acceptable to know T_s to ± 1 K, which is an order of magnitude smaller than the annual range (see Fig. 9.12), but larger than the interannual fluctuations (Barnett 1981*c*; Weare *et al.* 1976; Colebrook & Taylor 1979; Navato, Newell, Hsiung;

Billing & Weare 1981), except near the equator (Merle (1980, Rasmussen & Carpenter 1982). The salinity S_s need only be known to ± 1 kg/t, which is greater than the annual range in most locations (Taylor & Stevens 1980). We conclude that it is sufficient to know the annual mean salinity and the monthly mean temperature of the ocean mixed layer, plus the deviation from the monthly mean temperature near the equator in the winter, when the westerlies overlap the anomalies sufficiently to permit a large-scale dynamical response (Webster 1982). The accumulated data base of ship measurements has been analysed to give monthly mean temperature and annual mean salinity distribution (Robinson 1976; Robinson, Bauer & Schroeder 1979). The accuracy of the temperature measurements is probably not much better than ± 1 K (Saur 1963; Tabata 1978); the data coverage is inadequate in the southern hemisphere. Attempts have been made to construct global maps of SST from satellite IR radiance measurements (see Fig. 9.11): the accuracy of early measurements was not better than ± 2 K, but the methods are being improved and satellites are likely to provide the best source of ocean surface temperature data for short-term weather forecasting in the future.

Extended-range weather forecasting is based on persistence in certain patterns of atmospheric circulation, the boundary conditions being prescribed (Leith 1983, see Chapter 2 of this volume). The aim is to make forecasts for up to a month ahead, during which period the SST changes (according to the mean seasonal cycle) by as much as 2 or 3 K. Such large changes in SST cannot be ignored. The atmospheric circulation responds to the adjustment of the atmospheric boundary layer to the prevailing SST (Webster 1981). Four methods can be used to take account of SST change during extended-range weather forecasts:

(1) Assume that the SST follows the regular seasonal cycle, as described in standard analyses of data collected during the past century, mainly by merchant ships. Although broadly correct, even in the data-sparse southern hemisphere, the existing data base is probably not better than ± 2 K. The collection of a new SST climatology, based on satellite and drifting-buoy measurements with appropriate error control, could make a significant contribution to extended-range forecasting by this method.

(2) Assume that the *change* in SST occurs at the regular seasonal rate, starting from an initial global distribution of SST *observed* at the start of the forecast. This would have the advantage over (1) of taking account of known SSTAs, which may amount to several degrees. The requirement is for a global SST monitoring system (such as IGOSS) capable of establishing the global pattern of SST to an accuracy of ± 1 K every week.

(3) Assume that the SST follows its regular seasonal cycle with the exception of initial anomalies, which are assumed to decay at a rate pre-determined from analysis of past records, using the method of Frankignoul & Hasselmann (1977).

(4) Assume that the change in SST due to advection by geostrophic currents has the climatological mean value, but that changes due to solar heating and surface energy loss to the atmosphere and Ekman flow (including pumping) can be calculated deterministically during the period of the forecast using a diagnostic model of the OBL driven by the atmospheric model. The requirement is for monitoring of the initial conditions (as in (2) above) plus a model of the oceanic boundary layer that gives SST to better than ± 1 K. This ambitious aim has been made attractive by the successful simulation of SST changes observed at ocean weather stations by means of quite simple 'mixed layer' models of the OBL, (first by Denman & Miyake 1973).

Considerable effort has gone into designing OBL models capable of meeting these requirements (for recent reviews see Niiler 1977; Niiler & Kraus 1977; Kitaigorodskii 1979; Woods 1982). It is generally agreed that linear models of the boundary, such as the 'copper plate ocean' cannot meet the ± 1 K specification for extended-range weather forecasting. What is needed is a non-linear model in which the vertical transport processes change with surface fluxes and the temperature and velocity profiles. Comparisons of SST variations observed at ocean weather stations and the predictions of models run with the OWS meteorological observations (Denman & Miyake 1973); Thompson 1976; Gill & Turner 1976) show that such models can be tuned to meet the ± 1 K specification, at least at high latitudes where the large annual range of SST is due mainly to local heat storage. More work needs to be done before the models are equally satisfactory in the tropics, where Ekman flow and baroclinic waves give significant lateral redistribution of heat and upwelling. The neglect of geostrophic advection makes the boundary layer models unreliable at major currents, such as the Gulf Stream, but is not a serious problem over most of the ocean (Gill & Niiler 1973). A satisfactory compromise might be to include the climatological mean effect of geostrophic advection based on a steady circulation pattern and the mean seasonal cycle of temperature profile. Cushman-Roisin (1981) has shown that inclusion of the Ekman transport improves SST prediction.

9.3.2

Short-term climate prediction

As the time scale extends beyond a month, the climate prediction problem becomes increasingly influenced by inter-annual variability in the other components of the planetary climate system. Heat redistribution in the ocean is particularly important. It leads to SSTAs by advection from remote regions rather than from local anomalies in the weather. Such advectively induced anomalies in SST were specifically neglected in the OBL models discussed above in the context of extended-range weather forecasting. The reason that one is interested in such teleconnections inside the ocean lies both in the displacement of thermal anomalies, and in the delay between their creation at one location by the atmosphere and their subsequent emergence elsewhere as SSTAs. The displacement may feed a different part of the atmospheric circulation (e.g. the Westerlies) able to respond more vigorously, or with greater impact on the continents. The delay offers the promise of predictability. In this section we shall divide the processes responsible for oceanic teleconnections into two classes: advection by the general circulation, and propagation by waves. These will now be considered in turn.

9.3.2.1

The Gulf Stream

Historically, attention was focussed on teleconnections due to advection. It has long been conjectured that interannual variation in the advection of heat by the Gulf Stream, from the tropical source to a sink under the westerlies, may be responsible for interannual variation in the winter climate of western Europe

(Sabine 1846; Stommel 1979*b*). They can be produced either by anomalous upstream injection redistributed by the normal (seasonally varying) circulation, or by anomalies in the flow, or by a combination of both. There is evidence that the mass transport of the Gulf Stream does exhibit significant interannual variations and efforts are now being made to monitor them (see Wunsch 1983, Chapter 10 of this volume), but it has not yet been established whether they make a significant contribution to the observed patterns of emergent SSTAs in the North Atlantic. Woods (1980) has argued that turbulent diffusion will reduce the amplitude of SSTAs with the residence time of the thermal anomalies that provoke them; it seems unlikely that variations as large as 1 K can be achieved by the circulation. The observed SSTAs in the North Atlantic are only about 1 K (Weare 1977). Model studies have failed to reveal a significant atmospheric response to such modest anomalies (Webster 1981). The case is not strong for predicting short-term climate changes by monitoring the oceanic circulation of thermal anomalies, but interest in the idea continues, notably in the USSR (Marchuk 1979). The requirement is for systematic monitoring of:

(1) those upstream regions where the thermal anomalies enter the ocean (i.e. in the tropics);
(2) the major currents, such as the Gulf Stream; and
(3) energetically active zones, where the ocean most rapidly loses heat to the atmosphere.

The USSR 'Sections' programme is dedicated to testing this approach to short-term climate forecasting (Monin 1983).

9.3.2.2
Polar seas

In polar seas the interannual variation of sea ice cover dramatically modulates the local energy, and water fluxes the atmosphere, and can provoke large-scale atmospheric response (Goody 1980). Hibler (1979), Hibler & Walsh (1983) and Parkinson & Washington (1979) discuss progress in modelling seasonal and interannual fluctuations of sea ice. As with predicting the depth of winter convection in the subpolar seas, the problem involves using a boundary layer model forced both by the surface energy balance and by ocean currents (preconditioning), with significant horizontal inhomogeneity. It is likely that interannual variation of sea ice distribution results from fluctuations in both contributions. Lemke, Trinkl & Hasselmann (1980) have shown that much of the observed interannual variability can be explained as a Markov-like response to random fluctuations in the atmospheric forcing (i.e. to the weather); the decay time for sea ice anomalies created in that way is only a few months. Here we are interested in the possibility of significant sea ice response to advective changes of heat and fresh water (i.e. salinity deficit) of the kind described in the preceding paragraph. Indeed, one may speculate that such interannual variation of heat and water circulation may have a bigger impact in winter through its effect on sea ice than on the Westerlies. Studies of interannual and decadal changes of temperature and salinity in the northeast Atlantic (Smed, Meincke & Ellett 1982; Ellett 1982) are particularly relevant. Widespread atmospheric response to changes in sea ice cover have been found in modelling studies by Warshaw & Rapp (1973), Newson (1973), Herman & Johnson (1978).

9.3.2.3
El Niño and the Southern Oscillation

The tropical ocean seems more likely to contain teleconnections that might be exploited for short-term climate forecasting. Horel & Wallace (1981), Wallace & Gutzler (1981), Shukla & Wallace (1983) have detected, and Webster (1982) and Hoskins (1983) have explained, global atmospheric response to tropical SSTAs, which are observed to have amplitudes of several degrees in the western Pacific. The SSTAs represent a boundary layer response to dynamical depression of the thermocline. Considerable progress has been made in recent years in modelling the Kelvin–Rossby wave response to Trade Wind anomalies (O'Brien, Busalacchi & Kindle 1981 for the Pacific; Katz and Garzoli 1982 for the Atlantic) and the OBL response (Hughes 1980). The predictable delay time introduced by the Kelvin and Rossby wave propagation provoked by an initial disturbance to the Trade Winds in the western Pacific offers the promise of predicting the atmospheric climate response up to a year ahead. The requirement is for improved models of tropical dynamics and boundary layer processes, and for systematic monitoring of the tropical wind stress and components of the surface energy budget. Monitoring the upper ocean heat content and surface temperature will help to keep track of developing anomalies. The interval between major anomalies (El Niño events) is several years, so data needed to test ideas about them must extend over decades, even though there is no commitment to making such 'monitoring' permanent until the results of the research are available. Although case studies of individual events (e.g. Gill 1982, 1983; Rasmussen & Carpenter 1982) are stimulating, the real test of El Niño Southern Oscillation theories (e.g. McCreary 1982*a*, *b*) will come from prediction of a sequence of events. The logistics of collecting time series of observations over such a long period are likely to be similar to those needed eventually for operational forecasting. Satellite measurements of wind stress (by scatterometer), solar heating (cloud images), heat content anomalies (altimeter) and surface temperature (IR radiometer) will be particularly important when they become available in the late 1980s. Meanwhile, much can be done with *in situ* observations, including island tide gauges (Wyrtki 1979) and wind observations (Wyrtki & Meyers 1976). The USA research programme 'OACIS' is concerned with the study of the El Niño and Southern Oscillation in the Pacific; the French 'Focal' and USA 'Sequal' programmes are exploring the Tropical Atlantic Ocean; the international programme 'TOGA' is concerned with exploring the linkage between interannual anomalies in the three tropical oceans, and preparing the way for the new satellite observations.

9.3.3
Long-term climate prediction

Now we turn from the spontaneous variability of the ocean–atmosphere system on time scales less than a decade to its longer-term response to changes in the boundary conditions, concentrating on those due to mankind. The role of the ocean in climate response to CO_2 pollution of the atmosphere (Schiff 1981) and changes of river discharge was introduced in an earlier section. Here we consider how they influence the surface fluxes and the upper ocean, and identify requirements for future research.

9.3.3.1

Thermal consequences of carbon dioxide pollution

Ramanathan (1981) presents the following scenario (Fig. 9.23). An increase of atmospheric CO_2 (to double the present concentration) leads directly to a 3.5 W/m² increase of IR radiation from the atmosphere into the ocean, and a 0.6 W/m² decrease of the solar energy flux into the ocean. The planetary climate system reaches thermal equilibrium with the doubled CO_2 only when the ocean has warmed sufficiently for enhanced energy loss by IR, sensible and latent heat to balance the increased heating. Assuming no change in cloud cover or precipitation the increased evaporation raises the humidity[2] which has a larger influence on the IR heating of the ocean than the direct effect of CO_2 (That positive feedback had been neglected in some earlier studies.) In the final equilibrium for doubled CO_2 the IR heating of the ocean has increased by 15.5 W/m², and the SST by 2.2 K. The change in solar heating is unknown, being sensitive to changes in cloud cover the modelling of which remains uncertain. Regional variation may be much larger than the small decrease of global mean solar heating. All change of solar heating will be neglected in the discussion that follows of the upper ocean response.

The extra IR is absorbed in the skin of the ocean. The effect on the upper ocean is equivalent to a corresponding reduction in the IR loss to the atmosphere. The subsequent warming of the ocean is due to increased retention of solar energy absorbed below the surface. The reduced extraction of heat absorbed from the sun also affects convection in the OBL. It leads to a decrease in the thermal compensation depth during the day when the ocean is being warmed by the sun. As a result the diurnal thermocline extends closer to the surface. The increased stratification hinders the deepening of the mixed layer by turbulent entrainment; the heat absorbed each day is not mixed down so deeply during the next night. The surface marking the daily maximum depth of the mixed layer (i.e. the top of the seasonal thermocline) will become shallower as the atmospheric CO_2 pollution increases. When pCO_2 has doubled, an extra 1/3 MJ/m² will be stored each day above that surface. The extra heat absorbed in a year will be approximately 100 MJ/m². The effect of this extra energy income is best considered separately for the two regions divided by the line along which the annual energy fluxes through the sea surface sum to zero (Fig. 9.5). This line will slowly migrate poleward as the CO_2 pollution increases.

Equatorward of the line, the energy loss to the atmosphere is less than the annual energy income from the sun and atmosphere. The annual maximum depth of convective adjustment (D_0) normally is less than the maximum depth of solar heating (S_0, about 100 m). The water column holding the residual heat may be extended down below S_0 by turbulent mixing or by Ekman pumping, but seldom by more than 50 m in one year. In this heat storage region the change of SST for 100 MJ/(m² y) will be 0.25 K/yr, assuming $D = 100$ m.

Poleward of the line the atmosphere extracts each year *all*

[2] But, curiously, analysis of ice core data led Jouzel, Merlivat & Lorius (1982) to conclude that the relative humidity at the sea surface was higher during the last glaciation, when the CO_2 concentration was 59% of the present value (Delmas, Ascencio & Legrand 1980) and the SST was on average 5 K lower (CLIMAP 1976).

Fig. 9.23. Changes in the surface infrared flux resulting from doubled atmospheric CO_2, according to Ramanathan (1981). The early effect of $2 \times CO_2$ is a decrease of 3.5 W/m² or 100 MJ/(m² y) in the rate at which the ocean loses heat to the atmosphere. The surplus heat left in the ocean each year slowly raises the surface temperature by a few degrees. The resulting increase in absolute humidity over the ocean produces a much larger 'greenhouse' effect: 12.0 W/m² in Ramanathan's model. This large positive feedback makes it impossible to estimate climate response to CO_2 pollution by linear models of the surface fluxes.

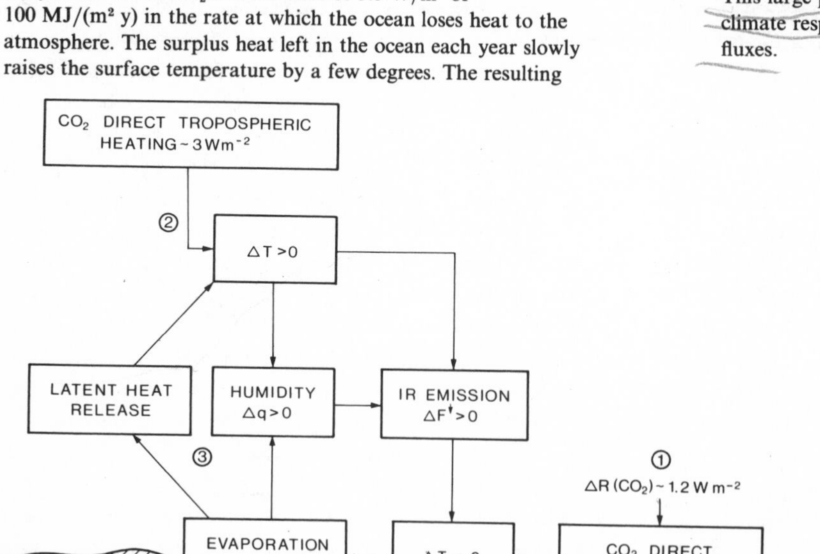

	Process ①	Process ②	Process ③	Total
Flux (Wm⁻²)	1.2	2.3	12.0	15.5
Percent	8.0	15.0	77.0	—
ΔT_S (Model dependent)	0.17	0.33	1.7	2.2

the energy received from the sun and the atmosphere, plus some more flowing in from the other region where there is a surplus. Here the annual maximum convection depth D_0 is greater than S_0. The addition of 100 MJ/m² from doubled CO_2 will reduce D_0 by an amount that depends on the vertical gradients of temperature and salinity introduced by advection (Woods 1982). Mixed layer deepening by turbulent entrainment is small compared with D and the slight reduction in it can be ignored. The convection depth D_0 will be further reduced by the weakening of the Westerlies as the meridional temperature gradient decreases with increasing CO_2 (Manabe & Weatherald 1980). So CO_2 pollution has two effects poleward of the zero line: (1) it reduces the depth of winter mixing, and (2) it reduces by 100 MJ/m² the cooling of the water advected into the water column whose lower boundary has depth D. The reduction in mixing depth may affect the ventilation of the interior of the ocean (Luyten, Pedlosky & Stommel 1983). The reduced cooling rate will produce a tiny change in SST; only 0.05K for a 100 MJ/m² input distributed over a water column 500 m deep.

Calculations with atmospheric GCMs suggest that the planetary climate system will achieve thermal equilibrium with doubled CO_2 when the SST is something like 2 K higher than at present (Manabe & Stouffer 1979, 1980; Bryan, Komro, Manabe & Spelman 1982). Taken as a whole the energy loss from the ocean surface will then have risen to match the extra IR input. The increased loss will be partitioned between IR radiation, sensible, and latent heat in roughly the present proportions, so the evaporation rate supplying the global hydrological cycle will increase by a few percent (Ramanathan 1981). The global distribution of the increased surface fluxes will not be uniform, it will depend on processes occurring in the upper ocean. We saw above that SST rises slowly and unevenly; it will take many years to reach the expected equilibrium. The problem for oceanographers is to calculate how long it will take, and what the global distribution of SST will be then, and during the interim. Solution of the problem involves a detailed understanding of the surface fluxes and boundary layer processes outlined in this chapter and of the circulation reviewed in the next chapter (Wunsch 1983). Although the processes involved are understood qualitatively, there has so far been no detailed calculation of SST response to CO_2 pollution. The problem has so far been simplified by treating the ocean as a set of thermal reservoirs with different time constants (Bretherton 1982*a*). The question is which reservoir is the pacemaker for the SST change. Brief comments on the principal candidates follow:

The solar heating zone. In the tropics (more accurately, equatorward of the zero line for net annual surface energy flux) the oceanic response to increased IR input is to store heat received from the sun in the top 100 m. The area for this region (about 200 (Mm)²) is over half that of the world ocean; the annual energy input due to CO_2 pollution will be about 2.10²² J/y. Treated in isolation from the rest of the world ocean, the thermal time constant for the layer is only six years. In reality, heat is swept out of the region horizontally to higher latitudes where $D_0 \geqslant S_0$. In equilibrium the heat flux divergence due to the circulation balances the annual storage. Calculations based on Bryden & Hall's (1980) analysis of the North Atlantic heat flux suggest that the leakage of heat from this store due to the geostrophic and Ekman flow would be less than 1% when the water column above S_0 has warmed by 2 K.

The warm water sphere. This is the lens of water extending down to the 8–10 °C surface, which outcrops at the polar fronts and reaches as deep as 1 km in places. Its volume is about 0.2 (Mm)³, roughly 10% of the total ocean. Much of the water passing through the solar heating zone circulates around the warm water sphere, being mixed down by winter convection in the extratropical region where D ⩾ S, and pumped down by Ekman convergence (Luyten, Pedlosky & Stommel 1983). The ventilation of the warm water sphere is not uniform (Worthington 1976), but if we treat it as a single isolated reservoir, the time constant to raise its temperature by 2 K, given 100 MJ/m² over an area of 250 (Mm)², is 64 years. That may be an underestimate since it assumes all of the tropical heat surplus is retained in the warm water sphere, and that there is no upward heat flow from the underlying cold water sphere.

The cold water sphere. Some of the water that has been heated in the solar heating zone and transported down to about 1 km in the warm sphere is subsequently carried horizontally by the ocean circulation into the subpolar regions where winter convection penetrates to great depths, providing the only link with the cold water sphere (Fig. 9.4(*c*)) (Killworth 1982). The rise of SST is limited by this access to the vast thermal reservoir of the cold water sphere. Assuming a local CO_2-induced energy input of 100 MJ/m² over 100 (Mm)², plus an advected input of, say, 10²² J/y (one-third of the warm water sphere income), feeding a volume of 1.2 (Mm)³, the time for a 2 K rise will be 500 years.

Sea ice. The annual energy input to the world ocean from doubled CO_2, approximately 3.6×10^{22} J, is comparable to the energy needed to melt all the sea ice of the Arctic and Antarctic, assuming a mean area of 20 (Mm)² and mean thickness of 2 m (Thorndike *et al.* 1975). The ice is sensitive to sea surface temperature. Parkinson & Kellogg (1979) estimate that summer temperatures only 4–5 K warmer than present would result in an ice-free Arctic ocean, at least in the summer. The atmospheric GCM calculations (Manabe & Weatherald 1980) suggest that the temperature rise will be 7–8 K in the Arctic. Such large changes could only be achieved if the cold water sphere ventilation were suppressed by oceanic boundary layer processes (Untersteiner 1983*a*).

9.3.3.2

Changes of continental run-off, salinity and sea ice

The sea ice cover of the Arctic Ocean is sensitive to the presence of a shallow halocline, created by the salt 'valve' effect of the annual freezing–thawing and sustained by 3 (Tm)³/y of fresh water from continents (Aagaard, Coachman & Carmack 1981). It has been conjectured that use of Siberian rivers for agriculture might weaken the halocline and reduce the ice cover (Budyko 1966, 1969; Donn & Shaw 1967; Flohn 1981). The response might be rapid; the residence time fresh water in the Arctic is only ten years (Aagaard & Coachman 1975).[3] The problem has been studied by Stigebrandt (1981), treating the Arctic as an estuary. The subject of sea ice physics is reviewed in the monograph edited by Untersteiner (1983*b*). Warshaw & Rapp (1973) and Newson (1973 have modelled the atmospheric response to a winter ice-free arctic Ocean.

[3] On the other hand, Untersteiner (1983*c*) points out that the planned reduction in river supply is less than the natural interannual variability and that circulation dynamics may limit its effect to coastal waters.

9.3.3.3

The carbon cycle

The ocean is the principal sink of anthropogenic CO_2 (Anderson & Malahoff 1977); so far it seems to have absorbed about half of the injection (Bolin, Degens, Kempe & Ketner (1979). The immediate question is whether the ocean will continue to absorb the same fraction in the future. Contemporary estimates, based on geochemical box models (Bolin 1981; Bolin *et al.* 1983), involve quite severe simplification of the OBL, processes that control not only the physical mixing of dissolved carbon, but also the biological processes essential to the accommodation of anthropogenic carbon in the ocean (Woods & Onken 1982). The present state of knowledge about these processes is inadequate. Recent studies of gas exchange at the sea surface are reviewed by Liss (1973), Deacon (1977) and Münnich *et al.* (1977), Hasse & Liss (1980). Broecker (1981) has reviewed the contribution made by geochemical tracer studies to the problem. Walsh (1980) has reviewed the role of primary production in the global carbon cycle, emphasizing the importance of the continental shelves. Considerable advances have been made in recent years in understanding the continental shelf boundary layer processes and the biological consequences (Swallow, Currie, Gill & Simpson 1981). Broecker (1982) has reviewed changes since the last glaciation, pointing out that 'the ocean contains about 60 times more carbon than the atmosphere; thus the atmospheric CO_2 content is on this time scale slave to the ocean's chemistry'; he emphasizes the role of nutrients.

9.3.3.4

Research priorities

In this section we have considered changes that occur over decades and longer. On these time scales the circulation of the ocean becomes increasingly important (Wunsch 1983), but in each of the examples considered here the role of the ocean depends critically on the regional and seasonal variation of the OBL. The development of improved boundary layer models, with particular regard to winter convection and sea ice has high priority for future research. More attention will have to be paid in future to the circulation of fresh water and the effect of salinity anomalies in the boundary layer. The direct monitoring of surface fluxes to the required accuracy (± 1 W/m²) is likely to be impossible, but efforts should be made to monitor SST (± 0.5 K), the depth of winter convection ($\pm 10\%$) and sea ice extent.

9.4

Surface fluxes – the present state of knowledge

The requirement for observing and predicting the GPFO, and inflow from the continents, was identified in the preceding section. Here we review the present state of knowledge about these fluxes and the methods available for measuring them. The subject has been presented in many excellent reviews and monographs published during the last decade. They include the following: Kraus (1972), Kitaigorodskii (1973), Businger (1975), Baumgartner & Reichel (1975), Paltridge & Platt (1976), Kraus (177), Favre & Hasselmann (1978), Dobson, Hasse & Davis (1980), Charnock (1981), Untersteiner (1983). No attempt will be made to repeat all the detailed information available in them. The aim here will be to provide a guide to the literature for the reader concerned with the climate problems introduced earlier in this paper. Reference will be made to some recent papers that are not quoted in the earlier reviews, and to new methods that hold promise for monitoring the surface fluxes during WCRP experiments.

Most of the requirements identified earlier would be satisfied if it were possible to monitor the fluxes globally with a temporal resolution of, say, one month, a spatial resolution of 1° (even 10° would be useful) and accuracies of ± 10 W/m² for each term in the energy flux (but ± 1 W/m² for the CO_2 problem), $\pm$ 0.1 m/month for precipitation and evaporation, and $\pm 10\%$ of the seasonal range for gas flux and continental run-off. At present, it is not possible to meet that specification. The accuracies can be achieved (with the exception of precipitation) in carefully controlled experiments such as JASIN (Guymer, Businger, Katsaros, Shaw, Taylor, Large & Payne, 1983) but they have not yet been achieved on the megametre scale of interest in climatology (Dobson *et al.* 1982). The problem is a combination of inadequate parameterization, inadequate measurement techniques and inadequate data coverage (Weare & Strub 1981).

Despite these limitations, the available climatological data have been used to produce maps of surface fluxes that have influenced our thinking about the role of air–sea interaction in climate. Famous examples include Jacobs (1951), Budyko (1956), Privett (1960), Wyrtki (1971), Zillman (1972), Hellerman (1967, 1968), Bunker (1976), Bunker & Worthington (1976), Saunders (1976*a*), Hastenrath & Lamb (1977, 1978, 1979), Leetma & Bunker (1978), Weare, Strub & Samuel (1980) and Hellerman (1980). River discharge analyses have been tabulated by UNESCO (1974). Precipitation and evaporation maps have been compiled by Baumgartner & Reichel (1975). Secondary compilations by the US Navy (1956, 1963), and Gorshkov (1976, 1979) have been widely used. Some of the data have been collected together on computer tapes, notably the US National climate centre tape TDF-11; others such as Bunker's are only now being made available in machine processable form (Bunker & Goldsmith 1979; Goldsmith & Bunker 1979).

These data have provided the basis for a number of climatological studies. Sverdrup (1957) and more recently Baumgartner & Reichel (1975), Stommel (1980) and Lamb (1981) estimated the global fluxes of heat and fresh water carried by ocean currents. The windstress data have been used as surface forcing in models of the oceanic circulation (Robinson & Baker 1979).

The present state of knowledge about the GPFO will now be reviewed in six brief sections: *solar and thermal radiation*, *evaporation and conduction*, *wind stress*, *precipitation and run-off*, *gases* and the *net annual budgets of energy and water*. The emphasis will be on practical methods used to calculate these components of the GPFO to space–time resolution appropriate for climatology, say one megametre and one month. The errors in present methods will be discussed. New methods based on satellite remote sensing of the ocean surface promise increased accuracy, systematic global coverage and some novel variables (e.g. vorticity flux) by the end of the century. As I write this article, there is intense activity in a number of centres around the world directed towards establishing these new methods, using data from prototypes such as Seasat-A and the GARP generation of meteorological satellites. The subject is advancing rapidly and any assessment of the final impact of satellite observations on the calculation of GPFO is premature. But the present indications are most encouraging, as the examples to be quoted in the following sections will show. Readers seeking more details of the state of the subject in the early 1980s should refer to the review by Stewart (1982) and the following monographs: Gower (1982), Bernstein (1982*a*), Kirwan *et al.* (1983), Stewart (1985).

9.4.1

Solar and thermal radiation

The problem is to calculate the solar irradiance incident on the sea surface, taking account of the astronomical cycles (Paltridge & Platt 1976; Berger 1981), and the influence of dust, water vapour and clouds (Kondratyev 1969; Paltridge & Platt 1976). Hinzpeter (1980) has reviewed methods of measuring the surface irradiance from ships. Climatologies based on ship observations of cloud cover using methods reviewed by Lumb (1964), Reed (1977) are available in the Bunker (1976), Hanson (1976), and the Hastenrath & Lamb (1978, 1979) atlases of the tropical Atlantic and Indian Oceans. There now seems to be a reasonable prospect of estimating the monthly mean surface irradiance from satellite images to ± 10 W/m² (Gautier 1982*a,b,c*; Gautier, Diak & Masse 1980; Bretherton 1982*b*); see Fig. 9.24. Payne (1972) and Cogley (1979) summarized measurements of the surface albedo.

Fig. 9.24 (*a*). Net incoming solar radiation at the sea surface in the Indian Ocean during 10–19 June 1979 measured by the VISSR on the Goes-1 satellite (From Gautier 1982*c*.) (*b*). Average wind stress on the Indian Ocean during 10–19 June 1979, calculated from measurements of low level cloud velocities (Goes-1 satellite). (From Wylie & Hinton 1982.)

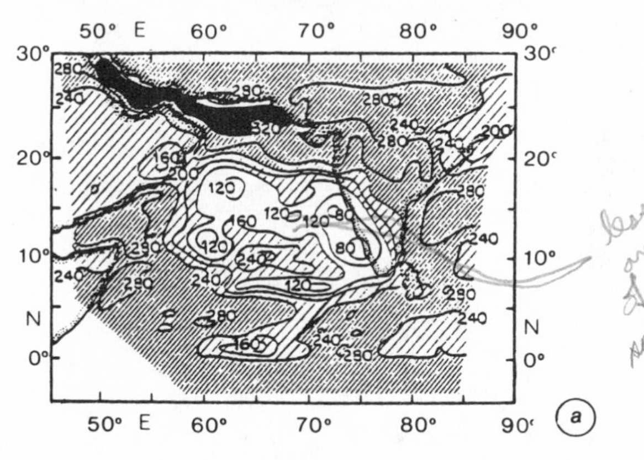

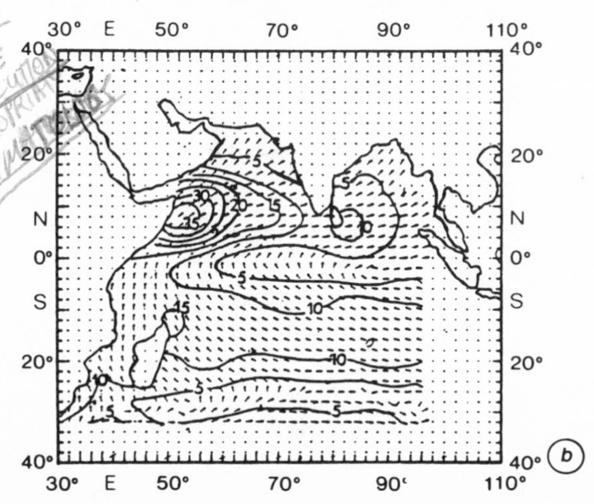

The longwave radiation from the ocean (IR_{out}) varies from 300 to 500 W/m² depending on the SST; a 1 K uncertainty of SST gives an error of 6 W/m² in IR_{out}. The long-wave radiation from the atmosphere into the ocean (IR_{in}) is on average 60 W/m² lower, and it is convenient to work with the net upward radiation (IR) which varies with humidity, cloud cover, the temperature difference between the mixed layers of the atmosphere and ocean, and the concentration of CO_2 and other radiatively active gases. Radiation models for calculation of IR in terms of those variables are available in Ramanathan & Coakley (1978). Techniques for measuring IR at sea are described by Hinzpeter (1980). Empirical formulae for calculating IR from merchant ship observations are discussed by Simpson & Paulson (1979). Climatologies based on such data have been published by Hastenrath & Lamb (1978, 1979), Hanson (1981), and are contained in the heat budget studies of Bunker (1976). Measurements during GATE show that the empirical formulae used in those climatologies overestimate IR by as much as 30 W/m² in the tropics (Fiegelson, Kontratyev & Prokovyev 1982; Kronfeld (1982), where the boundary layer can be so humid that changes of cloud have little affect on IR (the so-called 'super greenhouse effect'). New parameterizations were tested during JASIN (Lind & Katsaros 1982; Lind, Katsaros & Gube 1983). Much would be gained from equipping the weather ships to make high quality radiation measurements.

9.4.2

Evaporation and conduction

It is possible to make precise measurements (± 1 W/m²) of the sensible heat fluxes in the atmospheric boundary layer by the eddy correlation method, using data collected by aircraft (Nicholls 1978; Nicholls *et al.* 1982, 1983) flying on circuits of 100 km scale for a period of a few hours during experiments such as GATE (Volkov, Augstein & Hinzpeter 1982) and JASIN (Guymer *et al.* 1983). But such data are not available on a global basis, so the eddy correlation method cannot be used to calculate the GPFO. Climatologists therefore use much simpler bulk formulae, appropriate for the available global data base:

$$Q_L = L \cdot C_E \cdot (q_S - q_a) \cdot U \text{ (latent heat),}$$
$$Q_S = C_H \cdot (T_S - T_a) \cdot U \text{ (sensible heat),}$$

where C_L and C_H are transfer functions that parameterize the complicated processes by which moisture and heat flow from the mixed layer of the ocean to the mixed layer of the atmosphere. Various forms have been proposed for the transfer functions on the basis of the interfacial physics (Deardorff 1968; Kondo 1975; Friehe & Schmitt 1976; Liu, Katsaros & Businger 1979; Anderson & Smith 1981; Large & Pond 1982). Comparison of Q_L and Q_S estimated by the bulk formulae and the eddy correlation method during carefully controlled experiments at sea reveal the limits of the former. For example, Liu *et al.* (1979) found a latent heat flux error of ± 22 W/m² for their transfer function which takes account of the skin effect, and larger errors when C_E was assumed to be constant. Further work is needed to determine the form of the transfer function at high wind speeds, when the physics of the interface becomes more complicated (Ling & Kao 1976; Smith 1980; Charnock 1981; Smith & Katsaros 1981). Intermittency in the boundary layer fluxes due to clouds and spatial variability of ocean mixed layer temperature poses a serious sampling problem, even during well-conducted experiments. The problem becomes much more severe when one attempts to calculate the GPFO from archives of merchant ship data (Bunker & Goldsmith 1979; Goldsmith & Bunker 1979; Esbenson & Reynolds 1981). The transfer functions established during small-scale experiments may still be appropriate for estimating the GPFO (Bunker & Worthington 1976), or it may be more appropriate to tune them to the large-scale heat budget determined globally (Budyko 1974) or in an ocean basin, using oceanographic data to close the budget (Bunker, Charnock & Goldsmith 1982). The two methods (local physics, and large-scale budgets) give rather different values.

At present, it is not clear how much of the discrepancy between the different versions of the coefficients can be attributed to measurement errors and sampling biases (e.g. avoiding storms) in the merchant ship data base. There is an urgent need for a new global, systematic data base that can be used in its own right, and as a framework for correcting biases in the historical data. Satellite measurements will help by improving U (radar scatterometer) and T_S (infrared radiometer) both in accuracy and in unbiased global coverage. So will the use of atmospheric circulation models, providing they are reliable in predicting precipitation. Meanwhile, one should treat published GPFO distributions with caution: quoting the global mean evaporation rate to four significant figures (1.176 m/y; Baumgartner & Reichel 1975) is premature.

Calculation of evaporation and conduction in polar seas requires careful attention to the seasonally varying ice cover, and the open water regions (leads and polynyas) within the ice periphery (Goody 1980). Satellite monitoring of sea ice cover by radar and microwave radiometer will play an increasing role in future (Robin, Drewry & Squire 1983; Raney 1983). McBean (1983), Untersteiner (1983*b*), Herman (1983) and Barry (1983) have reviewed the subject in the monograph edited by Untersteiner (1983*b*), see also Chapter 8 of this volume (Untersteiner 1983*b*).

9.4.3
Wind stress

The momentum flux into the ocean, the wind stress τ, is a fundamental variable for models of the ocean's contribution to climate. It is needed for calculation of Ekman transport (τ/f), for Sverdrup transport (curl τ/β), for the deepening of the mixed layer by entrainment ($\tau^{3/2}$), and as a factor in the bulk formulae for surface fluxes of heat, water and gases ($\tau^{1/2}$). The principal method of calculating wind stress for climatological purposes has been to use the bulk aerodynamic formula

$$\tau = C_d \cdot \rho |\mathbf{U}_{10}| \mathbf{U}_{10},$$

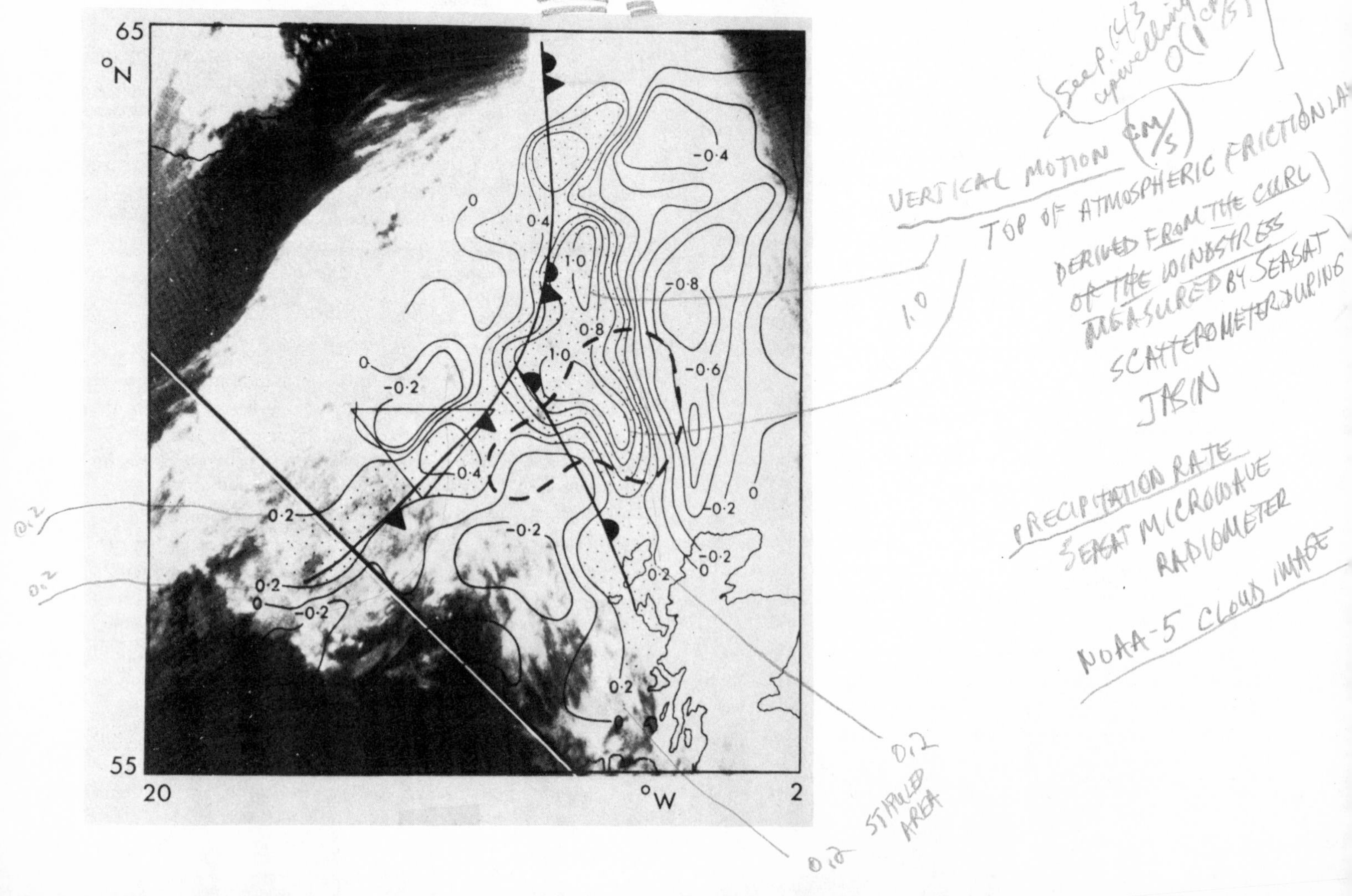

Fig. 9.25. The vertical motion (cm/s) at the top of the atmospheric friction layer derived from the curl of the windstress measured by Seasat scatterometer during JASIN, compared with precipitation rate estimated from Seasat microwave radiometer, and a NOAA-5 cloud image. (From Guymer *et al.* 1983, reproduced with permission.)

with U_{10} the wind at a standard height of 10 m derived from merchant ship observations, and the drag coefficient C_d determined empirically (see reviews by Garratt 1977; Charnock 1981). The role of surface waves in transferring momentum from the wind to the surface mixed layer of the ocean has been discussed by Richman & Garrett (1977). Climatological maps of wind stress and its derivatives have been published by Hantel (1970, 1972), Leetma & Bunker (1978), Hellerman (1967, 1976*a*,*b*, 1980), Han & Lee (1981). Saunders (1976*a*,*b*) has examined the sensitivity of such maps to the averaging scale in space and time. The distribution of merchant ship observations is seldom ideal and other data sources have to be used, including meteorological analysis (Hantel 1970; Holopainen 1967; Ellis, Vonder Haar, Levitus & Oort 1981) and cloud winds (Krishnamurti & Krishnamurti 1980; Wylie & Hinton 1982); see Fig. 9.24(*b*). Shaw, Watts & Rossby (1978) have estimated wind speed–stress from ambient noise measurements (a technique that can also, with suitable filtering, give rainfall rate). In the future, the standard method of measuring the wind stress will almost certainly be by satellite radar scatterometer (SASS), such as that carried on Seasat in 1978 (Born, Lame & Rygh 1981; Bernstein 1982; Lame & Born 1982; Guymer 1983). SASS algorithms have so far been developed for wind speed rather than stress, because the former offers more ground truth for tuning. Comparison of Seasat measurements with ship and buoy measurements during JASIN (ground truth errors $\pm$ 0.5 m/s, $\pm 5°$) suggest the SASS data will normally be accurate to ± 1.3 m/s and $\pm 16°$ (Liu & Large 1981; Jones *et al.* Guymer *et al.* 1981; Guymer 1983; Schroeder *et al.* 1982). There is some evidence that the SASS technique underestimates extreme winds at hurricanes (Jones *et al.* 1982). Comparison between SASS and merchant ship data show a much larger scatter, presumably because of errors in the latter. Chelton & O'Brien (1982) have compared maps of wind stress for the tropical Pacific derived from Seasat SASS, Seasat altimeter and routine ship observations. Chelton, Hussey & Park (1981) give the first global map of satellite-derived winds, based on Seasat altimeter data. Guymer *et al.* (1983) have compared SASS maps of wind stress with standard synoptic charts, and shown that the distribution of wind stress curl, which determines Ekman convergence and therefore vertical motion, correlates reasonably well with the distribution of precipitation rate (as estimated from Seasat microwave radiometer), a most exciting development in air–sea interaction research (Fig. 9.25).

9.4.4
Precipitation and run-off

Austin & Geotis (1980) and Mintz (1981) review the various methods used to measure precipitation over the ocean. The global distribution of annual precipitation has been mapped by Baumgartner & Reichel (1975) using what they describe as 'scanty and imprecise' ship measurements. Dorman (1982) draws attention to the dangers inherent in extrapolating island measurements out to sea. Korzun (1974), Hsu & Wallace (1976), Reed & Elliott (1979), Dorman (1981), Dorman & Bourke (1979, 1981) give seasonal, Jaeger (1976) gives monthly, distributions of precipitation over the ocean. Kidder & Vonder Haar (1977) have published seasonal oceanic precipitation data from Nimbus 5. During GATE, measurements of precipitation were made by ship radar (Hudlow & Patterson 1979) and satellite (Griffith *et al.* 1980). The Seasat scanning multichannel microwave radiometer indicated how satellite data will overcome the scantiness, but the question of precision still has to be resolved (Lipes 1982). Pessimists speak of the SMMR rainfall rate data being no better than within a factor of two; optimists speak of much higher accuracy (Atlas & Thiele 1981). It has also been proposed to estimate rainfall rate using a range-gated altimeter (Goldhirsh & Walsh 1981). See also Fig. 9.9.

Continental run-off presents fewer problems. The major rivers are routinely monitored, and the data are summarized by UNESCO (Fig. 9.8*c*). Icebergs are more difficult. Present methods of calculating calving rate, transport by ocean currents and melting rate are inadequate. Huppert & Turner (1978) have shown that double diffusion influences iceberg melting.

9.4.5
Gases

The flux of a gas such as CO_2 through the interface between the ocean and the atmosphere is limited by resistance in the water. It can be expressed by the following formula:

$$\text{Gas flux (mol/(m}^2\text{ s))} = F_g = W_s \cdot (C_s - A \cdot C_a),$$

where A is the solubility, C_s and C_a are the gas concentrations (mol/m³) in the mixed layers of the sea and air, and W_s is the gas transfer velocity (m/s). The gas solubility varies with temperature, so that CO_2 leaves the ocean in the tropics where the mixed layer is warm and enters the ocean at high latitudes where the mixed layer is cold and, incidentally but importantly, where it is deep in winter. The value of C_s varies regionally, reflecting the vertical motion through the base of the mixed layer (Fig. 9.3). The value of W_s depends on molecular diffusion of the gas through the laminar flow surface skin of the ocean. Two models have been proposed for W_s. The simplest, due to Whitman (1923) and called the stagnant film model, is based on the molecular diffusivity of the gas D (which varies significantly with temperature):

$$W_s = D/\delta,$$

where δ is the thickness of the viscous boundary layer. The second, due to Danckwerts (1970), is called the surface renewal model, and depends on the periodic replacement rate S of gas in the skin:

$$W_s = 1.13\ (D/S)^{\frac{1}{2}}.$$

The structure of the skin is changed by wind waves and therefore W_s is expected to increase with wind speed (Deacon 1980), which leads to the following development of the surface renewal model:

$$W_s = B \cdot (D/\nu)^n \cdot u_*,$$

where B is a constant, μ is the kinematic viscosity of sea water and u_* the friction velocity. Using a circular wind–water tunnel, Jähne, Münnich & Siegenthaler (1979) have measured a linear increase of W_s with wind speed, and determined $n = 0.41 \pm 0.09$. Hasse & Liss (1980) have argued that capillary waves are likely to be more important than gravity waves. At high speeds, disruption of the surface skin by wave breaking may become important (Wu 1982), although the magnitude of the change in W_s has not yet been measured. Broecker (1980) and Thorpe (1982) have considered the effect of bubbles created by wave breaking and mixed down by turbulence in the OBL before rising back to the surface, there to burst releasing their gas and ejecting droplets of skin water into the air.

Direct measurements of the flux of CO_2 have been made by the eddy correlation method in the atmospheric boundary layer (Jones & Smith 1977; Wesely, Cook, Hart & Williams 1982).

$$F_g = \overline{w'C_a'}.$$

Indirect estimates of W_s have been obtained from Broecker's radon deficit method in which the OBL concentration profile of ^{222}RN is measured. The radon has a radio-active decay constant of 5.52 days. Its concentration in the mixed layer is reduced relative to the equilibrium value in the underlying thermocline by loss to the atmosphere. Recent measurements by Peng, Broecker, Mathieu, Li & Bainbridge (1979) have failed to reveal the expected increase of W_s with wind speed. Kroemer & Roether (1982) have suggested that the change is masked by stronger variation in the mixed layer radon deficit due to spatial and temporal inhomogeneities in the exchange of water between the mixed layer and the thermocline, which were not measured during the radon deficit experiments. The exchange processes must have time scales of a few days for that to happen. Candidates include internal wave undulations (especially near the equator where f is small), tipping of isopycnals during mesoscale frontogenesis (MacVean & Woods 1980) and the undulation of particles flowing along frontal jet streaks distorted by meanders (Woods, Wiley & Briscoe 1977). Torgersen, Mathieu, Hesslein & Broecker (1982) sought to overcome some of these problems by measuring gas exchange in a small lake. They concluded their paper with the following remarks:

> 'The application of 'first principles' to complicated conditions of extreme and varied surface conditions is limited by our lack of understanding of the processes occurring in momentum, heat and mass transport. Until such times as these processes are clearly understood, simple single-parameter models are probably the best. This study supports the stagnant film model of gas exchange, which predicts a linear dependence of W_s on the molecular diffusivity of the gas.'

A comprehensive review of air–sea exchange of gases and particles is available in the monograph edited by Liss & Slinn (1983).

9.4.6 *The annual mean global pattern of fluxes of energy from the ocean*

Global circulation models of the atmosphere have climates that differ significantly from the observed climate. The models include parameterization schemes that can be tuned to reduce the discrepancy. An attractive climatic indicator for such tuning is the annual mean global pattern of surface fluxes of energy and water. It is attractive because the empirical annual energy and water budgets can be balanced everywhere on the continents. Interannual heat storage and transport are zero in the continents, so the energy budget (insolation minus net longwave radiation, sensible and latent heat) must balance everywhere. And, although there can be interannual variation of water storage (glaciers, lakes, ground water) and of transport (glaciers, rivers), they can be monitored and included in the annual water budget (precipitation and storage minus evaporation and run-off) to give a balance for each catchment. Campbell (1981), Burridge (1982) and Swinbank (1983) have used FGGE data to calculate the annual energy budget for continents and oceans (Table 9.1). The continental values should be zero, to within the experimental uncertainty in corresponding values for the external energy radiation budget, ± 10 W/m² (Barkstrom & Hall 1981). The errors are about ± 50 W/m², comparable with the basin-scale annual GPFO. Burridge (1982) has also plotted the regional variation of annual energy flux, revealing the existence of (erroneous) extrema with magnitudes exceeding ± 100 W/m² at megametre scale on the continents, comparable with the megametre-scale extrema in annual GPFO.

Let us suppose that further development of the models and improved atmospheric observations make it possible to reduce the annual energy budget to within 10 W/m² of zero everywhere on the continents. The challenge will then be to see how accurately the model simulates the observed pattern of annual mean energy and water fluxes over the ocean, where they are not zero, because the ocean circulation carries significant quantities of heat and water (order 1% of the throughput) from one ocean basin to another; and where the extended storage of heat in the ocean, which makes such transport possible, may exhibit significant interannual variation. It is necessary to have an independent measurement of the annual GPFO to ± 10 W/m². The discussion of individual components of the heat and water fluxes in earlier sections of this review indicated an uncertainty of ± 50 W/m² in the local annual budget. Although errors of this magnitude do not eliminate the megametre maxima in the annual GPFO, which exceed 100 W/m² (Fig. 9.9(*b*)), they are comparable with the basin-scale average fluxes deduced from direct oceanographic measurements (Fig. 9.7) (Bryden 1982).

There can be little confidence in the regional predictions of CO_2-induced climate change by atmospheric GCMs that have not been tuned to simulate the present annual GPFO, including the large megametre-scale variation. But, at present, the empirical description provides an inadequate basis for such tuning. How can it be improved? The most attractive method is to make a

Table 9.1. *Annual surface energy flux from radiation measurements and atmospheric flux divergence.*

		Campbell (1981) (W/m²)	Burridge (1982) (W/m²)
The continents			
North America		−7	−5
South America	(tropics)	+54	
	(extratropical)	+32	
Australia		−54	
Eurasia		−43	−38
Africa	(tropical and north)	−10	
	(extratropical, south)	−7	
The oceans			
Atlantic	(north)	−25	−28
	(tropical)	+75	
	(south)	+37	
Pacific	(north)	−28	+15
	(tropical)	+60	
	(south)	−21	
Indian	(tropical)	+29	
	(south)	−38	
Mediterranean		−18	

comprehensive network of measurements of the heat (and fresh water) transport by ocean currents. Wunsch (1983) discusses the prospects for such measurements; they provide one motivation for the World Ocean Circulation Experiment now being planned as a component of the WCRP (Houghton & Morel 1983). It may also be possible to improve the accuracy of direct calculations of surface fluxes by a combination of improved observations (from ships and satellites) and improved parameterizations, tuned to large-scale budget constraints following the method of Budyko (1974) and Bunker, Charnock & Goldsmith (1982). The interannual variation of oceanic heat content can be monitored relatively cheaply in the upper ocean by a ship-of-opportunity expendable bathythermograph (BT) programme (Bretherton, McPhaden & Kraus 1982), and in the deep ocean by acoustic tomography (Munk & Wunsch 1982). These new methods should begin to yield results by the end of the century. On the same time scale, we can look forward to the development of climate models in which the global circulations of the ocean and atmosphere are coupled, making the SST and surface fluxes free variables. Such coupled ocean-atmosphere models will also have to be tuned to yield realistic GPFO. Meanwhile, it is possible to calculate the oceanic transport of heat and fresh water with ocean GCMs based on prescribed boundary conditions (Bryan & Lewis 1979); the results are in fair agreement with the few available direct measurements (Bryan 1982; Bryden 1982).

All three methods of estimating the annual GPFO (measurement of component surface fluxes, measurement of the oceanic circulation, and ocean circulation modelling) are likely to be improved in the coming decades, but each will have uncertainties that require checking against independent assessments. One way is to check them against each other. A WCRP working group (Dobson *et al.* 1982) has examined the feasibility of a five-year case study in the North Atlantic, in which the annual GPFO will be calculated by all three methods, and compared with atmospheric GCM estimates. The project (the 'Cage' experiment) may become one element of the World Ocean Circulation Experiment (Woods 1984).

9.5 The boundary layer of the upper ocean

A model of the OBL is needed for all types of climate prediction. For extended-range weather forecasting it is used to calculate the modulation of the seasonal mean patterns of surface fluxes by weather systems, assuming that oceanic geostrophic flow does not change. For short-term climate prediction it is also used to calculate the change of surface fluxes resulting from oceanic processes; in particular, vertical motion associated with baroclinic Kelvin and Rossby waves in the tropics, and sea ice formation at high latitude. For long-term climate prediction it is also used to calculate the ventilation of the deeper layers of the ocean by winter convection and Ekman pumping, the response of the surface fluxes to slow changes in temperature and salinity of the non-turbulent interior of the ocean, and the biological–chemical processes that contribute to the accommodation of anthropogenic carbon. Modelling the OBL is as important for climate as modelling the circulation; it is important on all time scales. In this section, we review the present state of knowledge concerning the boundary layer, and the experience with modelling it. The material contained in the following review articles and monographs will be taken as read: Tully & Giovando (1963), Turner (1973), Niiler (1975, 1977), de Szoeke & Rhines (1976), Kraus (1977), Phillips (1977), Garwood (1979), Kitaigorodskii (1979), Woods (1982).

9.5.1. *Physics of the ocean boundary layer*

At present none of the models of the OBL proposed for climate research meets the specifications established in earlier references; at least, not for all seasons, all weathers and all over the world ocean (Woods 1982). That is largely because they have been based on an oversimplified view of the physics involved. So, before discussing the various models, it is worth considering the physics of the boundary layer in some detail. Of course, it is not practical to include all the details in climate models covering the world ocean. There must be parameterization. The emphasis here will therefore be on describing what is known of each aspect of OBL physics, then what proposals have been made to parameterize the results economically for climate modelling. Four aspects will be highlighted: solar heating, the upward heat flux (including the surface skin and the convection layer), turbulent mixing, and low-frequency vertical motion. We saw earlier that some aspects of the OBL will have to be monitored, to provide initial conditions for forecasts and, during the research phase, to test them. Monitoring the SST and upper ocean heat content and temperature profile will be briefly discussed at the end of this section.

9.5.1.1 *Solar heating profile*

The problem is to calculate how the solar irradiance decreases with depth, taking into account both particulate scattering and molecular absorption. Ivanoff's (1977) review provided an important stimulus for attempts to improve parameterizations of the vertical distribution of solar heating of the ocean. Paulson (1980) reviewed methods of *in situ* measurement. Raschke (1975), Preisendorfer (1976), and Gupta & Ghovanlou (1978) have treated the scattering problem theoretically. Woods (1980) calculated the solar heating close to the surface using pure water absorption coefficients (27 spectral bands) on the assumption that scattering can be neglected in much of the open ocean where biological activity is weak, and plankton congregate below the mixed layer (Woods & Onken 1982). A number of empirical parameterizations (model fits to measured irradiance profiles) have been proposed (Paulson & Simpson 1977; Zaneveld & Spinrad 1981; Simpson & Dickey 1981*a*, *b*). Woods, Barkmann & Horch (1984) have compared those parameterizations, proposed a robust three-exponential fit, and studied its sensitivity to cloud cover and sea water turbidity as a function of latitude and season. The development of an empirical climatology of oceanic solar heating for the world ocean as a function of season still lies in the future. The main difficulty at present is the inadequacy of existing statistical descriptions of clouds over the ocean (London 1957; Lumb 1964) and sea water turbidity (Jerlov 1976). The satellite climatologies will hopefully remedy that situation: surface irradiance based on clouds (ISCCP) and sea water turbidity (Smith & Baker 1978; Hovis *et al.* 1980). Useful progress is being made in relating satellite measurements of colour index to *in situ* measurements of irradiance profiles (Gienapp 1982; Robinson 1983).

9.5.1.2
The layer of upward heat flux

The heat flux in the uppermost layer of the ocean is always directed upward to supply the energy loss to the atmosphere. Here we consider how the depth H of that layer of upward heat flux varies regionally with time of day and season. In principle it is marked by a change in sign of the vertical temperature gradient, but in practice the gradient is normally too small to be measured, so H must be calculated using a diagnostic model. Within a few millimetres of the ocean, molecular viscosity keeps the flow laminar and the upward heat flux is carried by molecular conduction (Katsaros 1980). Street & Miller (1977) measured the thickness of the skin, across which there can be a temperature difference of several tenths of a Kelvin degree (Paulson & Simpson 1981). Katsaros *et al.* (1977) have modelled the temperature profile in the skin, taking account of the vertical distribution of heating by absorption of IR radiation from the atmosphere and (during the day) from the sun, and the emission of IR radiation from the ocean. The effective depth from which oceanic thermal radiation is emitted is a strong function of wavelength. McAllister, MacLeish & Corduan (1971) have used a two-wavelength radiometer to measure the skin temperature gradient and hence have calculated the upward heat flux. Satellite and aircraft radiometric measurements of ocean temperature have to be interpreted in terms of the effective sampling depth for the wavelength being used, and the skin temperature profile, calculated from the upward heat flux using a model such as that of Katsaros *et al* (1977).

During calm sunny days the surface cooling of the ocean can be supplied by solar heating inside the surface skin. The downward heat flux then starts very close to the surface. At night the surface heat loss is met by cooling the ocean through a water column the depth of which depends on the density profile. The Rayleigh number is large and the heat is carried up from depth H to the skin by buoyant convection (Turner 1973). The value of H can be calculated from the density profile and the surface buoyancy flux by the method of convective adjustment (Gill & Turner 1976). The density depends on temperature and salinity; the buoyancy flux on the surface energy loss and net water loss (evaporation minus precipitation). The forenoon decrease in H following sunrise and its initial increase in the afternoon can be calculated from the diurnal variation of the thermal compensation depth, which is independent of the density profile. Woods (1980) has exploited that simplification to explore the seasonal and meridional variation of daytime convection in the ocean boundary layer.

The density gradient controlling the depth of convection is determined primarily by temperature, but salinity can also be important. A halocline can limit the depth of convection, as in the Arctic Ocean (Carmack 1983) and after precipitation (Kronfeld 1982). Killworth (1983) has drawn attention to the contribution of horizontal T–S gradients in the formation of the very deep convection events that ventilate the deep ocean. Woods (1982) has suggested that in regions where the annual heat loss to the atmosphere exceeds the solar income, the depth of convection at the end of winter is influenced by buoyancy release from an unstable salinity gradient formed by the baroclinic advection that supplies the heat needed to balance the heat budget.

Table 9.2 *Turbulence in the ocean boundary layer.*

Measured rates of turbulent kinetic energy dissipation (ϵ)

Date	authors	ϵ/(mW m^3)	Comments
1962	Stewart and Grant	$\leqslant 4.2$	$z = 1$ m
1979	Gargett, Sanford & Osborn		diurnal variation
1980	Dillon & Caldwell	0.001	calm
		3	rough
1980	Oakey & Elliott	0.5–50	$\int \epsilon dz$ (mW/m^2)
1982	Lueck & Osborn	0.01–1	

Fig. 9.26 (*a*). Measured profiles of turbulent kinetic energy dissipation rate (ϵ) and stability frequency (N) in the ocean. (Based on Dillon & Caldwell 1980.) (*b*). An example of calculated profiles of the terms in the turbulent kinetic energy equation. (From the second-order closure model of Klein & Coantic 1981.)

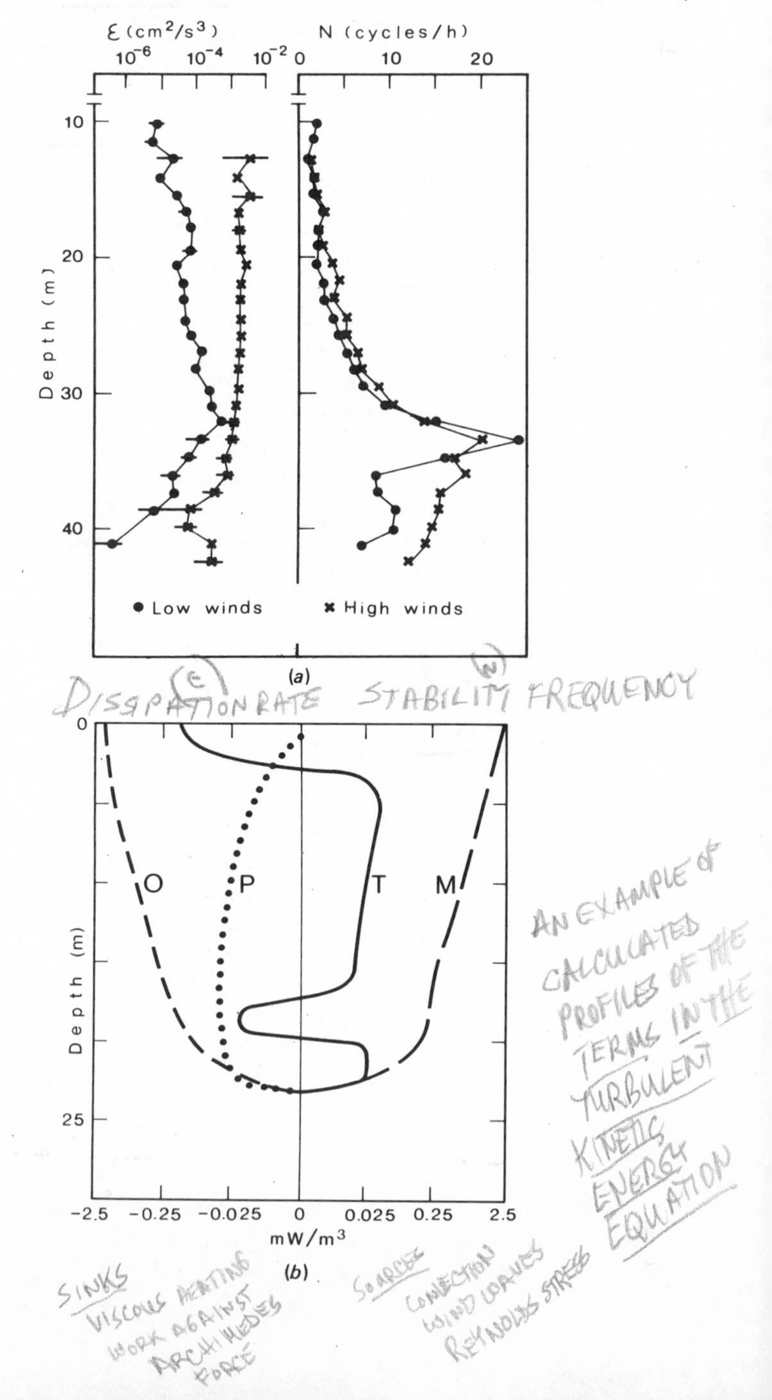

9.5.1.3

Turbulence

In recent years it has become possible to measure profiles of turbulent velocity and temperature fluctuations in the OBL (Table 9.2) and to model the turbulence with second- and third-order closure (Klein & Coantic 1981; Simpson & Dickey 1981*a,b*; Mellor & Yamada 1982). They both show values of order mW/m³ in the convection layer, decreasing to μW/m³ in the seasonal thermocline (Fig. 9.26). The depth of the turbocline is controlled by the density gradient. The models reveal the contributions of the various sources (convection, wind waves, Reynolds stress) and sinks (viscous heating and work against the Archimedes force), and reveal the role of vertical turbulent diffusion of TKE.

The three energy sources of turbulence occur at different levels in the OBL: wind waves supply energy to the top metre or so; convection throughout the mixed layer; current shear at all depths, but most strongly in the Ekman layer. In the extratropical winter, the mixed layer is normally much deeper than the Ekman layer and TKE from waves and current shear is dissipated by viscosity before diffusion carries it to the bottom of the mixed layer. But the release of potential energy by the upward fluxes of heat and fresh water provides a source of TKE deep in the mixed layer. Gill & Turner (1976) showed that only a tiny fraction of this convectively generated turbulence is consumed by work against the Archimedes force in entraining denser water into the mixed layer; most of the TKE is dissipated in viscous heating. Killworth (1982) has suggested that the entrainment fraction decreases as the mixed layer becomes deeper, becoming negligible when convection extends to the bottom of the deep ocean. In general, the influence of turbulence on the OBL is negligible in winter outside the tropics.

However, turbulent entrainment and the downward heat and fresh water fluxes below the mixed layer play a significant role in controlling OBL structure in the tropics and at higher latitudes in summer, when the mixed layer depth is comparable with both the Ekman and solar heating depths. Turbulence from wind waves loses most of its energy to viscous heating as it diffuses down to the pycnocline at $z = \mathrm{H}$. The small remainder is then consumed by work done against the Archimedes force in entraining dense (cold, salty) water into the mixed layer. The turbulence is quenched in the entrainment zone; none diffuses deeper. The horizontal entrainment front descends at a speed related to the rate of energy supply to the turbulence in the mixed layer by a parameter derived from laboratory experiments (Turner 1973; Phillips 1977). The third source of TKE, current shear, is unique in that it acts both in and below the mixed layer. The shear comes from the Ekman flow which occurs only in the mixed layer, geostrophic flow which increases towards the surface, and inertia–gravity waves at all depths. The stratification in the seasonal thermocline is sufficient to overwhelm the shear; the probability of subcritical Richardson number is small, and the flow is laminar except in short-lived billow turbulence events that have little effects on boundary layer structure (Woods 1968). But the stratification is much weaker in the diurnal thermocline, the no-man's-land lying between the mixed layer and the seasonal thermocline (Fig. 9.27), and it is often populated by weak continuous turbulence nourished solely by the geostrophic and inertial–gravity current shear. The stratification in the diurnal thermocline comes from a combination of solar heating and advection. Vertical diffusion of heat and salt by turbulence in the diurnal thermocline changes its potential energy, and thereby preconditions the OBL for the nocturnal descent of the mixed layer to the top of the seasonal thermocline. Such modification of mixed layer depth by turbulence generated below the mixed layer influences the SST in the climatologically sensitive east equatorial Pacific (Niiler 1981).

Fig. 9.27. Seasonal variation of upper ocean heat content and mixed layer temperature at ocean weather station Echo (35°N, 48°W). (From Gill & Turner 1976.) This type of diagnostic plot has proved to be valuable in assessing the performance of ocean boundary layer models (e.g. Bretherton 1982).

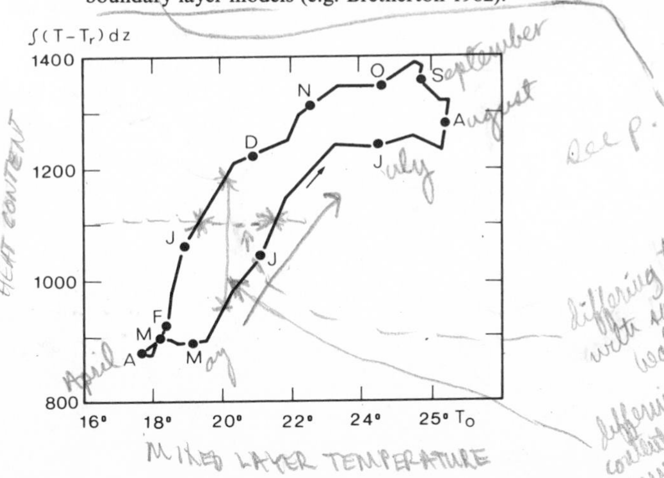

9.5.1.4

Sea surface temperature

The expression 'sea surface temperature' (SST) is widely used in the climatological literature; it needs careful definition in the context of air–sea interaction. It is necessary to distinguish between three different variables labelled SST: (1) the temperature at the interface between the ocean and atmosphere, (2) the temperature at some depth within the surface skin of the ocean, and (3) the temperature of the mixed layer of the ocean. The first of these has no practical value in the context of monitoring for climate research and prediction: the other two will be discussed below.

The SST specified by the WMO is the mixed layer temperature, sampled by bucket (the WMO issues standard buckets to selected merchant ships) or by thermometers mounted on the hulls of ships or buoys, or in ship cooling water intake. These methods of measuring SST give different results, and changes of measurement procedure have led to systematic shifts over the years (Saur 1963; Collins *et al.* 1975; Tabata 1978). Data collected over the past century are available in computer processable form at the US National Climatic Centre. The errors in the ocean mixed layer temperature data are a significant fraction of the air–sea temperature difference used in the bulk formula for surface heat flux. Difficulties arise in calm weather during the day when the mixed layer depth becomes shallower than the sampling depth. Sampling problems are caused by small-scale horizontal variability of mixed layer temperature, the amplitude of which increases by an order of magnitude during calm sunny days (Salmon 1978).

Techniques for estimating SST from satellite radiance measurements in the IR have advanced considerably during the past decade, and now offer the best method of global monitoring

(Bennett, Patzert, Webb & Bean 1979; see Fig. 9.11 above for a recent example). The main problem is to discriminate between clouds and the sea surface. This is normally achieved statistically, by the so-called histogram method (Brown 1981). The IR radiance measurement refers to the skin temperature, which is a few tenths of one Kelvin degree cooler than the mixed layer, the difference depending on the upward heat flux and sea state (Katsaros 1980). Although the IR wavelength chosen for SST measurements lies in the atmospheric window, there is a need to correct for absorption by the air; this is achieved by comparing radiances in two or more wavelengths (McClain 1982) and, in the future, byalong tracking scanning (Harris, Llewellyn-Jones, Minnett, Saunders & Zavody 1983). Tabata & Gower (1980) found a standard deviation of 0.5 K for the difference between satellite and ship measurements of SST. Radiance measurements in the microwave are less accurate, but have the advance of working through to clouds and therefore offering regular SST estimates at high latitudes where the cloud cover has only infrequent gaps through which IR radiance can be measured (Lipes 1982; Hofer, Njoku & Waters 1981; Bernstein 1982).

9.5.1.5

Temperature profile and heat content

Bathythermograph data banks comprise the primary source of our knowledge about the temperature profile in the upper ocean. The regional and seasonal coverage of the BT data are uneven. Each measurement is calibrated against a bucket sample of mixed layer temperature. BT accuracy lies in the range 0.1–1 K. The older, mechanical bathythermographs (MBT) seldom reached deeper than 100 m; the newer, expendable bathythermographs (XBT) reach 760 m, and a 1500-m version is being considered. The great advantage of the XBT is that a temperature profile can be obtained from moving ships of opportunity, without interfering with their normal work. Hydrographic stations by research ships (originally with Nansen bottles and reversing thermometers, nowadays with continuous CTD profilers) are much more expensive, but have no depth limit and provide salinity as well as temperature. Expendable CTDs are available, but not widely used. Towed undulating CTDs reaching to a few hundred metres are now capable of providing accurate, high-resolution transocean sections from ships moving at maximum speed. Thermistor chains attached to drifting buoys from which the Argos satellite system extracts position and data were used successfully during the Strex Experiment (Niiler 1982) and are now beginning to be used in the tropical Atlantic (J. Merle, personal communication). Looking to the future, there is a possibility of remote sensing of the temperature profile by lasers (Schwiesow 1980; Leonard 1980) and acoustic tomography (Munk & Wunsch 1982).

Our perception of the seasonal cycle of temperature profile in the ocean comes largely from empirical studies of the BT data base. Tulley & Giovando's (1963) study of the subArctic using OWS 'P' data, and Bathen's (1971) study of the tropical Pacific have been particularly influential. The monthly mean temperature distribution in the top 150 m has been mapped in the North Pacific and in the North Atlantic and Indian Oceans by Robinson (1976) and Robinson *et al.* (1979). Merle (1978) mapped seasonal (three months) average distributions of temperature and salinity in the tropical Atlantic using hydrographic station data. Gill (1974) analysed OWS BT data to describe the annual cycle of heat content down to a reference isotherm located below the maximum depth of solar heating. Gill & Turner (1976) introduced a plot of the annual cycles of heat content versus mixed layer temperature, which has a characteristic form useful in testing OBL models.

Our perception of the interannual variation of upper ocean temperature profile comes largely from BT time series collected at OWS (Colebrook & Taylor 1979; Yong 1982) and programmes of systematic XBT sampling by ships of opportunity operating along the shipping lanes (Barnett 1978). Systematic sampling by merchant ships using XBTs supplied for climate research has been a major success in the Pacific (Donguy & Henin 1982) and is being pursued in the tropical Atlantic. Bretherton, McPhaden & Kraus (1982) have examined the feasibility of monitoring long-term changes in North Atlantic content by a ship-of-opportunity XBT programme.

The thermal expansion of sea water leads to a change of sea level with water column heat content. Nagasaka & Nitani (1982) have calculated this steric change in sea level from the Japanese time series of hydrographic sections along 137°E, where the peak anomalies exceed 40 cm (Fig. 9.16). It contributes to long period changes in tide gauge records. Pattullo (1963) mapped the seasonal change of sea level around the world ocean, showing an extratropical range of order ± 10 cm. Large interannual changes are seen in tropical Pacific tide gauge records (Wyrtki 1977, 1979). The steric sea level changes are large enough to be monitored by satellite altimeter, which is expected to become the standard source of data for the global patterns of heat content variation by 1990. Mollo-Christenson & Masearenhas (1979) used Landsat data to infer heat storage changes.

9.5.2

Modelling the ocean boundary layer

A model of the OBL is needed for calculations of climate-related changes of surface fluxes and water mass conversion rates. It is a prerequisite for coupling GCMs of the atmosphere and ocean. The OBL model must be robust and accurate. It must permit calculation of the SST to within 10% of the annual range, and the mixed layer depth to significantly better than 10% of the monthly mean value at all locations around the world and at all seasons. The present state of OBL modelling does not meet that specification. The problem is not because the models have reached the limits of predictability (though they are in sight), but because we have not yet learnt how to parameterize the various physical processes controlling OBL structure. The relative importance of the different processes depends on location and season, and will vary with the climate. As with any boundary layer, the structure depends not only on the surface fluxes, but also on the character of the interior flow, yet we have scarcely begun to consider how the geostrophic circulation of the oceans influences the boundary layer and therefore the surface fluxes and water mass conversion rates. The immediate task is to establish in principle what processes are most important under different circumstances and to combine them economically in a model that gives acceptable results everywhere. The application of sophisticated mathematical techniques to squeeze out the last drop of predictability lies far in the future. The aim of this section is to assess the present state of OBL modelling and to highlight recent developments that promise future advance. The reader is referred to the following articles and monographs for further information: Niiler (1975, 1977) deSzoeke & Rhines (1976), Kraus (1977), Garwood (1979), Kitaigorodskii (1979) and Woods (1982).

9.5.2.1

Convection and turbulence

Kraus & Turner (1967) showed that the principal features of the mid-latitude seasonal cycle of mixed layer depth and temperature can be simulated by a model in which the forcing is the astronomical variation of daily mean solar heating, distributed exponentially with depth, constant surface cooling, and vertical heat redistribution by convective adjustment and turbulent entrainment. The entrainment velocity was calculated from the rate of TKE generated in the mixed layer by convection and the wind stress, with a parameter tuned to laboratory data (Turner & Kraus 1967). Denman & Miyake (1973) showed that when the entrainment rate is tuned to OWS data the model accurately describes the variation of mixed layer depth and temperature induced by changes in the weather at OWS 'P'. Gill & Turner (1976) tuned the model to simulate the relationship between seasonal heat content and mixed layer temperature, expressed in a curve (Fig. 9.27) that has since been widely used for assessing OBL model performance (Bretherton 1982*a*).

Thompson (1976) compared the seasonal cycle of mixed layer temperature predicted by such models with OWS data, concluding that various formulations all had errors of greater than ± 1 K (Fig. 9.28). Is that the limit of predictability of entrainment–parameterization models? They diagnose the change of mixed layer temperature in one step from the mean surface fluxes in that interval, so their predictability is limited only by the uncertainty in the initial conditions, the uncertainty in the surface fluxes, and the inadequacies of the model. Hofmann (1982) has investigated the sensitivity of the predicted depth and temperature of the mixed layer to uncertainty in the surface fluxes. Choosing random fluctuations in the following ranges (wind speed ± 2 m/s; surface cooling ± 30 W/m^2; cloud cover $\pm 50\%$) he showed that the error in mixed layer temperature is greatest in summer, but only reaches a few tenths of one Kelvin degree. The much larger errors detected by Thompson (1976) must be due to inadequacies in the models.

Fig. 9.28. The limited accuracy of mixed layer models of the upper ocean is revealed by the difference between observed and predicted mixed layer temperature in a one-year integration of the following models: DKT as Denman & Miyake 1973, Kraus & Turner 1967; PRT – as Pollard, Rhines & Thompson 1973; CON – assuming constant mixed layer depth. (Based on a diagram in Thompson 1976.)

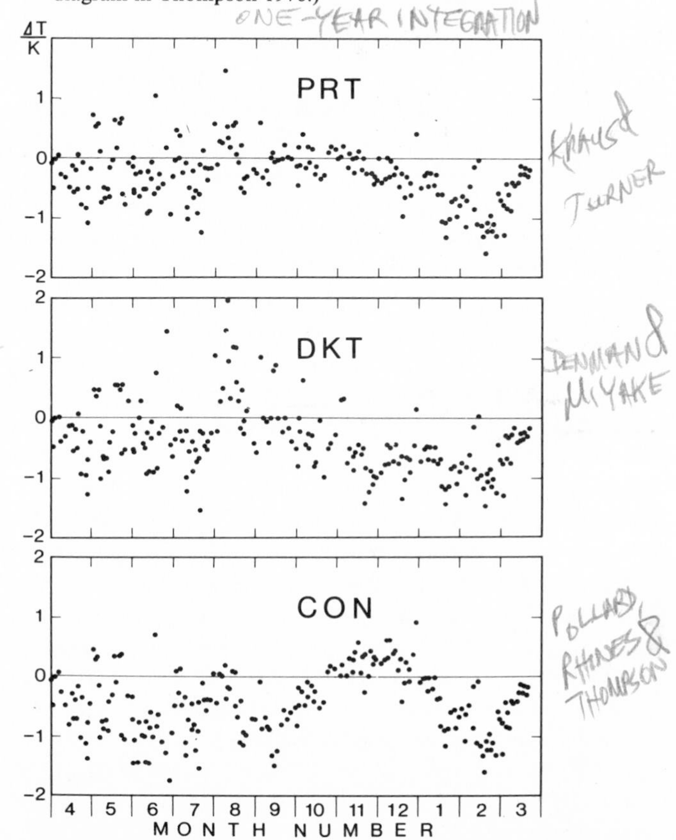

Attention focussed initially onto the parameterization of turbulent entrainment. Niiler (1977) and Stevenson (1979) attenuated the entrainment parameter with increasing depth to take account of viscous dissipation. That had the effect of eliminating turbulent entrainment in the extratropical winter when the mixed layer depth becomes much deeper than the Ekman depth (Wells 1979). In summer and in the tropics the mixed layer is shallow, giving more potential for SST changes, but then the viscous attentuation parameterization is not active. Others have chosen to abandon the Turner–Kraus entrainment parameterization in favour of an eddy diffusivity based on second- or third-order closure of the turbulence (Mellor & Durbin 1975; Marchuk *et al.* 1977; Warn-Varnas & Piacesk 1979; Kundu 1980; Blumberg & Mellor 1980; Klein 1980; Klein & Coantic 1981; Worthem & Mellor 1981; Warn-Varnas, Dawson & Martin 1981; Simpson & Dickey 1981*a*). The parameterization is based on laboratory measurements rather than field data, as was the case with earlier attempts to use diffusivity (Munk & Anderson 1949). The disadvantage is that models based on it are expensive in computer time. The advantage is a much more detailed treatment of the various terms in the TKE equation, including vertical diffusion of TKE and current shear below the mixed layer, both of which are absent from the entrainment parameterization (Mellor & Yamada 1982). One would expect these improvements to be most advantageous in modelling the OBL in the tropics and in the extratropical summer, when the mixed layer depth is sufficiently shallow to be strongly influenced by turbulent mixing.

9.5.2.2

Solar heating

If the full potential of the higher-order turbulence closure parameterization is to be realized, it will be necessary to improve the parameterization of solar heating in the models, taking into account the effects of clouds and sea water turbidity. Simpson & Dickey (1981*a*, *b*) have investigated the sensitivity of the temperature profile predicted by a diffusion–parameterization model to various empirical parameterizations of the solar heating profile. They conclude that a two-exponential parameterization is adequate unless one is concerned with variation in the top metre in calm weather. The entrainment model developed by members of my research group (e.g. Hofmann 1982; Fiekas 1982; Woods & Onken 1982) is based on a heating model with a 27-band spectrum (Woods 1980) to overcome that limit. Charlock (1982) showed that uncertainty in the sea water turbidity can produce summer temperature errors of ± 1 K in an entrainment model. It is more urgent to establish the world climatology of sea water turbidity and to include adequate parameterization of solar heating in OBL models than to replace the economical entrainment of turbulent mixing by the higher-order closure methods.

9.5.2.3

Ekman transport and upwelling–downwelling

Outside the tropics, the annual cycle of upper ocean heat content, and mixed layer temperature and depth is governed primarily by the annual cycle of solar heating and, secondarily, by the annual cycle of surface cooling. Dr J. Willebrand (private communication 1980) has shown that the annual cycle of SST observed at OWS can be simulated better by adding a term to represent the displacement of the mean horizontal temperature gradient by the Ekman transport. Cushman-Roisin (1981) has simulated the development of sub tropical fronts of large-scale Ekman convergence, following the ideas of Roden (1980). In the tropical Pacific and Atlantic, upwelling–downwelling due to divergence–convergence of the Ekman transport becomes a major factor controlling the structure of the OBL. The vertical motion must be introduced into the OBL models (de Witt & Leetma 1978; Hughes 1980, 1982). Shorter-period undulations due to internal waves (Hofmann 1982) and quasi-geostrophic motion (Voorhis, Schroeder & Leetma 1976; Stevenson 1980) affect the transfer of scalars between the mixed layer and the thermocline everywhere.

9.5.2.4

Geostrophic currents

Geostrophic advection is not taken into account in existing extra-tropical models of the OBL, yet it is crucial for the annual cycle of heat content, and it influences the temperature and salinity profiles and therefore the depth of convection, most noticeably in winter outside the tropics. It is surprising that the first-order effects of geostrophic advection have been neglected in favour of subtle refinements to parameterization of turbulence and solar heating. Perhaps it is because the best available data set for tuning OBL models came from OWS 'P' located in a corner of the subarctic Pacific where geostrophic flow is slow and contributes little to the annual heat budget (Fig. 9.5). Let us discuss the role of geostrophic advection in the annual cycle of the temperature profile at a fixed location outside the tropics. Consider a model integration starting at the moment at which the mixed layer reaches its annual maximum depth, about an hour after sunrise one day near the spring equinox. At that moment the rate of solar heating just balances the surface cooling: the temperature of the mixed layer is at its annual minimum. The value of that initial temperature of the OBL is controlled by the general circulation. Having painfully achieved its annual maximum depth, H_0, the mixed layer retreats smartly towards the surface (depth H) leaving most of the water column in adiabatic laminar flow (Fig. 9.4(*a*)). The speed of the decrease of mixed layer depth in spring was underestimated in the annual integration of Kraus & Turner (1967), who incorrectly argued that the vertical temperature gradient in the layer $H_0 \geqslant z \geqslant H$ is a solely fossil record of the temporal variation of mixed layer temperature during its apparently leisurely ascent. Observations and models with more detailed solar heating routines (Hofmann 1982) have since shown that the fossil contribution to the temperature gradient below the mixed layer is often no larger than those of solar heating, for $z \leqslant S_0$ (Charlock 1982) and geostrophic advection through the whole water column, $0 \leqslant z \leqslant D$ (Woods 1982). The latter arises from the vertical shear in the geostrophic current, which depends on the horizontal density gradient, or the baroclinicity (**B**) and the static stability (N):

$$\frac{\partial}{\partial t}\left(\frac{\partial T(z)}{\partial z}\right) = (N^2/f)\cdot \nabla_H T \times \mathbf{B}.$$

The vertical gradient of solar heating increases the static stability throughout the year in the seasonal thermocline ($H \leqslant z \leqslant S$), while baroclinicity is modulated seasonally by the meridional gradients of day length and Ekman pumping giving considerable regional variation in OBL structure (Fig. 9.29). The contribution of geostrophic shear to the growing temperature gradient therefore increases from zero at $z = D$ to a maximum in the mixed layer. The temperature profile below the mixed layer develops steadily through the year. Turbulent mixing keeps the vertical gradient very small near the surface; so advection of heat in the mixed layer produces a sharp step in the temperature profile at $z = H$. No work is done against the Archimedes force in the vertical redistribution of heat advected baroclinically into the mixed layer because geostrophic shear is orthogonal to the horizontal density gradient. The current introduces a vertical salinity gradient that compensates for density change due the warming. Because of western intensification, fast geostrophic currents mainly flow away from the tropics, giving (baroclinically) a stable temperature gradient and an unstable salinity gradient. In addition the temperature and salinity are changed independently by the barotropic component of the geostrophic flow in the OBL, i.e. by the velocity at $z = D$. The total change in heat and water contents of the water columns are given by the sums of the barotropic and baroclinic components:

$$\frac{\partial T(z)}{\partial t} = \nabla_H T\cdot \mathbf{U}(D_0) + \int_{D_0}^{z} [N^2(z)/f]\nabla_H T(z) \times \mathbf{B}(z)\,dz, \quad (9.1)$$

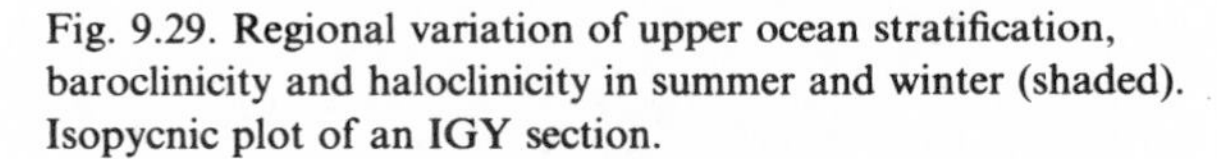

$$\frac{\partial S(z)}{\partial t} = \nabla_H S(z)\cdot \mathbf{U}(D_0) + \left(\frac{\alpha}{\beta}\right) \int_{D_0}^{z} [N^2(z)/f]\nabla_H T(z) \times \mathbf{B}(z)\,dz. \quad (9.2)$$

Fig. 9.29. Regional variation of upper ocean stratification, baroclinicity and haloclinicity in summer and winter (shaded). Isopycnic plot of an IGY section.

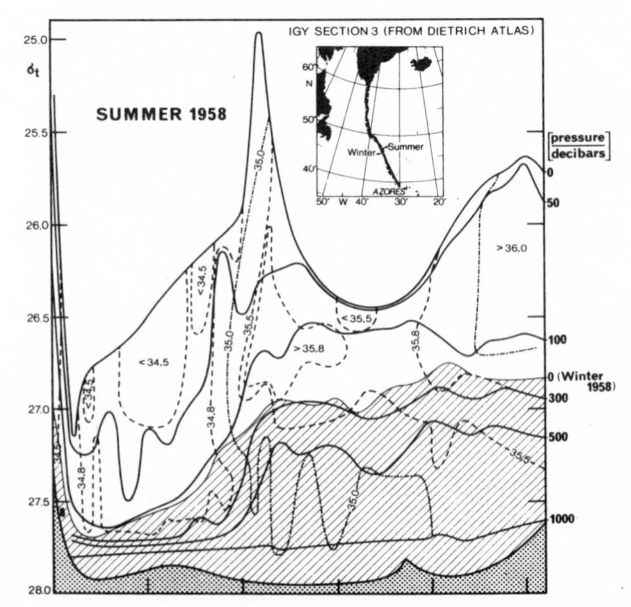

The changing heat and fresh water content is given by the vertical of Eqs. (9.1) and (9.2). For a steady seasonal annual cycle, the integral over one year must balance the annual values of surface cooling minus solar heating, and evaporation minus precipitation respectively.

Observations in the North Atlantic current show that the temperature and salinity gradients are quickly reestablished in the depth range $H \leqslant z \leqslant D$ after the vernal ascent of the mixed layer. The density gradient is also reestablished, even below the limit of solar heating ($z \geqslant S$). That must be due to the barotropic component bringing stable water from regions where D_0 is smaller:

$$\frac{\partial}{\partial T} N^2(z) = \mathrm{U}(D) \cdot \nabla_H N^2(z) + \cdot \frac{DN^2}{\partial z}.$$

The first term = 0, if $\nabla_H D_0 = 0$.

So geostrophic restratification depends on the upstream decrease in the annual maximum depth of the mixed layer. We note that D_0 increases downstream in the Gulf Stream and North Atlantic current in Fig. 9.4(*a*), as expected for a progressive increase in the net annual cooling. But the regional variation of D_0 cannot be calculated simply from the annual heat budget. Allowance must be made for the net annual water budget, and for the potential energy released in extracting both heat and water from the stable temperature but unstable salinity gradients introduced into the water column by baroclinic advection (Woods 1982). An OBL model incorporating advection and economic parameterizations of solar heating and turbulence is being developed by the author.

To summarize, after vernal retreat of the mixed layer, geostrophic advection contributes to restratification in the water column $0 \leqslant z \leqslant D_0$, and provides the only contribution below $z = S_0$, the maximum depth of solar heating. Geostrophic advection supplies the heat needed to balance the annual heat budget, but in doing so introduces a potentially unstable salinity gradient that influences D_0, the annual maximum depth of the mixed layer. Barotropic advection introduces a vertical density gradient, starting at $z = D_0$ and moving progressively shallower through the year until it reaches $z = H$, whereupon it begins to influence the value of H. The neglect of these effects of geostrophic advection is the principal reason why existing models of the OBL fail to describe the observed variation of boundary layer structure. The contributions to restratification of the OBL after it is mixed to D_0 at the end of the cooling season, are summarized in Figs. 9.30 and 9.31.

9.5.3 *Water mass transformation*

The boundary layer of the upper ocean is defined by the occurrence of powerful diabatic processes that alone can rapidly change the scalar properties of sea water. The adiabatic processes are absorption–emission of radiation and molecular diffusion, but the latter only becomes significant where three-dimensional turbulent strain enhances scalar gradients. Diabatic processes are assumed to change scalar properties much more slowly in the dark interior below the boundary layer where the turbulance is, on

Fig. 9.30. Elements of an ocean boundary layer model without advection. The diurnal and annual cycles of solar heating, mixed layer depth, and Ekman layer depth, expressed in terms of the instantaneous parameters (S', H', E'), and the daily maxima (S, H, E) and their annual maxima (S_0, H_0, E_0).

Fig. 9.31. Annual variation and summer solstice profiles of density, temperature and salinity stratifications in the upper ocean, at a subpolar site where winter mixing goes deeper than solar heating. After mixing reaches its annual maximum depth (D_0), three processes contribute to restratification of the annual thermostad: turbulent conduction (the fossil stratification established during the early spring decrease of mixed layer depth); solar heating at depths less than S_0; geostrophic advection. The older mixed layer models of the ocean are unsuitable for global climate research because they do not take account separately of temperature and salinity, and they neglect the contribution of geostrophic advection.

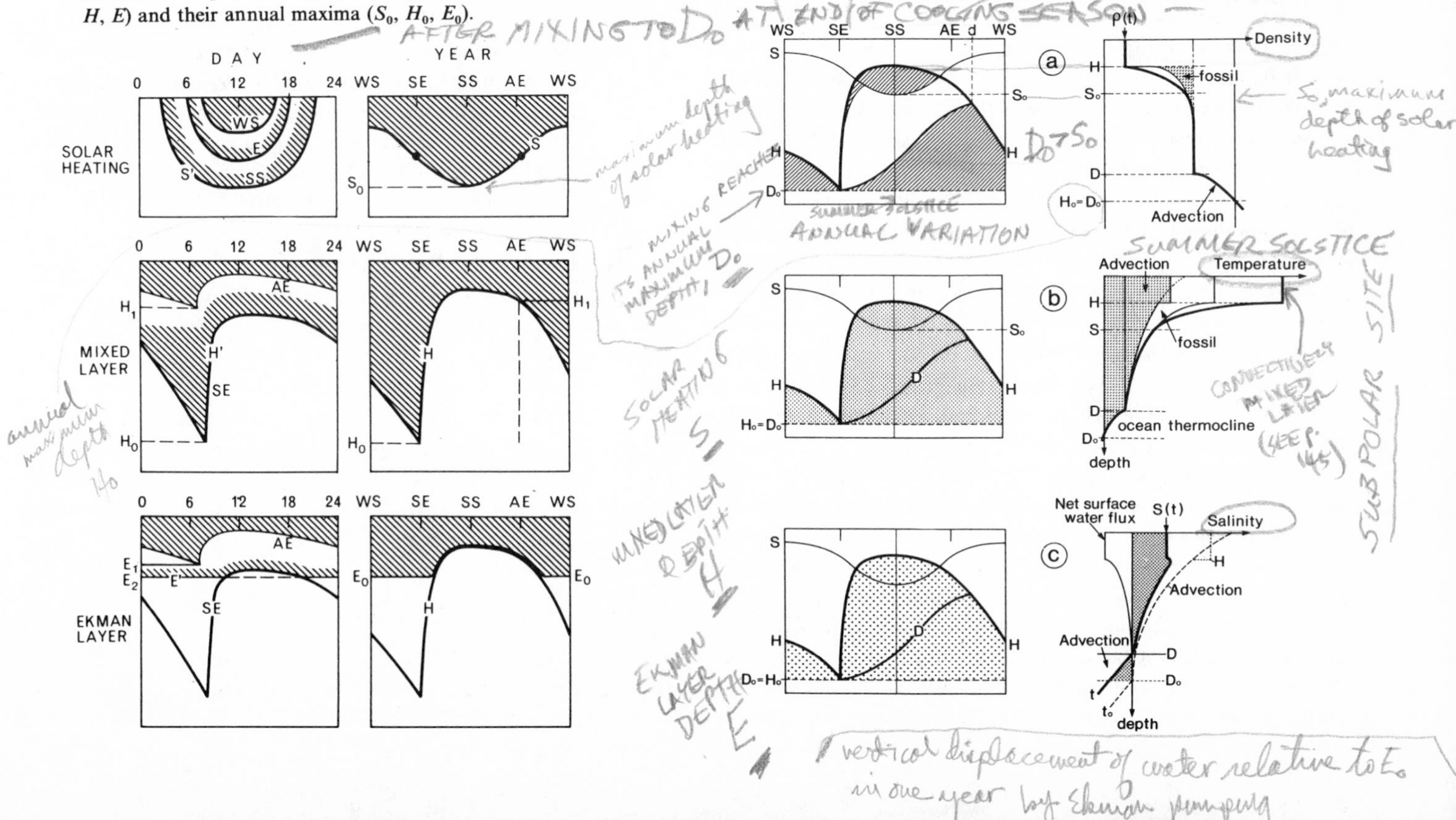

average, very weak (Garrett 1979). The absorption of solar radiation warms the water and, in the laminar flow region of the boundary layer, decreases the spacing between isopycnals. Weak turbulence in the diurnal and seasonal pycnoclines increases the spacing between isopycnals. Convective overturning in the mixed layer (where the buoyancy flux is directed upwards) changes the temperature and salinity of the water column uniformly in response to the surface fluxes of energy and water. The mean isopycnals rise vertically through the mixed layer (or, more precisely, they lean over backwards slightly in response to the upward buoyancy flux.

The volume of the boundary layer is only a tiny fraction of that of the world ocean, so only a tiny fraction of the ocean water is undergoing water mass transformation by diabatic processes in the boundary layer at any instant. The residence time of a water mass in the ocean is defined as the mean interval between successive passes through the boundary layer. Residence times vary from less than one day to over 1000 years. In order to describe the process of water mass transformation it is therefore necessary to consider what happens to a water particle when it makes one of its rather rare transits through the boundary layer. We consider what happens along the trajectory of a small parcel of water with known temperature, salinity and potential vorticity, moving in the upper ocean below the boundary layer. To simplify the problem, let us ignore the lateral and vertical displacements of the trajectory by transient motions (internal waves and quasi-geostrophic eddies and planetary waves) and assume that it coincides with a streamline of the mean circulation described by Wunsch in Chapter 10. So long as the water parcel lies below the mixed layer, its depth is changed only slowly by Ekman pumping. Luyten, Pedlosky & Stommel (1982) have studied the trajectories of such particles on the assumption that the flow is Sverdrupian. But once the parcel enters the mixed layer its depth at entry is rapidly forgotten. The water parcel is torn apart by the turbulence and its constituent water particles blended into a new water type whose temperature and salinity are those of the mixed layer. The incoming water parcel has lost its identity; its component particles have been dispersed throughout the water column; it is no longer meaningful to discuss its potential vorticity. The constituent particles leave the mixed layer independently from the bottom of the mixed layer in random order; up to the last minute they partake in the turbulent mixing that changes their depth rapidly (order one hour). The particles in the mixed layer are like members of a deck of cards[3] that is unceasingly shuffled, and dealt from the bottom (Woods & Onken 1982).

In order to calculate the mean rate at which parcels of water enter the boundary layer, the rates at which properties are changed there, and the rate at which water of newly acquired properties leaves the boundary layer, we need to discuss the seasonal variation of boundary layer properties. We saw in the last section that the seasonal variation of mixed layer depth properties depends not only on the surface fluxes, but also on advection by the geostrophic circulation. Also, that the depth distribution of solar heating depends on plankton concentration, which in turn is controlled by the mixed layer depth. Now we shall show that the circulation is itself affected by the diabatic processes in the boundary layer. A clear understanding of the symbiosis between boundary layer and circulation is a prerequisite for modelling the contribution of the ocean to climate. The general characters of the seasonal cycles of the diabatic and adiabatic processes in the upper ocean are shown schematically in Fig. 9.31. The regional variation depends largely on the relative depths of S_0, D_0 and L_0: the annual maximum depth of solar heating (taken conventionally to be 1 mK/d), the annual maximum depth of the mixed layer ($H_0 = D_0$), and the vertical displacement of water relative to E_0 in one year by Ekman pumping.

[3] The German name for the mixed layer is 'Deckschicht'.

9.5.3.1

The subtropical Sverdrup regime

Stommel (1979*a*) considered water mass conversion in the tropical Sverdrup regime of the North Atlantic where the three regional variables are similar ($S \approx D_0 \approx L_0$). By considering the annual cycles of S, D and L, he showed that the water leaves the boundary layer at a rate controlled by Ekman pumping, with the temperature and salinity of the mixed layer when it is deepest ($H = H_0 = D_0$). Defant (1936) had earlier sought a dynamical explanation for the observed variation of D_0, briefly entertaining the notion that D_0 might be related to the Ekman depth and therefore to $1/\sqrt{f}$. Luyten, Pedlosky & Stommel (1983) subsequently showed that, in the subtropical Ekman pumping region D_0 divided by the depth of the warm water sphere varies with latitude as (β/f). But at higher latitudes, where S and D exhibit large variation through the season cycle and where the net annual Ekman pumping is negative (see Fig. 9.3), the regional variation of D_0 is controlled thermodynamically as well as dynamically.

9.5.3.2

The subpolar regime

Let us consider the extratropical regions where $D_0 \geqslant S_0$. The date of the start of the heating season and therefore of the vernal retreat of the mixed layer is controlled by the increasing day length and exhibits little regional variation (the weather changes later, after the SST has begun to rise) but, during the cooling season, the mixed layer descends slowly to the maximum depth D_0 which varies regionally by an order of magnitude (100–1000 m) in the North Atlantic (Fig. 9.4). There is therefore a wide variation in the date at which the mixed layer descends through a given depth. A parcel of water continues to circulate adiabatically at depth $z_p \geqslant S_0$ until it finds itself in a region where $D_0 \geqslant z_p$ on a day when $H \geqslant z_p$. If it has the misfortune to be in the wrong place at the wrong time of year, the water parcel will become entrained by the mixed layer and blended into the new type that will be abandoned by the boundary layer at the start of the next heating season. Other water parcels circulating at the same depth z_p pass adiabatically through the danger zone (where $D_0 \geqslant z_p$) at a time of year when $H \leqslant z_p$. A parcel crosses the danger zone in a time $T_p(z_p) = L_p/U_g(z_p)$, where it moves at speed $U_g(z_p)$ along a streamline that crosses the zone with a path length L_p. If T_p is longer than the period T_a for which $H \geqslant z_p$ throughout the zone, no parcel approaching the zone at a depth z_p will escape entrainment. The fraction of water getting through unscathed is given by:

$$F(z_p) = (T_p - T_a)/(\text{one year}).$$

The rest of the water column from the surface down to depth z_p is blended into the new water type, the regional variation of which is indicated by the pattern of sea surface isotherms and isohalines on the date of the mixed layer's vernal retreat. The volume of water formed within a particular range of $(T \pm \Delta T,\ S \pm \Delta S$ depends on the surface area enclosed by the bounding isopleths and the mean depth D_0 in that area. The lower portion of that column of new water type will survive further transformation by diabatic processes in the boundary layer, provided its trajectory carries it below S_0 and successfully through the danger zones $(D_0 \geqslant z_p)$ along its path. The upper portion of the water type column and chunks caught by deep mixing events along the way are transformed to new T–S values. Calculation of the rate of formation of a particular water type each year is fairly straightforward, but calculation of the amount that survives unchanged by boundary layer processes each year thereafter presents problems that have not yet been resolved. Worthington (1976), who has studied the subject in detail, says 'it is one of the greatest areas of ignorance in physical oceanography'. Yet it is vital for climatology, being the process by which anomalies of heat and water in the OBL, created by anomalous surface fluxes, are carried into the ocean's interior. At present it is not possible to calculate the rate at which heat and water anomalies leaking into the ocean interior from the boundary layer return to the surface mixed layer. The solution lies in improved modelling of the OBL and the acquisition of an empirical climatology of the regional patterns of the mean and interannual variability of S_0, D_0 and L_0.

Diabatic processes in the OBL influence ocean circulation dynamics differently in the mixed layer $(z \leqslant H)$ and below it $(H' \leqslant z \leqslant D_0$ or S_0, whichever is greater). In the mixed layer:

(1) The vertical shear of the geostrophic current is independent of depth because the mean isopycnals rise vertically.

(2) Potential vorticity is not a rational variable, because turbulence stretches vortex tubes three-dimensionally and rapidly dispenses the particles of water parcels for which potential vorticity was defined before they entered the mixed layer.

Below the mixed layer:

(1) The vertical shear of the geostrophic current is nearly independent of depth in the diurnal thermocline $(H' \leqslant z \leqslant H)$, but varies strongly in the seasonal thermocline $(H \leqslant z \leqslant D)$ and in the seasonal ocean pycnocline $(D \leqslant z < D_0)$.

(2) Potential vorticity varies slowly with solar heating and divergence of the downward turbulent fluxes of heat and salt requiring a Lagrangian correction

$$gf\frac{\partial}{\partial z}\left[\frac{\partial}{\partial z}\left(\alpha K_T\frac{\partial T}{\partial z}+\beta K_S\frac{\partial S}{\partial z}\right)+\frac{\alpha}{C_p}\frac{\partial I}{\partial z}\right]. \qquad \text{(I)}$$

which is usually small.

The geostrophic restratification of the seasonal pycnostad is adiabatic and requires no correction to the potential vorticity equation, apart from the usual Eulerian advection terms (Stommel 1979*b*). Olbers, Willebrand & Wenzel (1982) have considered the problem of parameterizing the dynamical effects, advective and diabatic processes, occurring in the OBL in a diagnostic model of the oceanic circulation where the annual cycle is not resolved. They show that the dynamical effect of convection in the mixed layer, where potential vorticity is meaningless, is equivalent to a fictitious cooling of a laminar flow i.e. by a correction term similar to that due to solar heating in (I) above, but of opposite sign.

9.5.4

The ocean boundary layer in global climate models

Climate is sensitive over a wide range of time scales to non-linear processes occurring in the OBL. The specification for climate prediction demand rather accurate treatment of those processes in global climate models. The scientific problem is to design computationally economic but accurate parameterizations of the OBL processes suitable for inclusion in global models. The aim is to develop formulae for the rate of water mass formation and correction terms for ocean circulation dynamics with a spatial and temporal resolution appropriate to the climate prediction task.

Spatial averaging is similar for all global climate problems. It is required to resolve regional variability to, say, 5° × 5°, and to parameterize the effects of smaller scale variations. The oceanic synoptic- and meso-scale motions that provoke irreversible changes in OBL structure by undulating isopycnals have horizontal scales smaller than that climatological averaging scale, so their effects will be parameterized as a part of spatial averaging.

The temporal resolution of global climate prediction depends on the particular problem. For *extended-range weather forecasting* the time scale is determined by synoptic-scale variation in the atmosphere; for *short-term climate prediction* (El Niño, sea ice, etc.) the annual cycle must be resolved, but not the diurnal- synoptic-period changes; and for the *long-term climate prediction* (CO_2-pollution problem) it is not necessary to resolve the annual cycle, but secular changes in the OBL structure become important.

The detailed studies of OBL physics and modelling reviewed above suggest a strategy for parameterization of the temporal variation. The aim is to relate the rates of water mass transformation and dynamical correction to the depth parameters listed in Table

Table 9.3. *Principal parameters of the global ocean boundary layer.*

Process	Variation depends on	Depth	Daily maximum	Annual maximum (regional parameter)
Solar heating	Astronomy Plankton (αH)	S'	S	S_0
Convection	S', $\rho(z)$ Surface cooling Evaporation Freezing Advection	H'	H	H_0
Turbulence	Wind waves Current shear Convection	$H' \leqslant H^* \leqslant H$	H	H_0
Ekman transport (τ/f)	Wind stress (τ) Turbulence (A)	$E' =$ or H' whichever is less	E or H whichever is less	E_0 or H_0 whichever is less
Ekman pumping $\left(W_E = \frac{1}{\beta}\text{curl}\,\tau\right)$	Wind stress (τ)	—	L	L_0
			(measured relative to E, E_0)	

9.3, taking account of the fundamental differences in OBL structure arising from the relative magnitudes of the parameter in the daily or annual sets. The time scale of the climate prediction problem determines on which set (i.e. which column in Table 9.3) parameterization is based. Thus for El Niño prediction, in which the annual cycle will be resolved, OBL processes will be parameterized in terms of the daily set (S, H, E, L); whereas for predicting the consequences of CO_2-pollution over a period of decades it will be in terms of the annual set (S_0, H_0, E_0, L_0). Woods (1980) has drawn attention to the errors that arise from ignoring diurnal variation in the OBL in studies of the annual cycle, and suggests a method of parameterization.

No global climate model has been based on the full implementation of the 'SHEL' parameterization of OBL processes outlined above. But elements of it are incorporated in the Princeton model (Bryan & Lewis 1979; Manabe, Bryan & Spelman 1979; Bryan, Komro, Manabe & Spelman 1982). A number of groups at universities (e.g. Thompson & Schneider 1979; Wells 1979; Adamec, Elsberry, Garwood & Haney 1981) and at national meteorological centres are currently adding OBL routines to atmospheric GCMs for research directed towards climate prediction. The emphasis at the start of the 1980s is on inclusion of a mixed layer model à la Kraus & Turner (1967), which includes a simplified solar heating profile, weakly penetrative convections and wind stress entrainment, and a sea ice model à la Semtner (1976). The exclusion of oceanic geostrophic motion limits those models to extended-range forecasting. The addition of Ekman transport and Ekman pumping in the tropics will probably be needed before acceptable simulations of interannual variability can be achieved, and great care will be needed to specify lower boundary condition, i.e. the seasonally varying temperature and salinity distributions in the oceanic pycnocline, which are determined by geostrophic motion.

Oceanic teleconnections transmitted by the geostrophic currents are the subject of intense study at the basic research level (O'Brien, Busalacchi & Kindle 1981) and may be ready for incorporation in global GCMs designed to predict short-term climate changes within the next decade. The OBL problem is similar to that for extended-range weather forecasting, but the lower boundary condition now includes a deterministic calculation of motion associated with the oceanic teleconnections.

The development of global models for longer-term climate prediction (due, for example, to CO_2-pollution or river diversion) requires accurate treatment of water mass conversion, which depends on the symbiosis of the large-scale circulation and boundary layer processes of the upper ocean. Some indication of the complexity of the interaction was given in an earlier section of this paper. The problem is a long way from solution. We expect that changes in the OBL structure (in particular $H_0(x,y)$) will be one of the consequences of CO_2-induced increase in IR, and they will influence the mean ocean circulation in the upper ocean and the regional pattern of water mass formation, but we do not yet know how big the effect will be. Investigations of CO_2-induced climate change have so far been based on simplified treatments of the ocean circulation–boundary layer interaction (Hasselmann, Maier-Reimer, Müller & Willebrand 1981; Bryan, Komro, Manabe & Spelman 1982). It is worrying that we do not know enough about the OBL and the process of water mass formation to judge how sensitive such global models of longer-term climate change will be to their simplifications of the OBL. That is, of course, only one of many problems that need to be solved before the dream of truly coupled global atmosphere–ocean models of climate change can be realized.

Other difficulties that have been discussed in the recent years include the mismatch in time scales between the atmosphere and ocean. It seems unlikely that the expedient of running the ocean and atmosphere components of the model asynchronously (Manabe & Stouffer 1979) will prove satisfactory (Dickinson 1981; Ramanathan 1981). Much remains to be done, and there are many uncertainties, especially with regard to the OBL, but the prospects are good for major advances in the next decade. As issues become resolved and computer power increases, allowing better resolutions of critical elements of the climate system, there will be an increasingly urgent demand for a new data base to test the models. Global observations of the structure of the OBL will be among the highest priority objectives of WCRP experiments designed to collect the new data base.

9.6 Conclusion

Three themes have been discussed: the role of the ocean in the planetary climate system, the influence of air–sea interaction in climate prediction, and the present state of knowledge about ocean surface fluxes and the boundary layer of the upper ocean. The literature relevant to these themes is vast, the publications cited in this review are only one-tenth of those referred to in preparing it. In view of that evidence of massive research around the world, one cannot help wondering why the subject is advancing so slowly. As Goody (1980) wrote 'air–sea interaction is coming to be taken more and more seriously as the Achilles' heel of the short term climate problem'.

One can become depressed when assessing published maps of the GPFO in the light of error analysis of the methods used to produce them, let alone the patchy data base. Happily, new remote sensing methods promise significant improvements in both quality and coverage. Experience gained in analysing data from Seasat and the GARP generation of meteorological satellites lead us to anticipate that the new WCRP generation of satellites starting in the late 1980s will help to set our descriptions of GPFO on a sounder basis. But a word of caution comes from the Cage group (Dobson *et al.* 1982), who doubt whether it will be possible to measure the annual mean energy flux averaged over an area of several Mm^2 to an accuracy of ± 10 W/m^2; sufficient for studying the El Niño, which has interannual variability of up to 100 W/m^2, but not for CO_2-induced heating, at only a few W/m^2.

This review has emphasized the importance of the OBL in air–sea interaction. Our perception of the OBL is still rather limited, depending too heavily on bathythermograph data and mixed layer models. Here and elsewhere (Woods 1984) I have argued that the OBL models available at present are not accurate enough for climate research; the SST error exceeds ± 1 K; winter convection depth is unpredictable. Although they can be tuned to perform well at one site (e.g. OWS 'P'), the models do not then perform equally well at other locations. The problem is partly that the models are sensitive to the surface fluxes, which have large errors, and partly that they are based on too simple a perception of the OBL physics. In particular, no account has been taken of geostrophic advection, which restratifies the water column after deep winter convection, and determines the depth of that convection. There is scope for improvement, and the situation should improve rapidly over the next decade.

In summary, the present state of affairs is unsatisfactory, but new methods of observing the ocean from space, and improved parameterizations based on better understanding of the processes involved, encourage one to be optimistic about the future. Once the technical problems of calculating the surface fluxes and boundary layer structure have been overcome, we can look forward to exploiting the potential of the upper ocean for climate prediction; in particular, tropical teleconnections inside the ocean, sea ice fluctuation, and the ventilation of the warm and cold water reservoirs of the deep ocean. If air–sea interactions is the Achilles' heel of climate, then the upper ocean is the Achilles of climate prediction.

References

Aagaard, K. & L. K. Coachman (1975). 'Towards an ice free Arctic Ocean'. *EOS*, **56**, 484–6.

Aagaard, K., L. K. Coachman & E. C. Carmack (1981). 'On the halocline of the Arctic Ocean'. *Deep-sea Res*., **28**, 529–46.

Adamec, D., R. L. Elsberry, R. W. Garwood & R. L. Haney (1981). 'An embedded mixed-layer-ocean circulation model'. *Dynamics of Atmospheres and Oceans*, **6**, 69–96.

Anderson, N. R. & A. Malahoff (eds.) (1977). *The Fate of Fossil Fuel CO_2 in the Oceans*. Plenum, New York, 749 pp.

Anderson, R. J. & S. D. Smith (1981). 'Evaporation coefficient for the sea surface from eddy flux measurements'. *J. Geophys. Res*., **86**, 449–56.

Angstrøm, A. (1935). 'Teleconnections of climate changes in present time'. *Geografiska Annaler*, **17**, 242–58, Stockholm.

Anthes, R. A. (1982). *Tropical Cyclones: Their Evolution, Structure and Effects*. American Meteorological Society, Boston, 208 pp.

Atlas, D. & O. W. Thiele (eds.) (1981). *Precipitation Measurement for Space*. Rep., Goddard Laboratory for Atmospheric Sciences, NASA, Greenbelt, 358 pp.

Austin, P. M. & S. G. Geotis (1980). 'Precipitation measurements over the ocean'. *In: Air-Sea Interaction: Instruments and Methods* (F. Dobson, L. Hasse & R Davis, eds.). Plenum, New York.

Bach, W., J. Pankrath & W. Kellog (eds.) (1979). *Man's Impact on Climate*. Elsevier, Amsterdam, 325 pp.

Barkstrom, B. & J. B. Hall (1981). 'The Earth Radiation Budget Experiment (ERBE)'. *J. Energy* (November–December 1981).

Barnett, T. P. (1978). 'The role of the oceans in the global climate system'. *In: Climate Change* (J. Gribbin, ed.). Cambridge University Press.

Barnett, T. P. (1981*a*). 'Statistical prediction of North American air temperature from Pacific predictors'. *Mon. Wea. Rev*., **109** (5), 1021–41.

Barnett, T. P. (1981*b*). 'Statistical relations between ocean/atmosphere fluctuations in the tropical Pacific'. *J. Phys. Oceanogr*., **11**, 1043–58.

Barnett, T. P. (1981*c*). 'On the nature and causes of large-scale thermal variability in the central North Pacific Ocean'. *J. Phys. Oceanogr*. **11**, 887–904.

Barnett, T. P. & R. E. Davis (1975). 'Eigenvector analysis and prediction of sea surface temperature fluctuation in the Northern Pacific Ocean'. *In: Proc. Symp. Longterm Climatic Fluctuations*. WMO Technical Note No. 421. WMO Geneva, pp. 439–50.

Barnett, T. P., S. C. Patzert, W. C. Webb & B. R. Bean (1979). 'Climatological usefulness of satellite determinated sea surface temperature in the tropical Pacific'. *Bull. Am. Met. Soc*., **60**, 197–205.

Barry, R. G. (1983). 'Meteorological aspect of the seasonal sea ice zone'. *In: Air–Sea–Ice Interaction* (N. Untersteiner, ed.). Plenum, New York.

Bathen, K. H. (1971). 'Heat storage and advection in the North Pacific Ocean'. *J. Phys. Oceanogr*., **76**, 676–87.

Baumgartner, A. & E. Reichel (1975). *The World Water Balance*. Elsevier, Amsterdam, 179 pp.

Berger, A. (1977). 'Long-term variation of the earth's orbital elements'. *Celestial Mech*., **15**, 53–74.

Berger, A. (1981). 'The astronomical theory of paleoclimates'. *In: Climatic Variations and Variability: Facts & Theories* (A. Berger, ed.). Reidel, Dordrecht, pp. 501–25.

Berger, W. H. (1981). 'Paleoceanography: the deep-sea record'. *In: The Oceanic Lithosphere* (C. Emiliani, ed.) *The Sea*, vol. 7. Wiley-Interscience, New York, pp. 1437–1519.

Berner, W., H. Oeschger & B. Stauffer (1980). 'Information on the CO_2 cycle from ice core studies'. *Radiocarbon*, **22**, 227–35.

Bernstein, R. L. (ed.) (1982*a*). 'Seasat special issue, I: geophysical evaluation'. *J. Geophys. Res*., **87**, 3173–438.

Bernstein, R. L. (1982*b*). 'Sea surface temperature mapping with the SEASAT microwave radiometer'. *J. Geophys. Res*., **87**, 765–72.

Bjerknes, J. (1969). 'Atmospheric teleconnections from the equatorial Pacific'. *Mon. Wea. Rev*., **97**, 163–73.

Blackman, M. L., J. E. Geisler & E. J. Pritchard (1983). 'A general circulation model study of January climate anomaly patterns associated with interannual variation of equatorial Pacific sea surface temperature'. *Mon. Wea. Rev*. (in the press).

Blumberg, A.F. & G. L. Mellor (1980). 'A coastal ocean numerical model'. *In: Proc. Symp. Mathematical Modelling of Estuarine Physics*, J. Sundermann & K.-P. Holtz (eds.), Springer Verlag, Berlin, pp. 203–19.

Bolin, B. (ed.) (1981). *Carbon Cycle Modelling*. SCOPE 13. John Wiley, Chichester.

Bolin, B., E. T. Degens, S. Kempe & P. Ketner (eds.) (1979). *The Global Carbon Cycle*. SCOPE 16. John Wiley, Chichester.

Bolin, B., A. Björkström, K. Holmén & B. Moore (1983). 'The simultaneous use of tracers for oceanic studies'. *Tellus*, **35**, 206–36.

Born, G. M., D. B. Lame & P. J. Rygh (1981). 'A survey of goals and accomplishments of the Seasat mission'. *In: Oceanography from Space* (J. Gower, ed.). Plenum, New York, pp. 3–14.

Bretherton, F. P. (1982*a*) 'Ocean climate modelling'. *Progr. Oceanogr*., **11**, 93–129.

Bretherton, F. P. (1982*b*). 'On the interpretation of satellite cloud images'. Unpublished manuscript, *Rep. JSC XIII* (Dublin). WMO Geneva.

Bretherton, F., J. McPhaden & E. B. Kraus (1982). 'Heat storage measurements for Cage'. *In: The 'Cage' Experiment: a Feasibility Study* (Dobson *et al*., eds.). WCRP Rep. No. 22, WMO Geneva.

Broecker, H. C. (1980). 'Effect of bubbles on the gas exchange between atmosphere and ocean'. *In: Symposium on Capillary Waves and Gas Exchange* (L. Hasse, ed.). SFB 94 Rep. No. 17, University of Hamburg.

Broecker, W. S. (1981) 'Geochemical tracers and ocean circulation'. *In: Evolution of Physical Oceanography* (B. Warren & C. Wunsch, eds.). MIT Press, Cambridge, Mass. pp. 434–60.

Broecker, W. S. (1982). 'Glacial to interglacial changes in ocean chemistry'. *Progr. Oceanogr*., **11**, 151–98.

Brown, O. B. (1981). 'Observation of long term sea surface temperature variability during GATE'. *Deep-Sea Res*., **26** (GATE Suppl. II), 103–24.

Bryan, K. (1982). 'Poleward heat transport by the ocean: observations and models'. *Ann. Rev. Earth Planet Sci*., **10**, 15–38.

Bryan, K. & L. J. Lewis (1979). 'A water mass model of the world ocean'. *J. Geophys. Res*., **34**, 2503–17.

Bryan, K., F. G. Komro, S. Manabe & M. J. Spelman (1982). 'Transient climate response to increasing atmospheric carbon dioxide'. *Science*, **215**, 56–8.

Bryden, H. L. (1982). 'Ocean heat transport'. *In: Time Series of Ocean Measurements* (D. Ellett, ed.). WCP Rep. No. 21, WMO, Geneva.

Bryden, H. L. & M. M. Hall (1980). 'Heat transport by ocean currents across 25°N latitude in the Atlantic Ocean'. *Science*, **207**, 884–6.

Budyko, M. I. (1956). *The Heat Balance of the Earth's Surface*. (N. A. Stepanova, trans.). PB 131692 Office of Technical Services, Dept. of Commerce, Washington DC (1958), 259 pp.

Budyko, M. I. (1966). 'Polar ice and climate'. *In: Proc. Symp. Arctic Heat Budget and Atmospheric Circulation*. Meme. RM-5233-NSF, Rand Corp., pp. 3–22.

Budyko, M. I. (1969). 'Effects of polar radiation variations on the climate of the earth'. *Tellus*, **21**, 611–19.

Budyko, M. I. (1974). *Climate and Life*. Academic Press, New York, 508 pp.

Bunker, A. F. (1976). 'Computations of surface energy flux and annual air–sea interaction cycles of the North Atlantic Ocean'. *Mon. Wea. Rev.*, **104**, 1122–40.

Bunker, A. F. & L. V. Worthington (1976) 'Exchange energy charts of the North Atlantic Ocean'. *Bull. Am. Met. Soc.*, **57**, 670–8.

Bunker, A. F. & R. A. Goldsmith (1979). 'Archived time-series of Atlantic Ocean meteorological variables and surface fluxes'. *Technical Rep.* WHOI-79-3. Woods Hole Oceanographic Institution.

Bunker, A. F., H. Charnock & R. A. Goldsmith (1982). 'A note on the heat balance of the Mediterranean and Red Seas'. *J. Mar. Res.*, **40** (Suppl.), 73–84.

Burridge, D. (1982). 'Calculations of atmospheric energy flux divergence usng the ECMWF model'. Quoted by Dobson *et al.* 1982.

Busalacchi, A. J. & J. J. O'Brien (1981). 'Interannual variability of the equatorial Pacific in the 1960s'. *J. Geophys. Res.*, **86**, 10901–7.

Businger, J. (1975). 'Interaction between sea and air'. *Rev. Geophys. and Space Phys.*, **13**, 720–26; 817–22.

Campbell, G. G. (1981). Energy transport within the Earth's atmosphere–ocean system from a climatic point of view. Ph.D. Thesis, Colorado State University, Fort Collins.

Carmack, E. C. (1983). 'Circulation and mixing in ice-covered waters'. *In: Air-Sea-Ice Interaction* (N. Untersteiner, ed.). Plenum, New York.

Charlock, T. P. (1982). 'Mid-latitude model analysis of solar radiation, the upper layers of the sea, and seasonal climate'. *J. Geophys. Res.*, **87**, 8923–30.

Charnock, H. (1981) 'Air–sea interaction.' *In: Evolution of Physcial Oceanography* (B. A. Warren & C. Wunsch, eds.). MIT Press, Cambridge, Mass.

Chelton, D. B., K. J. Hussey & M. E. Parke (1981). 'Global satellite measurements of water vapour, wind speed and wave height'. *Nature*, **294**, 529–32.

Chelton, D. B. & J. J. O'Brien (1982). 'Satellite microwave measurements of surface wind speed in the tropical pacific'. *Tropical Ocean–Atmosphere Newsletter*, **11**, 2–4, Washington University, JISAO.

Chervin, R. M., J. E. Kutzbach, D. D. Houghton & R. G. Gallimore (1980). 'Response of the NCAR General Circulation model to prescribed changes in ocean surface temperatures, II midlatitude and subtropical changes'. *J. Atmos.* Sci., **37**, 308–22.

CLIMAP project members (1976). 'The surface of the ice-age earth'. *Science, 191'*, 1131–7.

Cogley, J. G. (1979). 'The albedo of water as a function of latitude'. *Mon. Wea. Rev.*, **107**, 776–81.

Colebrook, J. M. & A. H. Taylor (1979). 'Year-to-year changes in sea-surface temperature, North Atlantic and North Sea, 1948 to 1974'. *Deep-Sea Res.*, **26A**, 825–50.

Collins, C. A., L. F. Giovando & K. B. Abbott-Smith (1975). 'Comparison of Canadian and Japanese merchant ship observations of sea-surface temperature in the vicinity of present ocean weather ship 'P', 1927–1933'. *J. Fish. Board Canada*, **32**, 253–8.

Croll, J. (1875). *Climate and Time in their Geological Relations*. Appleton, New York.

Cushman-Roisin, B. (1981). 'Effects of horizontal advection on upper mixing: A case of frontogenesis'. *J. Phys. Oceanogr.*, **11** (10), 1345–56.

Danckwerts, P. V. (1970). *Gas-Liquid Reactions*. McGraw Hill, New York, 276 pp.

Davis, R. E. (1976). 'Predictability of sea surface temperature and sea level pressure anomalies over the North Pacific Ocean'. *J. Phys. Oceanogr.*, **6**(3), 249–66.

Davis, R. (1978). 'Predictability of sea level pressure anomalies over the North Pacific Ocean'. *J. Phys. Oceanogr.*, **8**, 233–46.

Deacon, E. L. (1977). 'Gas transfer to and across an air–water interface'. *Tellus, 29*, 363–74.

Deacon, E. L. (1980). 'Sea–air gas transfer: the wind speed'. *Boundary Layer Meteorology*, **21**, 31–7.

Deardorff, J. W. (1968). 'Dependence of air–sea transfer coefficients on bulk stability'. *J. Geophys. Res.*, **73**, 2549–57.

Defant, A. (1936). *Stratification and Circulation of the Atlantic Ocean: The Troposphere*. Scientific results of the German Atlantic Expedition of the Research Vessel 'Meteor' 1925–27, Vol. 6 Part 2. (English edn, 1982, Amerind, New Delhi.)

Delmas, R. J., J. M. Ascencio & M. Legrand (1980). 'Polar ice evidence that atmospheric CO_2 20000 yr BP was 59% of present'. *Nature*, **284**, 155–7.

Denman, K. L. & M. Miyake (1973). 'Upper layer modification at Ocean Station Papa: observations and simulation'. *J. Phys. Oceanogr.*, **3**(2), 185–96.

de Szoeke, R. A. & P. B. Rhines (1976). 'Asymptotic regimes in mixed layer deepening'. *J. Mar. Res.*, **34**, *111–16*.

De Witt, P. W. & A. Leetma (1978). 'A simple Ekman model for predicting thermocline displacement in the tropical Pacific'. *J. Phys. Oceanogr.*, **8**, 811–17.

Dickinson, R. E. (1981). 'Convergence rate and stability of ocean–atmosphere coupling scheme with a zero-dimensional climate model'. *J. Atmos. Sci.*, **38**, 2112–20.

Dietrich, G., K. Kalle, W. Krauss & G. Siedler (1980). *General Oceanography. An Introduction*, 2nd edn (translated from German). John Wiley, New York, 626 pp.

Dillon, T. M. & D. R. Caldwell (1980). 'The Batchelor spectrum and energy dissipation in the upper ocean'. *J. Geophys. Res.*, **85**, 1910–16.

Dobson, F., L. Hasse & R. Davis (eds.) (1980). *Air–Sea Interaction: Instruments and Methods*. Plenum, New York, 801 pp.

Dobson, F. E., F. P. Bretherton, D. M. Burridge, J. Crease, E. B. Kraus & T. H. Vonder Haar (1982). *The CAGE Experiment: A Feasibility Study*. WCP-22, WMO Geneva, 95 pp.

Donguy, J. R. & C. Henin (1982). 'An expendable bathythermograph and sea surface temperature experiment in the eastern and western Pacific'. *In: Time Series of Oceanographic Measurements* (D. Ellet, ed.). WCP-21, WMO, Geneva, pp. 123–34.

Donn, W. L. & D. M. Shaw (1967). 'The maintenance of an ice-free arctic ocean'. *Progr. Oceanogr.*, **4**, 105–13.

Dorman, C. E. (1983). 'Indian Ocean rainfall'. *Mon. Wea. Rev.* (in press).

Dorman, C. E. & R. H. Bourke (1979). 'Precipitation over the Pacific Ocean, 30°S to 60°N'. *Mon. Wea. Rev.*, **107**, 869–10.

Dorman, C. E. & R. H. Bourke (1981). 'Precipitation over the Atlantic Ocean, 30°S to 70°N. *Mon. Wea. Rev.*, **109**, 554–63.

Dorman, C. E. (1982). 'Comparison of ocean and island rainfall in the tropical Pacific: commments'. *J. Applied Meteor.*, **21**, 109–13; reply by R. K. Reed, pp. 114–15.

Douglas, R. & F. Woodruff (1981). 'Deep-sea benthic foraminifera'. *In: The Oceanic Lithosphere* (C. Emiliani, ed.), *The Sea*, Vol. 7. Wiley-Interscience, New York, 1233–327.

Drooger, C. W. (ed.) (1973). 'Messinian events in the Mediterranean'. *Geodynamics Sci. Rep.*, **7**, 1–272.

Duplessy, J. C. (1981). 'Oxygen isotope studies and quaternary marine climates'. *In: Climatic Variations and Variability: Facts and Theories* (A. Berger, ed.). Reidel, Dordrecht, pp. 181–92.

Ellett, D. J. (1982). 'Long-term water-mass changes to the West of Britain'. *In*: *Time Series of Oceanographic Measurements* (D. J. Ellet, ed.). WCP-21, WMO, Geneva, pp. 245–54.

Ellis, H. (1751). 'A letter to the Rev. Dr. Hales FRS from Captain Henry Ellis FRS dated Jan 7, 1750–51 at Cape Monte Africa, ship east of Halifax'. *Phil. Trans. Roy Soc., London*, **47**, 211–14.

Ellis, J. S., T. H. Vonder Haar, S. Levitus & A. H. Ooort (1981). 'Mean surface stress curl over the oceans as determined from the vorticity budget of the atmosphere'. *J. Atmos. Sci.*, **38**, 262–9.

Emiliani, C. (1954). 'Temperatures of Pacific bottom waters and polar superficial waters during the Tertiary'. *Science*, **119**, 853–5.

Emiliani, C. (1981). 'A new global geology'. *In: The Oceanic Lithosphere* (C. Emiliani, ed.), *The Sea*, Vol. 7. Wiley-Interscience, New York, pp. 1687–728.

Esbenson, S. K. & R. W. Reynolds (1981). 'Estimating monthly averaged air–sea transfers of heat and momentum using the bulk aerodynamic method'. *J. Phys. Oceanogr.*, **11**, 457–65.

Falkowski, P. G. (ed.) (1980). *Primary Productivity in the Sea.* Plenum, New York, 531 pp.

Favre, A. & K. Hasselmann (eds.) (1978). *Turbulent Fluxes through the Sea Surface, Wave Dynamics, and Prediction.* Plenum, New York, 677 pp.

Fiegelson, E. M., K. Ya. Kondratyev & M. A. Prokovyev (1982). 'Radiation processes and their parameterization'. Ch. 11 of *The GATE Monograph.* GARP Publication Series No. 25. WMO Geneva.

Fiekas, V. (1982). 'Untersuchungen mit einem eindimensionalen Modell des oberen Ozeans'. Diploma thesis, Kiel University.

Flohn, H. (1981). 'Scenarios of cold and warm periods of the past'. *In: Climate Variations and Variability: Facts and Theories* (A. Berger, ed.). Reidel, Dordrecht, pp. 689–98.

Fortak, H. G. (1979). 'Entropy and climate'. *In: Man's impact on climate* (W. Bach, J. Pankrath & W. Kellogg, eds.). Elsevier, Amsterdam, pp. 1–14.

Frankignoul, C. & K. Hasselmann (1977). 'Stochastic climate models. Part II. Application to sea-surface temperature anomalies and thermocline variability'. *Tellus*, **29**, 289–305.

Friehe, C. A. & K. F. Schmitt (1976). 'Parameterizations of air–sea interface fluxes of sensible and latent heat by bulk aerodynamic formulae'. *J. Phys. Oceanogr.*, **6**, 801–9.

Gargett, A., T. B. Sanford & T. R. Osborne (1979). 'Surface mixing in the Sargasso Sea'. *J. Phys. Oceanogr.*, **9**, 1090–111.

Garratt, J. R. (1977). 'Review of drag coefficients over ocean and continents'. *Mon. Wea. Rev.*, **105**, 115–29.

Garrett, C. (1979). 'Mixing in the ocean interior'. *Dynamics of Atmospheres and Ocean*, **3**, 239–66.

Gartner, S. & J. Keany (1978). 'The terminal Cretacous event: a geological problem with an oceanographic solution'. *Geology*, **6**, 708–12.

Garwood, R. W. (1979). 'Air–sea interaction and dynamics of the surface mixed layer'. *Rev. Geophys. and Space Science*, **17**, 1507–24.

GATE (1982). *The GARP Atlantic Tropical Experiment Monograph.* Garp Publication Series No. 25. World Meteorological Organisation, Geneva.

Gautier, C. (1982*a*). 'Daily shortwave energy budget over the ocean from geostationary satellite measurements'. *In: Oceanography from Space* (J. F. R. Gower, ed.). Plenum, New York, pp. 201–6.

Gautier, C. (1982*b*). 'Information content of the Seasat SMMR brightness temperatures for sea surface temperature retrieval'. *In: Oceanography from Space* (J. F. R. Gower, ed.). Plenum, New York, pp. 727–34.

Gautier, C. (1982*c*). 'Satellite measurements of insolation over the Tropical Oceans'. *Tropical Ocean–Atmosphere Newsletter*, **12**, 5–6.

Gautier, C., G. Diak & S. Masse (1980). 'A simple physical model to estimate incident solar radiation at the surface from GOES satellite data'. *J. Appl. Met.*, **19**, 1005–12.

Gienapp, H. (1982). 'Optical properties of seawater in the German Bight and the Mouth of the Elbe; the 1978 CZCS Prelaunch Experiment of the Deutsches Hydrographisches Institut': Part 1 *DHZ*, **35** (3); Part 2 *DHZ*, **35** (4), 136–67.

Gill, A. E. (1973). 'Circulation and bottom water formation in the Weddell Sea'. *Deep-Sea Res.*, **20**, 111–40.

Gill, A. E. (1974). 'The relationship between heat content of the upper ocean and the sea surface temperature'. *Norpax Highlights*, **2** (3), 1–4.

Gill, A. E. (1983). 'An estimate of sea-level and surface current anomalies during the 1972 El Niño and consequent thermal effects'. *J. Phys. Oceanogr.*, **13**, 586–606.

Gill, A. E. (1982). 'Changes in thermal structure of the equatorial Pacific during the 1972 El Niño as revealed by bathythermograph observations'. *J. Phys. Oceanogr.*, **12**, 1373–87.

Gill, A. E. & P. P. Niiler (1973). 'The theory of seasonal variability in the ocean'. *Deep-Sea Res.*, **20**, 141–77.

Gill, A. E. & J. S. Turner (1976). 'A comparison of season thermocline models with observation'. *Deep-Sea Res.*, **23**, 391–401.

Goldhirsh, J. & E. J. Walsh (1981). *Precipitation Measurements from Space Using a Modified Seasat Type Radar Altimeter.* Johns Hopkins University. Applied Physics Lab. No. S1R8IU-022, 77 pp.

Goldsmith, R. A. & A. F. Bunker (1979). '*Woods Hole Oceanographic Institution Collection of Climatology and Air–Sea Interaction Data*'. Woods Hole Oceanographic Institution. Technical Rep. No. WHOI-79-70.

Goody, R. (1980). 'Polar process and world climate (a brief overview)'. *Mon. Wea. Rev.*, **108** (12) 1935–42.

Gordon, A. L. (1981*a*). 'South Atlantic thermocline ventilation'. *Deep Sea Res.*, **28**, 1239–64.

Gordon, A. L. (1981*b*). 'Seasonality of Southern Ocean sea ice'. *J. Geophys. Res.*, **86**, 4193–7.

Gorshkov (ed.) (1976, 1979). *World Ocean Atlas:* I. *Pacific Ocean.* II. *Atlantic and Indian Oceans.* Pergamon, Oxford (302 and 306 plates).

Gower, J. F. R. (ed.) (1982). *Oceanography from Space.* Plenum, New York, 978 pp.

Griffith, C. G., W. L. Woodley, J. S. Griffith & S. C. Stromatt (1980). *Satellite-Derived Precipitation Atlas for GATE.* Environmental Research Laboratories, NOAA, 280 pp.

Gupta, J. N. & A. H. Ghovanlou (1978). 'Radiative transfer in turbid water'. *Ocean Optics*, **5**, 132.

Guymer, T. H. (1983). 'A review of Seasat-A satellite scatterometer data'. *Phil. Trans. Roy. Soc., London*, **A**, **309**, 399–414.

Guymer, T. H., J. A. Businger, W. L. Jones & R. H. Stewart (1981). 'Anomalous wind estimates from the Seasat scatterometer'. *Nature*, **294**, 737–9.

Guymer, T. H., J. A. Businger, K. B. Katsaros, W. J. Shaw, P. K. Taylor, W. G. Large & R. E. Payne (1983). 'Transfer processes at the air–sea interface'. *Phil. Trans. Roy. Soc., London*, **A 308**, 253–73.

Hall, M. M. & H. Bryden (1982). 'Direct estimates and mechanisms of ocean heat transports'. *Deep-Sea Res.*, **29**, 339–59.

Han, Y.-J. & S. W. Lee (1981). *A new analysis of Monthly Mean Wind Stress over the Global Ocean.* Rep. No. 26, Climate Research Institute, Oregon State University, 148 pp.

Haney, R. L. (1979). 'Numerical models of ocean circulation and climate interaction'. *Rev. Geophys. Space Sci.*, **17**, 1494–507.

Haney, R. L. & R. W. Davies (1976). 'The role of surface mixing in the seasonal variation of ocean thermal structure'. *J. Phys. Oceanogr.*, **6**, 504–10.

Hanson, K. J. (1976). 'A new estimate of solar irradiance at the Earth's

surface on zonal and global scales'. *J. Geophys. Res.*, **81**, 4435–43.
Hanson, H. P. (1981*a*). 'Infrared flux at the sea surface'. *Tropical Ocean–Atmosphere Newsletter*, July 1981. Joint Institute for Study of the Atmosphere and Ocean, University of Washington, Seattle.
Hanson, H. P. (1981*b*). 'Infrared flux at the ocean surface – disparity between data and formulae indicate need for reassessment'. *Tropical Ocean–Atmosphere Newsletter*, **7**, 4–7.
Hantel, M. (1970). 'Monthly charts of wind-stress curl over the Indian Ocean'. *Mon. Wea. Rev.*, **98**, 765–73.
Hantel, M. (1972). 'Wind-stress curl – the forcing function for oceanic motions'. *In: Studies in Physical Oceanography* (A. L. Gorden, ed.). Gordon and Breach, New York, pp. 121–36.
Harries, J. E., D. T. Llewellyn-Jones, P. Minnett, R. W. Saunders & A. M. Zavody (1983). 'Observations of sea surface temperature'. *Phil. Trans. Roy. Soc., London*, **A 309**, 381–95.
Hasse, L. & P. S. Liss (1980). 'Gas exchange across the air–sea interface'. *Tellus*, **32**, 470–81.
Hasselmann, K. (1981). 'Construction and verification of stochastic climate models'. *In: Climatic variation and variability* (A. Berger, ed.). Reidel, Dordrecht, pp. 481–500.
Hasselmann, K., E. Maier-Reimer, D. Müller & J. Willebrand (1981). *An Ocean Circulation Model for Climate Variability Studies*. Max Planck Institute for Meteorology, Hamburg, Research Rep. No. 104 02 612.
Hastenrath, S. & P. J. Lamb (1977). *Climatic Atlas of the Tropical Atlantic and Eastern Pacific Oceans*. University of Wisconsin Press, Madison, 105 pp.
Hastenrath, S. & P. Lamb (1978). *Heat Budget Atlas of the Tropical Atlantic and Eastern Pacific Oceans*. University of Wisconsin Press, 104 pp.
Hastenrath, S. & P. Lamb (1979). *Climatic Atlas of the Indian Ocean*. Part 1. *Surface Circulation and Climate*. Part 2. *The Oceanic Heat budget*. University of Wisconsin Press, 104 pp.
Hays, J. D., J. Imbrie & N. J. Shackleton (1976). 'Variations in the earth's orbit: pacemaker of the ice ages'. *Science*, **194**, 1121–32.
Held, I. M. (1982). 'Stationary and quasi-stationary eddies in the extra-tropical troposphere: theory'. *In: The Dynamics of the Extratropical Troposphere* (B. Hoskins, ed.). Academic Press, London.
Hellerman, S. (1967, 1968). 'An updated estimate of the wind stress of the world ocean'. *Mon. Wea. Rev.*, **95**, 607–26. (Correction: *Ibid*, **96**, 62–74.
Hellerman, S. (1980). 'Charts of the variability of the wind stress over the tropical Atlantic'. *Deep-Sea Res.*, **26**, (GATE suppl), Part 2, pp. 63–75.
Henin, C. & J. R. Donguy (1980). 'Heat content changes within the mixed layer of the equatorial Pacific Ocean'. *J. Mar. Res.*, **38**, 767–80.
Herman, G. F. (1983). 'Atmospheric modelling and air–sea–ice interaction'. *In: Air–Sea–Ice Interaction* (N. Untersteiner, ed.). Plenum, New York.
Herman, G. & W. T. Johnson (1978). 'The sensibility of the general circulation to Arctic sea ice boundaries'. *Mon. Wea. Rev.*, **106**, 1646–64.
Hibler, W. D. (1979). 'A dynamic-thermodynamic sea ice model'. *J. Phys. Oceanogr.*, **9**, 815–46.
Hibler, W. D. & J. E. Walsh (1983). 'On modelling seasonal and interannual fluctuations of Arctic Sea ice'. *J. Phys. Oceanogr.* (in press).
Hinzpeter, H. (1980). 'Atmospheric radiation instruments'. *In: Air–Sea Interaction: Instruments and Methods* (F. Dobson, L. Hasse & R. Davis, eds.). Plenum, New York, pp. 491–507.
Hofer, R., E. G. Njoku & J. W. Waters (1981). 'Microwave radiometric measurements of sea surface temperature from the Seasat satellite: first results'. *Science*, **212**, 1385–7.
Hofmann, K. (1982). 'Ein eindimensionales numerisches Modell zur Beschreibung der oberen Schichten des Ozeans'. Unpublished diploma thesis, University of Kiel.
Holopainen, G. O. (1967). 'A determination of wind-driven circulation from the vorticity budget of the atmosphere'. *PAGEOPH*, **67**, 156–65.
Horel, J. D. (1982). 'On the annual cycle in the tropical Pacific atmosphere and ocean'. *Mon. Wea. Rev.*, **110**, 1863–78.
Horel, J. D. & J. M. Wallace (1981). 'Planetary-scale atmospheric phenomena associated with the interannual variability of sea surface in the equatorial Pacific'. *Mon. Wea. Rev.*, **109**, 813–29.
Hoskins, B. J. (1983). 'Dynamical processes in the atmosphere and the use of models'. *Q. J. Roy. Met. Soc.*, **109**, 1–21.
Hoskins, B. J. & D. J. Karoly (1981). 'The steady linear response of a spherical atmosphere to thermal and orographic forcing'. *J. Atmos. Sci.*, **38**, 1179–96.
Hovis, W. A., D. K. Clark, F. Anderson, R. W. Austin, W. H. Wilson, E. T. Baker, D. Ball, M. R. Gordon, J. L. Mueller, S. Z. El-Sayed, B. Sturm, R. C. Wigley & C. S. Yentsch 1980). 'Nimbus 7 Coastal Zone Colour Scanner System description and initial imagery'. *Science*, **210**, 60–3.
Hsü, J. J. (1972). 'When the Mediterranean dried up'. *Scientific American*, December 1972.
Hsu, C. F. & J. M. Wallace (1976). 'The global distribution of the annual and semi-annual cycles in precipitation'. *Mon. Wea. Rev.*, **104**, 1093–101.
Hudlow, M. D. & V. L. Patterson (1979). *GATE Radar Rainfall Atlas*. NOAA Special Rep., Center for Environmental Assessment Services, EDIS, NOAA, 155 pp.
Hughes, R. L. (1980). 'On the equatorial mixed layer'. *Deep-Sea Res.*, **27**, 1067–78.
Hughes, R. L. (1982). 'Aspects of equatorial mixed layer dynamics'. *Tropical Ocean–Atmosphere Newsletter* (Jan. 1982), No. 9, pp. 4–5.
Huppert, H. E. & J. S. Turner (1978). 'On melting icebergs'. *Nature*, **271**, 46–8.
Højerslev, M. K. (1980). 'Water color and its relation to primary production'. *Boundary Layer Meteorology*, **18**, 203–20.
Imbrie, J. & K. P. Imbrie (1979). *Ice Ages: Solving the Mystery*. MacMillan, London, 224 pp.
Ivanoff, A. (1977). 'Oceanic absorption of solar energy'. *In: Modelling and prediction of the upper layers of the ocean* (E. B. Kraus, ed.). Pergamon Press, Oxford, pp. 47–71.
Jacobs, W. C. (1951). 'Large scale aspects of energy transformation over the ocean'. *In: Compendium of Meteorology* (T. F. Malone, ed.). American Meteorology Society, Boston, Mass. pp. 1057–70.
Jaeger, L. (1976). 'Monatskarten des Niederschlags für die ganze Erde'. *Berichte des Deutsches Wetterdienstes*, **18**, No. 139, Offenbach AM, 38 pp. + plates.
Jähne, B., K. O. Münnich & U. Siegenthaler (1979). 'Measurements of gas exchange and momentum transfer in a circular wind tunnel'. *Tellus*, **31**, 321–9.
Jerlov, N. G. (1976). *Marine Optics*. Elsevier, Amsterdam, 231 pp.
Jones, E. P. & S. D. Smith (1977). 'A first measurement of sea–air CO_2 fluxes by eddy correlation'. *J. Geophys. Res.*, **82**, 5990–2.
Jones, W. L., D. H. Boggs, E. M. Bracalente, R. A. Brown, D. Chelton & L. C. Schroeder (1981). 'Evaluation of the Seasat wind scatterometer'. *Nature*, **294**, 704–7.
Jones, W. L., L. C. Schroeder, D. H. Boggs, E. M. Bracalente, R. A. Brown, G. J. Dome, W. J. Pierson & F. J. Wentz (1982). 'The SEASAT-A Satellite Scatterometer: the geophysical evaluation of remotely sensed vector winds over the ocean'. *J. Geophys. Res.*, **87**, 3297–317.
Jouzel, J., L. Merlivat & C. Lorius (1982). 'Deuterium excess in an East Antarctic ice core suggests higher relative humidity at the oceanic surface during the last glacial maximum'. *Nature*, **299**, 68–69.
Julian, P. R. & R. M. Chervin (1978). 'A study of the Southern Oscillation and Walker circulation phenomenon'. *Mon. Wea. Rev.*, **106**, 1435–51.

Katsaros, K. B. (1980). 'The aqueous thermal boundary layer'. *Boundary Layer Meteorology*, **18**, 107–27.
Katsaros, K. B., W. T. Liu, J. A. Businger & J. E. Tillman (1977). 'Heat transport and thermal structure in the interfacial layer measured in an open tank of water in turbulent free convection'. *Tellus*, **29**, 229–39.
Katz, E. S. & S. Garzoli (1982). 'Response of the western equatorial Atlantic Ocean to an annual wind cycle'. *J. Mar. Res.*, **40** (suppl), 307–27.
Keshavamurty, R. N. (1982). 'Response of the atmosphere to sea surface temperature anomalies over the equatorial Pacific and the teleconnections of the Southern Oscillation'. *J. Atmos.* **Sci.**, **39**, 1241–59.
Kidder, S. Q. & T. H. Vonder Haar (1977). 'Seasonal oceanic precipitation frequencies from Nimbus 5 microwave data'. *J. Geophys. Res.*, **82**, 2083–6.
Killworth, P. D. (1982). 'Deep convection in the world ocean'. *Rev. Geophys. Space Phys.*, **21**, 1–26.
Kirwan, A. D., T. J. Ahrens & G. H. Born (eds) (1983). 'Seasat special issue II: scientific results'. *J. Geophys. Res.*, **88**, 1529–1952.
Kitaigorodskii, S. A. (1970, 1973). *Physics of Interaction of the Atmosphere and the Ocean.* Leningrad, Gidrometeoisdat. (English edn, 1973. IPST, Jerusalem.)
Kitaigorodskii, S. A. (1979). 'Review of the theories of wind-mixed layer deepening'. *In: Marine Forecasting* (J. C. J. Nihoul, ed.) Oceanography Series 25. Elsevier, Amsterdam, pp. 1–33.
Klein, P. (1980). 'A simulation of the effects of air–sea transfer variability on the structure of marine upper layer'. *J. Phys. Oceanogr.*, **10**, 1824–41.
Klein, P. & M. Coantic (1981). 'A numerical study of turbulent processes in the marine upper layers'. *J. Phys. Oceanogr.*, **11**, 850–63.
Kondo, J. (1975). 'Air–sea bulk transfer coefficients in diabatic conditions'. *Boundary Layer Meteorology*, **9**, 91–112.
Kondratyev, K. Ya. (1969). *Radiation in the Atmosphere.* Academic Press, New York, 912 pp.
Korzun, V. I. (ed.) (1974). *World Water Balance and Water Resources of the Earth.* UNESCO Press, Paris, 663 pp.
Kraus, E. B. (1972). *Atmosphere–Ocean Interaction.* Oxford University Press, New York.
Kraus, E. B. (1973). 'Comparison between ice age and present general circulations'. *Nature*, **245**, 129–33.
Kraus, E. B. (ed.) (1977). *Modelling and Prediction of the Upper Layers of the Ocean.* Pergamon, Oxford, 325 pp.
Kraus, E. B. & L. D. Leslie (1982), 'The interactive evolution of the oceanic and atmospheric boundary layers in the source regions of the trades'. *J. Atmos. Sci.*, **39**, 2760–72.
Kraus, E. B. & J. S. Turner (1967). 'A one-dimensional model of the seasonal thermocline. II. The general theory and its consequences'. *Tellus*, **19**, 98–105.
Krishnamurti, T. N. & R. Krishnamurti (1980). 'Surface meteorology over the GATE A-scale'. *Deep-Sea Res.*, **26** (GATE Suppl. 2), 29–61.
Kroemer, B. & W. Roether (1983). 'Field measurements of air–sea gas exchange by the radon deficit method during JASIN 1978 and FGGE 1979'. *Meteor Forschungsergebnisse A*, 55–75.
Kronfeld, U. (1982). 'Die Wärmebialanz der ozeanischen Deckschicht im GATE C-Gebiet. Eine Analyse des Batfish-Datensatzes'. Unpublished diploma thesis, University of Kiel.
Kundu, P. K. (1980). 'A numerical investigation of mixed layer dynamics'. *J. Phys. Oceanogr.*, **10**, 220–36.
Lamb, H. H. (1972). *Climate Past, Present and Future. Part I. Fundamentals and Climate Now.* Methuen, London, 613 pp.
Lamb, P. J. (1981). 'Estimate of annual variation of Atlantic Ocean heat transport'. *Nature*, **290**, 766–8.
Lame, D. B. & G. H. Born (1982). 'Seasat measurement system evaluation: achievements and limitations'. *J. Geophys. Res.*, **87**, 3175–8.
Large, W. G. & S. Pond (1981). 'Open ocean momentum flux measurements in moderate to strong winds'. *J. Phys. Oceanogr.*, **11**, 324–36.
Large, W. G. & S. Pond (1982). 'Sensible and latent heat flux measurements over the ocean'. *J. Phys. Oceanogr.*, **12**, 464–82.
Lau, K. M. (1982). 'A simple model of atmosphere–ocean ineraction during El Nino-Southern Oscillation'. *Tropical Ocean–Atmosphere Newsletter*, **13**, 1–2.
Leetma, A. & A. F. Bunker (1978). 'Updated charts of the mean annual wind stress, convergences in the Ekman layers and Sverdrup transports in the North Atlantic'. *J. Marine Res.*, **36**, 311–21.
Lemke, P., E. W. Trinkl & K. Hasselmann (1980). 'Stochastic dynamic analysis of polar sea ice variability'. *J. Phys. Oceanogr.*, **10**, 2100–20.
Leonard, D. A. (1980). 'The remote measurement of underwater temperature, salinity and attenuation profiles using Raman back scattering'. *In: Remote Measurement of Underwater Parameters* (H. Dolezalek, ed.). Royal Norwegian Council for Scientific and Industrial Research, Space Activity Division, Rep. SAD-91-T.
Levitus, S. (1982). *Climatological Atlas of the World Ocean.* NOAA professional paper 13. Rockville, Md., 173 pp. and 17 microfiches.
Li, Y. H., T. Takashi & W. S. Broecker (1969). 'The degree of saturation of $CaCO_3$ in the oceans'. *J. Geophys. Res.*, **74**, 5507–25.
Lighthill, M. J. (1969). 'Dynamic response of the Indian Ocean to the onset of the southwest monsoon'. *Phil. Trans. Roy. Soc., London*, **A 265**, 45–92.
Lind, R. J. & K. B. Katsaros (1982). 'A model of longwave irradiance for use with surface observations'. *J. Appl. Met.*, pp. 1015–23.
Lind, R. J., K. B. Katsaros & M. Gube (1983). 'Measurement of short and longwave radiation components during JASIN and their parameterization'. *J. Appl. Met.* (in press).
Ling, S. C. & T. W. Kao (1976). 'Parameterization of the moisture and heat transfer process over the ocean under white cap sea states'. *J. Phys. Oceanogr.*, **6**, 306–15.
Lipes, R. G. (1982). 'Description of Seasat radiometer status and results'. *J. Geophys. Res.*, **87**, 3385—95.
Liss, P. S. (1973). 'Processes of gas exchange across an air–water interface'. *Deep-Sea Res.*, **20**, 221–38.
Liss, P. S. & W. G. Slinn (eds.) (1983). *Air–Sea Exchange of Gases and Particles.* Reidel, Dordrecht, 561 pp.
Liu, W. T., K. B. Katsaros & J. A. Businger (1979). 'Bulk parameterizations of air–sea exchanges of energy and water vapour including the molecular constraints at the interface'. *J. Atmos. Sci.*, **36**, 1722–35.
Liu, W. T. & W. G. Large (1981). 'Determination of surface stress by Seasat–SASS: a case study with JASIN data'. *J. Phys. Oceanogr.*, **11**, 1603–11.
London, J. (1957). 'A study of atmospheric heat balance'. Final Report, Contract AF19 (122) 65. New York University.
Lorenz, E. (1975). *Climate Predictability.* GARP Publication Series No. 16, WMO Geneva, pp. 133–6.
Lovelock, J. E. (1979). *Gaia, a New Look at Life on Earth.* Oxford University Press.
Lueck, G. R. & T. R. Osborn (1982). *Dissipation Measurements from the Fronts – 80 Expedition.* Oceanography – The University of British Columbia, Manuscript Rep. No. 38.
Lumb, F. E. (1964). 'The influence of cloud on hourly amounts of total radiation at the sea surface'. *Q. J. Roy. Met. Soc.*, **90**, 43–56.
Luyten, J. R., J. Pedlosky & H. Stommel (1983). 'The ventilated thermocline'. *J. Phys. Oceanogr.*, **13**, 292 – 309.
MacVean, M. K. & J. D. Woods (1980). 'Redistribution of scalars during upper ocean frontogenesis: a numerical model'. *Q. J. Roy. Met. Soc.*, **106** (448), 293–311.
Manabe, S. & R. T. Wetherald (1975). 'The effects of doubling the CO_2 concentration on the climate of a general circulation model'. *J. Atmos. Sci.*, **32**, 3–15.

Manabe, S., K. Bryan & M. J. Spelman (1979). 'A global ocean atmosphere climate model with seasonal variation for future studies of climate sensitivity'. *Dyn. Atmos. Ocean*, **3**, 393–426.

Manabe, S. & R. J. Stouffer (1979). 'A CO_2 climate sensitivity, study with a mathematical model of the global climate'. *Nature*, **282**, 491–3.

Manabe, S. & R. T. Stouffer (1980). 'Sensitivity of a global climate model to an increase of CO_2-concentration in the atmosphere'. *J. Geophys. Res.*, **85**, 5529–54.

Manabe, S. & R. T. Wetherald (1980). 'On the distribution of climate change resulting from an increase in CO_2 content of the atmosphere'. *J. Atmos. Sci.*, **37**, 99–118.

Marchuk, G. I. (1979). Modelling of climate changes and the problem of long-range weather forecasting'. *Proc. World Climate Conference*. WMO, Geneva.

Marchuk, G. I., V. P. Kochergin, V. I. Klimock & V. A. Sukhorukov (1977). 'On the dynamics of the ocean surface mixed layer'. *J. Phy. Oceanogr.* **7**, 865–75.

Marcinek, J. (1964). 'Der Abfluß von den Landflächen der Erde'. *Mitt. Inst. Wasserwirtsch, Berlin*, **21**, 1–204.

Masuzawa, J. & K. Nagasaka (1982). 'Variations of oceanographic conditions and heat exchange through the sea surface in the western Pacific'. *In: Large Scale Oceanographic Experiments in the World Climate Research Programme* (A. Robinson, ed.). WMO Geneva.

Mason, B. J. (1976). Towards the understanding and prediction of climatic variations'. *Q. J. Roy. Met. Soc.*, **102**, 473–89.

McAllister, E. D., W. McLeish & A. Corduan (1971). 'Airborne measurements of the total heat flux from the sea during BOMEX'. *J. Geophys. Res.*, **86**, 4172–80.

McBean, G. A. (1983). 'The atmospheric boundary layer'. *In: Air–Sea–Ice Interaction* (N. Untersteiner, ed.). Plenum, New York.

McClain, E. P. (1982). 'Multiple atmospheric-window techniques for satellite-derived sea surface temperatures'. *In: Oceanography from Space* (J. F. R. Gower, ed.). Plenum, New York.

McCrea, W. H. (1975). 'Ice ages and the Galaxy'. *Nature*, **255**, 607–9.

McCreary, J. P. (1982). 'A linear model of tropical ocean–atmosphere coupling'. *Tropical Ocean–Atmosphere Newsletter*, **9**, 6–7.

McCreary, J. P. (1983). 'A model of tropical ocean–atmosphere interaction'. *Mon. Wea. Rev.* (in press).

McPhaden, M. (1982). 'Variability in the central equatorial Indian Ocean. Part II: oceanic heat and turbulent energy balances'. *J. Mar. Res.*, **40**, 403–19.

Mellor, G. L. & T. Yamada (1974). 'A hierarchy of turbulence closure models for planetary boundary layers'. *J. Atmos. Sci.*, **31**, 1791–1806.

Mellor, G. L. & P. A. Durbin (1975). 'The structure of the ocean surface mixed layer'. *J. Phys. Oceanogr.*, **5**, 718–28.

Mellor, G. L. & T. Yamada (1982). 'Development of a turbulence closure model for geophysical fluid problems'. *Rev. Geophys. Space Phys.*, **20**, 851–76.

Merle, J. (1978). *Atlas Hydrologique Saisonnier de l'Ocean Atlantique Intertropical*. ORSTOM, Paris, 184 pp.

Merle, J. (1980*a*). 'Seasonal variation of heat storage in the tropical Atlantic Ocean'. *Oceanologica Acta*, **3**, 455–63.

Merle, J. (1980*b*). 'Seasonal heat budget in the equatorial Atlantic Ocean'. *J. Phys. Oceanogr.*, **10**, 464–9.

Meyers, G. (1979). 'Annual variation of the slope of the 14 °C isotherm along the equator in the Pacific Ocean'. *J. Phys. Oceanogr.*, **9**, 885–91.

Milankovich, M. M. (1941). *Canon of Insolation and the Ice Age Problem*. Belgrade, Royal Serbian Academy, 484 pp. (English translation by IPST, NSF, Washington.)

Mintz, Y. (1981). A brief review of the present status of global precipitation estimates'. *In: Precipitation Measurements from Space* (D. Atlas & O. W. Thiele, eds.) NASA Goddard Laboratory.

Mollo-Christenson, E. & A. da S. Masearenhas, Jr (1979). 'Heat storage in the oceanic upper mixed layer inferred from Landsat data'. *Science*, **203**, 653–4.

Monin, A. S. (1975). *The Role of the Oceans In Climate Models*. GARP Publication Series No. 16. WMO, Geneva, pp. 201–5.

Moore, D. W. & S. G. H. Philander (1977). 'Modeling of the tropical ocean circulation'. *In: The Sea*, Vol. 6, *Marine modeling* (E. D. Goldberg, I. N. McCave, J. J. O'Brien & J. H. Steele eds.). Wiley, New York.

Moura, A. D. & J. Shukla (1981). 'On the dynamics of droughts in northeast Brazil: observations, theory and numerical experiments with a general circulation model'. *J. Atmos. Sci.*, **38**, 2653–75.

Münnich, K. L., W. B. Clarke, K. H. Fischer, D. Flothmann, B. Kromer, W. Roether, U. Siegenthaler, Z. Top & W. Weiss (1977). 'Gas exchange and evaporation studies in a circular wind tunnel, continuous radon-^{222} measurements at sea and tritium/helium-^{3} measurements in a lake'. *In: Turbulent Fluxes through the Sea Surface, Wave Dynamics and Prediction* (A Favre & K. Hasselmann, eds.). Plenum, New York.

Munk, W. H. & E. R. Anderson (1949). 'Notes on a theory of the thermocline'. *J. Mar. Res.*, **7**, 276–95.

Munk, W. & C. Wunsch (1982). 'Observing the ocean in the 1990s'. *Phil. Trans. Roy. Soc., London*, **A307**, 439–64.

Nagasaka, K. & H. Nitani (1982). 'Long-term oceanographic variations in the western north Pacific and Kuroshio regions'. *In: Time Series of Ocean Measurements* (D. Ellet, ed.). WCP-21, WMO, Geneva, pp. 101–22.

Namias, J. (1965). 'Short period climatic fluctuations. *Science*', **147**, 696–706.

Namias, J. (1969). 'Autumnal variations in the North Pacific and North Atlantic anticyclones as manifestations of air–sea interactions'. *Deep-Sea Res.*, **16** (Suppl.), 153–64.

Namias, J. (1976). 'Negative ocean–air feedback systems over the North Pacific in the transition between warm and cold seasons'. *Mon. Wea. Rev.*, **104**, 1107–21.

Namias, J. & D. R. Cayan (1981). 'Large-scale air–sea interactions and short-period climatic fluctuations'. *Science*, **214**, 868–76.

Navato, A. R., R. E. Newell, J. C. Hsiung, C. B. Billing & B. C. Weare (1981). 'Tropical mean temperature and its relationship to the oceans and atmospheric aerosols'. *Mon. Wea. Rev.*, **109**, 244–54.

Newson, R. L. (1973). 'Response of a general circulation model of the atmosphere to removal of the Arctic ice-cap'. *Nature*, **241**, 39–40.

Nicholls, N. & F. Woodcock (1981). 'Verification of an empirical long-range weather forecasting technique'. *Q. J. Roy. Met. Soc.*, **107**, 973–6.

Nicholls, S. (1978). 'Measurements of turbulence by an instrumented aircraft in a convective atmospheric boundary layer over the sea'. *Q. J. Roy. Met. Soc.*, **104**, 653–76.

Nicholls, S., W. J. Shaw & T. Hauf (1983). 'An intercomparison of aircraft turbulence measurements made during JASIN'. *J. Appl. Met.* (in press).

Nicholls, S., R. Brummer, F. Fiedler, A. L. M. Grant, T. Hauf, G. J. Jenkins, C. J. Readings & W. J. Shaw (1983). 'The turbulent structure of the atmospheric boundary layer'. *Phil. Trans. Roy. Soc., London*, **A308**, 291–309.

Niiler, P. P. (1975). 'Deepening of the wind-mixed layer'. *J. Mar. Res.*, **33**, 405–22.

Niiler, P. P. (1977). 'One-dimensional models of the seasonal thermocline'. *In: The Sea*, vol. 6, *Marine Modeling* (E. D. Goldberg, I. N. McCave, J. J. O'Brien & J. H. Steele, eds.). Wiley, New York, pp. 97–115.

Niiler, P. (ed.) (1981). *Tropical Pacific Upper Ocean Heat and Mass Budgets*. Hawaii Inst. of Geophysics Special Publication, Honolulu, Hawaii, 56 pp.

Niiler, P. P. (1982). 'Heat budgets of the tropical oceans'. *In: Large Scale Oceanographic Experiments in the World Climate Research Programme. Proc. CCCO-JSC Study Conference*, Tokyo 1982 (A. Robinson, ed.). WMO, Geneva.

Niiler, P. P. & E. B. Kraus (1977). 'One-dimensional models of the upper ocean'. *In: Modelling and Prediction of the Upper Layers of the Ocean.* Pergamon Press, pp. 143–77.

Oakey, J. A. & N. S. Elliott (1980). 'Average microstructure levels and vertical diffusion for phase III, GATE'. *Deep-Sea Res.*, **26** (GATE Suppl.), 273–94.

O'Brien, J. J., A. Busalacchi & J. Kindle (1981). 'Ocean models of El Nino. *In: Resource Management and Environmental Uncertainty: Lessons from Coastal Upwelling Fisheries* (M. H. Glantz & J. D. Thompson, eds.). Wiley-Interscience, New York, pp. 159–212.

Olbers, D., J. Willebrand & M. Wenzel (1982). 'The inference of ocean circulation parameters from climatological hydrographic data'. *Ocean Modelling*, **46**, 5–9.

Oort, A. H. & T. H. Vonder Haar (1976). 'On the observed annual cycle in the ocean–atmosphere heat balance over the northern hemisphere'. *J. Phys. Oceanogr.*, **6**, 781–800.

Opsteegh, J. D. & H. M. van den Dool (1980). 'Seasonal differences in the stationary response of a linearized primitive equation model: prospects for long range weather forecasting?' *J. Atmos. Sci.*, **37**, 2169–85.

Paltridge, G. W. & C. M. R. Platt (1976). *Radiative Processes in Meteorology and Climatology.* Elsevier, Amsterdam.

Pan, Y. H. & A. H. Oort (1983). 'Global climate variations connected with sea surface temperature anomalies in the equatorial Pacific for the period 1958–73'. *Mon. Wea. Rev.* (submitted).

Parkinson, C. L. & W. M. Kellogg (1979). 'Arctic sea ice decay simulated for a CO_2-induced temperature rise'. *Climatic Change*, **2**, 149–62.

Parkinson, C. L. & W. M. Washington (1979). 'A large scale numerical model of sea ice'. *J. Geophys. Res.*, **84**, 311–37.

Pattullo, J. G. (1963). 'Seasonal changes in sea-level'. *In: The Sea* (M. N. Hill, ed.), Vol. 2. Wiley-Interscience, London, pp. 485–96.

Paulson, C. A. (1980). 'Oceanic radiation measurements'. *In: Air Sea Interaction: Instruments and Methods* (F. Dobson, L. Hasse, R. Davis, eds.). Plenum, New York, pp. 509—21.

Paulson, C. A. & J. J. Simpson (1977). 'Irradiance measurements in the upper ocean'. *J. Phys. Oceanogr.*, **7**, 952–6.

Paulson, C. A. & J. J. Simpson (1981). 'The temperature difference across the cool skin of the ocean'. *J. Geophys. Res.*, **86**, 11044–54.

Payne, R. E. (1972). 'Albedo of the sea surface'. *J. Atmos. Sci.*, **29**, 959–70.

Peng, T. H., W. S. Broecker, G. G. Mathieu, Y.-H. Li & A. E. Bainbridge (1979). 'Radon evasion rates in the Atlantic and Pacific oceans as determined during the GEOSECS program'. *J. Geophys. Res.*, **84**, 2471–86.

Philipps, O. M. (1977). *The Dynamics of the Upper Ocean.* Cambridge University Press, 261 pp.

Pollard, R., P. Rhines & R. Thompson (1973). 'The deepening of the wind mixed layer'. *Geophys. Fluid Dynamics*, **3**, 381–404.

Preisendorfer, R. W. (1976). *Hydrologic Optics*, Vols. I–IV. US Dept. of Commerce Environmental Research Laboratories, Honolulu, Hawaii.

Privett, E. W. (1960). 'The exchange of energy between the atmosphere and the oceans of the southern hemisphere'. *Geophysical Memoir*, **13**, 104–65.

Quinn, W. H., O. Zopf, K. S. Short & R. T. Kuo Yang (1978). 'Historical trends and statistics of the Southern Oscillation, El Nino and Indonesian droughts'. *Fish. Bull.*, **76**, 663–78.

Ramage, C. S. (1981). 'On predicting El Nino'. *In: Resource Management and Environmental Uncertainty Lessons from Coastal Upwelling Fisheries* (M. H. Glantz & J. D. Thompson, eds.). John Wiley, New York, pp. 439–448.

Ramanathan, V. & J. A. Coakley, Jr (1978). 'Climate modelling through radiative–convective models'. *Rev. Geophys. Space Phys.*, **6**, 465–89.

Ramanathan, V. (1981). 'The role of ocean–atmosphere interaction in the CO_2 climate problem'. *J. Atmos. Sci.*, **38**, 918–30.

Raney, K. (1983). 'Synthetic aperture radar observations of ocean and land'. *Phil. Trans. Roy. Soc., London*, **A309**, 315–21.

Raschke, E. (1975). 'Numerical studies of solar heating of an ocean model'. *Deep-Sea Res.*, **22**, 659–65.

Rasmusson, E. M. & T. H. Carpenter (1982). 'Variations in tropical surface temperature and surface wind fields associated with the Southern Oscillation/El Nino'. *Mon. Wea. Rev.*, **110**, 354–84.

Reed, R. K. (1977). 'On estimating insolation over the ocean'. *J. of Phys. Oceanogr.*, **7**, 482–5.

Reed, R. K. & W. P. Elliott (1979). 'New precipitation maps for the North Atlantic and north Pacific Oceans'. *J. Geophys. Res.*, **84**, 7839–46.

Reid, J. L. (1981). 'On the mid-depth circulation of the world ocean'. *In: Evolution in Physical Oceanography*, Ch. 3. MIT Press, Cambridge, Mass. pp. 70–111.

Rennel, J. (1832). *An Investigation of the Currents of the Atlantic Ocean.* Rivington, London.

Revelle, R. (1981). 'Introduction'. *In: The Oceanic Lithosphere* (C. Emiliani, ed.), *The Sea*, Vol. 7. Wiley-Interscience, New York, pp. 1–17.

Revelle, R. & H. E. Suess (1957). 'Carbon dioxide exchange between the atmosphere and the ocean and the question of an increase of atmospheric CO_2 during the past decades'. *Tellus*, **9**, 18–27.

Reynolds, R. W. (1977). 'Sea surface temperature anomalies in the North Pacific Ocean'. *Tellus*, **30**, 97–103.

Richman, J. & C. Garrett (1977). 'The transfer of energy and momentum by the wind to the surface mixed layer'. *J. Phys. Oceanogr.* **7**, 876–81.

Robin, G. de Q., D. J. Drewry & V. Squire (1983). 'Observations of polar ice fields'. *Phil. Trans. Roy. Soc., London*, **A309**, 447–61.

Robinson, A. R. & D. J. Baker (eds.) (1979). 'Ocean models and climate models'. *Dynamics of Atmospheres and Oceans*, **3**, 81–526.

Robinson, I. S. (1983). 'Satellite observations of ocean colour'. *Phil. Trans. Roy. Soc., London*, **A309**, in press.

Robinson, M. K. (1976). *Atlas of North Pacific Ocean Monthly Mean Temperatures and Mean Salinities of the Surface Layer.* US Naval Oceanographic Office ref., Pub. 2., Washington DC, pp. 262.

Robinson, M. K., R. A. Bauer & E. H. Schroeder (1979). *Atlas of North Atlantic-Indian Ocean Monthly Mean Temperatures and Mean Salinities of the Surface Layer.* US Naval Oceanographic Office Ref., Pub. 18, Washington DC.

Robock, A. (1980). 'The seasonal cycle of snow cover, sea ice and surface albedo'. *Mon. Wea. Rev.*, **108**, 267–85.

Roden, G. I. (1980). 'On the variability of surface temperature fronts in the Western Pacific, as detected by satellite'. *J. Geophys. Res.*, **85**, 2704–10.

Roll, H U. (1965). *Physics of the Marine Atmosphere.* Academic Press, New York, 426 pp.

Rooth, C. (1982). 'Hydrology and ocean circulation'. *Progr. Oceanogr.*, **11**, 131–50.

Ruddiman, W. F. & A. McIntyre (1973). 'Time-transgressive deglacial retreat of polar waters from the North Atlantic'. *Quaternary Res.*, **3**, 117–30.

Sabine, E. (1846). 'On the cause of the remarkably mild winters which occasionally occur in England'. *Lond., Edin. and Dub. Phil. Mag. and J. Sci.* (April 1846).

Salmon, J. (1978). 'Temperature structure and variability in the diurnal thermocline'. Ph.D. thesis, University Southampton, Oceanography Dept.

Saunders, P. M. (1976*a*). 'Wind stress on the ocean over the eastern continental shelf of North America'. *J. Phys. Oceanogr.*, **7**, 555–66.

Saunders, P. M. (1976*b*). 'On the uncertainty of wind stress and calculations'. *J. Mar. Res.*, **34**, 155–60.

Saur, J. F. T. (1963). 'A study of the quality of seawater temperature reported in the logs of ships' weather observations'. *J. Appl. Met.*, **2**, 417–25.

Schiff, H. I. (1981). 'A review of the carbon dioxide green house problem'. *Planet. Space Sci.*, **29**, 935–50.

Schlesinger, M. E. & W. L. Gates (1981). *Preliminary Analysis of Four General Circulation Model Experiments on the Role of the Ocean in Climate.*, Climate Research Institute, Rep. No 25, Oregon State University, Corvallis, Oregon, 54 pp.
Schroeder, L. C., D. H. Boggs, G. Dome, I. M. Halberstam, W. L. Jones, W. J. Pierson & F. J. Wentz (1982). 'The relationship between wind vector and normalized radar cross-section used to derive Seasat-A satellite scatterometer winds'. *J. Geophys. Res.*, **87**, 3318–36.
Schwiesow, R. L. (1980). 'Temperature and current profiles by high resolution spectroscopy of scattered laser light'. *In: Remote Measurement of Underwater Parameter* (H. Dolezalel, ed.). Roy. Norwegian Council for Scientific and Industrial Research, Space Activity Division, Rep. SAD-91-T.
Semtner, A. J. (1976). 'A model for the thermodynamic growth of sea ice in numerical investigation of climate'. *J. Phys. Oceanogr.*, **6**, 379–89.
Shackleton, N. J. (1982). 'The deep-sea sediment record of climate variability'. *Progr. Oceanogr.*, **11**, 199–218.
Shackleton, N. J. & N. D. Opdyke (1977). 'Oxygen isotope and palaeomagnetic stratifigraphy of equatorial Pacific core V28-238: oxygen isotope temperature and ice volumes on a 10^5 and 10^6 year scale. *Quaternary Res.* **3**, 39–55.
Shaw, P. T., D. R. Watts & H. T. Rossby (1978). 'On the estimation of wind speed and stress from ambient noise measurements'. *Deep-Sea Res.*, **25**, 1225–33.
Shukla, J. (1975). 'Effect of Arabian sea-surface temperature anomaly on Indian summer monsoon: a numerical experiment with the GFDL model. *J. Atmos. Sci.*, **32**, 503.
Shukla, J. & B. M. Misra (1977). 'Relationship between surface temperature and wind speed over the central Arabian Sea, and monsoon rainfall over India'. *Mon. Wea. Rev.* **105**, 998–1002.
Shukla, J. & J. M. Wallace (1983). (Unpublished manuscript.) 'Numerical simulation of the atmospheric response to the Equatorial Pacific sea surface anomalies'.
Siegenthaler, U. (1983). 'Uptake of excess CO_2 by an outcrop–diffusion model of the ocean'. *J. Geophys. Res.*, **88**, 3599–608.
Simmons, A. J. (1982). 'The forcing of stationary wave motion by tropical diabatic heating'. *Q. J. Roy. Met. Soc.*, **108**, 503–34.
Simpson, J. J. & C. A. Paulson (1979). 'Mid-ocean observations of atmospheric radiation'. *Q. J. Roy. Met. Soc.*, **105**, 487–502.
Simpson, J. J. & C. A. Paulson (1980). 'Small scale surface temperature structure'. *J. Phys Oceanogr.*, **10**, 399–410.
Simpson, J. J. & T. D. Dickey (1981*a*). 'Alternative parameterization of downward irradiances and their dynamical significance'. *J. Phys. Oceanogr.*, **11**, 876–82.
Simpson, J. J. & T. D. Dickey (1981*b*). 'The relationship between downward irradiance and upper ocean structure'. *J. Phys. Oceanogr.*, **11**, 309–23.
Smed, J., J. Meincke & D. J. Ellett (1982). 'Time series of oceanographic measurements in the ICES area'. *In: Time Series of Ocean Measurements* (D. Ellet, ed.). WCP-21, WMO, Geneva, pp. 225–44.
Smith, R. C. & K. S. Baker (1978). 'The bio-optical state of ocean waters and remote sensing'. *Limnology and Oceanogr.*, **23**, 247–59.
Smith, S. D. (1980). 'Wind stress and heat flux over the ocean in gale force winds'. *J. Phys. Oceanogr.*, **10**, 709–26.
Smith, S. D. & K. B. Katsaros (1981). *HEXOS – Humidity Exchange Over the Sea.* Bedford Institute of Oceanography, Canada, Rep. Series BI-R-81-17, 133 pp.
Southam, J. R. & W. W. Hay (1981). 'Global sedimentary mass balance and sea level changes'. *In: The Oceanic Lithosphere* (C. Emiliani, ed.), *The Sea*, Vol. 7. Wiley-Interscience, New York, pp. 1617–84.
Stevenson, J. W. (1979) 'On, the effect of dissipation on seasonal thermocline models'. *J. Phys. Oceanogr.*, **9**, 57–64.
Stevenson, J. W. (1980). 'Response of the surface mixed layer to quasi-geostrophic oceanic motions'. Thesis, Harvard University, Cambridge, Mass.
Stewart, R. W. (1982). 'Oceanography from space'. *Proc. 33rd Congress of the International Astronomical Federation*, Paris, 1982.
Stewart, R. W. (1985). *Methods of Satellite Oceanography*, Scripps Studies in Earth and Ocean Sciences, La Jolla, California.
Stewart, R. W. & H. Grant (1962). 'Determination of the rate of dissipation of turbulent, energy near the sea surface in the presence of waves'. *J. Geophys. Res.*, **67**, 3177–80.
Stigebrandt, A. (1981). 'A model of the thickness and salinity of the upper layer in the Arctic Ocean and the relationship between ice thickness and some external parameters'. *J. Phys. Oceanogr.*, **11**, 1407–22.
Stommel, H. (1958). 'The circulation of the abyss'. *Scientific American*, July 1958.
Stommel, H. (1979*a*). 'Determination of water mass properties of water pumped down from the Ekman layer to the geostrophic flow below'. *Proc. Nat. Acad. Sci., USA., USA*, **76**, 3051–5.
Stommel, H. (1979*b*). 'Oceanic warming of Western Europe'. *Proc. Nat. Acad. Sci., USA*, **76**, 2518–21.
Stommel, H. (1980). 'Assymmetry of interoceanic fresh-water and heat fluxes'. *Proc. Nat. Acad. Sci., USA* **77**, 2377–81.
Street, R. L. & A. W. Miller (1977). 'Determination of the aqueous sublayer thickness at an air–water interface'. *J. Phys. Oceanogr.*, **7**, 110–17.
Sverdrup, H. U. (1957). 'Oceanography'. *In: Handbuch der Physik 48*, Springer Verlag, Berlin, pp. 609–70.
Swallow, J. C., R. I. Currie, A. E. Gill & J. H. Simpson (1981). *Circulation and Fronts in Continental Shelf Seas.* The Royal Society, London, 177 pp.
Swinbank, R. (1983). 'A comparison of energy budget calculations based on ECMWF, and Meteorological Office FGGE analyses'. Met.0.20 Technical Note II/23, Bracknell.
Tabata, S. (1978). 'An evaluation of the quality of sea surface temperatures and salinities measured at station P and line P in the northeast Pacific Ocean'. *J. Phys. Oceanogr.*, **8**, 970–86.
Tabata, S. & J. F. R. Gower (1980). 'A comparison of ship and satellite measurements of sea surface temperature off the Pacific coast of Canada'. *J. Geophys. Res.*, **85**, 6636–48.
Taylor, A. H. & J. A. Stephens (1980). 'Seasonal and year-to-year variations in surface salinity at the nine North Atlantic Ocean weather stations. *Oceanology Acta*, **3**, 421–30.
Thompson, R.O.R.Y. (1976). 'Climatological numerical models of the surface mixed layers of the ocean'. *J. Phys. Oceanogr.*, **6**, 496–503.
Thompson, S. L. & S. H. Schneider (1979). 'A seaonal zonal energy balance climate model with an interactive lower model'. *J. Geogr. Res.*, **84**, 2401–14.
Thorndike, A. S., D. A. Rothrock, G. A. Maykut & G. Colony (1975). 'The thickness distribution of sea ice'. *J. Geogr. Res.*, **80**, 4501–13.
Thorpe, S. A. (1982). 'On the clouds of bubbles formed by breaking wind-waves in deep water, and their role in air–sea gas transfer'. *Phil. Trans. Roy. Soc. London*, **A304**, 155–210.
Torgersen, T., G. Mathieu, R. H. Hesslein & W. S. Broecker (1982). 'Gas exchange dependency on diffusion coefficient: direct ^{222}Rn and ^{3}He comparisons in a small lake'. *J. Geophys. Res.*, **87**, **Cl**, 546–56.
Tully, J. P. & L. F. Giovando (1963). 'Seasonal temperature structure in the eastern subarctic Pacific Ocean'. *In: Marine Distributions* (M. J. Dunbar, ed.). Roy. Soc. Canada Publ. No. 5, University of Toronto Press, pp. 10–36.
Turner J. S. (1973). *Buoyancy Effects in Fluids.* Cambridge University Press, 367 pp.
Turner, J. S. & E. B. Kraus (1967). 'A one-dimensional model of the seasonal thermocline I. A laboratory experiment and its interpretation'. *Tellus*, 19, 88–97.
UNESCO (1974). *Discharge of Selected Rivers of the World.* Vol. III, *Mean Monthly and Extreme Discharges (1969–72).* UNESCO Press, Paris, p. 124.
Untersteiner, N. (1983*a*). 'Surface heat and mass balance'. *In: Air–Sea–Ice Interaction.* Plenum, New York.

Untersteiner, N. (ed.) (1985). *Geophysics of Sea Ice*. Plenum, New York.

US Navy (1956, 1963). *Marine Climatic Atlas of the World*. Vols. 1–6. NAVAER, SO-IC-528-533, US Govt Print Office, Washington.

Vincent, E. & W. H. Berger (1981). 'Planktonic foramifera and their use in palaeoceanography'. *In: The Oceanic Lithosphere* (C. Emiliani, ed.) *The Sea*, Vol. 7. Wiley-Interscience, New York, pp. 1025–119.

Volkov, Yu.A., A. Augstein & H. Hinzpeter (1982) 'Boundary Layer Phenomena. Ch. 10 of the *GATE Monograph*, GARP Publication Series No. 25. WMO, Geneva.

Vonder Haar, T. H. & A. H. Oort (1973). 'New estimates of annual poleward energy transport by northern hemisphere oceans'. *J. Phys. Oceanogr.*, **2**, 169–72.

Voorhis, A. D., E. H. Schroeder & A. Leetma (1976). 'The influence of deep mesoscale eddies on sea surface temperature in the North Atlantic subtropical convergence'. *J. Phys. Oceanogr.*, **6**, 953–61.

Walker, G. T. (1923). 'Correlation in seasonal variations of weather. VIII. A preliminary study of world weather. *Mem. Indian Met. Dept.*, **24**, 75–131.

Wallace, J. M. & D. S. Gutzler (1981). 'Teleconnections in the geopotential height field during the northern hemisphere winter'. *Mon. Wea. Rev.*, **109**, 784–812.

Walsh, J. E. & C. M. Johnson (1979). 'An analysis of arctic sea ice fluctations, 1953–1977'. *J. Phys. Oceanogr.*, **9**, 580–91.

Walsh, J. E. & J. E. Slater (1981). 'Monthly and seasonal variability in the ocean-ice-atmosphere system of the north Pacific and the North Atlantic'. *J. Geophys. Res.*, **86**, 7425–45.

Walsh, J. J. (1980). 'Marine photosynthesis and the global carbon cycle'. *In: Primary Productivity in the Sea* (P. G. Faulkowski, ed.). Plenum, New York, pp. 497–506.

Warn-Varnas, A. C. & S. A. Piacsek (1979). 'An investigation of the importance of the third order correlations and choice of length scale in mixed layer modelling'. *Geophys. Astrophys. Fluid Dynamics*, **17**, 63–85.

Warn-Varnas, A. C., G. M. Dawson & P. T. Maring (1981). 'Forecast and studies of the oceanic ixed layer during the MILE experiment'. *Geophys. Astrophys. Fluid Dynamics*, **17**, 63–85.

Warren, B. A. (1981). 'Deep circulation of the world ocean'. *In Evolution of Physical Oceanography* (B. A. Warren & C. Wunsch, eds.) MIT Press, Cambridge, Mass., pp. 6–41.

Warshaw, M. & R. R. Rapp (1973).'An experiment on the sensitivity of a global circulation model'. *J. Appl. Met.*, **12**, 43–9.

Weare, B. C. (1977). 'Empirical orthogonal analysis of Atlantic Ocean surface temperatures'. *Q. J. Roy. Met. Soc.*, **103**, 467–78.

Weare, B.C., A. R. Navato & R. E. Newell (1976). 'Empirical orthogonal analysis of Pacific sea temperature'. *J. Phys. Oceanogr.* **6**, 671–8.

Weare, B. C., P. T. Strub & M. D. Samuel (1980). 'Marine climate atlas of the tropical Pacific Ocean'. *Contributions to Atmospheric Science*, No. 20, University of California at Davis.

Weare, B. C. & P. T. Strub (1981). 'The significance of sampling biases on calculating monthly mean oceanic surface heat fluxes'. *Tellus*, **33**, 211–24.

Weare, B. C., P. T. Strub & M. D. Samuel (1981). 'Annual mean surface heat fluxes in the tropical Pacific Ocean'. *J. Phys. Oceanogr.*, **11**, 705–17.

Webster, P. J. (1981). 'Mechanisms determining the atmospheric response to sea surface temperature anomalies'. *J. Atmos. Sci.*, **38**, 554–71.

Webster, P. J. (1982). 'Seasonality in the local and remote atmospheric response to sea surface temperature anomalies'. *J. Atmos. Sci.*, **39**, 41–54.

Wells, N. C. (1979). 'A coupled ocean–atmosphere experiment: the ocean response'. *Q.J. Roy. Met. Soc.*, *105*, 355–70.

Wessely, M. L., D. R. Cook, R. L. Hart & R. M. Williams (1982). 'Air-sea exchange of CO_2 and evidence for enhanced upward fluxes'. *J. Geophys. Res.*, **87**, 8827–32.

Wetherald, R. T. & S. Manabe (1975). 'The effect of changing the solar constant on the climate of a general circulation model'. *J. Atmos. Sci.*, **32**, 2014–20.

Whitman, W. G. (1923). 'The two-film theory of gas absorption'. *Chem. Met. Eng.*, **29**, 146–8.

Willebrand, J. (1980). Personal communication.

Woods, J. D. (1968). 'Wave-induced shear instability in the summer thermocline'. *J. Fluid Mech.*, **32**, 791–800.

Woods, J. D. (1980). 'Diurnal and seasonal variation of convection in the wind-mixed layer of the ocean'. *Q. J. Roy. Met. Soc.*, **106**, 379–94.

Woods, J. D. (1982). 'Climatology of the upper boundary layer of the ocean'. *In: Large Scale Oceanographic Experiments in the World Climate Research Programme*. WCRP Publication Series 1, 147–79, WMO Geneva.

Woods, J. D. (1985). 'The world ocean circulation experiment'. *Nature*, **314**, 501–11.

Woods, J. D., R. L. Wiley & M. G. Briscoe (1977). 'Vertical circulation at fronts in the upper ocean'. *In: A voyage of Discovery* (M. Angel, ed.), suppl. to *Deep-Sea Res.*, **24**, 253–75.

Woods, J. D. & R. Onken (1982). 'Diurnal variation and primary production in the ocean – preliminary results of a Langrangian ensemble model'. *J. Plankton Res.*, **4**, 735–56.

Woods, J. D., W. Barkmann & A. Horch (1984). 'Solar heating of the Ocean'. *Q. J. Roy. Met. Soc.*, **110**, 633–56.

Worthem, S. & G. Mellor (1981). 'Turbulence closure model applied to the upper tropical ocean'. *Deep-Sea Res.*, **26** (GATE Suppl.), 237–72.

Worthington, V. (1976). *On the North Atlantic Circulation*. Johns Hopkins University Press, Baltimore & London, 110 pp.

Worthington, L. V. (1981). 'Water masses of the world ocean: some results of a fine-scale census'. *In: Evolution of Physical Oceanography* (B. A. Warren & C. Wunsch, eds.). MIT Press, Cambridge, Mass.

Wu, Jin (1982). 'Sea spray: a further look'. *J. Geophys. Res.*, **87**, 8905–12.

Wylie, D. P. & B. B. Hinton (1982). 'The wind stress patterns over the Indian Ocean during the summer monsoon of 1979'. *J. Phys. Oceanogr.*, **12**, 186–9.

Wyrtki, K. (1971). '*Oceanographic Atlas of the International Indian Ocean Expedition (1963–1965)*. National Science Foundation, Washington, DC, 531 pp,

Wyrtki, K. (1975). 'El Nino the dynamic response of the equatorial Pacific Ocean to atmospheric forcing'. *J. Phys. Oceanogr.*, **5**, 572–84.

Wyrtki, K. (1977). 'Sea level during the 1972 El Nino'. *J. Phys. Oceanogr.*, **7**, 779–87.

Wyrtki, K. (1979). 'The response of sea level topography to the 1979 El Nino'., *J. Phys. Oceanogr.*, **9**, 1223–31.

Wyrtki, K. & G. Meyers (1976). 'The trade field over the Pacific Ocean'. *J. Appl. Met.*, **15**, 698–704.

Yong, Q. Yang (1982). 'Temperature fluctuations in the upper 300 metres of the North Atlantic'. (Unpublished manuscript, DAMPT, University of Cambridge.)

Zaneveld, J. R. V. & R. W. Spinrad (1981). 'An arctangent model of irradiance in the sea'. *J. Geophys. Res.*, **85**, 4919–22.

Zillman, J. W. (1972*a*). 'A study of some aspects of the radiation and heat budgets of the southern hemisphere oceans'. *Meteorological Study No. 26*. Australian Govt. Publ. Service, Canberra, 562 pp.

Carl Wunsch†
Department of Applied Mathematics and Theoretical Physics, University of Cambridge, Silver St., Cambridge, CB3 9EW

The ocean circulation in climate

Abstract

The three-dimensional circulation of the ocean is an integral part of the overall climate system of the earth. Atmospheric climate is, to some as yet unclear degree, controlled through the interaction of air and sea. Existing knowledge of the ocean circulation is based upon a combination of fragmentary, possibly badly aliased observations (hydrography), and a number of highly plausible theoretical ideas which have rarely been tested directly. Such basic concepts as Ekman flux and divergence, Sverdrup balance, the physics of western boundary currents and the structure of the thermocline have never been demonstrated quantitatively. Existing numerical models, while superficially becoming very realistic, also to a great extent rest upon this fragile foundation. The role of the ocean in climate and climate change is unlikely to be demonstrated and understood until observations become meaningful in terms of the physics being tested. Because the ocean is a three-dimensional global turbulent fluid, with much of its variability on long time scales, a meaningful observation system also must be global, three-dimensional and of long duration. The only observations now conceivable that are capable of addressing the longest time scales are chemical tracers. Other means exist for oceanic sampling for periods of days to years. Combined with suitable numerical models, a system could be deployed to begin to answer the questions addressed by meteorologists to oceanographers concerning the climate role of the ocean.

10.1 Introduction

It is widely believed that the ocean plays a major role in the earth's climatic system. The thermal capacity of sea water is larger (about four times) than that of air. The ocean stores heat in the upper layers in summer, and releases it back into the atmosphere in winter, thus ameliorating the seasonal extremes of atmospheric temperature. Much of the attention given to the ocean circulation in the past decade has been stimulated, however, from work of Vonder Haar & Oort (1973), and Oort & Vonder Haar (1976), who proposed that the ocean actually *transports* half the total global heat flux from equator to pole, at least in the region equatorward of about 30° latitude.

The ocean also has a more subtle role in climate (we believe). It acts as a source and transporter of water vapour, important in the heat budget of the atmosphere; and provides salt nuclei for the formation of raindrops. It provides a 'memory' for previous atmospheric states. And it is a sink of carbon dioxide – important as the atmospheric content of CO_2 increases. The ocean contains a myriad of time scales – it probably has a continuum of them – ranging from seconds to thousands of years. Because 'climate' means so many things to different people (depending upon their particular interest), it is difficult to rule out any component of oceanic motion as being unimportant in the climatic regime.

But the ocean should not be regarded as a mere appendage of the atmosphere. In a fundamental sense, finding the climatic state of the sea is as important a problem as determining that of the atmosphere. The immediate impact upon human life may not be quite so obvious but, together with the atmosphere, the cryosphere, the hydrological cycle, etc., knowledge of how the ocean 'works'

† On leave from Department of Earth and Planetary Sciences and Department of Meteorology and Physical Oceanography, Massachusetts Institute of Technology, Cambridge, MA 02139, USA.

on the average is an important component of our understanding of the earth's climate as a whole.

Here I wish to summarise something of what is known of the ocean circulation, its role in climate and, more important, what is not known, or only guessed at.

A great deal *is* known about the ocean circulation; the papers by Reid (1981), Warren (1981), and Veronis (1981) conveniently summarise the observations (the first two), and theory (the latter). As the reader will see, the major difficulties we face are that much of this information is qualitative (we are not sure exactly what we have got right), and we are unable to put firm error bars on many of the quantitative statements that are made about the ocean. We are often unable to make real comparisons between theory and observation, so as to test critical ideas.

One cannot define precisely what is meant by the ocean circulation and ocean climate. Because the ocean is a non-linear physical system, its time-average properties are probably not physically realisable. We also do not know the extent to which the ocean fluctuates on time scales of years to decades and longer. In this chapter, I will operationally define the ocean circulation as encompassing all those motions exceeding spatial scales of order of the baroclinic Rossby radius of deformation (about 50 km), and with time scales exceeding a few days, but I will emphasise averages that might be supposed to apply over a period of a few years.

Most of the attention to the role of the ocean in climate has been addressed to sea-surface thermal anomalies and their advection by the general circulation (e.g. Woods, chapter 9 of this volume). The 'teleconnection' hypothesis, the Southern Oscillation, and the evident impact upon numerical models of the atmosphere of tropical anomalies in sea-surface temperature has led to much

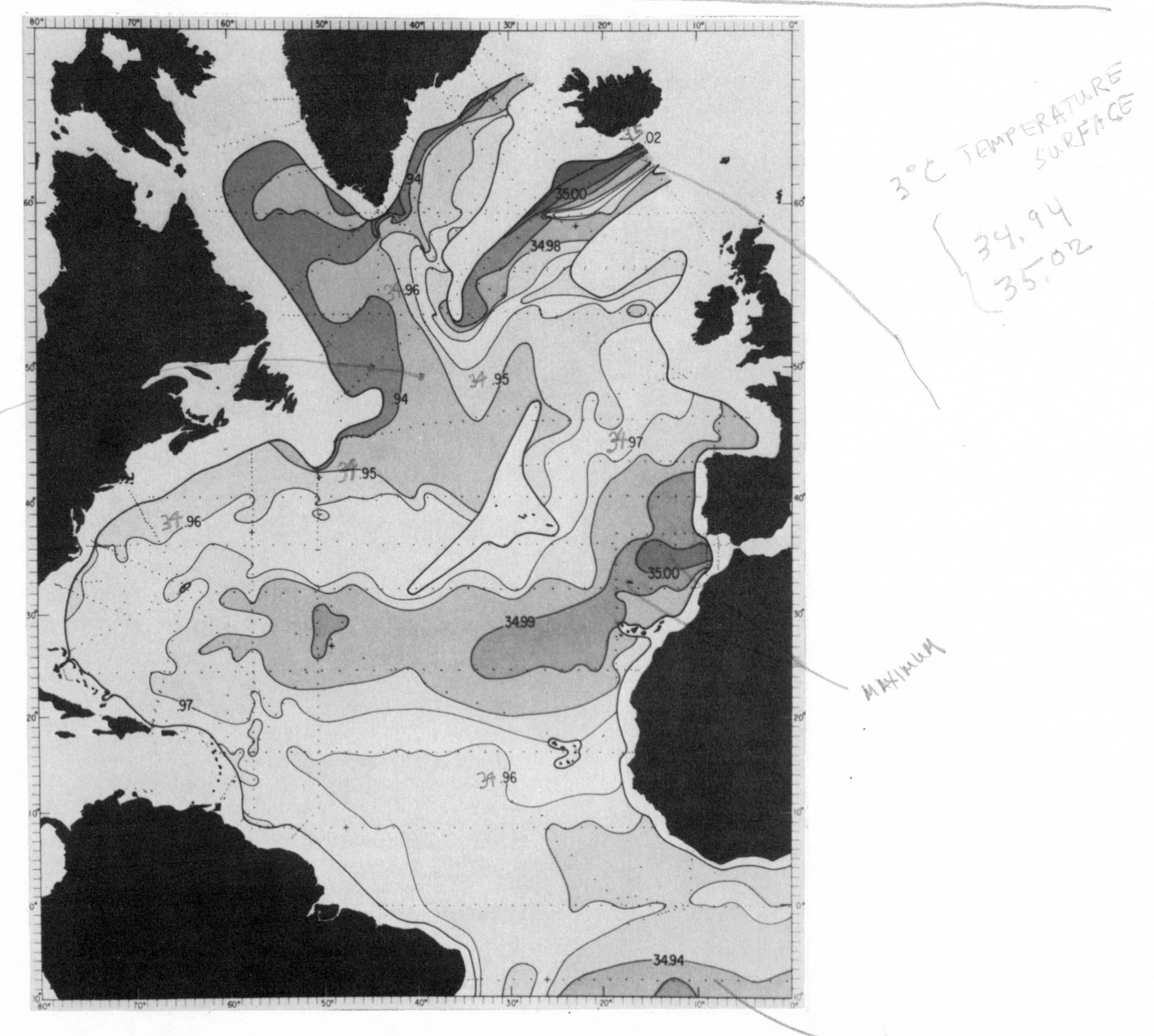

Fig. 10.1. Salinity on a temperature surface (3 °C) (from Worthington and Wright 1970). Note high salinity tongue emanating from the Mediterranean Sea on the east, and the (comparatively) low salinity tongue in the northwest emanating from the overflow regions. These gross properties are permanent features of the North Atlantic.

of this activity. Because this is the 'fashionable' role of the ocean in climate, and because it is the focus of chapter 9, I will devote this chapter instead to the problem of determining and understanding the climatic state of the ocean itself – not its superficial role in forcing the atmosphere. The job of oceanographers is to describe and understand the ocean, not merely to provide a set of black box boundary conditions to meteorologists.

10.2
The largest time and space scales

The ocean probably changes on all time and space scales (Wunsch 1981, Robinson 1983), although direct evidence for significant fluctuations on decadal and longer periods is slight. That the ocean does, however, contain elements of a non-zero time average flow is compelling. Consider Fig. 10.1, taken from

Fig. 10.2. (*a*) Worthington's (1981) compilation of the temperature–salinity classes in which the global water masses are found. Most of the world ocean lies within narrow limits. Upper panel is an expanded version of the cold water classes of lower panel. (*b*) Worthington's (1981) chart of adequate hydrography in the deep ocean (for purposes of constructing Fig. 10.2 (*a*)). (Black areas are not sampled properly.)

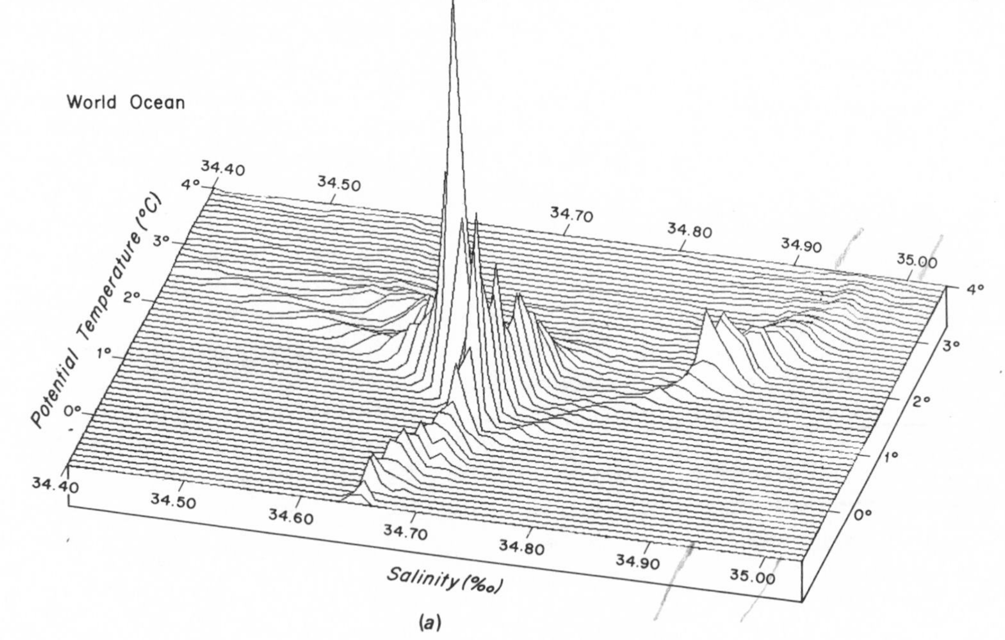

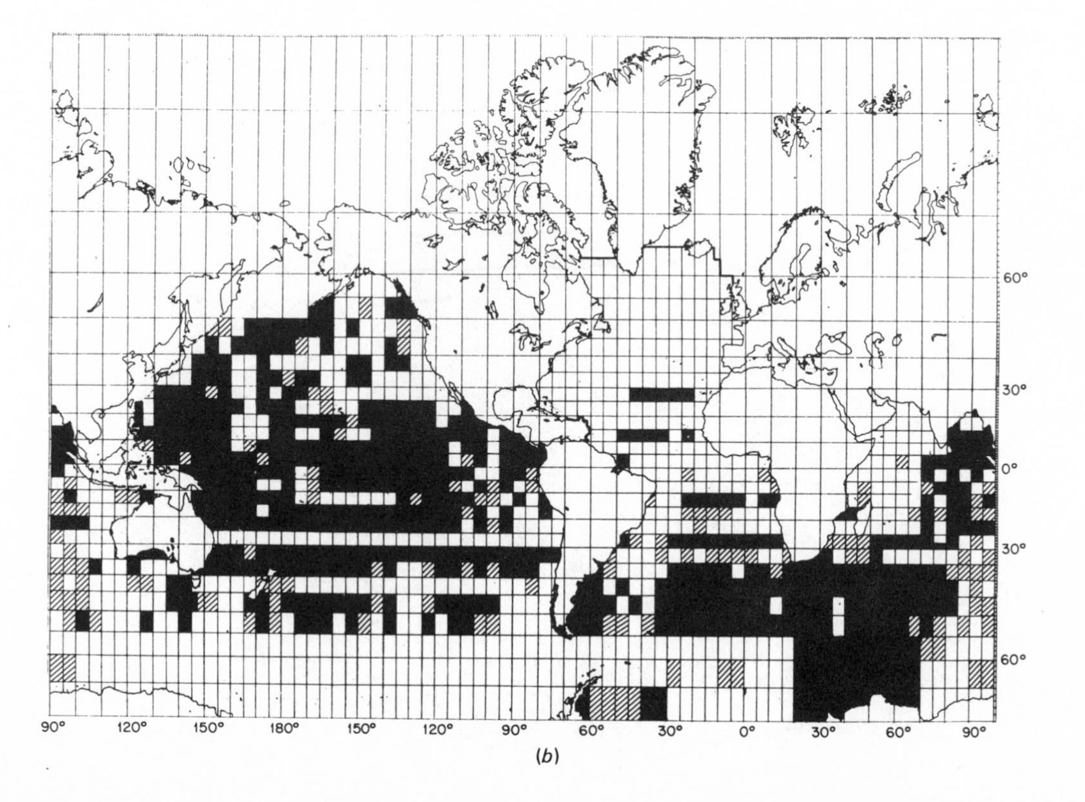

Worthington & Wright (1970). It shows the salinity on a surface of constant potential temperature in the North Atlantic. The eye sees the enormous maximum corresponding to the flow of highly saline water out of the Mediterranean Sea (the Mediterranean 'salt-tongue'). A secondary feature is the lower salinity 'tongue' corresponding to the fresher water moving down the north western boundary from the 'overflow' regions to the north. As far as we are able to tell (the data base extends backwards in time only about 100 years), these gross property distributions are permanent (although the detailed convolutions, etc., are not). No ship leaving New England to the south, crossing 35°N in the Atlantic, has ever failed to find the Gulf Stream, and we see a density structure in the Gulf Stream as determined by the Challenger (about 1873), indistinguishable, except in details, from that measured by a ship today. The great boundary currents of the world ocean appear to be permanent features of a large-scale time-average circulation, and the overall distribution of salt and heat also appears to be permanent, at least on a human time scale.

Fig. 10.2 is taken from Worthington (1981), and shows his estimates of the present volumetric temperature–salinity properties of the world ocean. Worthington makes explicit his belief that these *T–S* proportions will provide a future climatological 'baseline' against which future generations will be able to test the hypothesis that the oceanic climate has changed.

Fig. 10.3 from Dietrich *et al.* (1980) shows, schematically, the water masses of the North and South Atlantic in meridional section. Each water mass is defined by a particular range of temperature and salinity, seemingly also permanent features of the circulation and suggesting 'source' regions for each water type.

Figs. 1–3 attempt to give a feeling for the gross thermohaline structure of the world ocean. Details (but what exactly constitutes a 'detail' is not known) are unimportant.

10.3

Methods of deduction

10.3.1

The dynamic method

Oceanic water movements on time scales exceeding a pendulum day are nearly geostrophic both in theory (Phillips 1963, Pedlosky 1979), and in observation (Bryden 1977, Swallow 1977). Since Bjerknes and colleagues developed the dynamic method, the thermal–wind relationship has been used to make inferences about the large-scale oceanic circulation; Reid (1981) discusses some of the history of such attempts. There are two major issues in this use of the dynamic method. They are

(i) the lack of a reference level velocity (the level of no motion problem); and

(ii) the lack of synoptic data.

Usually, only the first problem is discussed. Typically, investigators have assumed a deep level of no motion and then contoured the circulation in the thermocline above. Fig. 10.4 is a not untypical shear profile in the open ocean. Because the shears in the upper 1000 m are much larger than in the lower ocean, vertical movement of the reference level, if it is kept anywhere between about 1500–3000 m, makes little change in the water *velocities* in and above the thermocline. Thus all discussions based upon deep reference levels of the circulation of the upper layers of the ocean resemble each other. Of course, a 1 cm/s difference in estimated velocity integrated over an area of ocean 100 km across and 1000 m deep results in a transport of 10^6 m^3/s (10^9 kg/s). If extended across an ocean basin, massive amounts of water are involved, with corresponding effects on the oceanic heat flux.

These massive transport uncertainties, owing to the uncertainty in the reference level, become paramount in discussing the deep circulation of the ocean. A reference level at the bottom gives an entirely different picture of the deep circulation than does one at 1500 m. Indeed, discussions of the deep circulation were comparatively rare (Defant 1941 is a notable exception) until Worthington (1976) made vivid the difficulty. Another exception is the work on deep western boundary currents; see Warren (1981).

Fig. 10.3. Schematic diagram (Dietrich, Kalle, Kraus and Siedler, 1980) of water masses of the Atlantic.

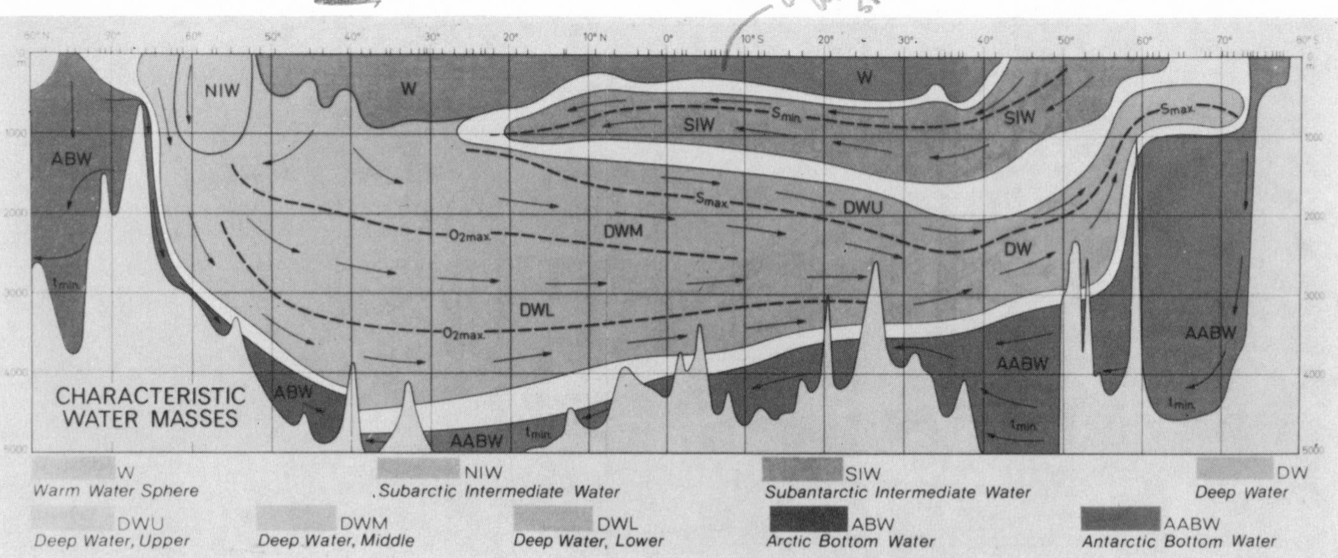

I have elsewhere (Wunsch 1982) discussed the issue of the lack of a non-synoptic data set. For want of synoptic data, oceanographers (myself among them) have used data obtained in different seasons of different years as though it were simultaneous and as though it represented 'average' conditions. We know that the ocean circulation is highly variable, at least on the mesoscale. Fig. 10.5 is the direct transport float measurement of Niiler & Richardson (1973), taken in the Florida Current (as replotted by R. Molinari). The Current appears to have extreme values between 19 and 39×10^9 kg/s. Assuming these are true extremes (they might not be), then which value of the transport is the appropriate one to combine with some particular hydrographic stations, perhaps made in 1956?

The mean of the measurements in Fig. 10.4, about 30×10^9 kg/s, is one of the most quoted 'known' quantities relevant to the ocean circulation. Fig. 10.5 may plant seeds of doubt in the reader's mind – the 'annual cycle' was determined from different seasons of different years, and the scatter about the mean is very large.

From the standpoint of conventional sampling theorems, we have only a potentially badly aliased representation of the ocean; this sampling would be regarded as intolerable by anyone used to dealing with ordinary time-series measurements of a fluctuating quantity. We still have the hope that the small wavenumber components of the circulation are largely time invariant – and there is always a Gulf Stream. But we cannot quantify that hope, and undoubtedly the low wavenumber components are not strictly steady; their change may be important in climate dynamics.

In a time-varying fluid, because of the non-linearity of the density conservation equation, the time-average circulation must be obtained (if it is of direct significance) as an average of instantaneous circulations, *not* from the time average of the hydrography. The absence of basin-scale synoptic data sets means that we cannot determine what the ocean circulation looks like at any given time, much less determine its average properties.

The dynamic method combined with our existing data base thus leads to an ambiguous picture of the ocean circulation. Figs. 10.6 and 10.7 display two very different interpretations of the ocean circulation. They *do* have many features in common, but they differ in many important details. It is unlikely that such ambiguities will be resolved with present techniques and data.

10.3.2

Core layer and related methods

Pictures such as Figs. 10.1 and 10.3 led Wüst (1935) to develop his 'core-layer' method for describing the ocean circulation. In essence, it amounts to saying that water masses move ('spread') away from their sources, and toward their sinks. 'Sources' are regions where the *T–S* properties are modified, usually through intense air–sea interactions. Thus in Fig. 10.1 we would argue that the salty Mediterranean water must, on the average, move from the source region (identifiable in part, as the eastern Mediterranean, a region of intense evaporation) into the North Atlantic as a whole. In the absence of a detailed budget, the rate of movement is unknown. Generally speaking, Wüst confined himself to discussions of directions rather than of rates. But there is little doubt that

Fig. 10.4. Profile of horizontal velocity from a station pair in the South Pacific Ocean. The zero is purely arbitrary, but note that vertical shear is comparatively small between about 1.5 and 3 km.

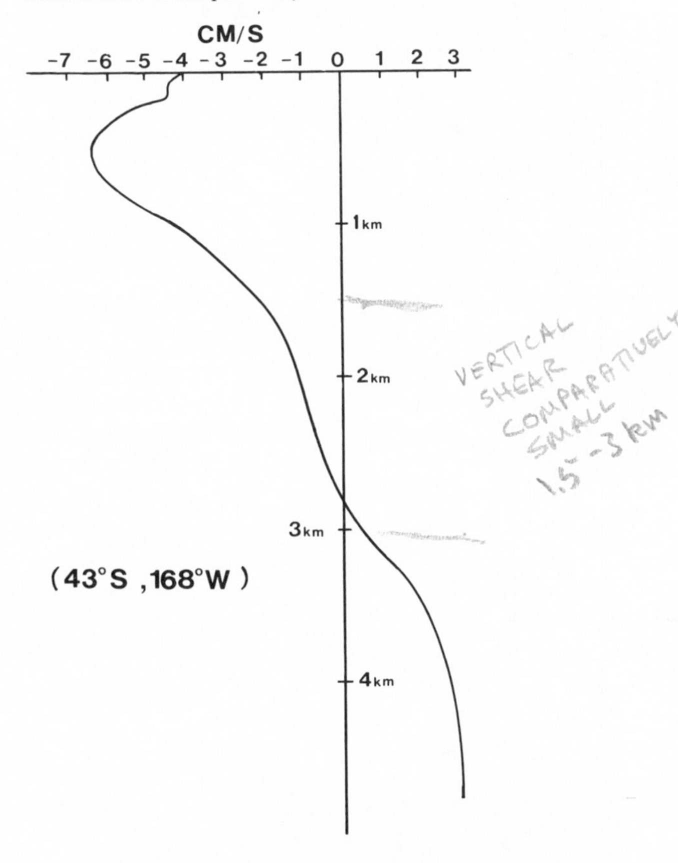

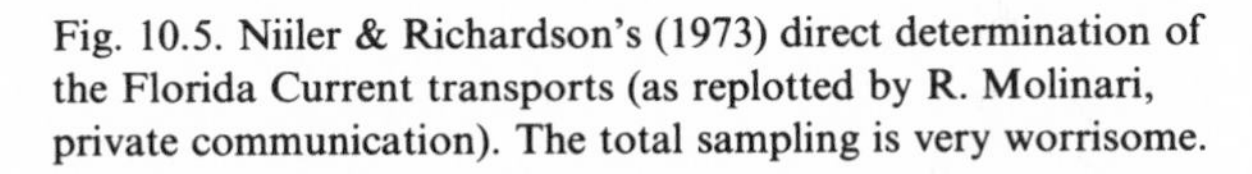
Fig. 10.5. Niiler & Richardson's (1973) direct determination of the Florida Current transports (as replotted by R. Molinari, private communication). The total sampling is very worrisome.

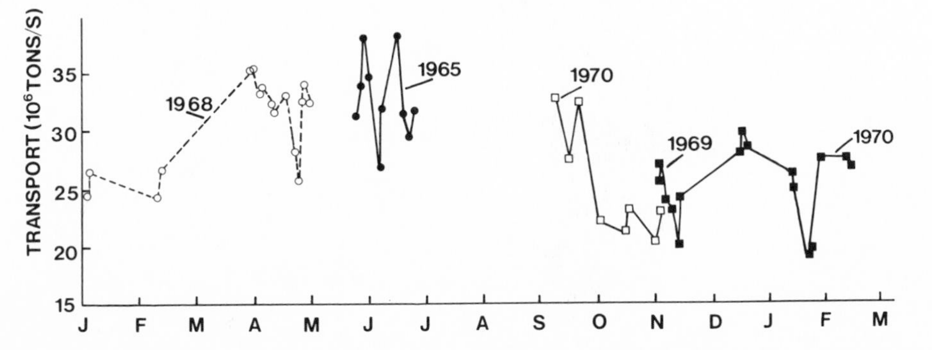

Fig. 10.6. Worthington's (1976) top-to-bottom estimate for the flow in the North Atlantic (this scheme does violate geostrophic balance over large parts of the ocean). (Units 10^6 tons/sec.)

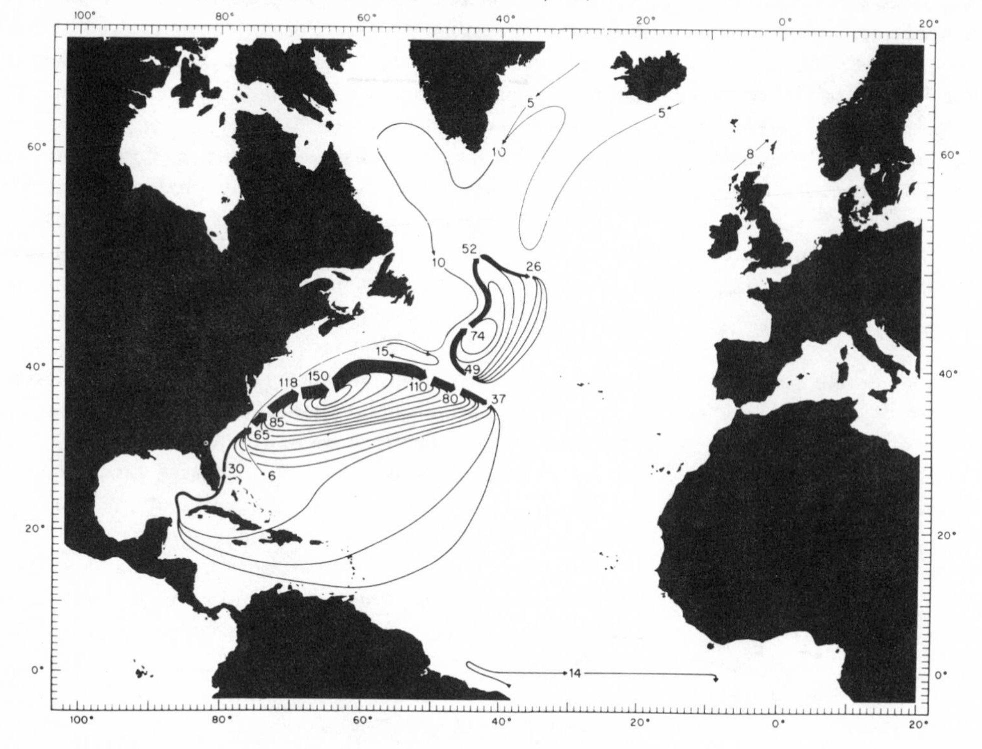

Fig. 10.7. Wunsch & Grant's (1982) estimate of the top-to-bottom flow in the North Atlantic (compare with Fig. 10.6). (Contour interval 10^7 tons/sec.)

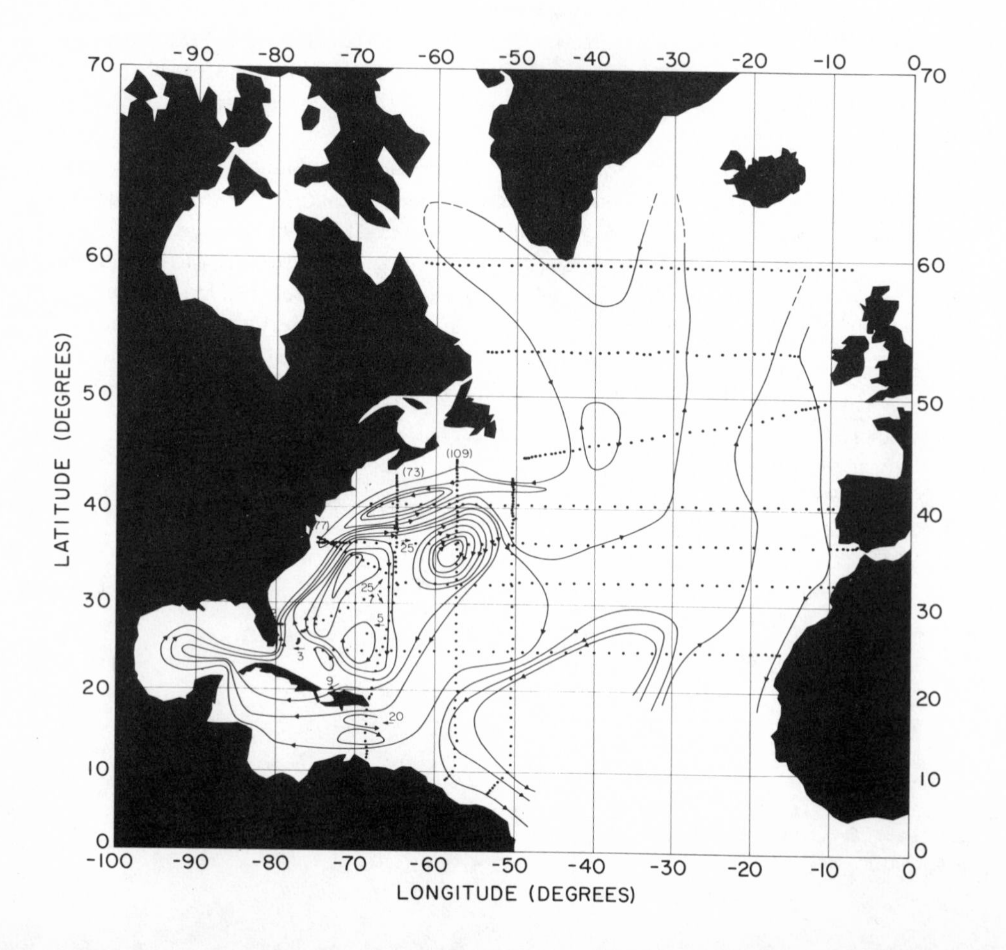

Fig. 10.3 does describe something real in the ocean circulation, if only because of thermodynamics (water masses must flow, on the average, from their sources to their sinks). Wüst (1935) used the term 'ausbreitung' (spreading) to describe what was happening – it is neither flow nor mixing; perhaps it is both. But, the rate of movement of the water can only be inferred if something is known quantitatively about the rate of water-mass formation – and here again we have only qualitative ideas. Yet these pictures are probably the firmest information we now have about the ocean circulation.

A consequence of the semi-qualitative nature of these pictures is that they have a somewhat 'geological' air to them. One speaks of 'layers', 'tongues', and 'cores', etc., as though they were geological strata, and the verbal descriptions that go with them are more in the nature of 'scenarios' than recognisable pieces of scientific deduction from first principles.

There have been attempts to combine core-layer methods with geostrophic calculations. This is a risky procedure because of the aliasing problem: a hydrographic section need not have been acquired at a time when the time-variable flow was moving in its time-average direction. Forcing it to do so may greatly distort reality (Wunsch 1982).

10.3.3 *The forcing functions*

The ocean is forced by the wind and by the thermodynamic coupling to the atmosphere (e.g. Kraus 1972). We make inferences about the ocean circulation from estimates of the wind stress, heat and moisture transfer, combined with our knowledge of dynamics.

Fig. 10.8 displays an analysis by Han & Lee (1981) of the global-wind field; this picture is based upon ship reports of surface winds. Previous such efforts have been published by Hellerman (1967), and for the North Atlantic alone, by Leetmaa & Bunker 1978). The number of reporting ships is extremely variable (Fig. 10.9), differing by orders of magnitude in different areas. It is very difficult to obtain quantitative error estimates for pictures such as Figure 10.8. In large parts of the world, there is a strong annual cycle in the wind field, rendering estimates of the mean even more uncertain. The ocean circulation as a whole depends directly upon the curl of the wind stress – involving derivatives of the wind field and hence greatly exaggerating the observation noise. Saunders (1976) has discussed some of the problems.

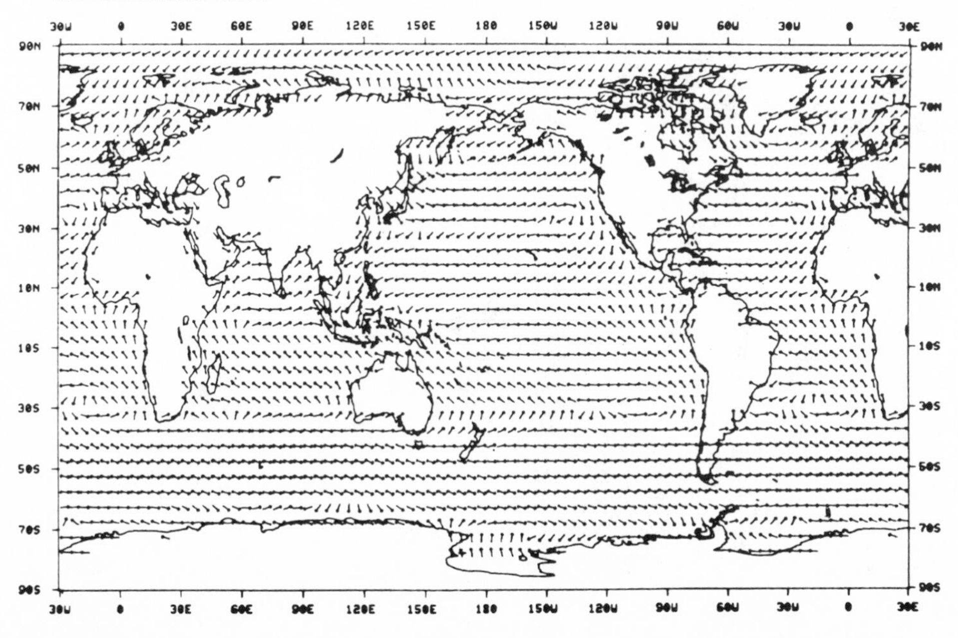

Fig. 10.8. Global wind field analysis (Han & Lee 1981). Vectors are annual mean stress.

Generally speaking, our knowledge of the wind field over the sea is inadequate; we are not even sure of the magnitude of the errors in what we now have.

There is strong thermodynamic coupling of the ocean and the atmosphere through the exchange of sensible and latent heat. Studies of the evaporation, precipitation and sensible heat exchange over the ocean have been published by Budyko (1963), Hastenrath & Lamb (1977), Bunker (1976), and others. Most of these rely upon bulk aerodynamic formulae and cannot be considered very reliable (Dobson *et al.* 1982 critically reviewed these methods). Fig. 10.10 displays the Budyko (1963) and the Bunker & Worthington (1976) estimates for the North Atlantic.

The net heat flux obtained from integrating Bunker's (1976) values disagrees considerably with that determined from direct oceanographic observations (Hall & Bryden 1982. Wunsch 1980, Roemmich 1980) and in the South Pacific, Hastenrath's (1980) bulk formulae estimates differ greatly from some of the oceanographic ones (Bennett 1978, Wunsch *et al.* 1982). Much is still uncertain.

10.4 Analytical understanding

There are some cornerstones of dynamical understanding of the circulation, that almost all models, irrespective of their additional complexities, take as given.

All circulation models start with the assumption that the time mean of the wind stress τ, drives a net water-mass motion to the right (left) of the stress vector in the northern (southern) hemisphere of magnitude

$$|\mathbf{V}| = |\tau|/f. \qquad (10.1)$$

Eq (10.1) comes from the vertical integral of Ekman's (1905) famous spiral and does not depend upon the details or magnitude of the eddy coefficient in the theory (see Stern 1975). What is the evidence for the validity of Eq. (10.1)? Ekman spirals have been clearly observed in the laboratory, but the most interesting aspect of their physics is their instability (Faller 1981). Hunkins (1966) observed a well-defined Ekman spiral under an ice sheet but, in general, spirals are not seen. The difficulties of deducing the relationship between τ and $\mathbf{V}$ may be inferred from the meteorological experience (Holton 1979, p. 111). But since Eq. (10.1) does not depend upon the details of the vertical stress distribution, one might hope that it would have application beyond the validity of a spiral. However, evidence for this true cornerstone of all ocean circulation theories is almost all indirect. The sub-tropical gyres, with warm water heaped up into the centres, do coincide roughly with the expected convergence zones of the wind stress; surface fronts occur where variations in stress suggest they should occur, and coastal upwelling exists where Ekman theory qualitatively suggests it should. But it must be admitted that there is no direct demonstration of Eq. (10.1) in the sense that most experimental scientists would prefer and the indirect tests are only qualitative, partly because τ and its time average are poorly known.

Over most of the ocean the linear vorticity balance

$$\beta v = f\frac{\partial w}{\partial z} \tag{10.2}$$

is believed to apply. Direct tests of Eq. (10.2) are very few; among them are Bryden's (1980) and Schott & Stommel's (1978). If

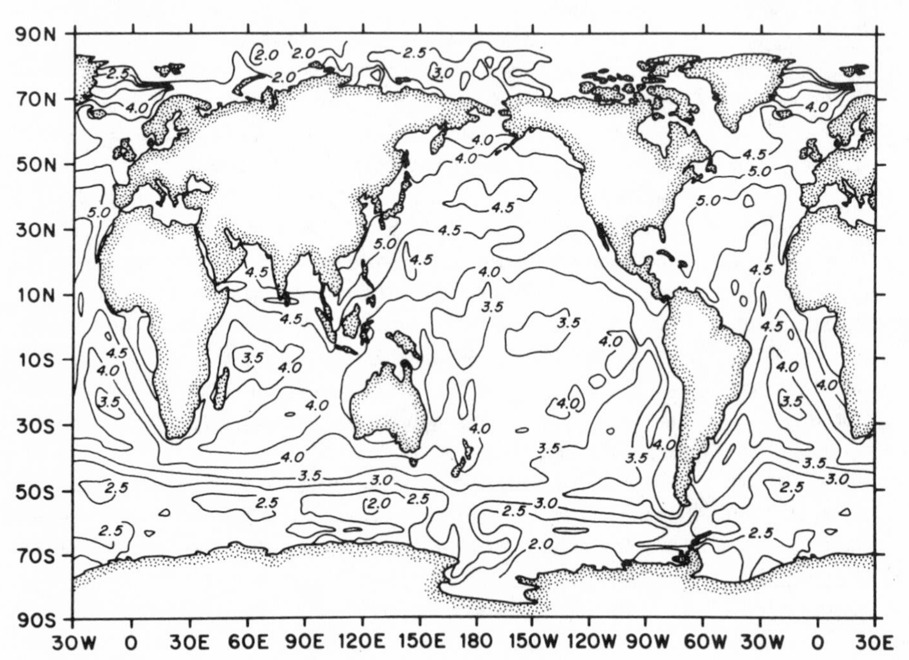

Fig. 10.9. Logarithm to the base 10 of number of ship reports used to construct Fig. 10.8 (Han & Lee 1981). Note enormous variation.

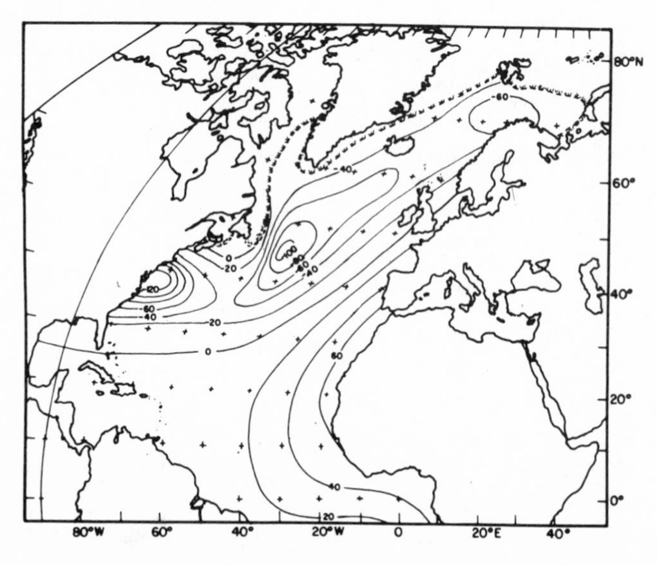

Fig. 10.10. Budyko's (1962) chart (left) of net heat gain in North Atlantic and that of Bunker & Worthington (1976) (right) in units of kcal/cm² per yr.

integrated from the base of the Ekman layer to some depth d' where v is supposed to vanish, we obtain the famous Sverdrup relationship

$$\beta V = f\mathbf{k} \cdot \nabla \times \tau \tag{10.3}$$

between the curl of the wind-stress and V, the meridional transport between the surface and d'. (Some authors have confused Eqs. (10.2) and (10.3), claiming them both to be the Sverdrup relationship. But Eq. (10.2) implies (10.3) only under very special circumstances). Eq. (10.3) has been almost an article of faith since Sverdrup (1947) first developed it. It was used, for example, by Bryan (1962) and Bennett (1978) to determine the oceanic heat flux from observational data. What are the tests that have been made of this relationship?

Leetmaa *et al.* (1977) compared the North Atlantic wind-stress curl estimates of Leetmaa & Bunker (1978) with the inferred geostrophic flow across 24°N in a hydrographic section obtained in 1958, and with the estimate of Niiler & Richardson (1973; see Fig. 10.5), of 30×10^9 kg/s flowing northward in the Florida Current.

But the simple relationship applies only to an ocean in which there is no Ekman divergence (or 'bottom torques') of any significance owing to the bottom slopes. But all the evidence (Ivers 1975, Wunsch & Grant 1982, Schmitz 1980) is that very substantial flows occur everywhere in the deep sea and the possible effects of permitting such flows is made apparent in the diagnostic model results of Holland & Hirschman (1972); see Holland (1973). Indeed, Bryden (1980) found in a test of Eq. (10.2) that in at least one location, it was dominated not by the contribution from $\nabla \times \tau$, but by the divergence of the bottom flow. Furthermore (Leetmaa *et al.* 1981, Meyers 1980), it appears that even Sverdrup's (1947) original comparison of wind and currents was a lucky accident. Undoubtedly, there is a relationship between the wind-stress curl and the meridional flux, but it is hardly likely to be as simple as Eq. (10.3) implies. Averages over large areas *may* obey Eq. (10.3), but this has never been demonstrated (see Welander 1981).

The third idea fundamental to ocean circulation theory is that westward intensification dominates the meridional flows of the sea. Patently, the Gulf Stream, Kuroshio, Agulhas Current and many bottom currents (Warren 1981) exist. But the detailed balances (friction – Stommel 1948, Munk 1950; inertia – Charney 1955, Morgan 1956) remain fuzzy and unquantitative. And what does one make of the extremely time-variable East Australian Current?

An example of the type of controversy that has occurred may be illuminating. Beginning with Munk (1950), it was noticed that Gulf Stream models and available wind-stress data did not produce a Gulf Stream transport as large as that 'observed'. There ensued discussions of the adequacy of the determination of drag coefficients, the ability to determine the wind-stress curl, etc. It is now generally accepted that the apparent high Gulf Stream transports arise from the intense recirculations occurring on both sides of the Stream, flows not directly related to the overall sub-tropical gyre; the existing wind-stress data do seem to give the right order of magnitude of the net Gulf Stream 'throughput', but the comparisons remain qualitative because of the absence of good estimates of τ, the sampling problem already described and the great time-varying complexity of the western boundary current regions.

In the early 1960s, it appeared that an acceptable zero-order understanding of the Gulf Stream had been obtained; attention began to turn from the basic existence of westward intensification to the meandering and breakdown of the Stream as it reentered the interior ocean. Moore's (1963) paper appeared to be the first in a series of efforts that would elucidate this process, with considerable attention being paid to the bottom control of meandering proposed by Warren (1963). Twenty years later, on the heels of the direct measurements of Luyten (1977), Hendry (1982) and Schmitz (1978), showing complicated small-scale time-average features, the discovery (or rediscovery) of the intense recirculations (Worthington 1976, Wunsch 1978, Reid 1981), the very intense associated eddy field (Schmitz 1978), it is reasonable to conclude that we have no theory of the Gulf Stream east of Cape Hatteras. (We do have numerical models which appear superficially realistic – but we don't really know if they have got the basic physics right.)

Perhaps the greatest flowering of analytical modelling took place in the late 1950s, with the production of the non-linear similarity solutions for the 'thermocline' (Robinson & Stommel 1959, Welander 1959). These are connected to a fourth fundamental notion – that the thermohaline circulation as we observe it is a subtle interplay of wind driving and thermal forcing. Sinking of cold water occurs in comparatively small, isolated regions of the world ocean (Stommel 1962, and compare Fig. 10.3), as convective 'chimneys' and must return vertically elsewhere. In the sub-tropical gyres, the diffusive balance theories yield the thermocline as a boundary layer sandwiched between the downwelling warm water driven by the Ekman convergence and the upwelling cold water from below. Apart from the existence of the convective chimneys (Killworth 1982), and the requisite deep western boundary currents (Warren 1981), a demonstration of the overall scheme has proved elusive. Munk (1966) found that the vertical profiles of chemical properties in the deep ocean were consistent with this picture, although he called his results 'recipes' in recognition that they were not purely deductive. In a schematic form, these ideas led Stommel & Arons (1960) to the only extant global semi-analytical picture of the complete circulation (see Stommel 1966). Although efforts have continued (see especially Welander 1971), these analytical models have not yet proved especially fruitful: it became clear that the similarity solutions would never satisfy realistic boundary conditions, the role of vertical mixing proved very difficult to elucidate from the solutions themselves, and the discovery of the intense eddy field cast doubt upon the entire assumption that a steady ocean circulation model could realistically model a time-average non-linear ocean.

10.5 Numerical models

Until the early 1970s, numerical models of the ocean circulation tended to be extensions of the laminar analytical ones. These models have been reviewed by Veronis (1973, 1981), Holland (1979), and others. The models range from those intended to study a particular simplified physics (e.g. the effects of non-linearity upon the Gulf Stream, as in Bryan 1963), to complete global, multi-layer thermodynamic models forced by the best estimates of global available wind, e.g. Bryan & Cox (1968), Bryan & Lewis (1979). These latter models, whether truly global or confined to a single ocean basin, are, in turn, of two types, usually denoted as 'prognostic' and 'diagnostic' respectively. Prognostic models are computed using only imposed boundary conditions (wind stress and heat flux), with the oceanic interior permitted to adjust as it wishes given the boundary conditions and governing physics. Diagnostic models take the observed, three-dimensional density field of the ocean and compute the resulting flow field. Examples are Sarkisiyan (1977), Holland & Hirschman (1972), and Sarmiento & Bryan (1982).

The physics underlying these models is the same as in the analytical ones – Ekman and Sverdrup dynamics, westward intensification, friction of various types, wind and thermal driving, etc. They do permit the exploration of complex geometrical situations (e.g. bottom topography) that the analytical models cannot deal with. Some components of the models, specifically the parameterisation of friction in the form of eddy coefficients, is clearly artificial. The resemblance between the flows computed in the basin and larger-scale models, and that observed, is reassurance that the underlying physics is qualitatively correct. The degree to which it is quantitatively correct is much less clear, and in some cases it is known to be quantitatively wrong. We have not yet reached the stage where numerical prediction of a specific current would carry much weight in plans to measure oceanic hydrography.

With the demonstration (The MODE Group 1978) of the existence of an energetic field of mesoscale variability in the ocean, and the specific supposition that the variability might play a crucial role in determining the ocean circulation, a number of so-called eddy resolving general circulation models (EGCMs) was developed (see Rhines & Holland 1979, and Harrison 1980). These models had adequate spatial resolution to begin to resolve mesoscale variability which manifests itself as a strong time dependence. Because of the required spatial resolution, none of these models as yet occupies even a full ocean basin. Rather, they are restricted to substantial fractions of one or two oceanic gyres (e.g. Semtner & Holland 1978). All these models have been prognostic.

The EGCMs have led to a shift in outlook – they made clear that the general circulation has to be obtained from the average of many realisations of the instantaneous time-dependent flow field. Unless some as yet simple parameterisation of the eddy field is found, the relationship between steady models of the ocean circulation, and the time average of time-dependent models is going to remain remote, and perhaps unbridgeable.

Schmitz & Holland (1982) have discussed the relationship between the results of one EGCM and observations. A safe summary of that work seems to be that there are many points of similarity, but also many remaining differences. The rate of progress is such that one anticipates that within a decade there should exist models, occupying full ocean basins, which successfully mimic many of the features of the ocean circulation. At some point one will encounter both the inadequacy of knowledge of the forcing functions (wind, heating) since even a perfect model, if imperfectly forced, will yield imperfect results, and the paucity of adequate observations to test such models.

Diagnostic models are interesting because they can be thought of as maximising the deductive power of real data sets (see Davis 1978). Perhaps the highest form of this type is the meteorologists 'assimilation' and 'initialisation' (of the global weather data sets twice per day) into complex atmospheric models (see Bengtsson, Ghil & Källén 1981). The chief criticism that one can level against the oceanographic diagnostic models is that they have paid little or no attention to the realities of data – all data contain errors, systematic and random. Many of the complicated features observed in models are undoubtedly owing to slight inconsistencies in the data being fed into non-linear equations of motion. The results may be misleading, but one has no way of knowing which parts of the model are believable (if any) and which are not. Some possible ways out of this dilemma are discussed by Sarmiento & Bryan (1982), and Wunsch & Minster (1982).

10.6 The heat flux problem

The meridional flux of heat by the ocean is often separated out for special discussion because of its obvious relationship to the climate problem. Much of the recent attention to the role of the ocean in climate comes from the work of Oort & Vonder Haar (1976), who concluded that at low latitudes the ocean was carrying at least half the poleward flux of heat. Bryan (1982) and Hall & Bryden (1982) have reviewed aspects of the problem and we will thus confine ourselves to some general remarks.

Generally speaking, the flux of heat (or any other property carried by the ocean) can be written as the sum of four terms A_i, $i = 1,4$ as follows: the component of the flux owing to

$i = 1$, the thermal wind relative to some reference level;
$i = 2$, the (non-zero) reference level velocity;
$i = 3$, non-geostrophic (e.g. Ekman) fluxes;
$i = 4$, undetermined components.

Wunsch, Hu & Grant (1983) have discussed the problems with each of these terms. In A_1, if we wish to interpret the result as a mean flux, how can we determine the error bar owing to having used only a single section obtained in one month of one year?

In A_2 if, for example, the reference level velocity has been determined (say) from the Sverdrup relationship, which wind field should we use (the instantaneous one, the time average, the average for that season)? What is the error bar?

For A_3, we have the same questions about the wind, and in addition – over what depth should we distribute the flux?

Term A_4 represents what inverse methods denote by the null-space. If the reference level velocity has been determined instead by the Sverdrup relationship, it represents all components of flux yielding no net meridional flux, but correlated with the thermal field. How big are these components?

Terms A represent the direct calculation of heat transport by the ocean. Other methods exist – the direct calculation through bulk formulae already mentioned above of the air–sea–heat transfer at the interface, and the meteorological residual calculation. In some cases, we seem to have a direct conflict between the different methods. For example, in the southern hemisphere, Hastenrath (1980) and Trenberth (1980) find a net poleward flux of heat by the ocean; but direct estimates in the ocean (Bennett 1978, Fu 1981, Wunsch, Hu & Grant 1983) suggest that the net oceanic flux of heat may be equatorward. Until these different methods are brought into reconciliation, there will be little assurance that we really understand the role of the ocean in climate, much less whether that role is changing.

The flux of heat in the ocean is intimately connected

(Stommel & Csanady 1980) with the flux of water vapour in the ocean and atmosphere. Consider a zonal section across an ocean basin closed in the poleward direction. Then the mass flux across the section can be written

$$\iint \rho v \, \mathrm{d}x \, \mathrm{d}z = -F, \tag{10.4}$$

where v is the area total ocean water velocity, and F is the net rate of precipitation (or evaporation if negative) poleward of the section. Because the atmosphere provides no significant pathway for salt, we have for the salt flux

$$\iint \rho v S \, \mathrm{d}x \, \mathrm{d}z = 0. \tag{10.5}$$

We can write the heat flux H across the section as

$$c_\mathrm{p} \iint \rho v T \, \mathrm{d}x \, \mathrm{d}z = H, \tag{10.6}$$

where c_p is the heat capacity. Let S_0 be the mean salinity of the section. Then multiplying Eq. (10.4) by S_0 and subtracting from Eq. (10.5), we have

$$\iint \rho v S' \, \mathrm{d}x \, \mathrm{d}z = F S_0, \quad S' = S - S_0. \tag{10.7}$$

We have seen (e.g. Fig. 10.2) that there is a relationship between the temperature and salinity of the oceanic water masses. If F should vanish, and if the T–S relationship were linear (it is often nearly so), then the heat flux, Eq. (10.6), would also have to vanish. More generally (see Stommel & Csanady 1980) there is a very close connection between our ability to quantify F and our ability to estimate oceanic heat fluxes. It may be optimistic to assert that we know F to an order of magnitude in the global oceans.

Even though heat and moisture are transferred from ocean to atmosphere only at the sea surface, it does not mean that an understanding of the upper ocean alone will suffice to understand what the ocean is doing. Heat and fresh water fluxes can only be computed in closed systems, i.e. mass conserving systems, and these involve the entire ocean circulation from top to bottom, Because of the linkages between temperature, salinity, heat and water vapour fluxes, and the varying dynamical elements of the oceanic general circulation, the problem must be viewed as an entirety – i.e. one needs to determine and understand the general circulation of the ocean, its relationship to water-mass 'conversion' (modification of T–S relationships), mixing rates, forcing and back-coupling to the atmosphere; it is dangerous to assert that any element can be ignored safely.

10.7 What needs to be done

The picture we have painted of knowledge of the ocean circulation is somewhat bleak. But the definition of the problem helps a great deal toward finding a solution.

It seems clear that there are certain fundamental issues we now need to face before we can even begin to quantitatively discuss the role of the ocean in climate.

We understand many of the specific elements that must make up the ocean circulation – vorticity dynamics, mixing processes, structure of boundary layers, and the like. What we lack is an understanding of how these disparate and competing physical processes fit together. The major reason for this lack of understanding is that our observation base does not permit a quantitative sorting out of various possibilities. The models, both analytical and numerical, are quite sophisticated enough (Veronis, 1981) to carry most of the relevant physics. But over the past decade one can see a decline in attention to the large-scale ocean circulation (a vigorous field in the 1950s and early 1960s), at least partly because the models became increasingly isolated from the physics that oceanographers in the field could test.

The ocean circulation is a large-scale phenomenon – forced by global winds and global thermodynamic processes. Until we know the global-wind field and the air–sea transfer processes, even hypothetically 'perfect' models (with infinite resolution and all possible physics) will produce incorrect results.

Until we are able to observe the ocean on the large scale synoptically, there will be little hope of quantitatively defining what the ocean circulation *is*, much less of explaining its causes and effects.

These comments are in some sense trivial and obvious – that until we can observe a physical system we cannot hope to describe or understand it. Obvious as this is, it has tended to be obscured and not discussed, as oceanographers made what they could of the data they did have available. What can one do about it?

Three major problems seem to stand out: (i) to understand the degree of time aliasing our existing and future conventional observations; (ii) to find observables that represent true temporal averages; and (iii) to find methods to observe the ocean on basin-wide and global scales.

10.7.1 *(i) The aliasing problem*

This is perhaps the easiest problem. Something as simple as a sensible (as described and prescribed below) hydrographic sampling program could settle a number of outstanding issues – Is the seasonal cycle of baroclinicity confined to upper levels at high latitudes? Is there significant interannual variability at all latitudes? If sustained over a few years, do direct transport measurement programmes such as that of Niiler & Richardson (1973) show a stable, unchanging annual mean and cycle? Are we entitled to assume that small wavenumber components of the flow are unchanging? These are all fundamental sampling issues that can be addressed on a regional basis. In a few years, working in a few carefully chosen areas, we could understand the extent to which this represents a serious issue.

10.7.2 *(ii) What represents the time average?*

We know we can observe some quantities assuredly representing the time-average ocean circulation. These are the large-scale property and chemical distributions described above. We may not yet understand what maintains these distributions, or by what complex hydrodynamical paths they achieved their forms, nor even precisely which details are permanent; nonetheless, they represent a datum that any model purporting to describe the time-average circulation must reproduce. It is difficult to make that statement about any other observable. Yet we still cannot make global maps of oxygen, silicon, tritium, or anything besides temperature and salinity, and even there (see Fig. 10.2) major gaps exist.

Chemical tracers are widely regarded as the best hope for understanding the time-average ocean circulation. A truly passive tracer (e.g. tritium) satisfies an equation like

$$\frac{\partial C}{\partial t} + \mathbf{v}\cdot\nabla C = \frac{\partial}{\partial z}\left(K_\mathrm{v}\frac{\partial C}{\partial z}\right) + \frac{\partial}{\partial x}\left(K_\mathrm{h}\frac{\partial C}{\partial x}\right) + \frac{\partial}{\partial y}\left(K_\mathrm{h}\frac{\partial C}{\partial y}\right) - Q, \tag{10.8}$$

where Q involves the sources (which may be a function of time) and sinks (radioactive decay, chemical biological processes for some tracers) and boundary conditions. If enough is known about Q, $\mathbf{v}$, K_h, K_v we can integrate

$$C(x,y,z,t) = \int_{-\infty}^{t} \mathrm{d}t' \iiint_{\text{ocean}} G(\mathbf{x}-\mathbf{x}', t-t') \times Q(\mathbf{x}', t')\, \mathrm{d}t'\, \mathrm{d}\mathbf{x}', \quad (10.9)$$

where G is the appropriate causal Green's function, which results in some observed distribution $C(x,y,z,t)$. The observed distribution is a result of an integral from the infinite past, and our observations of C are a measure of the entire history of the trajectory and diffusion of C and do not represent (as does a current metre measurement) a purely instantaneous snapshot. This has long been recognised – tracers are telling us about the history of the ocean circulation, something intimately related to a weighted time-average flow. But the weighting implied by the integral, Eq. (10.9), is a complex one and we have a difficult inverse problem to determine $\mathbf{v}$, K_h, K_v.

We do not as yet have a general theory for the use of even passive, stable, conserved oceanic tracers; most studies are based on *ad hoc* simplifications; it often becomes difficult to understand what the deductions are sensitive to – whether it is the model used (i.e. the simplication of Eq. (10.8) or Eq. (10.9)), or the data (i.e. $C(\mathbf{x}, t)$, or the boundary conditions (i.e. $Q(\mathbf{x}, t)$).

Much of the difficulty is that the tracer data base is so sparse. Broecker (1981) discusses the tracers determined during the Geosecs – the most modern geochemical study available of the sea. The stations are widely scattered. Reid (1981) succeeded in mapping oxygen and silica at certain depths in the world ocean, and others (e.g. Needell 1980) have produced similar maps. The difficulties in working with inadequate data, however, have produced the observation that chemical oceanographers made one-dimensional models when they had isolated station data; with Geosecs they obtained some sections so they discovered two-dimensional models. The implication is that one day they will discover the virtues of three-dimensional modeling.

What the chemists have done is perfectly defensible. One works with the data one can obtain – keeping in mind its limitations. But sometimes the limitations have been forgotten; and one-, two-, and three-dimensional tracer models tend to give qualitatively different pictures of what the ocean is doing. There is no substitute for adequate data.

10.7.3
(iii) How can one obtain a global view?

This is a subject discussed at some length by Munk & Wunsch (1982). Satellites and acoustic techniques hold great promise for giving a global (in the former case) and basin-scale (in the latter) data set that would very strongly constrain any sensible ocean model. Other important possibilities exist too, perhaps involving the large-scale deployment of large numbers of floats and drifters tracked by satellite system.

Satellite measurements of many of the forcing functions, on a global scale, are feasible in the next few years. Stewart (1983) has reviewed the possibilities; wind stress is the most promising measurement. Because the bulk formulae for heat and moisture transfer to the atmosphere depend critically upon the accuracy of the wind measurement, all estimates of oceanic forcing by the atmosphere would greatly improve the situation.

Altimetric measurements can be made globally, too. Here one determines directly one of the dynamical boundary conditions (the pressure field) governing the general circulation. Munk & Wunsch (1982) have described how the wind and altimetric measurement could be coupled together to make inferences about the interior ocean.

Acoustic tomography is another large-scale observational system described by Munk & Wunsch (1982). Here one uses acoustic signals to determine large-scale averages of important dynamical quantities (heat content, vorticity, vertical velocity,...). Because of the geometric growth of information with instrument numbers, comparatively modest numbers of instruments provide very large numbers of integral constraints upon the ocean circulation.

The final component of any global view of the ocean is the development of suitable models to integrate the measurements with known kinematics and dynamics. Despite the somewhat bleak existing state of observations, we know a good deal about the fluid behaviour that permits one to relate local and regional measurements to large-scale behaviour. For example, the simple fact that the ocean conserves mass is an important global kinematical constraint permitting us to relate diverse measurements in diverse locations. Much more interesting and subtle relationships are possible through numerical models. Inevitably, any global view of the ocean will be obtained through model making in which the global observations are 'assimilated' into the models.

10.8
A global ocean circulation programme

Our major *in situ* data base is going to continue to depend on shipboard surveys; we have found no substitute for the ship for hydrographic and chemical sampling. New methods exist for handling these types of data, and further ones are under study. But we very badly need a new global survey, on the scope of that conducted during the International Geophysical Year in the Atlantic (Fuglister 1960, Dietrich 1969).

The immediate climatic interaction of ocean and atmosphere (our primary concern here) manifests itself partially in the addition to, and extraction from, the ocean of heat and moisture. This interaction is visible in the ocean directly as convection (e.g. the MEDOC Group 1978, Killworth 1982) or through the presence of modified water masses (e.g. the North Atlantic 18° water and other 'mode-waters' (Worthington 1959, Warren 1972, McCartney 1977). Oceanographers have devoted much attention to the study of mixing processes (see Garrett 1980, for a review). This work involves detailed measurements of quantities such as local Richardson numbers, velocity and temperature microstructure, etc. These studies are important for complete understanding where, why and how the ocean mixes.

But for understanding the rate of oceanic mixing, and its climatic consequences on a global scale, an indirect approach is necessary.

Consider the zonally integrated North Atlantic flow shown in Fig. 10.11. It shows that about 15×10^9 kg/s of warm salty water is flowing poleward above about 5 °C and is returning equatorward, colder and fresher, below that temperature. Because the North

Atlantic is nearly closed off to the north, we can infer that the northward-moving warm salty water must be converted to the southward-moving cold fresh water. Whatever mixing process is actually occurring poleward of 24°N, it must produce the observed balance. In fact Wunsch (1980), Roemmich (1980), and Bryden & Hall (1980), inferred from the results shown in Fig. 10.11 that poleward of 24°N, about 10^{15}W was being lost to the atmosphere.

The corresponding South Pacific flow, also shown in Fig. 10.11, carries much less heat than that of the North Atlantic – consistent with independent inferences about deep water formation in the Pacific. (Why are these oceans so different?)

There are two points to be made here: first, to determine the rate of air–sea heat transfer we do not need to actually understand in detail how it occurs (much as we would like to do so). The second point is that the inference made about the 24°N section (and at other sections where similar calculations have been made) involves examination of the deep ocean just as much as it does the upper ocean. One cannot close and compute heat budgets (or water vapour, etc.) *without examining a closed, i.e. mass-conserving, system*. One cannot *understand* the role of the ocean in climate, even on the short time-scales, simply by focussing upon the upper several hundred meters of the water column, although one might conceivably determine its effects that way.

The continuing use of the old Meteor stations (Wüst 1935), the IGY sections, and the Scorpio sections (Stommel *et al.* 1973) coupled with new methods like inverse techniques, show what is required: high quality CTD measurements, on near-mesoscale resolution, taken from top to bottom and continent to continent. The last two requirements may not be obvious – they permit us to do a number of things not otherwise possible. Land-to-land, top-to-bottom sections permit use of mass and other conservation constraints – important kinematic requirements. They permit the calculation of zonal and meridional integrals of quantities of direct climatic interest like heat and salt fluxes which require overall mass balance. The fragmentary ('dangling') sections often made are much less useful than the trans-oceanic ones.

Many chemical tracers can be obtained in the process of doing hydrography (some cannot – carbon-14 requires large water bottle samples which interfere with hydrographic sampling). Conventional nutrients, tritium, probably freon, etc., ought to be obtained globally and on the same scales as the hydrographic measurements. The availability of such chemical data will not only strongly constrain the ocean circulation and provide tests of our understanding of the circulation, but will shed much light on the chemical–biological processes governing these distributions.

The introduction of these global baseline surveys into the rapidly evolving and improving circulation models of the ocean, along with the near-global satellite and acoustic data sets described by Munk & Wunsch (1982) would greatly improve knowledge of all aspects of the ocean circulation, including its role in the climate system.

Fig. 10.11. Zonally integrated meridional flows in the South Pacific (left) from Wunsch, Hu & Grant (1983) and in North Atlantic (right) from Wunsch & Grant (1982). North Atlantic shows a single overturning cell, carrying much heat; South Pacific contains a multi-cellular structure carrying little heat.

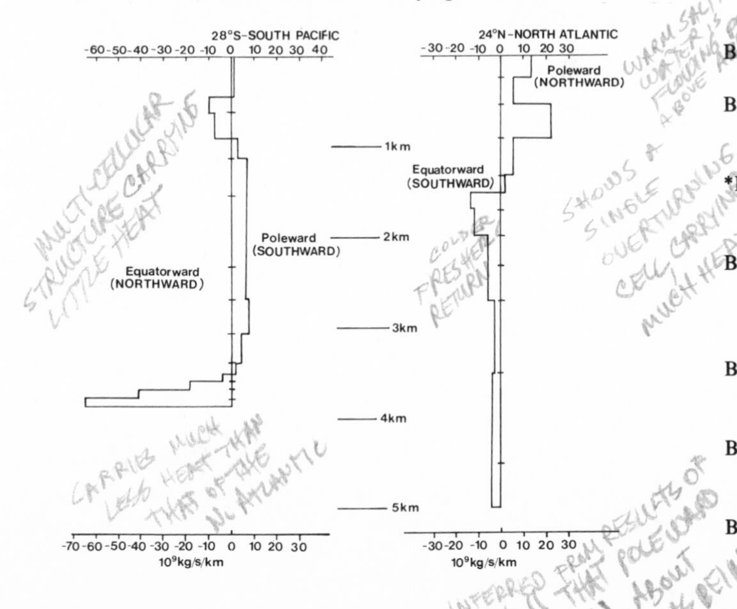

Acknowledgement

The hospitality of the Department of Applied Mathematics and Theoretical Physics, University of Cambridge, where this essay was prepared is gratefully acknowledged. I was supported in part, 1981–2, by funds from the John Simon Guggenheim Foundation, the Fulbright Programme, and the UK Natural Environmental Research Council.

References

*Recommended review article for additional background.

Barnett, T. P. (1981). 'Statistical relations between ocean/atmosphere fluctuations in the tropical Pacific'. *Journal of Physical Oceanography*, **11**, 1043–58.

Bengtsson, L. M. Ghil & E. Källén (1981). *Dynamic Meteorology: Data Assimilation Methods*. Springer Verlag, New York, 330 pp.

Bennett, A. F. (1978). 'Poleward heat fluxes in southern hemisphere oceans'. *Journal of Physical Oceanography*, **8**, 785–98.

Broecker, W. S. (1979). 'A revised estimate for the radiocarbon age of North Atlantic deep water'. *Journal of Geophysical Research*, **87**, 3218–26.

*Broecker, W. S. (1981). 'Geochemical tracers and ocean circulation'. *In: Evolution of Physical Oceanography, Scientific Surveys in Honor of Henry Stommel*, B. A. Warren & C. Wunsch, eds. pp. 434–60. The MIT Press, Cambridge.

Bryan, K. (1962). 'Measurements of meridional heat transport by ocean currents'. *Journal of Geophysical Research*, **67**, 3403–14.

Bryan, K. (1963). 'A numerical investigation of a non-linear model of a wind-driven ocean'. *Journal of the Atmospheric Sciences*, **20**, 594–606.

*Bryan, K. (1982). 'Poleward heat transport by the ocean: observations and models'. *Annual Review of Earth and Planetary Sciences*, **10**, 15–38.

Bryan, K. & M. D. Cox (1968). 'A non-linear model of an ocean driven by wind and differential heating: Part 1, Description of the three-dimensional velocity and density fields'. *Journal of the Atmospheric Sciences*, **25**, 945–67.

Bryan, K. & L. J. Lewis (1979). 'A water mass model of the world ocean: observations and models'. *Journal of Geophysical Research*, **84**, 2503–17.

Bryden, H. (1977). 'Geostrophic comparisons from moored measurements of current and temperature during the Mid-Ocean Dynamics Experiment'. *Deep-Sea Research*, **24**, 667–81.

Bryden, H. (1980). 'Geostrophic vorticity balance in midocean'. *Journal of Geophysical Research*, **95**, 2825–8.

Bryden, H. & M. M. Hall (1980). 'Heat transport by currents across 25°N latitude in the Atlantic Ocean'. *Science*, **207**, 884–6.

Budyko, M. I. (1963). *Atlas of the Heat Balance of the Earth* (in Russian). Globnaia Geofiz. Observ., 69 pp.

Bunker, A. F. (1976). 'Computations of surface energy flux and annual air–sea interaction cycles of the North Atlantic Ocean'. *Monthly Weather Review*, **104**, 1122–40.

Bunker, A. F. & L. V. Worthington (1976). 'Energy exchange charts of the North Atlantic Ocean'. *Bulletin of the American Meteorological Society*, **57**, 670–8.

Charney, J. G. (1955). 'The Gulf Stream as an inertial boundary layer'. *Proceedings of the National Academy of Science of the USA*, **41**, 731–40.

Davis, R. E. (1978). 'Techniques for statistical analysis and prediction of geophysical fluid systems'. *Geophysical and Astrophysical Fluid Dynamics*, **8**, 245–77.

Defant, A. (1941). 'Die absolute topographie des physikalishen meerensiniveaus und der druckflachen sowie die wasserbewegungen im raum des Atlantischen Ozeans'. *Wissenshaftliche Ergenbisse der Deutschen Atlantischen Expedition auf dem Forchungs-und Vermessungsschift 'Meteor'*, **6**, 191–260.

Dietrich, G. (169). *Atlas of the Hydrography of the Northern North Atlantic Ocean*. Conseil International pour L'Exploration de la mer. Copenhagen. pp. 140.

Dietrich, G., K. Kalle, W. Kraus & G. Siedler (1980). *General Oceanography, An Introduction*. Wiley Interscience, New York 626 pp. and 8 plates.

*Dobson, F. W., *et al.* (1982). *The 'Cage' Experiment: a Feasibility Study, Final Report, Commissioned by the JSC/CCCO Liaison Panel*. UNESCO, Paris, 134 pp.

Ekman, V. W. (1905). 'On the influence of the earth's rotation on ocean-currents'. *Arkiv för Matematik, Astronomi och Fysik*, **2** (11), 52 pp.

Faller, A. J. (1981). 'The origin and development of laboratory models as analogues of the ocean circulation'. *In: Evolution of Physical Oceanography. Scientific Surveys in Honor of Henry Stommel*, B. A. Warren & C. Wunsch, eds, pp. 462–79. The MIT Press, Cambridge.

Fu, L.-l (1981). 'The general circulation and meridional heat transport in the subtropical South Atlantic determined by inverse methods'. *Journal of Physical Oceanography*, **11**, 1171–93.

Fuglister, F. C. (1960). *Atlantic Ocean Atlas of Temperature and Salinity Profiles and Data from the International Geophysical Year of 1957–1958*. Woods Hole Oceanographic Institution Atlas Series 1: 209 pp.

*Garrett, C. (1980). 'Mixing in the ocean interior'. *Dynamics of Atmospheres and Oceans*, **3**, 239–65.

Hall, M. M. & H. Bryden (1982). 'Direct estimates and mechanisms of ocean heat transport'. *Deep-Sea Research*, **29**, 339–59.

Han, Y-J & S-W Lee (1981). *A New Analysis of Monthly Mean Wind Stress over the Global Ocean*. Report No. 26, Climate Research Institute, Oregon State University, 148 pp.

*Harrison, D. E. (1980). 'Eddies and the general circulation of numerical model gyres; an energetic perspective'. *Reviews of Geophysics and Space Physics*, **17**, 969–79.

Hastenrath, S. (1980). 'Heat budget of tropical ocean and atmosphere'. *Journal of Physical Oceanography*, **10**, 159–72.

Hastenrath, S. & P. Lamb (1977). *Climate Atlas of the Tropical Atlantic and Eastern Pacific Oceans*. University of Wisconsin Press, 104 pp.

Hellerman, S. (1967). 'An updated estimate of the wind stress on the world ocean'. *Monthly Weather Review*, **95**, 607–26 (and see correction in *Monthly Weather Review*, **96**, 63–74, 1978).

Hendry, R. (1982). 'On the structure of the deep Gulf Stream'. *Journal of Marine Research*, **40**, 119–42.

Holland, W. R. (1973). 'Baroclinic and topographic influences on the transport in western boundary currents'. *Geophysical Fluid Dynamics*, **4**, 187–210.

Holland, W. R. (1979). 'The general circulation of the ocean and its modelling'. *Dynamics of Ocean and Atmospheres*, **3**, 111–42.

Holland, W. R. & A. D. Hirschman (1972). 'A numerical calculation of the circulation in the North Atlantic Ocean'. *Journal of Physical Oceanography*, **2**, 336–54.

Holton, J. R. (1979). *An Introduction to Dynamic Meteorology*, 2nd edn. Academic Press, New York, 391 pp.

Hunkins, K. (1966). 'Ekman drift currents in the Arctic Ocean'. *Deep-Sea Research*, **13**, 607–20.

Ivers, W. D. (1975). 'The deep circulation in the northern North Atlantic with special reference to the Labrador Sea'. Ph.D. Thesis, University of California at San Diego, 179 pp.

*Killworth, P. D. (1982). 'Deep convection in the world ocean'. *Reviews of Geophysics and Space Physics*, **21**, 1–26.

Krauss, E. B. (1972). *Atmosphere–Ocean Interaction*. Oxford University Press, 275 pp.

Leetmaa, A. & A. F. Bunker (1978). 'Updated charts of the mean annual wind stress, convergence in the Ekman layers and Sverdrup transports in the North Atlantic'. *Journal of Marine Research*, **36**, 311–22.

Leetmaa, A., J. P. McCreary, Jr & D. W. Moore (1981). 'Equatorial currents: observations and theory'. *In: Evolution of Physical Oceanography. Scientific Surveys in Honor of Henry Stommel*, B. A. Warren & C. Wunsch, eds., pp. 184–96. The MIT Press, Cambridge.

Leetmaa, A., P. Niiler & H. Stommel (1977). 'Does the Sverdrup relation account for the mid-Atlantic circulation?' *Journal of Marine Research*, **35**, 1–10.

Luyten, J. R. (1977). 'Scales of motion in the deep Gulf Stream and across the continental rise'. *Journal of Marine Research*, **35**, 49–74.

McCartney, M. S. (1977). 'Subantarctic mode water'. *In: A Voyage of Discovery: George Deacon 70th Anniversary Volume*, M. V. Angel, ed., pp. 103–19. Pergamon Press, Oxford.

MEDOC Group, (1970). 'Observation of formation of deep water in the Mediterranean Sea, 1969'. *Nature*, **227**, 1037–40.

Meyers, G. (1980). 'Do Sverdrup transports account for the Pacific North Equatorial Countercurrent'? *Journal of Geophysical Research*, **85**, 1073–5.

MODE Group, The (1978). 'The Mid-ocean dynamics Experiment'. *Deep-Sea Research*, **25**, 859–910.

Moore, D. W. (1963). 'Rossby waves in ocean circulation'. *Deep-Sea Research*, **10**, 735–47.

Morgan, G. W. (1956). 'On the wind-driven ocean circulation'. *Tellus*, **8**, 301–320.

Munk, W. (1950). 'On the wind-driven ocean circulation.' *Journal of Meteorology*, **7**, 79–93.

Munk, W. (1966). 'Abyssal recipes'. *Deep-Sea Research*, **13**, 707–30.

*Munk, W. & C. Wunsch (1982). 'Observing the ocean in the 1990's'. *Philosophical Transactions of The Royal Society of London A*, **307**, 439–64.

Needell, G. J. (1980). 'The distribution of dissolved silica in the deep western North Atlantic Ocean'. *Deep-Sea Research*, **27**, 941–50.

Niiler, P. P. & W. S. Richardson (1973). 'Seasonal variability of the Florida Current'. *Journal of Marine Research*, **31**, 144–67.

Oort, A. H. & T. H. vonder Haar (1976). 'On the observed annual cycle in the ocean–atmosphere heat balance over the northern hemisphere'. *Journal of Physical Oceanography*, **6**, 781–800.

Pedlosky, J. (1979). *Geophysical Fluid Dynamics*. Springer Verlag, New York 624 pp.

Phillips, N. A. (1963). 'Geostrophic motion'. *Reviews of Geophysics*, **1**, 123–76.

*Reid, J. L. (1981). 'On the mid-depth circulation of the world ocean'. *In: Evolution of Physical Oceanography. Scientific Surveys in Honor of Henry Stommel*, B. A. Warren & C. Wunsch, eds. pp. 70–111. The MIT Press, Cambridge.

Rhines, P. B. & W. Holland (1979). 'A theoretical discussion of eddy-driven mean flows'. *Dynamics of Atmospheres and Oceans*, **3**, 289–325.

Robinson, A. R. (ed.) (1982). *Eddies in Marine Science*. Springer-Verlag, New York, 609 pp.

Robinson, A. R. & H. Stommel (1959). 'The oceanic thermocline and the associated thermohaline circulation'. *Tellus*, **3**, 295–308.

Roemmich, D. (1980). 'Estimation of meridional heat flux in the North Atlantic by inverse methods'. *Journal of Physical Oceanography*, **10**, 1972–83.

Sarkisiyan, A. S. (1977). 'The diagnostic calculations of a large-scale oceanic circulation'. *In: The Sea: Ideas and Observations on Progress in the Study of the Seas*, Vol. 6: *Marine Modeling*, E. D. Goldberg, I. N. McCave, J. J. O'Brien & J. H. Steele, eds., pp. 363–459. Wiley, New York.

Sarmiento, J. L. & K. Bryan (1982). 'An ocean transport model for the North Atlantic'. *Journal of Geophysical Research*, **87**, 394–408.

Saunders, P. (1976). 'On the uncertainty of wind stress curl calculations'. *Journal of Marine Research*, **24**, 155–60.

Schmitz, W. J., Jr (1978). 'Observations of the vertical distribution of low-frequency kinetic energy in the western North Atlantic'. *Journal of Marine Research*, **36**, 295–310.

Schmitz, W. J., Jr (1980). 'Weakly depth-dependent segments of the North Atlantic circulation'. *Journal of Marine Research*, **38**, 111–35.

Schmitz, W. J., Jr & W. R. Holland (1982). 'A preliminary comparison of selected numerical eddy-resolving general circulation experiments with observations'. *Journal of Marine Research*, **40**, 75–117.

Schott, F. & H. Stommel (1978). 'Beta spirals and absolute velocities in different oceans'. *Deep-Sea Research*, **25**, 961–1010.

Semtner, A. J. & W. R. Holland (1978). 'Intercomparison of quasi-geostrophic simulations of the western North Atlantic circulation with primitive equation results'. *Journal of Physical Oceanography*, **8**, 735–54.

Stern, M. E. (1975). *Ocean Circulation Physics*. Academic Press, New York, 246 pp.

*Stewart, R. H. (1983). *Methods of Satellite Oceanography*. University of California Press, to appear.

Stommel, H. (1948). 'The westward intensification of wind-driven ocean currents'. *Transactions of the American Geophysical Union*, **29**, 202–6.

Stommel, H. (1962). 'On the smallness of sinking regions in the ocean'. *Proceedings of the National Academy of Sciences of the USA*, **48**, 766–72.

Stommel, H. M. (1966). 'The large-scale oceanic circulation'. *In: Advances in Earth Science*, P. Hurley, ed., pp. 175–84. The MIT Press, Cambridge.

Stommel, H. & A. B. Arons (1960). 'On the abyssal circulation of the world ocean: Part 2, an idealized model of the circulation pattern and amplitude in ocean basins'. *Deep-Sea Research*, **6**, 217–23.

Stommel, H. & G. T. Csanady (1980). 'A relation between the *T-S* curve and global heat and atmospheric water transports'. *Journal of Geophysical Research*, **85**, 495–501.

Stommel, H., E. D. Stroup, J. L. Reid & B. A. Warren (1973). 'Transpacific hydrographic sections at lats. 43°S and 28°S: The SCORPIO Expedition-I. Preface'. *Deep-Sea Research*, **20**, 1–7.

Sverdrup, H. U. (1947). 'Wind-driven currents in a baroclinic ocean; with application to the equatorial currents of the eastern Pacific'. *Proceedings of the National Academy of Sciences of the USA*, **33**, 318–26.

Swallow, J. C. (1977). 'An attempt to test the geostrophic balance using minimode current measurements'. *In: A Voyage of Discovery: George Deacon 70th Anniversary Volume*, M. Angel, ed., pp. 165–76. Pergamon Press, Oxford.

Trenberth, K. E. (1980). 'Mean annual poleward energy transport in the oceans in the southern hemisphere'. *Dynamics of Atmospheres and Oceans*, **4**, 57–64.

Veronis, G. (1973). 'Large-scale ocean circulation'. *Advances in Applied Mechanics*, **13**, 1–92.

*Veronis, G. (1981). 'Dynamics of large-scale ocean circulation'. *In: Evolution of Physical Oceanography. Scientific Surveys in Honor of Henry Stommel*, B. A. Warren & C. Wunsch eds. pp. 140–83. The MIT Press, Cambridge.

Vonder Haar, T. H. & A. H. Oort (1973). 'New estimates of annual poleward energy transport by northern hemisphere oceans'. *Journal of Physical Oceanography*, **2**, 169–72.

Warren, B. A. (1963). 'Topographic influence on the path of the Gulf Stream'. *Tellus*, **15**, 167–83.

Warren, B. A. (1972). 'Insensitivity of subtropical mode water characteristics to meteorological fluctuations'. *Deep-Sea Research*, **19**, 1–20.

*Warren, B. A. (1981). 'Deep circulation of the world ocean'. *In: Evolution of Physical Oceanography, Scientific Surveys in Honor of Henry Stommel*, B. A. Warren & C. Wunsch eds., pp. 6–41. The MIT Press, Cambridge.

Welander, P. (1959). 'An advective model of the ocean thermocline'. *Tellus*, **11**, 309–18.

Welander, P. (1971). 'Thermocline problem'. *Philosophical Transactions of The Royal Society of London*, **A270**, 69–73.

Welander, P. (1981). 'A note on the overall vorticity balance in wind-driven ocean circulation'. *Dynamics of Atmospheres and Oceans*, **6**, 125–30.

Worthington, L. V. (1959). 'The 18° water in the Sargasso Sea'. *Deep-Sea Research*, **5**, 297–305.

Worthington, L. V. (1976). 'On the North Atlantic circulation'. The Johns Hopkins Oceanographic Studies, **6**, 110 pp.

Worthington, L. V. (1981). 'The water masses of the world ocean; some results of a fine-scale census'. *In: Evolution of Physical Oceanography, Scientific Surveys in Honor of Henry Stommel*, B. A. Warren & C. Wunsch, eds. pp. 42–69. The MIT Press, Cambridge.

Worthington, L. V. & W. R. Wright (1970). *North Atlantic Ocean Atlas of Potential Temperature and Salinity in the Deep Water Including Temperature, Salinity and Oxygen Profiles from the Erika Dan Cruise of 1962*. Woods Hole Oceanographic Institution Atlas Series 2, 24 pp. and 58 plates.

Wunsch, C. (1978). 'The North Atlantic general circulation west of 50°W determined by inverse methods'. *Reviews of Geophysics and Space Physics*, **16**, 583–620.

Wunsch, C. (1980). Meridional heat flux of the North Atlantic Ocean'. *Proceedings of the National Academy of Sciences of the USA*, **77**, 5043–7.

Wunsch, C. (1981). 'Low-frequency variability of the sea'. *In: Evolution of Physical Oceanography. Scientific Surveys in Honor of Henry Stommel*, B. A. Warren & C. Wunsch, eds., pp. 342–74. The MIT Press, Cambridge.

Wunsch, C. (1982). 'Comments on the problem of determining the general circulation of the ocean'. Unpublished manuscript.

Wunsch, C. & B. Grant (1982). 'Towards the general circulation of the North Atlantic Ocean'. *Progress in Oceanography*, **11**, 1–59.

Wunsch, C., D. Hu & B. Grant (1983). 'Mass, heat, salt and nutrient fluxes in the South Pacific Ocean', *Journal of Physical Oceanography*, **13**, 725–53.

Wunsch, C. & J-F Minster (1982). 'Methods for box models and ocean circulation tracers – mathematical programming and nonlinear inverse theory'. *Journal of Geophysical Research*, **87**, 5617–62.

Wüst, G. (1935). 'Schichtung and Zirkulation des Atlantischen Ozeans. Die Stratosphare'. *In: Wissenschaftliche Ergebnisse der Deutschen Atlantischen Expedition auf den Forschungs- und Vermessungsschiff*, 'Meteor' 1925–1927, Part 1, Vol. 2, 180 pp.

M. N. Koshlyakov and A. S. Monin
PP Shirshov Institute of Oceanology,
Academy of Sciences of the USSR,
Moscow, USSR

11 *Strategy of ocean monitoring for climate research*

Abstract

Ocean monitoring for climate research should be directed at:

(1) studies of ocean circulation structure and variability;
(2) studies of ocean circulation effects on the distribution and variability of heat content in the active ocean layer;
(3) studies and exploration of specific phenomena in the ocean–atmosphere interaction system;
(4) global monitoring of climatically significant fields in depths and at the ocean surface, as well as in the lower atmosphere.

A number of national and international programmes relevant to these objectives have been undertaken in recent years or are planned for the next decade.

The world ocean exerts, in many ways, a decisive effect on earth's climate and its changes by transporting enormous quantities of heat, moisture and carbon dioxide through the ocean–atmosphere interface. The ocean accumulates heat mostly in the tropical latitudes, due to an intensive influx of short-wave solar radiation and the distribution of thus-accumulated heat within the active ocean layer by wind mixing. The accumulated heat is transported by a system of ocean currents to the temperate and high latitudes where it is given up to the atmosphere, mainly in winter, by evaporation of ocean water, effective long-wave ocean radiation to the atmosphere and contact heat exchange. Thus, ocean monitoring for climate research should imply both measurements of the exchange processes themselves at the ocean–atmosphere interface (when possible) and observations of the fields in the ocean and the atmosphere, which affect these processes directly or indirectly. These processes and fields may be grouped together in the following way:

(1) ocean surface temperature and heat content of its active layer – the leading fields of paramount influence on heat, moisture and gas exchange between the ocean and the atmosphere;
(2) ice cover at the ocean surface;
(3) ocean water circulation, governing, in many respects, temperature spatial distribution and variability in the upper layer, large-scale currents, synoptic eddies, temperature and motion of ocean water (upwellings and downwellings), and convective and wind mixing in the upper ocean layer;
(4) atmospheric fields, together with the ocean surface temperature determining the intensity of energy exchange between the ocean and the atmosphere – temperature and humidity of the atmospheric boundary layer, wind speed near the ocean surface, quantity and composition of clouds;
(5) direct and reflected fluxes of short- and long-wave radiation near the ocean surface.

At the present stage in the studies of ocean effects on climate, observations of the above-mentioned oceanic and atmospheric fields should probably be of a research character, i.e. they should be arranged in such a way that the maximum amount of data could be gained to construct the ocean–atmosphere interaction models, which, in turn, should give a basis for long-range weather and climate change forecasts.

Among the above-mentioned oceanic and atmospheric fields, the general circulation of the ocean is most difficult to observe. The large-scale experiments in the ocean, carried out in the two recent decades, as well as those now in progress or planned for the near future, which are aimed at monitoring climatically important processes and fields in the ocean and the adjacent atmospheric layer, can conveniently be classified in the following way:

(1) studies of ocean circulation structure and variability;
(2) studies of ocean circulation effects on the distribution and variability of heat content in the active ocean layer;
(3) studies and explanation of specific phenomena in the ocean–atmosphere interaction system;
(4) global monitoring of climatically significant fields in depths and at the ocean surface, as well as in the lower atmosphere.

In what follows we shall consider some experimental studies of the ocean. This classification is clearly conventional and in a brief paper like this we shall be able to mention only part of the numerous experiments associated with the studies of ocean effects on climate.

11.1 Studies of ocean circulation structure and variability

Among the field studies of the ocean, carried out in the recent decades, a prominent place belongs to the studies of synoptic variability of the ocean. Synoptic disturbances of the ocean circulation, usually called synoptic eddies, can conveniently be subdivided into two categories: eddies of the western boundary currents (rings) and open-ocean eddies. Among the eddies in the first category, the most intensive research was made of the eddies in the Gulf Stream, the Kuroshio, the East Australian and the Agulhas Currents and, to some extent, of those in the Antarctic Circumpolar Current (Kamenkovich, Koshlyakov & Monin, 1982).

As observations show, the rings representing rather concentrated solitary formations result from the cutoff of meanders of the western boundary jet currents, with only cyclonic rings formed to one side of the current and only anticyclonic rings to the other side. Due to the frontal character of the western boundary currents, the cyclonic rings contain cold subpolar water and, as a result of their formation, they appear in the warm subtropical water regions, whereas the anticyclonic rings, on the contrary, are characterized by warm cores and intrude into the cold subpolar water.

It is easy to establish that the formation of five pairs of rings in the Gulf Stream region implies heat exchange across the Gulf Stream of some 6×10^{21} joules/year (Mintz, 1979). However, this value cannot be taken as an estimate of the integral annual heat transport across the Gulf Stream related to its synoptic activity: as the observations of the recent years show (Richardson, 1980), all the Gulf Stream anticyclones and most of its cyclones, after several weeks or months of drift in the ocean, come into secondary contact with the Gulf Stream, which is accompanied, as a rule, by an intensive water and heat exchange between the eddy and the Gulf Stream and, in many cases, terminates in a reverse absorption of eddies by their generating current.

The discovery of synoptic eddies of the open ocean was one of the most significant events in oceanography during the post-war years. First indications of the existence in the ocean of intensive synoptic disturbances of currents were obtained through the late 1950s to the early 1960s, both through measurements of deep currents with the neutrally buoyant floats (Swallow, 1971), and as a result of the accomplishment of long-term series of measurements of the vertical profiles of temperature and salinity at fixed points of the ocean (Yasui, 1961). The spatial patterns of the open-ocean synoptic eddies, well resolved in the horizontal plane, were first obtained by processing the data of the two hydrological surveys made by the Soviet oceanographic expedition in the Arabian Sea in 1967 (Koshlyakov, Galerkin & Truong Din Hien, 1970).

Finally, the first comprehensive direct measurements of the open ocean synoptic currents were made by the Soviet expedition Polygon-70 at a network of 17 moorings maintained in the tropical zone of the North Atlantic during six months in 1970 (Koshlyakov & Grachev, 1973). Three powerful cyclonic and anticyclonic eddies were recorded to pass, one after the other, across the polygon with the velocities of currents much in excess of the velocity in the main North Equatorial Current. The eddies were penetrating an ocean depth of at least 2000 m. The relation between the horizontal dimensions of these disturbances (200–300 km) and the velocity of their movement westward–southwestward (5–6 km/day) was in good agreement with the dynamics of baroclinic Rossby waves by means of which the physical nature of these formations was revealed.

During MODE-1 (Mid-Ocean Dynamics Experiment) carried out by oceanographers from the United States in the Sargasso Sea in 1973 (MODE Group, 1978), several open-ocean eddies resembling in many parameters, Polygon-70 eddies, were recorded.

A logical continuation and development of Polygon-70 and MODE is known as the POLYMODE, an international experiment carried out from 1974–79 in the North Atlantic, with its most intensive period, a synoptic experiment, accomplished in 1977–78 in the southwestern part of the Sargasso Sea. The backbone of this experiment was the Soviet network of 19 moorings at which continuous measurements of velocities and water temperatures in the 100–1400 m ocean layer were made over 13 months, supplemented by 58 temperature and density surveys of the area. The experiment confirmed the high degree of the development of synoptic currents in the Sargasso Sea and revealed the structure of these currents which consist of the coexistence of strong concentrated jets of synoptic currents, well-pronounced cyclonic and anticyclonic eddies and regions with weak currents between the intensive jets and eddies (Koshlyakov, Grachev & Enikeev, 1980). The high level of energy of the synoptic currents caused strong turbulence in the current field, which resulted in intensive water, energy and vorticity exchange between neighbouring eddies and jets (Fig. 11.1), as a result of which the lifetime of individual eddies was comparable to the time of their passing through fixed points in the ocean.

The reliably measured vertical phase shift of velocity oscillations at depths above and below the main thermocline was indicative of eddy generation due to baroclinic instability of the large-scale current (Enikeev, Kozubskaya, Koshlyakov & Yaremchuk, 1982). The corresponding energy flux from the large-scale current to eddies turned out to be rather variable in time, which entailed sharp changes in the synoptic current structure. During the periods of high energy fluxes the eddies had well-pronounced baroclinic structure and were characterized by diameters of some 150 km, whereas during the periods of weak energy fluxes, sharp barotropization of the synoptic velocity field took place and the eddy diameters almost doubled. Such evolution of the velocity field is characteristic, as is generally known, of two-dimensional turbulence (Rhines, 1977; Mirabel & Monin, 1980).

The American Local Dynamics Experiment carried out, within the framework of POLYMODE from 1978–79, in the area in the north immediately adjacent to the Soviet experimental area, included measurements of currents at moorings and, with the aid of the SOFAR floats, temperature and density surveys from ships and some other measurements (Ebbesmeyer, Taft, Cox, McWilliams, Owens, Sayles & Shen, 1978).

As a whole, the 1970s and early 1980s were marked by a rapid development of the field studies of synoptic eddies in different areas of the world ocean (Kamenkovich, Koshlyakov & Monin, 1982). As a result of these studies it has been established that synoptic eddies are of an ocean-wide distribution and contain the major part of its kinetic energy. Eddies are usually essentially baroclinic, their horizontal dimensions are most often comparable to the Rossby deformation radius $L_R = (gh\delta\rho/\rho_0)^{\frac{1}{2}} f^{-1}$, where g is the acceleration of gravity, f the Coriolis parameter, ρ_0 the mean ocean water density, $\delta\rho$ the density difference across the main thermocline, and h the depth of the main thermocline core. Their mean energy, as a rule, is higher in the regions where the energy of the mean large-scale currents is higher. All these features, according to the synoptic currents theory, are indicative of baroclinic instability of the large-scale currents as the basic eddy generation mechanism.

It can probably be assumed that eddies contribute essentially to the mean meridional heat fluxes in the polar front area in the southern ocean (Sciremammano, 1980) and, maybe, in the polar front areas in the North Atlantic and North Pacific. In the equatorial and subtropical zones of the ocean the meridional heat

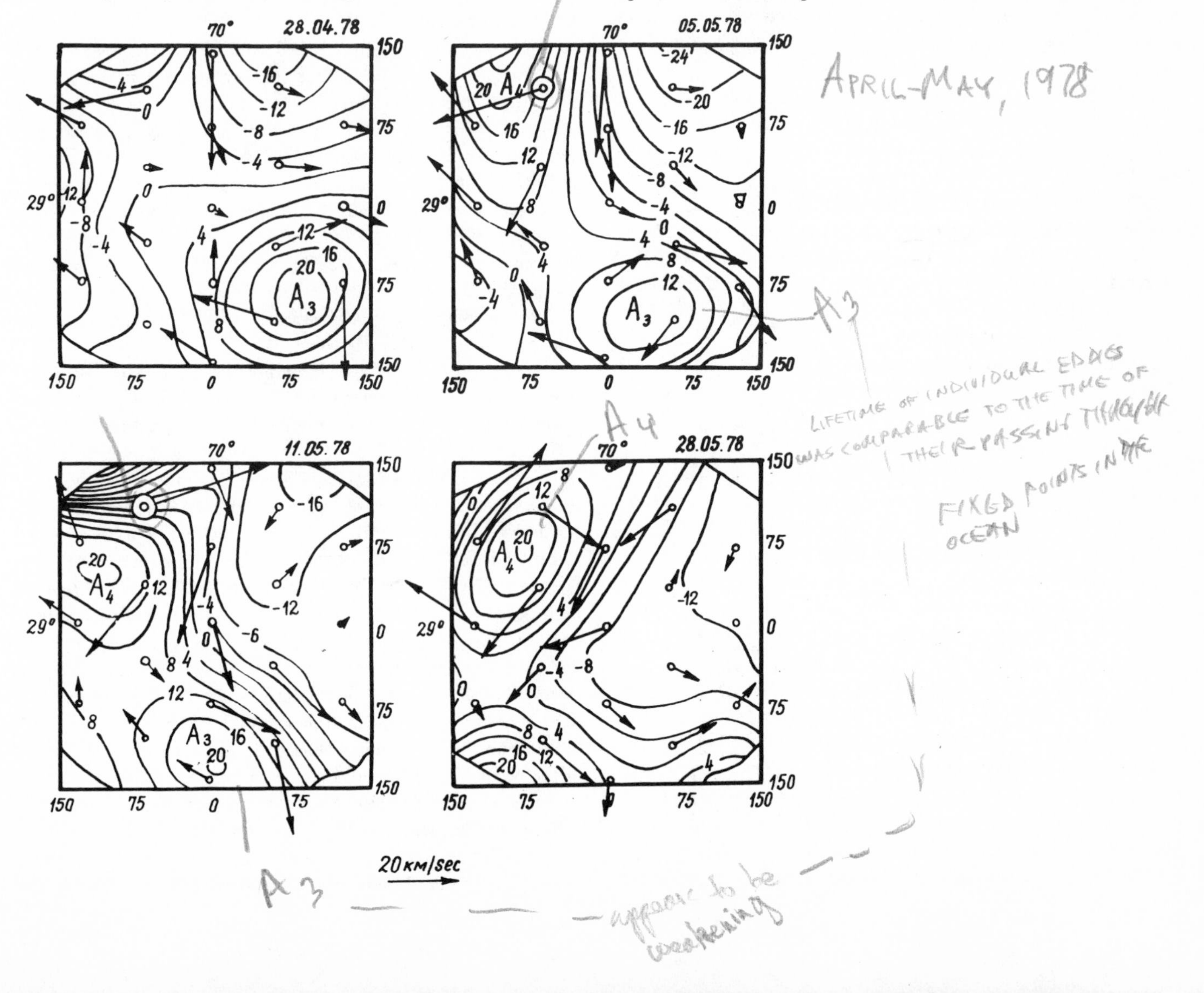

Fig. 11.1. Velocity vectors and stream lines of synoptic eddies at a 700 m depth in the POLYMODE area in April–May, 1978. Stream function of 10^7 cm^2 × sec^{-1}. Circles stand for mooring positions. Numbers near the frames denote distances in kilometres. A_3 and A_4 are individual anticyclonic eddies. Water exchange between separate eddies and streams is evident. The double circle designates the station through which a sharp velocity front passed in the first 10 days of May, 1978.

fluxes associated with eddies are apparently small and, besides, are directed from the poles towards the equator (Bryden & Hall, 1980).

11.2

Studies of ocean circulation effects on heat transport in the ocean and on heat exchange between the ocean and the atmosphere

It is generally known that heat and moisture exchange between the ocean and the atmosphere is, first, usually of a seasonal character and, second, is most intensive in some definite regions which can be called the energy-active zones of the ocean (EAZO). This is the basis of the Soviet project Sections – a grand programme of oceanographic and meteorological studies in five regions of the world ocean (Norwegian, Newfoundland, Gulf Stream and Kuroshio, and the Atlantic tropical region). In all these regions, investigations are planned to be carried out in all the four seasons for at least five years beginning in 1981, aimed at both direct measurements of the intensity of energy exchange between the ocean and the atmosphere including changes of this intensity in time, and studies of the factors responsible for these changes – primarily, ocean circulation variability.

In each of the seasons and at each of the polygons, the work is mainly conducted from two ships and consists of two successive surveys of the region, more-detailed observations in the sections running across the main currents, and many-day observations in fixed points inside the polygon (Fig. 11.2). At all these stages a wide complex of oceanographic studies (primarily temperature and density soundings of the ocean), meteorological, actinometric and aerological measurements are made. In addition to this, sea current

Fig. 11.2. Scheme of work in the Norwegian polygon under the Sections programme. Circles and dots designate the locations of hydrological stations, taken under the main and supplementary programmes. Double lines are hydrological sections with a more-frequent spacing of stations. Also seen are the locations of moorings with submerged buoys and current meters, weather ships, continental and island aerological stations and ice edge in winter and summer.

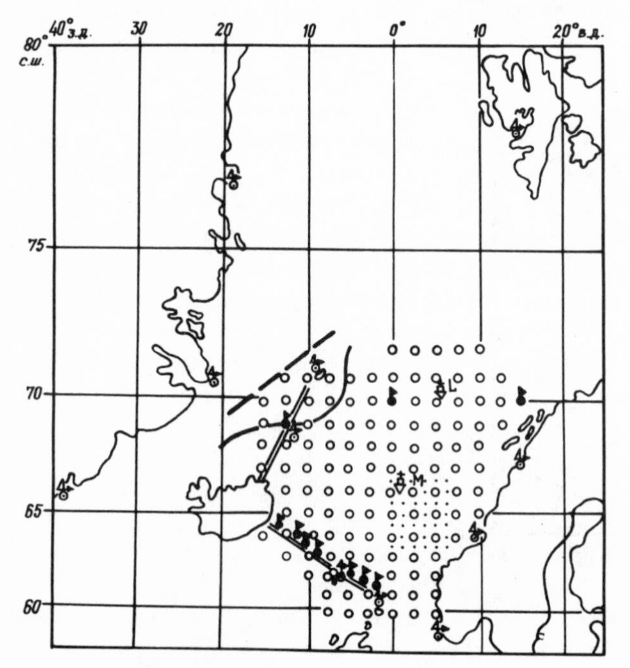

and water temperature measurements are made at moorings. The work results in estimates of heat content of the active ocean layer and heat content variability in time, heat transport by large-scale currents and synoptic eddies, heat and moisture transport in the atmospheric circulation system, and radiation heat and moisture exchange fluxes at the ocean–atmosphere interface. The Sections programme also includes a complex of satellite observations – ocean surface temperature measurements, particularly important for detecting oceanic fronts, measurements of the major jets of the frontal currents and synoptic eddies, as well as observations of clouds, wave and wind measurements and direct measurements of the radiation budget components in the upper ocean layer and the atmosphere.

The Soviet Sections programme may give material for calibrating all the available climatic maps of the ocean–atmosphere interaction, as well as monitoring of the annual seasonal variability.

In its scientific objectives the Sections programme is close to several national and international oceanographic programmes (Summary Report of CCCO, 3rd Session, 1982), among which should, first of all, be made of the French experiment FOCAL (French Equatorial Atlantic Ocean Climate Program) and the US SEQUAL (Seasonal Equatorial Atlantic Experiment), the proposed international (mainly, the USA, Canada, Great Britain) Cage Experiment and the American programme of acoustic tomography of the ocean. FOCAL and SEQUAL, together with the Soviet investigations under the Sections programme at the Atlantic tropical polygon, are naturally combined into a large programme for monitoring the Atlantic equatorial zone, the main objective of which consists of the studies of seasonal and year-to-year variability of currents and water heat content as dependent on wind field variability. Both the French and the American experiments are planned for 1982–85 and envisage the accomplishment of temperature and density sections across the ocean, a complex of meteorological observations, current measurements at moorings and drifting buoys, near-bottom pressure oscillation measurements and a number of other observations.

The proposed Cage Experiment (Bretherton, Burridge, Crease, Dobson, Kraus & Vonder Haar, 1982) is aimed at obtaining estimates of the current generated heat fluxes (and their changes by seasons and from year to year) across the southern and the northern boundaries and across some inner sections of the Atlantic Ocean region, limited in the south by the 25° parallel and in the north by the lines connecting France, Great Britain, Iceland, Greenland, Baffin Island and Labrador. It is proposed that estimates be obtained by three independent methods: (a) direct measurements of heat fluxes in the ocean; (b) heat content measurements in different parts of the area under study (mostly to 1500 m depth) and overall measurements of heat exchange through the upper interface of the area; (c) measurements of radiative heat fluxes at the upper boundary of the atmosphere, heat and moisture fluxes in the atmosphere through its lateral boundaries, and heat content of the ocean. To obtain all these estimates it would be necessary to undertake a wide complex of oceanographic, meteorological, aerological and actinometric observations, which would include repeated temperature and density sections along the routes of passing and research vessels, current measurements at moorings and, with the aid of drifting buoys, meteorological, aerological and actinometric measurements from passing and research vessels and weather ships, as well as at island and shore meteorological stations; radiation measurements from the earth's artificial satellites; satellite measurements of the ocean surface level variations (satellite altimetry); and some other observations. The feasibility study (Bretherton *et al.*, 1980) points out the difficulties of making all measurements required with the necessary accuracy to make a useful inference of the heat fluxes in the ocean. A similar project has been outlined for the North Pacific.

The acoustic tomography project (Munk & Wunsch, 1982) envisages deployment in the northeastern Pacific (the tropical Pacific and the northeastern Atlantic are the alternatives) of a series of acoustic stations along the perimeter of a polygon of about 2000 m width. Each station consists of a chain of acoustic transmitters and receivers spaced in ocean depth. Accurate measurement of acoustic signal transmission and reception times at different stations makes it possible to determine (as a function of time and ocean depth) water temperature averaged over different ranges of depth along the straight lines connecting the station locations. The resulting information shows the distribution in depth and variability in time of the following most important parameters of the ocean heat content inside each of the triangles formed by intersections of the above-mentioned straight lines; water flows and heat fluxes across the triangle sides; and the potential vorticity values integrated over the area of each triangle.

11.3 Studies of the large-scale ocean–atmosphere interaction

Striking examples of the empirical studies of specific phenomena in the large-scale ocean–atmosphere interaction system are represented by the oceanographic and meteorological observations of interrelated year-to-year changes of the oceanic and atmospheric circulation in the equatorial and northern parts of the Pacific Ocean. By the empirical and theoretical studies, among which those by Bjerknes (1969) were pioneering, the following chain of phenomena has been rather reliably established. The weakening of the southeastern trade wind in the equatorial Pacific, recorded every three to seven years, results in the weakening of the South Equatorial Current and in cold water upwellings in this zone, which, in turn, cause the appearance of a positive (2–3 °C) ocean surface temperature anomaly. At the same time the Equatorial Surface Countercurrent intensifies and the warm equatorial water sets up near the Peruvian coasts (the El Nino phenomenon).

The intensified heat flux from the ocean to the atmosphere, associated with the positive ocean temperature anomaly, strengthens the tropical cell of the atmospheric circulation, which leads, in particular, to intensification of the northeastern trade wind and meridional temperature and pressure gradients in the upper atmospheric layers in the subtropical zone of the northern hemisphere. The latter causes intensification of eddy activity in the temperate latitude atmosphere and, in the end, intensification of the eastern transport in the atmosphere above the North Pacific. Strengthening of near-surface wind leads to intensification of the subtropical water gyre, thereby influencing heat emission from the ocean to the atmosphere in winter. It is obvious that a relation should exist between the atmospheric and oceanic circulations in both hemispheres of the earth, resulting in the regeneration of a strong southeastern trade wind.

The links of this chain of phenomena were studied in a number of the atmospheric and oceanic experiments of the recent years and, in particular, in the Norpax and Westpac programmes commenced nearly two decades ago and still being carried out. In 1978–80, within the framework of Norpax, a very interesting experiment known as Shuttle was accomplished. The main objective of the experiment was to study changes of the large-scale zonal currents in the equatorial–tropical parts of the Pacific Ocean and the corresponding variability of ocean water thermal structure (Wyrtki, Firing, Halpern, Knox, McNally, Patzert, Stroup, Taft & Williams, 1981). The experiment consisted of 15 density sections, made from ships, and 35 temperature sections, performed from aircraft along the meridians of 150, 153 and 158° W between 20°N and 17°S, a complex of shipboard meteorological, aerological, actinometric and hydrochemical measurements, as well as direct ocean current measurements with the aid of drifting buoys, and shipborne acoustic velocity profilometers at moorings deployed along the equator. These observations were supplemented by standard meteorological measurements and ocean surface level measurements on islands, as well as by satellite observations of wind from cloud motions. The main result of the work was a reliable confirmation of the idea advanced previously by Wyrtki (1974), that the intensification of the South Equatorial Current and the Equatorial Undercurrent caused by the strengthening of the southeastern trade wind was accompanied by the weakening of the Equatorial Surface Countercurrent and the North Equatorial Current and vice versa. A well-pronounced solitary character of synoptic eddies in the Equatorial Surface Countercurrent area was an interesting result.

The studies under the Norpax and Westpac programmes are being carried out intensively by oceanographers from the Soviet Union, Japan, USA, Canada, France, Great Britain, Australia, China and some other countries, and embrace practically the whole of the northern and equatorial parts of the Pacific Ocean and the western half of its southern part (Final Report of the Joint WMO/IOC Regional IQOSS Implementation Co-ordination Meeting in Westpac and Norpax Regions, Tokyo, 9–13 November 1981, unpublished document). The principal ones among these studies are regularly repeated temperature sections of the ocean (Fig. 11.3), unique in their scope, carried out primarily by passing ships and particularly frequent in the area adjacent to Japan, in the Sea of Japan and the East China Sea, as well as standard density sections in the western part of the ocean (mainly the Soviet Union and Japan), repeated four times a year. During these studies standard meteorological and actinometric measurements aboard ships are carried out.

In addition to the above, Norpax and Westpac include aircraft temperature sections of the ocean along the Aleutians–Hawaii and Hawaii–Tahiti routes; current and water temperature measurements at moorings, mainly in the areas adjacent to the USA, Canada and Japan; meteorological observations and measurements of level variations of the ocean surface on islands (mostly those in the central and western parts of the tropical ocean), as well as at continental stations, and satellite observations of clouds and ocean surface temperature. Ocean current measurements by drifting buoys with satellite tracking are also planned.

The main objective of all these studies is to investigate seasonal and year-to-year heat content oscillations in separate Pacific Ocea regions, the relation between these oscillations and the oceanic circulation, and their effect on heat exchange with the atmosphere and on climate changes, as well as reverse effects of the atmospheric circulation disturbances on ocean currents.

11.4 Global monitoring of climatically significant ocean fields

Up to the present, studies of the ocean–atmosphere interaction in separate, though sometimes vast, areas of the earth have been dealt with most thoroughly. Inevitably, however, the observational system in the ocean, aimed at studying the role of the ocean in the formation of the earth's climate and its anomalies, should be of a global character (Monin, 1969), this because of a close interrelation between thermo- and hydrodynamic processes in different parts of the globe.

The basis of the global monitoring of the ocean should, undoubtedly, be formed by satellite observations because the earth's artificial satellites are the only means of making global-scale, continuous observations. The major types of ocean measurements from satellites, which can be used for climate research, will evidently be the measurements of the following geophysical parameters:

(1) short-wave and long-wave radiation fluxes at the upper atmospheric boundary;
(2) ice coverage of the ocean (especially ice-edge position), being, perhaps, one of the major indicators of climate changes;

Fig. 11.3. Network of regular shipboard (expendable bathythermographs) temperature sections across the Pacific Ocean under the Norpax–Westpac programme. The hatched regions are the regions of dense temperature surveys. The dashed lines show the planned sections.

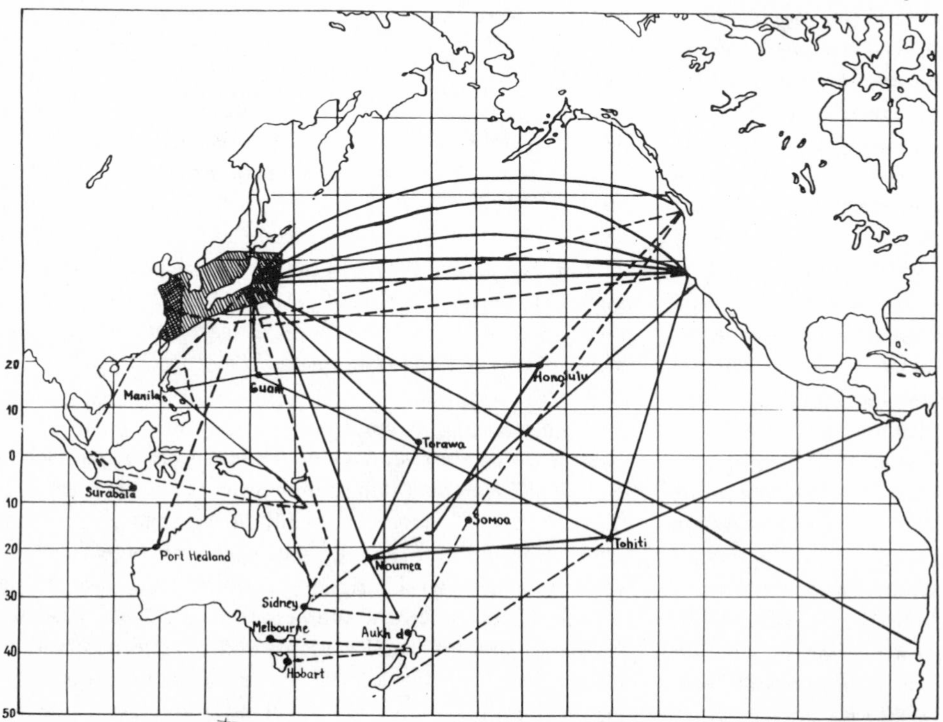

(3) ocean surface temperature (including the position of the oceanic thermal fronts);
(4) velocity and direction of ocean currents (from their reflection in the temperature field and by determining the positions of drifting surface buoys);
(5) ocean surface level variations related to the oceanic circulation (satellite altimetry);
(6) ocean surface waves (and through them, near-surface wind);
(7) quantity and composition of clouds;
(8) vertical temperature and humidity profiles in the atmosphere.

Measurements from satellites should be supported by a system of observations in the ocean itself, aimed mainly at obtaining systematic information on the vertical distributions of temperature and density in the ocean; heat content of the active ocean layer and its variability in time; the oceanic circulation structure and variability; and heat transport by ocean currents and heat and moisture fluxes at the ocean surface. Of climatic significance may also be year-to-year temperature fluctuations of the waters of polar origin in the greatest (ultraabyssal) deeps of the ocean, reflecting the climate state in the earth's polar regions.

According to the existing plans (Summary Report of CCCO, 3rd Session, 1982), the observational system, which in principle should embrace the entire world ocean and should be most detailed in the regions of the most intensive interaction between the ocean and the atmosphere, is to include the following elements.

(1) Regular trans-oceanic temperature (expendable bathythermographs) and, in future, density (expendable bathythermosalinometers) sections of the ocean are to be carried out, mainly by ships of ships of opportunity, and are to be supplemented by frequent temperature and salinity measurements of the surface water, together with meteorological observations.

(2) Regular detailed density sections of the ocean, mostly running across the major ocean currents, are to be carried out by

research ships. Good examples of such sections are represented by the standard section along the 'Kola meridian' in the Barents Sea, which has been carried out by Soviet oceanographers since 1929 (with the forced break in 1941–45), and several sections regularly repeated by Japan and the Soviet Union in the Kuroshio area. The density sections are to be accompanied by a complex of meteorological, actinometric and aerological measurements and, in future, also by the measurements of currents with velocity profilometers.

(3) Acoustic tomography of the ocean with the aid of several networks of acoustic stations is to be deployed in different world ocean areas.

(4) Aircraft temperature (with the aircraft modification of the expendable bathythermograph) and actinometric sections are to be established along some aircraft routes across the ocean.

(5) Ocean level measurements, as well as meteorological, actinometric and aerological measurements are to be taken on islands and at continental shore stations.

(6) Measurements of ocean current velocities and directions, water temperature and air pressure are to be carried out by drifting surface buoys with satellite tracking. During the last years observations of this kind yielded rich material for the study of oceanic circulation in the southern Ocean.

(7) Measurements of ocean current velocity and water temperature vertical distributions are to be supplemented, possibly by measurements of some meteorological elements at moorings deployed in key climate research areas of the ocean.

Studies of the physical mechanism of climate change require special experiments to be undertaken with the aim of more-detailed research of different phenomena in the global ocean–atmosphere interaction system. An experiment of this type, to be undertaken internationally, is the World Ocean Circulation Experiment (WOCE) which will be carried out (Summary Report of CCCO, 3rd Session, 1982) in the late 1980s and the early 1990s (see also Chapters 1 and 10 of this volume). The basis of this experiment is the global survey of the ocean dynamic level (i.e. ocean surface deviations from the geoid equilibrium surface related to the ocean current system), its surface temperature, surface wind speed (in terms of wave parameters), and clouds from several simultaneously operating earth's artificial satellites. The survey is to be made continuously over at least five years, will cover practically the entire world ocean and will be so detailed and precise as to be sufficient for resolving the ocean synoptic processes.

The measurements from satellites will be supplemented by the following observations: about thirty trans-oceanic temperature and density sections regularly repeated in different world ocean areas, accompanied by meteorological observations; measurements of ocean currents and water temperature with drifting surface buoys traced by satellites; ocean level measurements, meteorological, actinometric and aerological measurements on islands and at continental shore stations; and ocean current and water temperature measurements at a limited network of self-contained moorings.

The accomplishment of such a programme, particularly of the measurements from satellites, will obviously yield unique material for study of the oceanic circulation structure and variability, long-period oscillations of the large-scale ocean currents, and statistical parameters of synoptic eddies in different ocean areas. Simultaneous measurements of currents and water temperature in the upper ocean layer will make it possible to evaluate separately the contributions of the large-scale currents and synoptic eddies to the mean horizontal heat transport in different ocean areas and, at the same time, will yield extremely important information about the character and intensity of heat transformation of the current transported water masses under the influence of heat and moisture exchange with the atmosphere, clouds and vertical mixing in the upper ocean. Apart from supplying oceanographers and meteorologists with original data, such research would be very important for the improvement of the ocean–atmosphere interaction models aimed, in the end, at the creation of methods for forecasting climatic anomalies.

References

Bjerknes, J. (1969). 'Atmospheric teleconnections from the equatorial Pacific'. *Monthly Weather Review*, **97** (3), 163–72.

Bretherton, F. P., Burridge, D. M., Crease, J., Dobson, F. W., Kraus, E. B. & Vonder Haar, T. H. (1982). '*The 'Cage' Experiment*: a Feasibility Study'. World Climate Research Programme document. WMO, Geneva.

Bryden, H. L. & Hall, M. H. (1980). 'Heat transport by currents across 25°N in the Atlantic Ocean'. *Science*, **207**, 884–6.

Ebbesmeyer, C., Taft, B., Cox, J., McWilliams, J., Owens, B., Sayles, M. & Shen, C. (1978). 'Preliminary maps from the Polymode Local Dynamics Experiment – first half'. *Polymode News*, **54**. Unpublished manuscript.

Enikeev, V. Kh., Kozubskaya, G. I., Koshlyakov, M. N. & Yaremchuk, M. I. (1982). 'On the dynamics of synoptic eddies in the POLYMODE area'. *Doklady Akademii Nauk SSSR*, **262** (35), 573–7.

Final Report of the Joint WMO/IOC regional IQOSS Implementation Co-ordination Meeting in Westpac and Norpax Regions, Tokyo, 9–13 November 1981. Unpublished document.

Kamenkovich, V. M., Koshlyakov, M. N. & Monin, A. S. (1982). *Synoptic Eddies in the Ocean*. Leningrad, Gidrometeoizdat.

Koshlyakov, M. N., Galerkin, L. I. & Truong Din Hien (1970). 'On the mesostructure of the open-ocean geostrophic currents'. *Okeanologiya*, **10**, 805–14.

Koshlyakov, M. N. & Grachev, Yu. M. (1973). 'Meso-scale currents at a hydrophysical polygon in the tropical Atlantic'. *Deep-Sea Research*, **20** (6), 507–26.

Koshlyakov, M. N., Grachev, Yu. M. & Enikeev, V. Kh. (1980). 'Kinematics of the open-ocean synoptic eddy field'. *Doklady Akademii Nauk SSSR*, **252** (3), 573–577.

Mintz, Y. (1979). *On the Simulation of the Oceanic General Circulation*. Report of the JOC Study Conference on climate models: performance, intercomparison and sensitivity studies. V. 11. *GARP Publication Series No. 22*, pp. 607–87. Geneva: WMO.

Mirabel, A. P. & Monin, A. S. (1980). 'Geostrophic turbulence'. *Izvestiya Akademii Nauk SSSR. Fizika Atmosphery i Okeana*, **16** (10), 1011–23.

Monin, A. S. (1969). *Weather Forecasting as a Problem in Physics*. Moscow, Nauka, 183 pp.

The MODE Group (1978). 'The Mid-Ocean Dynamical Experiment'. *Deep-Sea Research*, **25** (10), 859–910.

Munk, W. & Wunsch, C. (1982). 'Observing the ocean in the 1990s'. *Philosophical Transactions of the Royal Society of London*, **A307**, 439–64.

Rhines, P. B. (1977). 'The dynamics of unsteady currents'. *In: The Sea*, eds. E. D. Goldberg, I. N. McCane, J. J. O'Brien & J. H. Steele, vol. 6, pp. 189–318. New York, Wiley.

Richardson, P. L. (1980). 'Gulf Stream ring trajectories'. *Journal of Physical Oceanography*, **10** (1), 90–104.

Sciremammano, F. (1980). 'The nature of the poleward heat flux due to low-frequency current fluctuations in Drake Passage'. *Journal of Physical Oceanography*, **10** (6), 843–52.

Summary Report of CCCO, 3rd Session, 1–5 March, 1982. Unpublished document.

Swallow, J. C. (1971). 'The 'Aries' current measurements in the western North Atlantic'. *Philosophical Transactions of the Royal Society of London*, **A270** (1206), 451–60.

Wyrtki, K. (1974). 'Equatorial currents in the Pacific 1950 to 1970 and their relations to the trade winds'. *Journal of Physical Oceanography*, **4** (3), 372–80.

Wyrtki, K., Firing, E., Halpern, D., Knox, R., McNally, G. L., Patzert, W. C., Stroup, E. D., Taft, B. A. & Williams, R. (1981). *Science*, **211** (4477), 22–8.

Yasui, M. (1961). 'Internal waves in the open ocean (an example of internal waves progressing along the oceanic frontal zone)'. *Oceanographic Magazine*, **12**, 157–67.

Biogeochemical processes and climate modelling

Bert Bolin
Department of Meteorology,
University of Stockholm

Abstract

Some basic steady state features of the global biogeochemical cycles are reviewed, particularly with regard to turnover times for carbon in terrestrial and marine reservoirs, and the means for their determination. The processes of evapo-transpiration, the dependence of surface albedo on type of vegetation and moisture supply, photosynthesis and gaseous exchange between the atmosphere and plants are analysed as is their possible role in modelling long-term changes of climate. Similarly, the role of the nutrient balance of the sea for climate change is discussed.

It is clear that regional studies of major biomes are required to formulate dynamic models of these subsystems and arrive at adequate methods for including biogeochemical processes in climate models. A record of land surface changes due to man will also be required in the analysis of future climate changes and their causes. It is proposed that a data bank be developed that includes both data with regard to such direct interventions by man, and simple dynamic models describing the response of the dominating ecosystems to climatic change.

12.1
Introduction

The biosphere is the thin, outer shell of the earth, within which life exists. It comprises most of the atmosphere, the oceans, the top layers of the soil, lakes and rivers. The climatologist calls this part of the earth the climate system. His attempts to describe, understand and model the climate of the earth may thus be considered as attempts to develop a consistent theory of how the physical state of the biosphere varies, depending on incident solar radiation, when the distribution of land and sea and the orography of the earth's surface is given. When concerned with the quasi-steady state of the present climate, this can be done without explicitly dealing with the life processes on earth. We may assume that the composition of the atmosphere is given and not changing. We prescribe the characteristic features of the earth's surface which are of importance in this context, such as surface roughness and albedo. In this way the general features of the present climate have been simulated rather well. This implies that the physical features of the biosphere, which serve as boundary conditions in determining the motions of the atmosphere and the sea, to a first approximation, have been described adequately. Such models can also be used to study climatic variations that depend on internal changes of dominating circulation pattern, e.g. interannual climatic variability. They cannot, however, tell us how the present climate has evolved. With this aim in mind we must also consider how the features of the earth's surface have changed with time.

We know that the composition of the atmosphere has changed in the course of the earth's evolution and that its present features, to a considerable degree, are the result of the development of life on earth. The oxygen content, and thus the presence of ozone and the concentration of carbon dioxide in the atmosphere, are related to the life processes. Similarly, the albedo of the earth's surface, surface roughness and other surface characteristics depend on the terrestrial vegetation which thus is another link between life processes on earth and climate. Lovelock (1979) has even advanced the hypothesis that the evolution of life on earth has established close to optimum climatic conditions for life to prevail. He refers to the remarkable observation that the changes of climate since life developed on earth (about $4 \cdot 10^9$ years ago) has been small in comparison with the increase of the flux of solar energy, estimated to have been about 30%. Also when concerned with climate variations on a shorter time scale, i.e. over periods of from hundreds to millions of years, which are well documented by quarternary geologists, we need to assess the role played by the changing terrestrial vegetation.

Studies of this time-dependent problem are much more difficult than attempts to simulate the present steady state. Feedback mechanisms play a key role and must be carefully analysed. The inertia of soils and terrestrial vegetation, and of the oceans, implies that slight but consistent departures from an equilibrium, may lead to a considerable change over longer time periods. As has already been indicated, however, the climatic variations have been rather small in comparison with solar radiation, which implies that negative feedback mechanisms have been of importance. In this sense the climate system seems to have been rather stable.

More rigorous and systematic studies of the mechanisms that govern the earth's climate have come rather late. The complexity of the system requires a systematic development of models, starting with rather simple ones to establish the general features of the interplay of the most fundamental mechanisms. The large general circulation models (GCM), which now are becoming available and which simultaneously consider transfer processes in the atmosphere and the sea in some detail, also offer possibilities to study the importance of biological and chemical processes for the past, present climate regimes on earth. Which are the key processes to be first included in such studies? Taking present knowledge of the biogeochemical cycles and their interaction as our starting point, we shall show the importance of dealing with the biological processes on earth in a manner that allows for a proper dynamic interaction between the life processes, on one hand, and the physical processes that, so far, have been emphasized primarily, on the other. Such an approach is also necessary when we wish to analyse in a consistent manner how a changing climate may change conditions for the life of man on earth.

12.2
Some general features of the major biogeochemical cycles

Many overviews of the major biogeochemical cycles and their interactions have appeared during the last few years. We shall not make another survey of this kind but refer to some such more-detailed presentations of the carbon cycle (Bolin *et al.*, 1979), the nitrogen cycle (Rosswall, 1981), the sulphur cycle (Ivanov *et al.*, 1983) and the interactions of the major biogeochemical cycles (Likens, 1981; Bolin & Cook, 1983). It is important for the following treatments, however, to emphasize some general features of these cycles, particularly with regard to their response to external disturbances, i.e. climatic change or man's interventions. Since carbon plays a key role in all biological processes and since the amount of CO_2 in the atmosphere is of prime importance for the radiative transfer through the atmosphere, we choose the carbon cycle as the basis for such a discussion.

12.2.1
Vegetation and soil

Fig. 12.1 shows the general features of the carbon cycle in terms of the amount of carbon (in 10^9 ton = Pg) in the major reservoirs and the rate of transfer (in 10^9 ton yr^{-1} = Pg yr^{-1}) between them. Table 12.1 presents a more detailed picture of the partitioning of carbon between the major terrestrial biomes and their net primary production (NPP), i.e. the difference between photosynthesis and respiration. In a steady state NPP equals the annual detritus formation and return of CO_2 to the atmosphere by bacterial decomposition of organic matter in the soil.

We note that the amount of carbon in the atmosphere (as CO_2) is about the same as the total amount of carbon in living matter on land, while the compartment of dead organic matter in the soil contains twice as much. The circulation rate between these three reservoirs, 50–60 Pg yr^{-1}, means a turnover rate of carbon in the atmosphere and in terrestrial biota of 12–15 years and in the soil of 25–30 years. These figures are, however, misleading when concerned with characteristic response times of the global terrestrial ecosystem. Only about 30% of the organic compounds formed by plants becomes structural matter, primarily cellulose in the tree trunks of the forests, which contain about 90% of all carbon in living matter (see Table 12.1). The average turnover time of carbon in forest trees is, therefore, rather, 40 years. Conditions vary, however, from one type of forest to another. The turnover time of carbon in the northern boreal forests (taiga) is about 100 years, but merely 20–30 years in tropical forests.

Conditions are similar in the soils. Two-thirds of the primary production is in the form of leaves, needles, grass and fine roots, which die and decay within a few years. Only a small fraction of the detritus is incorporated into the soil as slowly degradable organic matter (Schlesinger, 1977). The renewal time of carbon in the soil, therefore, is of the order of several hundred years. It is obviously important to realize the difference between turnover time and age (the latter, for example, determined with the aid of ^{14}C analyses), see Bolin & Rodhe (1973). There is, furthermore, a difference between the podsols and peat bogs in northerly latitudes, where the turnover time may be 1000 years or more, and tropical soils, some of which have a turnover time much of less than 100 years. Because of the more rapid decay in warm and humid climate, the latter soils also contain much less organic matter (see Table 12.1). A more detailed analysis of this kind of interplay between the atmosphere, vegetation and the soil has been presented by Bolin (1981).

The brief analysis given above is, however, not applicable if we are concerned with evolutionary processes, where, for example, weathering also becomes of importance. The process of soil formation on a barren site takes thousands of years. The podsols, which today are found in regions covered by ice 10000 years ago (e.g. in Scandinavia and Canada), have been formed since the ice disappeared; and the top layer containing carbon is usually 10–20 cm thick. A succession of biomes have contributed to this evolution. In this process the availability of nutrients is crucial. Plants that can provide their own nitrogen supply by fixation from atmospheric nitrogen are obviously at an advantage in the early phase of the development, but airborne nitrogen compounds may also play a role. Phosphorus on the other hand has no volatile compounds and is supplied by weathering, primarily of apatite. As evolution progresses, more species appear on the site and gradually a circulation pattern of the nutrients develops, which is almost closed with small losses in terms of gas emissions to the atmosphere and run-off to the sea.

The features of the terrestrial ecosystems summarized above illustrate the important fact that when concerned with changes of climate on time scales of 100 years or more, these ecosystems must not be considered as static, but, rather, their response characteristics must be analysed and possible feedback mechanisms searched for.

12.2.2
The sea

As shown by Fig. 12.1, more than 90% of the carbon that circulates rather rapidly in the biosphere is found in the sea. Its role in the global biogeochemical cycles is crucial. The biogeochemical cycles in the sea are also of fundamental importance for both present climate and climate evolution.

Life in the sea is essentially limited to the photic zone, i.e. the upper part of the mixed layer above the thermocline, which is about 75 m deep, except in polar regions. Here photosynthesis takes place at a rate of about $40 \cdot 10^{15}$ g C yr^{-1} for the oceans as a whole. About 90% of the organisms die, decay and dissolve within the mixed layer and at any one time only about $3 \cdot 10^{15}$ g C are present in the form of living organic matter (de Vooys, 1979). The turnover time of carbon within the surface layer is thus merely about one month.

The limiting factor for phytoplankton growth is lack of

Fig. 12.1. Major features of the carbon cycle. Reservoir sizes are given in 10^{15} g (= Pg = G ton) and fluxes between reservoirs 10^{15} g yr^{-1}. For a discussion of uncertainties in these estimates, reference is made to Bolin *et al.*, (1979).

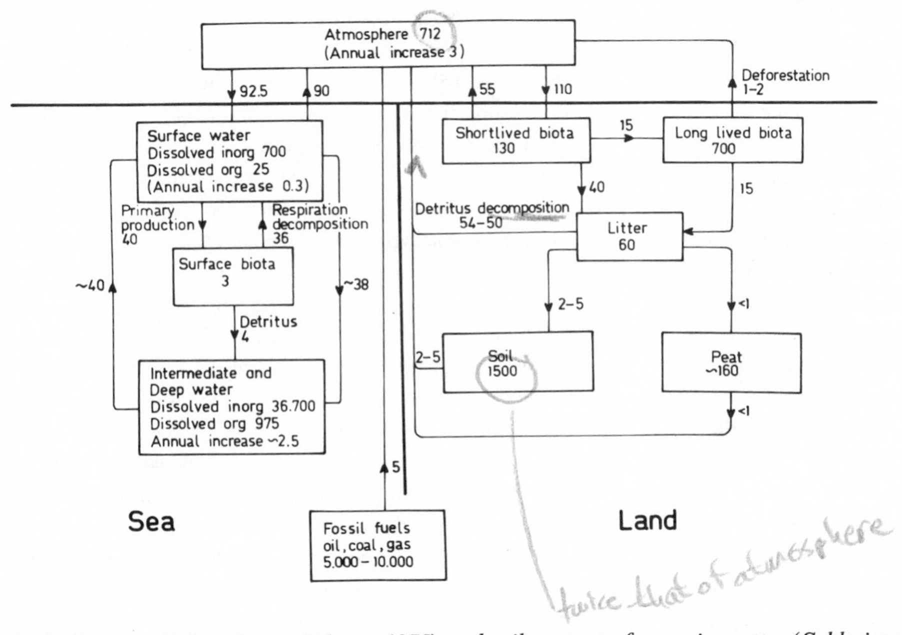

Table 12.1. *Biomass* (*Whittaker & Likens, 1975*) *and soil content of organic matter* (*Schlesinger, 1977*)

Ecosystem type	Biomass: Dry matter: Normal range kg m^{-2}	Biomass: Dry matter: Mean kg m^{-2}	Biomass: Dry matter: Total 10^9 ton	Biomass: Carbon: Total 10^9 ton	Detritus, soil carbon: Mean kg m^{-2}	Detritus, soil carbon: Total 10^9 ton	Net primary production: Dry matter mean kg m^{-2} yr^{-1}	Net primary production: Carbon total 10^9 ton yr^{-1}
Tropical rain forest	6–80	45	765	344 }	11.7	288	2.2	16.8
Tropical seasonal forest	6–60	35	260	117 }			1.6	5.4
Temperature forest	6–200	32	385	174	13.4	161	1.3	6.7
Boral forest	6–40	20	240	108	20.6	247	0.8	4.3
Woodland, shrubland	2–20	16	50	23	6.9	59	0.7	2.7
Savannah	0.2–15.0	4	60	27	4.2	63	0.9	6.1
Temperate grassland	0.2–5	1.6	14	66	18.9	170	0.6	2.4
Tundra, alpine	0.1–30	0.6	5	2	20.4	163	0.14	0.5
Desert, semidesert shrub	0.1–40	0.7	13	6	5.8	104	0.09	0.7
Extreme desert	0–0.2	0.02	0.5	—	0·2	4	0.003	0.0
Cultivated land	0.4–12.0	1	14	6	7.9	111	0.7	4.1
Swamps, marshes	3–50	15	30	14	72.3	145	3.0	2.7
Lakes and streams	0–0.1	0.02	0.5	—	—	—	0.4	0.4
Total		12.2	1837	827	10.2	1515		52.6

nutrients, particularly phosphorus and nitrogen, even though nitrogen compounds may become available by nitrogen fixation. When organisms die they settle out of the photic zone before much bacterial decomposition and dissolution occurs. For this reason the uppermost part of the mixed layer usually has low nutrient concentrations. Vertical mixing and upwelling within the mixed layer are necessary processes to maintain phytoplankton growth in the photic zone. It follows that, on the average, the nutrient content of this uppermost layer is renewed once a month. More precise determinations of the vertical mixing and overturning of the mixed layer can be made by more-careful study of the biological processes in the surface layers. About 10% of the inorganic particulate matter (carbonate and silicon shells) and dead organic matter settle out of the mixed layer into the intermediate and deep waters of the sea, in the case of carbon $3\text{–}5 \cdot 10^{15}$ g C yr^{-1}. Most of these particles dissolve before reaching the bottom of the sea. Direct measurements of this particle flux are difficult (Fiadeiro, 1983), but possibly 10%, i.e. $0{\cdot}5 \cdot 10^{15}$ g C reaches the bottom, where it serves as an energy source for the sparse bottom fauna. Only a small fraction is incorporated into the bottom sediment.

The particle flux into the thermocline region and the deep sea also brings nutrients with it. Carbon, phosphorus, nitrogen, silicon and many other elements are therefore enhanced in these deeper layers. The decomposition of organic matter further requires oxygen, whereby its concentration is diminished. The quasi-steady distributions of all these elements, as observed below the thermocline, represent a balance of the particulate flux and the decomposition processes on one hand, and the transfer of the dissolved compounds by water motions on the other (Bolin, 1983; Bolin *et al.*, 1983). The rate of this slow turnover of the oceans, which depends on climate, determines the vertical gradients of all chemical elements involved in the life processes. Fig. 12.2 shows the distribution of total dissolved inorganic carbon (DIC) in the oceans. We note the 15% difference between surface concentrations and those of the deep sea. If the vertical turnover of the sea were slower, the rate of nutrient supply by upwelling would be reduced and thus also the rate of photosynthesis. The vertical carbon profile (Fig. 12.2) would probably not change much (Keeling, 1973). If, on the other hand, more nutrients were supplied to the sea without any change of vertical turnover, the rate of photosynthesis would increase and the quasi-steady state now maintained might be disturbed. We then recall that the percentage increase of the CO_2 partial pressure in sea water increases by a factor of 10–15 more quickly than the amount of DIC because of the chemical characteristics of the carbon system of the sea (see Keeling, 1973). Thus the prevailing concentration of atmospheric CO_2 is fundamentally dependent on this dynamic balance. An understanding of these interconnections is also important when interpreting climatic records in marine sediments, e.g. those of ^{14}C (see Broecker *et al.*, 1977; Bolin, 1981).

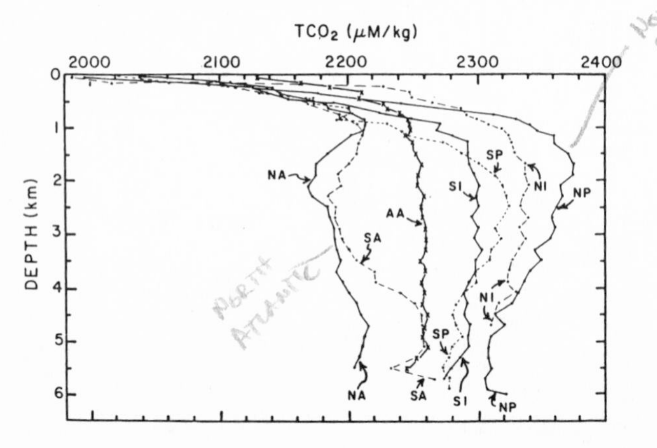

Fig. 12.2. The mean vertical distribution of total dissolved carbon in seven regions of the world oceans. NA = North Atlantic, SA = South Atlantic, NP = North Pacific, SP = South Pacific, NI = North Indian, SI = South Indian and AA = Antarctic region. (From Takahashi *et al.*, 1981.)

The circulation of the deep sea is not well known. Direct measurements are difficult because of the slow and sluggish motions, even though neutrally buoyant floats have been used successfully in recent years (Rossby, 1983). Data are, however, inadequate to verify prevailing theories for deep ocean circulation (Veronis, 1978; Wunsch, this volume) in more than their broad outline. We may, however, ask ourselves, in which way the quasi-steady distributions of a large number of tracers can be used to gain information about the circulation and turbulent processes that establish them. Such attempts are of course not new, but have long been employed in regional studies of water transfer and exchange.

There are basically two approaches that can be used. Most directly the general circulation models of the oceans, as developed by Bryan (1975) and others, can be supplemented by a series of continuity equations for the chemical compounds concerned, in which the biogeochemical processes appear as sources and sinks. These latter equations are integrated in parallel with those that account for changes of momentum, heat, water and salt until a steady state is reached. Assuming that a quasi-steady state prevails in reality, we may compare the computed distributions with those observed. A principle difficulty is of course that the biogeochemical processes are not well known and are therefore difficult to incorporate properly into such computations. It should be noted, however, that a detailed understanding may not always be needed. In a first attempt, one might use the hypothesis advanced by Redfield (1958) that the approximate proportions in which basic elements are incorporated into organic tissue and also released in the decomposition process, can be specified as constants, 'Redfield ratios'. Fractionation processes, that certainly occur may, in a first approximation, be neglected. A series of computations can be made with different assumptions about the biochemical processes and their spatial distribution to achieve optimum agreement with the observed distribution of the trace elements. There is of course no standard method for how to proceed. Attempts of this kind do not necessarily converge and may in any case be very time consuming.

As an alternative method we may address this inverse problem more directly, i.e. attempt to solve for the distribution of water circulation, rate of turbulent water exchange and biological transfer, which is consistent with the steady state spatial distributions of a set of tracer elements (see Bolin *et al.*, 1983; Wunsch & Mercier, 1982). To illustrate the method, we consider a finite difference formulation of the continuity equations for k tracers. Let us divide the ocean into a set of m reservoirs in exchange with each other across altogether n internal surfaces between these reservoirs.

Fig. 12.3. 12-reservoir model of the world oceans (*a*) Tracer concentrations used in the computations – First row: dissolved inorganic carbon (mol m^{-3}); $\Delta^{14}C$ (relative to $\Delta^{14}C$ standard). Second row: alkalinity (eq); phosphorus (m mol^{-1} m^{-3}); oxygen (mol m^{-3}). Third row: volume (10^{15} m^3). (*b*) Deduced fluxes for model version (*a*) reference case – Between boxes: advective (→) and turbulent (↔) fluxes of water (10^{15} m^3 yr^{-1}). In boxes: upper figure, net organic detritus formation and loss of carbon (−) or decomposition and gain of carbon (+) in 10^{15} mol C yr^{-1}; lower figure, net carbonate formation and loss of carbon (−) or dissolution and gain of carbon in 10^{15} mol C yr^{-1}; bottom figure, turn-over time for water in the box. (See further, Bolin *et al.*, 1983.)

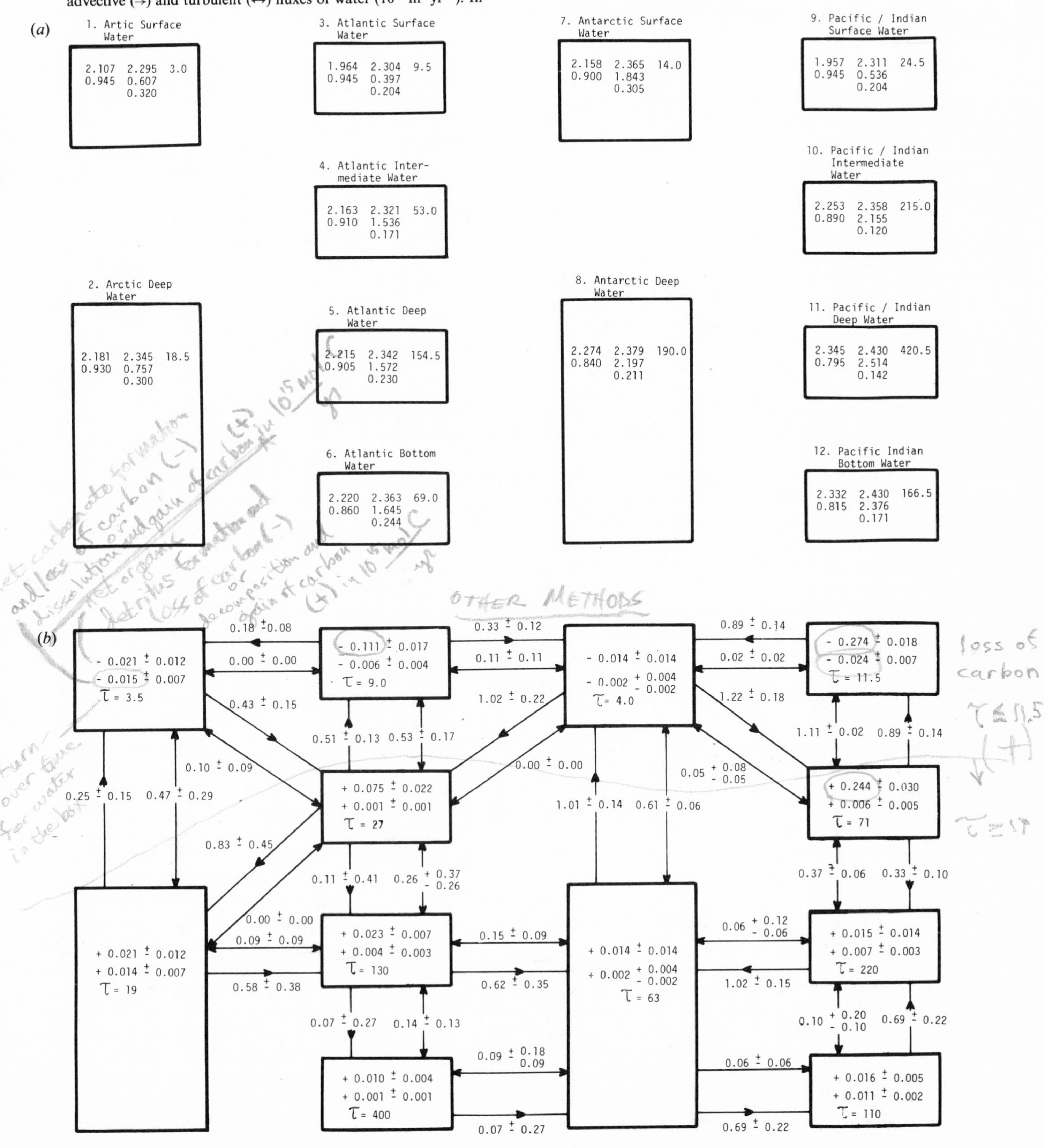

We further define a mean advective and a turbulent flux of water across each such surface, which become our $2n$ unknown variables that we wish to determine. In addition, we introduce the rates of *net* production or decomposition of organic tissue, carbonate and silicate in each one of the reservoirs as $3m$ more unknown variables. We assume that the sources and sinks of other tracer elements can be related to the processes of primary production and decay as expressed by sources and sinks of organic tissue, carbonate and silicate, by using Redfield ratios. With the aid of the k tracer distributions we can formulate $k \cdot m$ equations based on the condition that the amount of tracer material for each tracer and for each reservoir is conserved. These equations can be used to determine the $2n+3m$ unknowns that describe the water motions and the biochemical processes. If $k \cdot m \leqslant 2n+3m$, constraints may have to be introduced to derive a solution. We can also apply a principle of minimization, e.g. ask for that minimum vector with the $2n+3m$ unknowns as components, which satisfies our set of equations. In the case $k \cdot m > 2n+3m$, we may instead ask for the solution that minimizes the errors in satisfying the $k \cdot m$ conservation equations. Methods of matrix inversion are available to solve large sets of equations, but practical difficulties arise when $k \cdot m$ becomes large.

The method has been preliminarily tested by using a 12-box model of the ocean as shown in Fig. 12.3. In addition to water continuity, five tracers were used: total dissolved inorganic carbon, alkalinity, ^{14}C, oxygen and phosphorus (see Bolin *et al.*, 1983). Water motions between the reservoirs were described by 21 advective velocities and turbulent exchange by 21 exchange rates. Silicate transfer was not accounted for and there were thus altogether $2n+2m = 66$ unknowns. In addition to the $6m = 72$ conservation equations, we demanded that the detritus formation in each surface reservoir was balanced by decomposition (of organic matter) or dissolution (of carbonate) in the reservoirs below. This yielded an additional eight equations. Even though the spatial resolution is crude and the errors in the finite difference formulation thus considerable, we obtain spatial patterns of advective and turbulent motions and detritus flux, which, in general, agree with what has been deduced by other methods (Fig. 12.3 (*b*)).

We have a diagnostic tool which can be further developed. Thus the incorporation of some dynamic constraints and the inclusion of heat and salt transfer is obviously desirable.

Even though the results described briefly above are tentative, some further considerations are of interest. Having derived a set of advective and turbulent transfer rates for an ocean model, which in their gross features depict the real ocean reasonably well (for further details see Bolin *et al.*, 1983) we may ask the question: How effective is this model of the oceans as a sink for the CO_2 emissions that have been occurring during the last 100 years? The transient response of the model to such an emission has been analysed (Bolin *et al.*, 1983). We find that the part of the total emissions that remain in the atmosphere, i.e. the airborne fraction, using the present model, is close to 70%. It is clear, however, that the resolution as adopted is not sufficient to determine this value accurately. A more detailed consideration of the thermocline region would most likely reduce this figure. It is of interest in this context to compare this result with a similar analysis by Siegenthaler (1983). He has used an ocean model in which he considers both vertical diffusion from the ocean surface through the mixed layer into deeper layers and turbulent transfer along isopycnic surfaces from the Arctic and Antarctic surface waters. If, assuming that the latter process is infinitely rapid, the rate of transfer of CO_2 into the ocean is primarily dependent on the size of that part of surface ocean water that communicates directly and rapidly with deep ocean waters and the rate of air sea exchange in this area. On the basis of his model, Siegenthaler determines the minimum value for the airborne fraction to be 62%. It seems likely, however, that a model with better resolution might even yield a smaller value for the airborne fraction (see Broecker et al., 1980). We note, further, that if only considering emissions due to fossil fuel combustion, the observed airborne fraction during the last 20 years at Mauna Loa has been 55% (Bacastow & Keeling, 1981). However, a very considerable net transfer of CO_2 to the atmosphere has also occurred due to deforestation and expanding agriculture (Moore *et al.*, 1981). Thus the airborne fraction of anthropogenic emissions to the atmosphere is considerably less than 55%. We conclude that our understanding of the global carbon cycle is still inadequate to permit a more accurate assessment of the CO_2 increase in the atmosphere, due to a given scenario of the future use of fossil fuel. The possible increased photosynthesis as a result of the enhanced atmospheric CO_2 concentrations and the sedimentation and burial of organic matter in lakes and the marine coastal zone should be analysed more carefully. In doing so we must consider the interactions between the carbon cycle and the nutrient cycles in terrestrial, freshwater and marine coastal ecosystems (see Bolin & Cook, 1983).

Finally, it is important to emphasize that the processes of advection and mixing in the oceans, which account for the transfer of excess CO_2 into deeper layers of the sea, also transfer heat in the oceans and thus determine the thermal response of the ocean to a changing pattern of heat transfer in the atmosphere. Also in this regard, the role of the thermocline region has not yet been properly considered.

12.3 Some key biogeochemical processes of importance in climate modelling

Present general circulation models used in climate research, resolve atmospheric and oceanic motions down to scales of 300–500 km in the horizontal and one or a few km in the vertical, somewhat better in the boundary layers between the atmosphere and the sea. This seems necessary to account properly for the transfers of energy and momentum both in the atmosphere and the sea, moisture in the atmosphere and salt in the sea, which all are fundamental in any climate model. Attempts to model the global biogeochemical processes have not progressed much beyond problems of the overall global balances. It should be clear from the previous section that further progress now requires the consideration of the regional features of the biogeochemical cycles. Also, most of the processes that might be of importance in climatic modelling need be considered with proper spatial resolution. We shall consider some of these in more detail.

12.3.1
Terrestrial processes
12.3.1.1
Heat and moisture exchange

In most climate models heat and moisture transfer in the soil is considered and usually made dependent on atmospheric conditions and some bulk soil properties, such as thermal conductivity and moisture storage capacity, which, however, usually are assumed not to vary in space and time. For this reason, the response of the earth's surface to changing atmospheric conditions is rather restricted. Reality is, of course, more complex.

Except in desert areas, the soil is covered by vegetation which, to a very considerable degree, determines its characteristics with regard to heat and moisture transfer. Because of the fact that trees are tens of metres high and roots generally extend one or sometimes several metres into the soil, the boundary surface between the atmosphere and the ground is transformed into a transition zone, the characteristics of which are determined by the vegetation (see Section 12.3.1.3). Since the vegetation is climate dependent, its response to climate change should be considered, particularly when concerned with long-term changes of climate. Water storage capacity of the soil should include the water that can be stored in the vegetation and in the soil down to the depth that is reached by the roots. This capacity is obviously very different from one biome to another. It would be of interest to make some sensitivity analyses of how much changes in areal extension of the present biomes would change the pattern of potential evapo-transpiration. Some recent experiments by Shukla & Mintz (1982) that concern rather extreme conditions for the moisture exchange between the ground and the atmosphere, indicate that a proper description of the mutual interplay between climate and distribution of vegetation on land is indeed of importance in predicting climate change.

12.3.1.2
Radiative processes at the earth's surface

It is well known that the albedo of the land surface varies considerably in space and time (Kondratiev, 1969; Kukla, 1981). The obvious difference depending on whether land is covered by snow and ice or not is accounted for in most present climate models. Some other differences depending on the vegetation cover have usually not been considered and certainly not in terms of a dynamic interplay between climate and vegetation.

Table 12.2. *Albedo (in %) for bare soil and different vegetation types. (After Kondratiev, 1969)*

		Dry	Moist
Savanna, semidesert	24		
Decidous forest / Tops of oak	18		
Coniferous forest / Tops of pine	14		
Tops of fir	10		
Fallow fields		10	6
Black earth		14	8
Grey earth		25–30	10–12
Blue clay		23	16
Snow on open fields	70–80		
Forest, stable snow cover	45		
Forest, unstable snow cover, fall	30		
Forest, unstable snow cover, spring	25		
Forest, without snow (Coniferous forest)	14		

The albedo values shown in Table 12.2 (from Kondratiev, 1969) have been arranged in order to bring out three characteristic features that are of interest in this context. Analyses of the very large data source from satellites certainly could bring out such systematic differences more accurately.

(1) The albedo of vegetation is less at high than at low latitudes. It seems likely that this is a result of environmental adaptation. At high latitudes it is important for plants to be able to use efficiently the sparse solar radiation during a short growing season, while at low latitudes it is rather important to strike a balance between the need for solar radiation in the process of photosynthesis and the minimization of evapo-transpiration, which increases with temperature

(2) Vegetation and bare soil has a larger albedo under dry conditions than when wet.

(3) In winter the albedo of a forest is much less than that of an open field. It also differs depending on whether snow is covering the branches or the trees have lost their snow cover due to thawing.

It might not be necessary to account in detail for the daily variations of the albedo due to rain or snow, even though this is, in principle, possible and sometimes done with regard to snow. The duration of periods with dry or wet vegetation much depends, however, on temperature. At low temperatures (but still above freezing) modest amounts of precipitation may keep the ground constantly wet, while at high temperatures even abundant rains are quickly absorbed into the ground or evaporated, whereby periods of wet ground are reduced to short spells.

Climatic changes on time scales of a few decades or more could lead to changes of the areal extension of different types of vegetation and also frequencies of dry or wet ground. A warmer climate at middle and high latitudes in the northern hemisphere permits the advance northward of the boreal forest into regions at present covered by tundra. At its southern boundary, on the other hand, the decidons forest might invade areas of boreal forests. The rapidity with which changes might occur also depends on the rate of soil changes that would be associated with such biome changes. It is difficult to judge the importance of the changes of albedo that would result. It is interesting in this context to note that the boreal forest in Asia primarily consists of larch which is not an evergreen. What would a change to pine or spruce with a smaller albedo in winter time imply for the regional or even possibly for the global climate?

12.3.1.3
Photosynthesis

Photosynthesis in plants requires energy (solar and diffuse radiation) water, carbon dioxide and nutrients. The process proceeds differently in C3 and C4 plants. The latter are more efficient and also more optimally tuned to the atmospheric CO_2 concentration of about 300 ppm (v) that prevailed before the recent increase due to fossil fuel burning began. Corn is the most well-known C4 plant, but a systematic inventory of whether a plant is of one kind or the other is not available. Very little is known about the large variety of tropical plants.

Even though plants under optimum conditions are able to utilize up to a few per cent of the incoming solar radiation, much less is actually being transformed globally into chemical energy and stored in the growing plants. Land only constitutes about 30% of the earth's surface, 25% of the land is covered by desert, i.e. the dry deserts in the sub-tropics, cold deserts in polar regions and at sufficient high elevation in mountains, and semi-deserts occupy another 10%. Only about 50% of the total land surface is covered by grassland, farmland and forests, which contribute significantly to photosynthesis on earth. Due to inadequate water supply or nutrient deficiencies, on the average less than 50% of the optimum photosynthetic capability is utilized in large parts of these areas. Finally, about half of the energy captured in the process of primary production is required for respiration, in which a transformation from the aldehydes to more-complex organic compounds takes place. Only about 0.1% of the solar radiation coming to the earth is actually stored as chemical energy in plants.

Water in plants is required both as a transport agent and in the biochemical processes. Sophisticated mechanisms have evolved whereby the plants can regulate their evapo-transpiration within wide limits. The stomata, through which carbon dioxide and some gaseous nutrients are diffused into the chloroplasts, open and close, depending primarily on incoming short-wave radiation and the internal water supply. Maximum rate of photosynthesis occurs rather early during the day when the plant is close to saturation and the organic products can be quickly transferred from the site of photosynthesis into the plant. During the assimilation processes, water is lost through the open stomata. If this loss exceeds the uptake from the roots, the stomata are partly closed. If sufficiently narrow, the rate of CO_2 diffusion will be reduced and photosynthesis decreased. Obviously a higher ambient CO_2 pressure would, under such circumstances, increase the rate CO_2 diffusion through the stomata. If the nutrient supply is adequate, the rate of photosynthesis may then be larger for a given rate of water loss. Such an increased water use efficiency may permit greater plant survival in dryer regions than would be the case at a lower ambient CO_2 pressure. A possible change of this kind cannot be detected in natural ecosystems because of the large spatial and temporal variability. An increase in the rate of photosynthesis in agriculture may well have occurred, particularly where water is a limiting factor for growth. Again it would be difficult to detect because of constantly changing farming practices and variations in yield due to interannual climatic variations. We are therefore not yet able to assess the magnitude of such possible changes.

Different biomes behave very differently to climatic change with regard to the processes of photosynthesis. The determination of the global changes therefore requires the proper consideration of spatial distributions of biomes, which so far has not been the case in modelling any of the biogeochemical cycles.

12.3.1.4
Exchange of other gases between terrestrial biota and the atmosphere

The stomata play a role for transfer of gases other than water vapour and CO_2. Gaseous emissions also emerge from the soils. Both processes are important for the maintenance of present concentrations in the atmosphere. This is so for CO, SO_2, H_2S, N_20, NO, NO_2, NH_3 and a number of organic compounds. As an illustration we shall consider SO_2 and H_2S.

Sulphur is needed in small amounts in the formation of some proteins. Already, atmospheric concentrations of a few ppb provide enough sulphur for the plant, but light is required for the process of uptake (Hällgren *et al.*, 1982). No SO_2 transfer occurs during the night. The sulphur transfer begins in the morning when radiation increases, but soon falls behind the maximum possible, which would be expected on the basis of the estimated diffusive transfer of water vapour out of the plant. This difference is particularly noticeable in highly polluted air with SO_2 concentration well above what is required for the production of organic matter in the plant. A careful analysis shows that the SO_2 flux is not much reduced, but the plant, rather, emits reduced sulphur compounds, presumably produced by sulphate reduction, which diminish the net inward flux of sulphur. It seems that this is a protective measure because it increases when atmospheric SO_2 concentrations increase. At levels above one or a few hundred ppb concentrations the rate of photosynthesis is reduced and damage to leaves and needles appears.

The transfer processes and associated biochemical processes described above show the active chemical role that plants play, which need to be considered in assessing the biological sources and sinks for atmospheric constituents. The occurrence of nitrogen oxides in the atmosphere is similarly a result of biological processes, primarily in the soil, and is associated with dinitrification in which leakage to the atmosphere of nitrogen oxides takes place. Soil moisture plays an important role in determining the partitioning between N_2O and N_2 in the emission to the atmosphere.

12.3.2
The role of the sea: the nutrient balance

Nitrogen and phosphorus are two most crucial nutrients for both terrestrial and marine ecosystems. It has almost without exception been assumed that pre-industrial conditions were characterized by steady state N and P cycles. It is, however, not known to what degree this has been the case. It is interesting that this question was raised about 100 years ago, when the processes of dinitrification had recently been discovered. In an address to the Royal Society in England Sir William Crookes proposed that fertilization with nitrate seemed necessary to avoid a gradual decrease of the food production of the world (Bolin, 1972).

Broecker (1981) and McElroy (1982) have recently analysed some implications of the long-term variations in the phosphorus and nitrogen cycles respectively, particularly with regard to the role of the oceans. Changes of temperature and salinity of the oceans that were associated with the transition from glacial to interglacial conditions 13000–8000 years ago disturbed the distribution of carbon between the atmosphere and the sea. Those changes would, however, be rather small and less than the change from about 210 ppm to 350 ppm that analyses of air in Antarctic ice indicate have taken place. Broecker (1981) points out, however, that sediment records reveal other simultaneous changes: (1) the amount of carbon in living matter increased significantly as shown by the change of the $^{13}C/^{12}C$ ratio; (2) deposition of $CaCO_3$ probably took place maintaining an approximately constant alkalinity of the sea in spite of transfer of a sizeable amount of carbon from the sea to organic matter on land; (3) the phosphorus content of the sea decreased as indicated by different changes of the $^{13}C/^{12}C$

ratio for bentic and planktonic organisms in the sea.

Broecker (1981) considers that the rise of sea level by more than 100 m during this period, and a considerable increase of the shelf area, is an important reason for these changes. The phosphorus decrease was presumably caused by some (slight) increase in sedimentation and burial or organic matter in the coastal zone.

The changes in the nitrogen cycle as discussed by McElroy (1982) are similar in nature and would be associated with cyclic imbalance of nitrification and dinitrification. In this case the transfer of nutrients from land to sea, and biological processes in estuaries and the coastal zone, are of prime importance.

The suggestions made by Broecker (1981) and McElroy (1982) should be considered as plausible working hypotheses that emphasize the role of biological processes that might be of importance for climatic change. Man is today markedly influencing the global biogeochemical cycles. The production of nitrogen oxides and nitrate in combustion and fertilizer manufacturing is presently more the 50% of the rate of nitrification in the global ecosystem. An understanding of the complex interactions among the biogeochemical cycles that are associated with both climatic change and man's increasing activities requires a considerable modelling effort in order to derive likely changes in a consistent manner. Such work has merely just begun.

12.4 Dynamic modelling of ecosystems for studies of climatic change

12.4.1 *Modelling ecosystem processes*

The behaviour of an ecosystem basically depends on physical, chemical and biological processes and their interaction from the molecular scale up to any scale less than that of the total system itself. It is obviously impossible by successive syntheses, starting at the molecular level, to develop a model for the system as a whole. Nevertheless, the basic processes must be properly considered. We need to describe in some simplified manner how they interact to bring about changes of importance for the overall characteristics of the system, for example a biome, which we are studying. In the present case we are also interested in the possible effects of a changing ecosystem for the global climate. This requires an understanding of those *integral* properties of the subsystem that are of importance in this regard, as qualitatively outlined in Section 12.3. Some hypotheses about the relative significance of a number of possible effects should preferably serve as a basis of such modelling attempts and comparisons between model computations and observations of reality are important to progress in an orderly and consistent manner.

It is of some principal interest to analyse more closely the way we might describe the behaviour of a biome as dependent on abiotic parameters such as temperature, humidity and atmospheric CO_2 concentration. Greenhouse experiments and perhaps also field experiments, might yield relations between the net primary production and abiotic factors. We may be able to simulate both the daily and annual cycles but may still not have learned much about the *long-term* dependence of the biome on these environmental parameters. On the basis of such experiments, it has, for example, been concluded that the rate of photosynthesis might increase proportionally to some power, β, of the atmospheric CO_2 concentration (Keeling, 1973; Revelle & Munk, 1977). Relevant questions from the point of view of climate modelling are rather:

- Will there be an accumulation of carbon in the terrestrial ecosystem and thus a net withdrawal of carbon dioxide from the atmosphere?
- Will soil and root characteristics of the biomes change in a manner that will modify the potential evapo-transpiration?
- Will the species composition of the biome change and affect the albedo?

The answer to the first question is dependent on: (1) whether the ecosystem is close to climax conditions or not; (2) the way in which climax depends on abiotic factors or, rather, is a function of the species of which it is composed; (3) how rapidly such changes may occur as dependent on soil characteristics and their response to change. We are not able to give good answers to such questions at present. New methods of ecosystem analysis need to be developed.

12.4.2 *Spatial variations within a given biome*

There are marked spatial variations of characteristic features of a biome. In the boreal forest, close to the tundra, for example, trees are far apart and thus biomass comparatively small, while

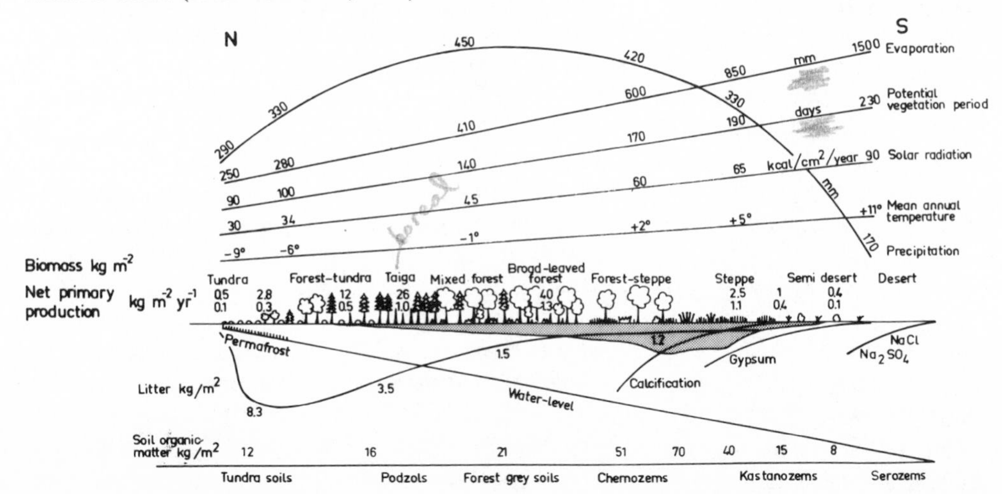

Fig. 12.4. North–south succession of biomes in the USSR from tundra to desert. (From Bolin *et al.*, 1979.)

dead organic matter in the soil has accumulated in large amounts. Conditions are reversed when we approach a more temperate climate, see Fig. 12.4. The present distribution is a result of an evolution over centuries and millenia, and a steady state may not yet have been reached. We ask the question if the present distributions of biomes, their spatial characteristics and their changes during the last few thousand years, as revealed by paleoecological studies, can be used to deduce their response characteristics to change of abiotic factors. The problem is, of course, complex because of the dependence of ecosystem features on the soil and geological substrate. Serious consideration should, however, be given to the way this source of information could possibly be used for modelling of ecosystem dynamics and how to include such processes into climate models.

12.4.3 *A data bank for land surface processes*

It is clear that land surface processes are of importance in studies of climate and climate change. In present models, land surface features are usually described by a few parameters such as albedo, roughness, soil water storage capacity, etc, which are assumed to be constant. In the light of the previous discussion it seems important to develop methods to describe the interactions between the atmosphere and the land surface in a more dynamic way. We will, to begin with, have to use some simple models in which subscale processes are parameterized. Because of the marked spatial variations, it may be necessary to adopt different models as dependent on soil features, biome characteristics, prevailing climatic conditions, etc. A systematic approach is needed because a careful analysis for the globe as a whole needs to be conducted with a resolution corresponding to what is commonly being used in climate models. It would obviously be desirable to agree on a standard grid to be used for the development of such a data bank.

Man today is obviously modifying land surface characteristics significantly. There are attempts to describe those well enough to obtain information on ongoing regional and global changes. Careful mapping of the changes of the extension of tropical forests during the last 25 years by UNEP (United Nations Environment Programme) are in progress. Moore *et al.* (1981) have attempted to assess the total release of carbon dioxide to the atmosphere, due to deforestation and expansion of agriculture. In none of these attempts have climatic parameters been much used in order to obtain a more internally consistent description of these changes. The combination of such attempts to map the land surface changes due to man and those needed for climatic modelling is obviously desirable.

To assess the consequences, for man, of climatic change, it is obviously necessary to analyse the global distribution of such changes and to carefully study their impacts on the terrestrial biomes. Also in this context, it is necessary to have available a data bank that can be used to deduce the ecosystem response in as much detail as is possible.

12.5 Concluding remark

The role of biological process in climatic change has so far been inadequately considered. Important feedback mechanisms may therefore have been overlooked, some of which may possibly cause natural climatic oscillations. The climate changes that may be caused by anthropogenic effect such as an increasing atmospheric CO_2 concentration have so far been deduced without consideration of feedbacks from terrestrial or marine ecosystems. It seems urgent to analyse some of the most obvious possibilities in this regard to better ascertain the uncertainties of the results deduced so far. The development of simple ecosystem models for this purpose should be given high priority.

References

Bacastow, R. B. & Keeling, C. D. (1981). 'Atmospheric carbon dixoide concentration and the observed air borne fraction'. *In: Carbon Cycle Modelling*, SCOPE 16, ed. B. Bolin, pp. 103–16. John Wiley, Chichester.

Bolin, B. (1972). 'Atmospheric chemistry and environmental pollution'. *In: Meteorological Challenges; A History*, ed. D. P. McIntyre, pp. 237–66. Canadian Meteorological Service.

Bolin, B. (1981). 'Steady state and response characteristics of a simple model of the carbon cycle'. *In: Carbon Cycle Modelling*, SCOPE 16, ed. B. Bolin, pp. 316–2. John Wiley, Chichester.

Bolin, B. (1983). 'Changing global biogeochemistry'. *In: Oceanography; The Present and Future*, ed. P. Brewer, pp. 305–26. Springer Verlag, Berlin.

Bolin, B. & Rodhe, H. (1973). 'A note on the concepts of age distribution and transit time in natural reservoirs'. *Tellus* **25**, 58–62.

Bolin, B., Degens, E. T., Kempe, S. & Ketner, P. eds. (1979). *The Global Carbon Cycle*, SCOPE 16. John Wiley, Chichester.

Bolin, B. & Cook. R. B. eds. (1983). *The Major Biogeochemical Cycles and their Interactions*, SCOPE 21. John Wiley, Chichester.

Bolin, B., Björkström, A., Holmén, K. & Moore, B. (1983). 'The use of tracers for oceanic studies and particularly for determining the role of the oceans in the carbon cycle'. *Tellus*, **358**, 206–36.

Broecker, W. S. (1981). 'Glacial to interglacial changes in ocean and atmospheric chemistry'. *In: Climate Variations and Variability: Facts and Theory*, ed. A. Berger, pp. 111–21. NATO Advanced Study Institute Series. D. Reidel, Dordrecht.

Brocker, W. S. & Takahashi, T. (1977). 'Neutralization of fossil fuel CO_2 by marine calcium carbonate'. *In: The Fate of Fossil Fuel CO_2 in the Oceans*. eds. N. R. Andersen & A. Malakoff, pp. 213–42. Plenum Press. New York.

Broecker, W. S., Peng, T.-H. & Engh, R. (1980). 'Modeling the carbon system'. *Radiocarbon*, **22**, 565–98.

Bryan, K. (1975). 'Three-dimensional numerical models of the ocean circulation'. *In: Numerical Models of the Oceans*, pp. 94–106. Nat. Ac. Sci., Washington DC.

de Vooys, C. G. N. (1979). 'Primary production in aquatic environments'. *In: The Global Carbon Cycle*, SCOPE 16, eds. B. Bolin, E. T. Degens, S. Kempe & P. Ketner, pp. 259–92. John Wiley, Chichester.

Fiadeiro, M. (1983). 'Physical–chemical processes in the open sea'. *In: The Major Biogeochemical Cycles and their Interactions*, SCOPE 21 eds. B. Bolin & R. B. Cook, pp. 461–76. John Wiley, Chichester.

Hällgren, J. E., Linder, S., Richter, A., Troeng, E. & Granat, L. (1982). 'Uptake of SO_2 in shoots of Scots pine: field measurements of net flux of sulfur in relation to stomatal conductance'. *Plant. Cell and Environment*, **5**, 75–83.

Ivanov, M. V. & Freney, J. R. ed. (1983). *The Global Biogeochemical Sulfur Cycle*. John Wiley, Chichester, in press.

Keeling, C. D. (1973). 'The carbon dioxide cycle. Reservoir models to depict the exchange of atmospheric carbon dioxide with the oceans and the plants'. *In: Chemistry of the Lower Atmosphere*. ed. S. Rasool, pp. 251–329. Plenum Press, New York.

Kondratiev, K. Ya. (1969). *Radiation in the Atmosphere*, Academic Press, New York.

Kukla, G. (1981). 'Surface albedo'. *In: Climate Variations and Variability: Facts and Theory*, ed. A. Berger, pp. 85–110. NATO Advanced Study Institute Series. D. Reidel, Dordrecht.

Likens, G. E., ed. (1981). *Some Perspectives of the Major Biogeochemical Cycles*. SCOPE 17, John Wiley, Chichester.

Lovelock, J. E. (1979). *Gaia, A New Look at Life on Earth*. Oxford University Press, Oxford.

McElroy, M. B. (1982). Presentation at the COSPAR Symposium, Ottawa, May, 1982.

Moore, B., Boone, R. D., Hobbie, J. E., Houghton, R. A., Melillo, J M., Peterson, B. J., Shaver, G. R., Vörösmaety, C. J. & Woodwell, G. M. (1981). 'A simple model for analysis of the role of terrestrial ecosystems in the global carbon budget'. *In: Carbon Cycle Modelling*, SCOPE 16, ed. B. Bolin, pp. 365–85. John Wiley & Sons, Chichester.

Oeschger, H., Siegenthaler, U., Schotterer, U. & Quegelman, A. (1975). 'A box diffusion model to study the carbon dioxide exchange in nature'. *Tellus*, **27**, 168–92.

Redfield, A. C. (1958). 'The biological control of chemical factors in the environment' *Am. Sci.*, **46**, 206–26.

Revelle, R. & Munk W. (1977). 'The carbon dioxide cycle and the biosphere'. *In: Energy and Climate*. Studies in Geophysics, Nat. Ac. Sci., Washington DC.

Rossby, H. T. (1983). 'Eddies and ocean circulation'. *In: Oceanography; The Present and Future*, ed. P. Brewer, pp. 137–62. Springer Verlag, Berlin.

Rosswall, T. (1981). 'The biogeochemical nitrogen cycle'. *In: Some Perspectives of the Major Biogeochemical Cycles*, SCOPE 17, ed. G. Likens, pp. 25–49. John Wiley, Chichester.

Schlesinger, W. H. (1977). 'The world carbon balance in terrestrial detritus'. *Am. Ecol. Syst.*, **8**, 51–81.

Shukla, J. & Mintz, Y. (1982). 'Influence of land-surface evapo-transpiration on the earth's climate'. *Science*, **215**, 1498–1501.

Siegenthaler, U. (1983). 'Uptake of excess CO_2 by an outcrop–diffusion model of the ocean'. *J. Geoph. Res.*, **88**, 3599–608.

Takahashi, T., Broecker, W. S. & Bainbridge, A. E. (1981). 'The alkalinity and total carbon dioxide concentration of the world ocean'. *In: Carbon Cycle Modelling*, SCOPE 16, ed. B. Bolin, pp. 271–86. John Wiley, Chichester.

Veronis, G. (1978). 'Model of world ocean circulation: III Thermally and wind-driven. *J. Marine Res.*, **36**, 1–44.

Whittaker, R. L. & Likens, G. E. (1975). 'The biosphere and man'. *In: Primary Productivity of the Biosphere*. Ecol. Studies 14, eds. H. Leith & R. H. Whittaker, pp. 305–28. Springer Verlag, Berlin.

Wunsch, C. & Minster, J.-F. (1982). 'Method for box models and ocean circulation tracers: mathematical programming and non-linear inverse theory, *J. Geoph. Res.*, **87**, 5647–62.

K. Ya Kondratyev and
N. I. Moskalenko
Laboratory of Remote Sensing, Institute for Lake Research, Leningrad, USSR

The role of carbon dioxide and other minor gaseous components and aerosols in the radiation budget

Abstract

Calculations are presented of the greenhouse effect, which arises from different atmospheric constituents, in particular from CO_2, H_2O, O_3, CH_4, NH_3, nitrogen oxides and freons. Estimates are made of the mean surface temperature at different epochs during the evolution of the earth's atmosphere. The effect of man's changing activities (especially those resulting in increases in CO_2 and freons) are shown to be significant, although our lack of knowledge of different feedbacks including, for instance, the oceans in the climate system means that reliable estimates of these effects cannot yet be made. The presence of aerosol can lead to net heating or cooling, dependent on its type and distribution. In the ozone layer in the stratosphere, complex interactions occur between radiation, photochemistry and dynamics.

13.1 **The greenhouse effect**

The problem of the impact of radiative factors on atmospheric circulation and climate is central to studies of global energetics of the atmosphere. The radiative regime of the planet is substantially dependent on the chemical composition and turbidity of the atmosphere. As the earth has evolved, the chemical composition and structure of its atmosphere has changed substantially [5, 7, 15, 18]. Climatic changes at present and in the near future can be, to a significant degree, of anthropogenic origin [1–3, 12, 21, 23–35] and therefore the problem of the growing climatic impact of atmospheric pollution is an important aspect of global ecology.

Variations in the concentration and chemical composition of the optically active gaseous and aerosol components of the atmosphere can substantially affect the atmospheric radiation budget and, consequently, the atmospheric circulation. The main mechanism for the climatic impact of radiative factors is *the greenhouse effect*, in that the atmosphere, being comparatively transparent to solar radiation, acts as a screen for thermal emission from the planetary surface.

In this chapter, we shall summarise some recent results from studies of the greenhouse effect and its impact on climate; some more-detailed aspects of studies accomplished by 1980 may be found in the references [1–3, 12, 21, 31, 33].

The greenhouse effect may be defined by the difference in temperatures

$$\Delta T = T_s - T_r, \tag{13.1}$$

where T_s is the temperature of the planetary surface and T_r is a radiative temperature defined by

$$T_r = \left[\frac{F^{\uparrow}}{\sigma}\right]^{1/4}, \tag{13.2}$$

where $F^{\uparrow}$ is the outgoing thermal emission flux at the top of the atmosphere and σ is the Stefan Boltzmann constant. In the absence of the atmosphere, $T_r = T_s$ and $\Delta T = 0$.

The globally averaged thermal emission of the planet is characterised by the equilibrium temperature, T_e. In the absence of interior heat sources, the average value of T_r is the equilibrium temperature, T_e, which is determined from the balance averaged over the planet between the absorbed solar energy and outgoing thermal emission, i.e.

$$T_e = [q_0\,(1-A)/4\sigma]^{\frac{1}{4}}, \tag{13.3}$$

where q_0 is the solar constant for the planet and A is its total albedo.

Calculations of the atmospheric greenhouse effect are presented below. For these calculations the absorption of radiation by atmospheric gases was deduced both from the results of complex measurements of absorption spectra and from direct line-by-line calculations, taking account of the fine structure of the absorption spectra [11]. Optical characteristics of atmospheric aerosols and cloudiness (spectral coefficients of absorption, scattering, and phase function) have been calculated for models of aerosol distributions, both with regard to the chemical composition and the multi-modal size distribution of the aerosol [13, 22].

At present, the radiative regime of the cloudless atmosphere is mainly determined by its water vapour, carbon dioxide, ozone and aerosol, the effect of water vapour in the greenhouse effect being most important.

In between the main absorption bands of CO_2, H_2O and O_3 are 'window' regions where the atmosphere is partially transparent. The most important 'window region' situated near the peak of the black-body curve at terrestrial temperatures is in the region 8–13 μm in wavelength, where continuous absorption due to water vapour occurs and also where absorption bands of minor constituents such as NH_3 and the freons occur. Although these gases are only present in very small quantities, the absorption bands lead to a noticeable greenhouse effect.

Many processes taking place in the atmosphere are mutually correlated. For instance, an increase in tropospheric temperature is followed by increased water content. The major absorbing components – carbon dioxide and water vapour – exhibit a strong temperature dependence of their spectral transmission functions, and with growing temperature their absorptivity strongly increases. Thus, a rise in the temperature of the troposphere and the surface is followed by intensification of the greenhouse effect mechanism [13].

Calculations show that for a standard model atmosphere, the total greenhouse effect amounts to 33.2 K, with the following contributions from optically active gaseous components: H_2O – 20.6 K; CO_2 – 7.2 K; N_2O – 1.4 K; CH_4 – 0.8 K; O_3 – 2.4 K; NH_3 + freons + NO_2 + CCl_4 + O_2 + N_2 – 0.8 K.

13.2 Evolution of the greenhouse effect in the earth's atmosphere

It is of interest to follow the evolution of the greenhouse effect and climate on the earth, by modelling the process of radiative heat exchange at different stages of the evolution of the atmosphere's chemical composition.

For these calculations, the data on the vertical ozone profile have been taken from Hart [5], Morss & Kuhn [18], and the vertical profiles for NH_3 have been drawn from the data on photochemical reactions [12]. The vertical profiles of water vapour concentration have been inferred from that of temperature on the assumption that the atmospheric relative humidity does not exceed 50% at $T_s > 298$ K. Pressure-induced absorption by CO_2, O_2, N_2, NH_3 were taken from the data of Kondratyev *et al.* [11].

In the early stage of the earth's evolution, its atmosphere contained a large amount of methane and other hydrocarbons reaching 10^6 atm cm per vertical column. The measurement data of Moskalenko & Parzhin (see [11]) obtained with a multi-path cell with an optical path up to 1 km, have been used in calculations of the spectral transmission functions for such amounts of hydrocarbons. All the absorption bands for methane in the 0.1–25 μm region have been taken into account in calculations.

The mean-global temperature T_s of the planetary surface and the globally averaged greenhouse effect ΔT are given in Table 13.1 for the models of the chemical composition of the atmosphere given in the same table. According to these models and assuming no variation in the solar constant, the earth's surface was considerably warmer at various stages in its evolution than it is currently, largely due to the increased greenhouse effect arising from substantial

Table 13.1. *Evolution of the greenhouse effect,* ΔT, *of the earth's atmosphere from the moment of the planet's formation*

Time, 10^9 years	0	0.2	0.8	1	2	2.5	3.5	4	4.5[a]	5[b]
A	0.15	0.5	0.53	0.53	0.43	0.25	0.29	0.28	0.3	0.33
T_e(K)	235	216	218	219	225	246	253	258	256	253
$C(O_2)$	0	0	$.2\,\bar{4}$	$.2\,\bar{3}$	$.2\,\bar{2}$	$.2\,\bar{1}$	0.04	0.05	0.2	0.33
$C(N_2)$	0.03	0.1	0.04	0.04	0.40	0.95	0.95	0.95	0.78	0.65
$C(NH_3)$	$.3\,\bar{5}$	$.1\,\bar{4}$	$.3\,\bar{4}$	$.25\,\bar{4}$	$.3\,\bar{4}$	$.5\,\bar{6}$	$.25\,\bar{7}$	$.2\,\bar{8}$	$.1\,\bar{9}$	$.3\,\bar{9}$
$C(CO_2)$	0.9	0.5	0.1	0.08	0.04	0.02	$.5\,\bar{2}$	$.1\,\bar{2}$	$.32\,\bar{3}$	$.32\,\bar{2}$
$C(CH_4)$	0.05	0.4	0.83	0.85	0.57	$.2\,\bar{2}$	$.1\,\bar{5}$	$.18\,\bar{4}$	$.13\,\bar{5}$	$.13\,\bar{5}$
$W_\perp(O_3)$ (atm cm)	0	0	$.3\,\bar{6}$	$.3\,\bar{4}$	$.3\,\bar{3}$	$.3\,\bar{2}$	$.3\,\bar{1}$	0.17	0.3–0.56	0.15
$C(N_2O)$	0	0	0	0	0	$.1\,\bar{7}$	$.1\,\bar{6}$	$.2\,\bar{6}$	$.3\,\bar{6}$	$.6\,\bar{6}$
$C(CO)$	0	0	$.3\,\bar{3}$	$.8\,\bar{3}$	$.4\,\bar{3}$	$.2\,\bar{3}$	$.3\,\bar{4}$	$.8\,\bar{5}$	$.1\,\bar{5}$	$.1\,\bar{4}$
$W_\perp(H_2O)$ (g/cm²)	2	8.1	22	16	7.1	2	1.4	1.2	1.4	4
P_s (atm)	0.4	1	1.4	1.32	0.73	0.75	0.78	0.8	1	1.18
T_s (K)	296	318	336	328	316	298	288	285	282	306
ΔT (K)	63	102	116	109	91	52	35	27	32	53

$\cdot 2\,\bar{4}$ stands for $0{\cdot}2 \times 10^{-4}$; C is the volume concentration; $W_\perp$ is the content of a component in vertical air column; P_s and T_s are the surface pressure and temperature; T_e is the effective temperature of the planet.

(a) Conditions as at present.

(b) A hypothetical model, 5, corrected for possible anthropogenic effect.

quantities of CH_4, NH_3 and H_2O. The last column of Table 13.1 considers the greenhouse effect due to a hypothetical atmosphere with a substantial increase in CO_2 and a decrease of O_3.

Substantial variations in the greenhouse effect may be expected to occur during periods of volcanic activity arising from changes in gaseous composition and in aerosol content. In studies of evolution of the atmospheric chemical composition during the geological past, considerable temporal variations in carbon dioxide have been discovered connected with volcanic activity [27]. Also, in high volcanic activity conditions it is natural to expect a heightened concentration in the atmosphere of such volatiles as SO_2, H_2S and the presence of a dense haze layer of H_2SO_4-solution droplets. SO_2 is known to have strong absorption bands in the longwave spectral region, coinciding with the atmospheric transparency windows [11] and, consequently, intensifying the greenhouse effect of the atmosphere. At the same time, the presence of the dense sulphate haze at altitudes of 20–30 km with temperatures 200–220 K lowers the radiative temperature of the planet and intensifies the greenhouse effect. The SO_2 strong absorption bands in the UV spectral region could have led to considerable temporal variations in the structural characteristics of the atmosphere in the periods of 'sporadic explosions' of volcanic activity, during which the aerosol of sulphate water solutions had considerably affected the planetary albedo.

13.3 Anthropogenic changes in gaseous constituents and the greenhouse effect

The problem of anthropogenic effects on climate is known to be rather urgent [1–3, 6–10, 12, 16, 17, 21, 23, 27–33]. Most attention has been given to the effect of increased CO_2 concentration in the atmosphere, due to the burning of fossil fuel [4, 8, 17, 19, 24, 26, 28]. However, growing pollution of the atmosphere causes an increase in concentrations of such components as sulphur dioxide, carbon monoxide, halocarbons, nitrogen oxides, nitric acid, hydrocarbon compounds, etc. Also of great importance is aerosol – both injected directly to the atmosphere and resulting from chemical transformations of gaseous components to solid ones – the effect of aerosol will be considered in the next section. All these components have absorption bands in the IR spectral region and all make some contribution to the greenhouse effect of the atmosphere [3, 6, 12, 16, 21, 29, 31, 33, 34].

First, considering changes of CO_2, calculations (the results of which are shown in Table 13.2) have been carried out with a 1-D model of the atmosphere (assuming radiative–convective equilibrium) of the change of surface temperature, ΔT, which would occur for increasing CO_2 concentration. Two different models have been used, one with fixed water vapour and one allowing the water vapour content W g cm^{-2} of the troposphere to change with surface temperature, T_s, according to the expression:

$$W = 1.4 \exp 0.07\,(T_s - 288). \tag{13.4}$$

Within the troposphere the relative humidity was assumed to be independent of height.

In Table 13.2, it is of interest to note the absence of greenhouse 'saturation' effect, even in the case of a ten-fold CO_2 increase.

Fig. 13.1 shows two vertical temperature profiles calculated for the mean global atmosphere of the earth (curve 1) and an atmosphere with the ten-fold increase of the CO_2 content (curve 2) as compared to that of the present day. Curve 2 refers to the combined greenhouse effect caused by increased concentrations of CO_2 and water vapour. Of interest in Fig. 13.1 are the stratospheric temperature decrease and the growing altitude of the tropopause due to the CO_2 greenhouse effect.

The calculations of Table 13.2 have not taken into account the feedback effect of clouds or the effect of atmospheric or oceanic circulation which will be referred to in Section 13.8.

Calculations have also been carried out on the greenhouse effect due to other gaseous constituents. Doubled concentration of nitrous oxide N_2O leads to a 0.7 K rise in the mean temperature, and when concentrations of ammonia and methane are doubled, the surface temperature rises, respectively, by 0.1 and 0.3 K. A 20-fold increase of freons can lead to the greenhouse effect reaching

Fig. 13.1. Calculated vertical temperature profiles: (1) for a mean global model of an atmospheric vertical structure and a mean chemical composition; (2) for an atmosphere with a ten-fold increase of CO_2 concentration.

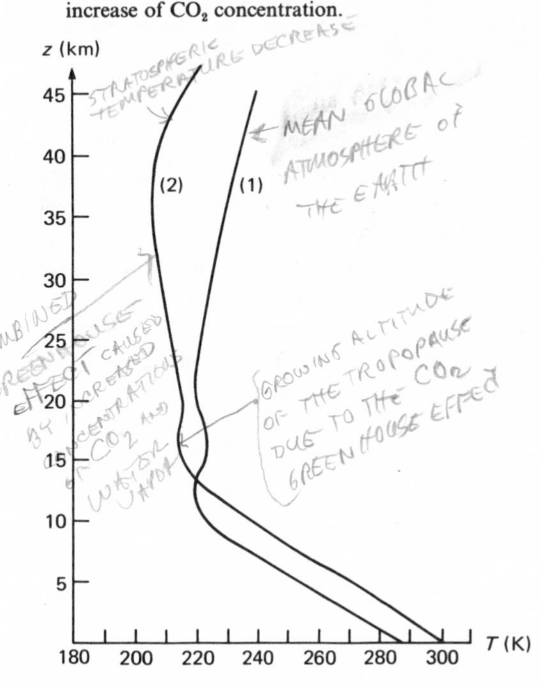

Table 13.2. *Surface temperature increase ΔT v. n-fold increase of CO_2 concentration*

n	1	1.5	2	3	4	5	6	7	8	10
$\Delta T'$	0	1.6	2.4	4.1	5.1	5.7	6.3	6.7	7.2	8.1
$\Delta T''$	0	2.9	4.4	7.6	9.4	10.4	11.2	11.9	12.6	13.1

$\Delta T'$ and $\Delta T''$ are values of ΔT obtained, respectively, with fixed water-vapour and with varying water-vapour content.

0.6–1 K and the total greenhouse effect due to doubled concentrations of N_2O, CH_4, SO_2 and HNO_3 reaches 1.2 K.

The greenhouse effect is also significant in the stratosphere. A doubling of the water-vapour content there leads to a warming by 1 K, and a decrease of O_3 concentration by 25% would lead to a reduction of temperature by 0.4–0.5 K.

The above estimates of the effect of variations in concentrations of minor gaseous components on the temperature, T_s, agree well with the results of other authors [16, 29] and confirm the conclusion that anthropogenic changes in the concentration of a number of minor components, the effect of which on climate had previously been considered negligible, would lead to a possible climate warming in the future. It is therefore important to monitor global trends of the concentration of these minor components.

On the basis of a 1-D global radiative–convective model, similar to our own, Lacis *et al.* [34] have made calculations of surface warming due to the greenhouse effect of anthropogenic changes in the components CO_2, chlorofluorocarbons, methane, and nitrous oxide for two cases:

(1) observed rates of injections for the decade 1970–80;

(2) doubled concentrations.

Their results are given in Table 13.3 which shows that the total contribution of methane, chlorofluorocarbons and nitrous oxide is equal to 50–100% of the CO_2 contribution. The equilibrium warming for arbitrary changes may be fitted to an analytical expression:

$$T_e\,(K) = 0.57\,(CH_4)^{0.5} + 2.8\,(N_2O)^{0.6} - 0.057 \times CH_4 \times N_2O + 0.15 \times CCl_3 + 0.18 \times CCl_2F_2 + 2.5 \ln[1 + 0.005\,(\Delta CO_2) + 10^{-5}\,(\Delta CO_2)^2], \quad (13.5)$$

where the abundances are in ppm except for CCl_3F and CCl_2F_2, which are in ppb, and the CO_2 amount is in ppm above a reference value of 300 ppm. This equation fits the model results to better than 5% for abundances $CH_4 < 5$ ppm, $N_2O < 1$ ppm, $CCl_2F < 2$ ppb, $CClF_3 < 2$ ppb and $\Delta CO_2 < 300$ ppm; the third term corrects for overlap of CH_4 and N_2O bands.

Lacis *et al.* [34] point out that their estimates do not take into account the atmosphere-ocean interaction, which slows down the climate warming due to the greenhouse effect [see 4, 17]. In the case of the interaction of the atmosphere with the upper mixed oceanic layer of about 100 m depth, the surface temperature rise by 1970 would be half i.e. ≈ 0.1 °C, or ≈ 0.1–0.2 °C if atmospheric pollution before 1970 is taken into account. Since such a level of temperature increase is close to the natural mean-global temperature fluctuations on decadal timescales, the anthropogenic impact on climate cannot yet be observed with any certainty. But because total warming for the 1970s, 1980s and 1990s may reach 0.2–0.3 °C it may become observable by the year 2000.

Considering possible climatic changes due to anthropogenic variations in the chemical composition of the atmosphere, one should bear in mind the interrelationship of various climatological factors. For instance, an increase in the chlorofluorocarbon content can substantially change the atmospheric ozone concentration. Therefore, atmospheric temperature variations have been calculated [12] taking account of the vertical profile of ozone resulting from photochemical reactions between ozone and chlorofluoromethanes.

It is important to take into account ozone variations not only in the stratosphere, but also in the troposphere, since variations in tropospheric ozone often lead to opposite effects as compared to those arising from stratospheric disturbances of ozone concentration. Observations exist indicating that upper tropospheric ozone concentration has increased 20–25% during the past 15 years (Attmanspacher 1982, unpublished result) so that special attention should be given to the tropospheric ozone problem.

13.4 The effect of aerosol

In considering the effect of aerosol on the radiative regime of the atmosphere, the mechanisms for radiation scattering and absorption by aerosol have to be taken into account [3, 13]. Depending on the aerosol optical density, its size distribution and chemical composition, and also depending on the surface albedo, conditions may appear when radiation processes due to aerosol can either raise or lower the albedo of the earth–atmosphere system. The presence of aerosol can lead to an increase of surface temperature due to a positive greenhouse effect or to a lowering of the surface temperature (referred to as an anti-greenhouse effect).

Figure 13.2 shows the vertical profile of atmospheric aerosol optical density for a mean-global atmosphere assumed in the calculations. A fine disperse background aerosol consists of 80% particles of gas-to-particle conversion origin and 20% small-sized dust particles. The tropospheric aerosol contains 40% mineral dust, 20% water solution of sulphates, 20% industrial aerosols, and 20% sea salt. The stratospheric aerosol layer is represented by the 75% H_2SO_4 water solution particles. Optical characteristics of

Table 13.3. *Greenhouse effects of several trace gases*

	Arbitrary change			1970–80 change		
Species	a_0(ppb)	Δa(ppb)	ΔT_e (°C)	a_0(ppb)	Δa(ppb)	ΔT_e (°C)
CH_4	1600	1600	0.26	1500	150	0.032
N_2O	280	280	0.65	295	6	0.016
CCl_3F	0	2	0.35	0.045	0.135	0.020
CCL_2F_2	0	2	0.36	0.125	0.190	0.034
CO_2	300000	300000	2.9	325000	12000	0.14

a_0 (Δa): initial (increase of) concentration.
ΔT: temperature rise.

aerosol formations have been considered by Moskalenko *et al.* [22]; examples are given in Table 13.4 of ω_0, the single scattering albedo in the wavelength region 0.3–1.7 μm for industrial, background and mineral aerosols.

For small-sized background aerosol particles in the IR spectral region, the absorption coefficient σ_a exceeds the scattering coefficient σ_s; for wavelengths $\lambda > 1$ μm the extinction coefficient is mainly determined by aerosol absorption. Therefore, small-sized aerosol warms up the atmosphere due to solar radiation absorption and also emits in the longwave spectral region, thereby screening thermal emission from the surface. The small-sized background aerosol, like an absorbing gaseous component, intensifies the greenhouse effect of the atmosphere, raising the surface temperature by 3 K for mean-global conditions.

The above estimates of the greenhouse effect for the small-sized fraction of the atmospheric aerosol are partially compensated for by the anti-greenhouse effect of the coarse fraction of the tropospheric aerosol which amounts to $\Delta T = -1.4$ K for the model of Fig. 13.2. The anti-greenhouse effect ($\Delta T_s = -0.5$ K) is also typical of the stratospheric sulphate aerosol layer. However, for the planet on the average, the greenhouse effect due to the strongly absorbing small-sized aerosol fraction probably dominates the anti-greenhouse effect of coarse aerosol particles.

Fig. 13.2. The vertical structure of the optical density for (1) the atmospheric aerosol; and (2) background aerosol; (3, 4) the models of the optical density of cirrus clouds, assumed to cover $\frac{1}{10}$ and $\frac{1}{20}$ of the sky with the upper level clouds.

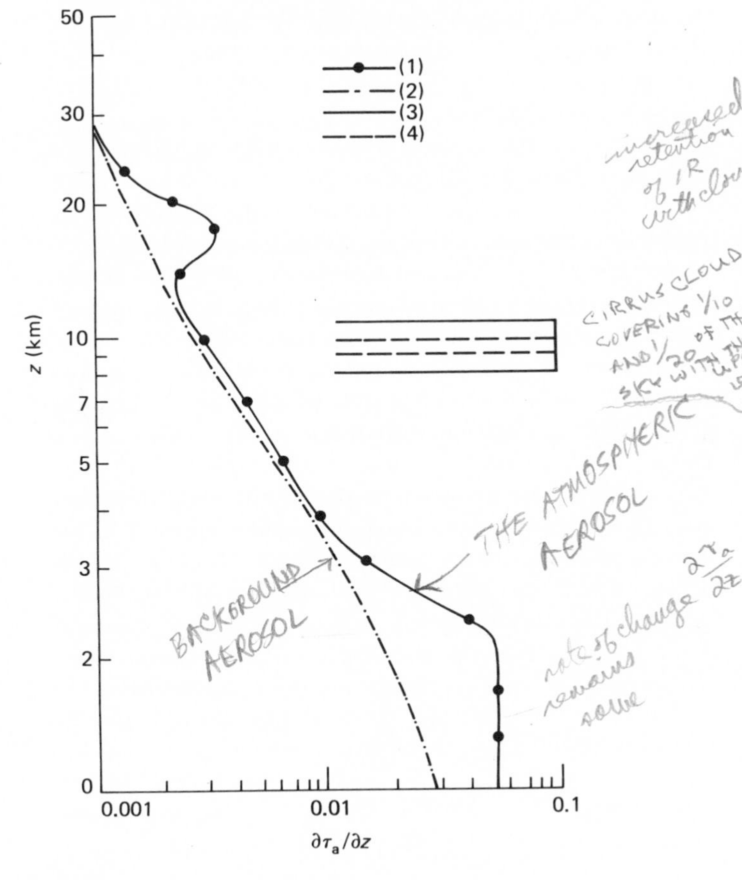

Table 13.4. *Single scattering albedo* ω_0 *for different types of aerosols*

Wavelength (μm)	0.3	0.5	0.69	1.06	1.7
Industrial aerosol	0.621	0.603	0.592	0.660	0.640
Mineral aerosol	0.776	0.864	0.892	0.952	0.956
Background aerosol	0.660	0.720	0.560	0.260	0.090

Atmospheric pollution due to man's activity raises the concentration of strongly absorbing aerosol particles (Table 13.3), which does not lead to substantial variations in the planetary albedo but increases the screening of the outgoing thermal emission of the surface. An increase by a factor of 1.5 of the tropospheric aerosol concentration due to man's activity will warm the surface by 1.7 K through the greenhouse effect. A doubled optical thickness of the stratospheric background aerosol (through an increasing number of free carbon-containing particles) will raise the temperature, T_s, by 0.8 K.

13.5 The effect of cloud cover

Clouds are the most important factor governing the radiative regime of the earth. Growing cloudiness raises the albedo and reduces the solar radiation flux reaching the earth's surface. On the other hand, the presence of cloudiness lowers the radiative temperature corresponding to the outgoing radiation, leading to a positive greenhouse effect. At night, this positive greenhouse effect is important but during the daytime, over surfaces that are not snow covered, it is likely that the cooling due to increased albedo dominates over the greenhouse effect. However, in winter conditions (the earth's surface is covered with snow) the presence of clouds does not substantially change the albedo, and the greenhouse effect in these conditions is dominant.

Due to anthropogenic causes, the structural and optical characteristics of the earth's cloud cover can be changed substantially. In particular, flights of supersonic aircraft can lead to an increase in the upper-level cloud cover and hence to radiative cooling. At present it is difficult to foresee the scale of the greenhouse effect due to this factor.

Some calculations have been carried out of variations in the surface temperature, T_s, caused by increasing upper-level cloud amount (upper troposphere, lower stratosphere). The results depend strongly on selected models of the cloud cover (Fig. 13.2), and lie within the range -0.7 to 3 K. The value -0.7 K corresponds to clouds of 1 km thickness with their base at 9 km and the value -3 K corresponds to clouds of 3 km thickness with their base at 8 km.

13.6 The earth–atmosphere system albedo

The earth–atmosphere system albedo is known to depend substantially on latitude and is determined by surface reflectivity, cloud amount, and absorbing and scattering properties of the atmospheric gas phase. Different types of the surface have drastically different albedos. For instance, the sea surface albedo, A, is about 0.1, the vegetation cover albedo is close to 0.4, and the snow albedo can reach 0.8–0.9. Albedos of different types of clouds vary strongly depending on the location of the sun in the sky. Table 13.5 lists the dependences of the albedo, A, on the solar zenith angle calculated using the Eddington approximation for different values of the surface albedo and optical thickness of mineral aerosols. In these tables, reflectivity $A = 0.1$ corresponds to the sea surface, $A = 0.4$ to the vegetation cover, and $A = 0.7$ to dense clouds or snow-covered surface.

For the sea surface, the aerosol raises the albedo, A, of the surface–atmosphere system but in the case of the snow-covered surface or the overcast sky, the albedo decreases considerably. If the underlying surface is the vegetation cover, then with increasing atmospheric turbidity the albedo becomes lower at solar zenith angles $0 < 60°$ and becomes larger when solar zenith angles are high. Table 13.6 illustrates the effect of industrial aerosol on the system albedo. For most solar zenith angles, the growing atmospheric turbidity due to industrial aerosol, causes a decrease of the surface–atmosphere system albedo.

For a cloudless atmosphere in the absence of dust clouds, the albedo over water bodies grows if the imaginary part of the complex index of refraction $x < 0.015$. The situation, however, changes drastically in the presence of clouds over water bodies. First, the marine aerosol, as a rule, is located below the upper boundary of the cloud cover, and its optical effect is screened by cloudiness. Second, the albedo of the sea surface–cloud system increases drastically. Because of the latter, the presence of aerosol in the troposphere above the cloud cover lowers the albedo. The albedo decreases still lower if the aerosol absorbs shortwave radiation. The latter occurs in a real atmosphere, since the above-cloud aerosol is an ensemble of dry particles of the mean-global salt and mineral dust.

The above-cloud dust aerosol with an optical thickness $\tau_a = 0.2$ lowers the albedo of the earth–atmosphere system by 3%. Over the oceans, the effect of the atmospheric aerosol on the albedo is determined by the cloud amount. In the atmosphere, with overcast cloudiness, the presence of aerosol always decreases the albedo, regardless of the surface type. A similar conclusion holds for a cloudless atmosphere and snow-covered surface.

For the continents, when the land is covered with vegetation, and there are no clouds, the effect of aerosol on the total albedo is practically absent in the range of optical thicknesses of the atmospheric aerosol 0.05–0.5. The latter means that only the vertical distribution of the shortwave radiation balance changes. With growing atmospheric turbidity, absorption of shortwave radiation by the surface decreases, and by the atmosphere increases.

For industrial regions with surface reflectivity of 0.3, the industrial aerosol, characterised by the albedo of single scattering $\tau_s/\sigma_a = 0.6$ and optical thickness $\tau_a = 0.4$, lowers the albedo, A, by 20% in the absence of clouds.

It is of interest to estimate the effect of dust transport from deserts on the radiative regime. Calculations have shown that a dust cloud decreases the planetary albedo, on the average, due to

Table 13.5. *The dependence of the surface–atmosphere system albedo, A, on its determining factors at the 0.55 μm wavelength for dust aerosol.*

$\Theta°$	$A_s = 0.1$ $\tau = 0.2$	$A_s = 0.1$ $\tau = 0.5$	$A_s = 0.1$ $\tau = 1.0$	$A_s = 0.7$ $\tau = 0.2$	$A_s = 0.7$ $\tau = 0.5$	$A_s = 0.7$ $\tau = 1.0$	$A_s = 0.4$ $\tau = 0.2$	$A_s = 0.4$ $\tau = 0.5$	$A_s = 0.4$ $\tau = 1.0$
	$A(\Theta)$								
0	0.12	0.14	0.18	0.65	0.60	0.51	0.37	0.35	0.33
30	0.15	0.17	0.23	0.66	0.59	0.52	0.39	0.37	0.36
60	0.18	0.22	0.30	0.68	0.60	0.53	0.41	0.40	0.40
70	0.22	0.29	0.36	0.69	0.62	0.55	0.44	0.44	0.46
80	0.30	0.42	0.48	0.70	0.64	0.60	0.50	0.52	0.52
85	0.41	0.50	0.51	0.72	0.67	0.63	0.55	0.60	0.56
90	0.52	0.56	0.58	0.74	0.72	0.66	0.64	0.64	0.62

Here $\sigma^s_a|\sigma_a = 0.9$; A_s is the surface albedo; τ is the optical thickness; Θ is the solar zenith angle.

Table 13.6. *The dependence of the surface–atmosphere system albedo, A, on its determining factors at the 0.55 μm wavelength for industrial aerosol.*

$\Theta°$	$A_s = 0.1$ $\tau = 0.2$	$A_s = 0.1$ $\tau = 0.5$	$A_s = 0.1$ $\tau = 1.0$	$A_s = 0.7$ $\tau = 0.2$	$A_s = 0.7$ $\tau = 0.5$	$A_s = 0.7$ $\tau = 1.0$	$A_s = 0.4$ $\tau = 0.2$	$A_s = 0.4$ $\tau = 0.5$	$A_s = 0.4$ $\tau = 1.0$
	$A(\Theta)$								
0	0.09	0.08	0.07	0.57	0.44	0.28	0.33	0.26	0.20
30	0.10	0.09	0.09	0.57	0.41	0.27	0.34	0.26	0.21
60	0.11	0.10	0.10	0.55	0.38	0.26	0.35	0.27	0.23
70	0.13	0.13	0.14	0.54	0.37	0.27	0.36	0.28	0.25
80	0.18	0.20	0.20	0.53	0.36	0.28	0.38	0.32	0.29
85	0.23	0.26	0.24	0.51	0.42	0.35	0.41	0.40	0.36
90	0.30	0.32	0.28	0.49	0.47	0.45	0.44	0.46	0.40

Here $\sigma^s_a|\sigma_a = 0.7$; A_s is the surface albedo; τ is the optical thickness; Θ is the solar zenith angle.

shortwave radiation absorption. If a dust cloud with an optical thickness $\tau_a = 0.4$ covered the whole planet, its albedo would have decreased by 16%. Since the dust cloud can cover about 0.1 of the planetary area, the planetary albedo can decrease by 1.6%. The latter is equivalent to a rise of the radiative temperature of the planet by about 0.8 K. Calculation of surface temperature change in the approximation of radiative–convective equilibrium, using a scheme suggested by Kondratyev & Moskalenko [15] gives an increase of the mean global temperature for this case of 2.1 K. Thus the growing dust load of the troposphere is an important climate-forming factor not only on the regional scale, but also on the planetary scale.

13.7 Feedback factors

The studies which have been mentioned so far have employed simple 1-D models and have not taken into account dynamical processes, the influence of the oceans, or feedback processes such as those which occur through cloud–radiation interaction. Predictions with a 3-D general circulation model of the atmosphere of the effect of a doubling of the CO_2 content of the atmosphere have been carried out by Manabe & Wetherald [8] who find an increase in surface temperature of about 2 K in the tropics, 3 K in mid-latitudes and up to 8 K in polar regions.

Mitchell *et al.* [35] have undertaken numerical climate modelling for cases of normal (320 ppm) and doubled CO_2 concentration on the basis of the Lawrence Livermore National Laboratory Statistical Dynamical Model. They have made estimates of the sensitivity of the earth radiation budget and its components, namely absorbed solar radiation (S), outgoing longwave radiation (F) and net total flux (R) to changes in surface temperature (T), precipitable water vapour (r), total cloud cover fraction (C), surface albedo (A), and CO_2 concentration. Table 13.7 shows the changes to be expected in the various parameters when the CO_2 increases; Table 13.8 shows the changes in S, F and R which would occur due to the change in each parameter. Warmer temperatures lead to an increase of F but do not change S, while the increase in water vapour absorbs more of the solar and longwave radiation, increasing S but reducing F.

Table 13.7. *Hemispheric and globally averaged changes in various parameters (doubled-CO_2 run minus control run)*

Parameter	Control case	Change
Surface temperature T (K)		
NH	285.2	2.333
SH	289.1	0.514
G	287.2	1.424
Precipitable water vapor r (g cm^{-2})		
NH	2.1	0.251
SH	2.3	0.190
G	2.2	0.220
Total cloud cover fraction C		
NH	0.56	−0.005
SH	0.54	0.027
G	0.55	0.011
Surface albedo A		
NH	0.23	−0.007
SH	0.13	−0.002
G	0.18	−0.004
CO_2 concentration (ppmv)		
NH	320.0	320.0
SH	320.0	320.0
G	320.0	320.0

Of great interest is the strong hemispheric difference in the effect of precipitable water on solar radiation in spite of the symmetry for longwave radiation. This situation is determined by the fact that solar radiation is absorbed primarily by water vapour in the lower troposphere, its content being larger in the northern hemisphere (NH) than in the southern hemisphere (SH). Since CO_2 doubling creates some decrease of cloud amount in the NH and some increase in the SH, consequent changes of absorbed solar radiation take place.

An important fact in the longwave radiation changes is the spatial (latitudinal and altitudinal) redistribution of cloudiness. Surface albedo variations are due only to soil moisture changes. The impact of ice albedo feedback does not show itself, because sea ice extent has not been allowed to change in the model.

Mitchell *et al.* proceed in their paper to discuss the effect of multiple feedbacks i.e. changes in more than one parameter at a time. It is clear from their study that further study of the various feedback processes is required before more-definite estimates can be made of the effect of changes in atmospheric CO_2. In particular, much-better understanding is required of the feed back effect of clouds on the various radiation components and of the influence of the ocean circulation on the various changes. For the latter studies, coupled atmospheric ocean models will be required.

13.8 Interactions in the stratosphere between radiation, photochemistry and dynamics

The problem of the interaction between radiation, photochemistry and dynamics is relevant to climate studies, especially in the stratosphere. In the pioneering paper by Bojkov [9], the interaction between the patterns of ozone distribution and of the stratospheric circulation have been studied. Since that time, a large number of investigations have been carried out on various feedback relationships. To describe these in detail is outside the scope of this chapter; many of them have been reviewed in monographs and

Table 13.8. *Hemispheric and globally averaged changes (doubled CO_2 minus control) in net downward solar flux (δS), net upward longwave flux (δF) and in net total flux (δR) at the top of the atmosphere due to each of the five factors listed in the column headings. All values in W m^{-2}*

	T	r	C	A	CO_2
δS					
NH	0.000	0.569	1.906	0.501	0.000
SH	0.000	0.120	−3.048	0.285	0.000
G	0.000	0.344	−0.571	0.393	0.000
δF					
NH	4.359	−1.823	1.570	0.000	−2.454
SH	2.700	−1.657	−0.041	0.000	−2.640
G	3.530	−1.740	0.764	0.000	−2.547
δR					
NH	−4.359	2.392	0.335	0.501	2.454
SH	−2.700	1.777	−3.007	0.285	2.640
G	−3.530	2.085	−1.336	0.393	2.547

papers [3, 12, 14, 21]. However, a brief description will be given of recent work by Haigh & Pyle [20].

Using a 2-D general circulation model at Oxford University, Haigh & Pyle modelled, numerically, various possible anthropogenic impacts on the ozone layer and the associated climate changes. Their model provides for chemical, radiative and dynamical interactive processes in the stratosphere and reproduces, for example, the feedback effects between temperature (through temperature dependence of the chemical reaction rates), changes in circulation and variations in the ozone concentration. Since the upper stratospheric ozone content is negatively correlated with temperature, stratospheric cooling arising from CO_2 increase will be greater than in the absence of ozone.

The model extends from the surface to approximately 80 km. Zonal mean temperatures, wind vector components and concentrations of gas constituents are held as functions of time, latitude and altitude (log pressure) with resolutions of four hours, 9.47° of latitude and 0.5 in. log pressure (about 3.5 km). At each time step, the meridional circulation necessary to preserve the geostrophic balance against the perturbing effects of heating and eddy transports of momentum and heat is calculated.

The photochemical scheme in the model includes 50 reactions describing the oxygen, hydrogen, nitrogen, and chlorine cycles (the natural chlorine cycle is excluded). Component concentrations for the following families are calculated: (O^1D), O, and O_3; N, NO, NO_2 and $ClONO_2$; HNO_3; H_2O_2; H, OH and HO_2; Cl, ClO, $ClONO_2$, and HCl; $CFCl_3$; CF_2Cl_2 (within each group photochemical equilibrium is assumed). Invariant profiles of H_2O, CH_4, H_2, N_2O and CO are specified independent of latitude. Although some of the rate constants have since been updated, the values assumed in the model were adequate for examining the coupling between the radiation, photochemistry and dynamics.

Particular attention was paid to providing an adequate radiation parameterisation scheme. Between the tropopause and 25 km, radiative equilibrium was assumed because of the impossibility of making sufficiently accurate calculations of the heating rates in this region where large terms of opposite sign nearly cancel. Fixed heating rates are used in the troposphere.

Four model runs were made:

(A) control (chlorine cycle excluded, CO_2 mixing ratio of 320 ppmv assumed);

(B) CO_2 volume mixing ratio increased up to 625 ppmv;

(C) chlorofluorocarbon (CFC) injections taken at four times the average 1973–76 production rate;

(D) simultaneous doubling of CO_2 content and CFC injection.

Comparing the control run with observational data demonstrated a sufficient agreement to give plausibility to the model, although there are significant differences (e.g. too low winter polar temperatures, overestimated zonal winds, high total ozone at the equatorial minimum).

The (B) run was performed in three steps with CO_2 mixing ratios, respectively, of 400, 500 and 625 ppmv considered appropriate for the years 2000, 2020 and 2040 AD. Doubling of CO_2 brings about considerable (up to 10–12 K) decrease in the upper stratospheric temperatures, coupled with a 20–25% increase in O_3 concentration. The effect on zonal circulation is also pronounced; however, the meridional transport remains practically unperturbed.

The total O_3 suffers strong increase (about 20%) near the north pole in summer, and a weaker one of 14% in southern hemisphere high latitudes, this difference reflecting the asymmetry in the global ozone field. The global increase in total O_3 is around 9%, considerably larger than any of the values predicted by the 1-D models which fail to account for the dynamics of the lower stratosphere. The major source region of NO_x is at around 30 km.

Injections of CFCs (run C) reduce the temperatures by up to 8–10 K and, contrary to the effect of doubling the CO_2, simultaneously deplete O_3 concentration by as much as 35% around 40 km, with a corresponding decrease in the total O_3. The joint perturbation (CO_2+CFCs) results presented in Table 13.9 show that their separate effects on the total O_3 are not additive. For example, addition of those effects would lead to the total O_3 depletion by only 4% around 2040, while the more accurate calculation for the joint effect produces the figure of 8.1%. Qualitatively, the results are similiar for ozone concentrations at different levels (i.e. interactive drop is considerably less than additive).

The interactive temperature decrease, peaking in excess of 20 K in the upper stratosphere, leads to substantial changes in the zonal circulation. Non-linearity of the interactive effect is due to the varying temperature dependences of the ozone concentration for the different models, namely: $d \ln (O_3)/d(T^{-1}) = 1188$ K(A); 1172 K(B); 557 K(C); 461 K(D). The data in Table 13.10 characterise the relative importance of each chemical cycle in depleting the total ozone at the equator near 40 km in April 2045. These data testify to the importance of the chlorine cycle.

13.9 Conclusions

This paper has considered the present status of our knowledge regarding the impact of atmospheric composition changes on the earth's radiation budget and on climate. Further research is

Table 13.9. *Percentage change in total O_3 for various CO_2 and CFC contents*

CO_2 content (ppmv)	zero CFC	CFC increases predicted for year (2000)	(2020)	(2040)
320	0	−3.2	−8.3	−12.8
400	+3.0	−0.6		
500	+5.9		−4.1	
625	+8.8			−8.1

Table 13.10. *Percentage of total ozone destruction due to various cycles at 40 km, equator, April 2045*

Cycle	Run (A)	(B)	(C)	(D)
pure oxygen	26.3	24.4	6.6	4.5
HO_x	17.9	19.6	9.5	8.7
NO_x	55.8	56.0	30.0	25.9
ClO_x	0	0	53.9	60.9

required in two key areas:

(1) the prediction of changes of composition including the spatial–temporal variations of different components through increased knowledge and understanding of biogeochemical cycles (see Chapter 12);

(2) adequate modelling of detailed atmospheric processes including the interactions between radiation, chemistry and dynamics.

The pursuit of research in these areas is an important component of the World Climate Research Programme.

References

[1] Borisenkov, E. P. 'Climate studies and their application aspects'. *Meteorol. and Hydrol., 1981*, **6**, 32–48 (in Russian).

[2] Budyko, M. I. *Climate in the Past and Future.* Leningrad, Gidrometeoizdat, 1980, 351 pp. (in Russian).

[3] Kondratyev, K. Ya. *Radiative Factors of the Present-Day Global Climate Change.* Leningrad, Gidrometeoizdat, 1980, 279 pp. (in Russian).

[4] Cess, R. D. & Goldenberg, S. D. 'The effect of ocean heat capacity upon global warming due to increasing atmospheric carbon dioxide'. *J. Geophys. Res.*, 1981. **C86** (1), 498–502.

[5] Hart, M. A. 'The evolution of the atmosphere of the Earth'. *Icarus*, 1978, **33** (1), 23–39.

[6] Houghton, J. T. 'Greenhouse effects of some atmospheric constituents'. *Phys. Trans. Roy. Soc., London*, 1979, **A290**, (1376), 515–21.

[7] Kondratyev, K. Ya. & Hunt G. E. *Weather and Climate on Planets.* Pergamon Press, Oxford, 1982, 755 pp.

[8] Manabe, S. & Wetherald, R. T. 'On the distribution of climate change resulting from an increase in CO_2 content of the atmosphere'.*J. Atmosph. Sci.*, 1980, **37** (1), 99–118.

[9] Bojkov, R. D. 'Planetary study of the ozone heating of the stratosphere'. *Bull. of the Cairo Meteorol. Department*, 1979, **1**, 193–244.

[10] Fels, S. B., Mahlman, J. D., Schwarzkopf, M. D. & Sinclair, R. W. 'Stratospheric sensitivity to perturbations of ozone and carbon dioxide: radiative and dynamical response'. *J. Atmosph. Sci.*, 1980, **37** (10), 2265–97.

[11] Kondratyev, K. Ya. & Moskalenko, N. I. *Thermal Emission of Planets.* Leningrad, Gidrometeoizdat, 1977, 263 pp. (in Russian).

[12] Kondratyev, K. Ya. *Stratosphere and Climate. Progress in Sci. and Technol., Meteorol. and Climatol.*, Moscow, VINITI, 1981, Vol. 6, 223 pp. (in Russian).

[13] Kondratyev, K. Ya., Moskalenko, N. I., Terzy V. F. & Skvortfsova, S. Ya. 'Modeling of the optical characteristics of the atmospheric aerosols'. *In: Polar Aerosol, Extended Cloudiness, and Radiation.* FGGE Ser., Vol. 2, Leningrad, Gidrometeoizdat, 1981, pp. 130–53. (in Russian).

[14] Hartmann, D. L. 'Some aspects of the coupling, between radiation, chemistry and dynamics in the stratosphere'. *J. Geophys. Res.*, 1981, **C86** (10), 9631–40.

[15] Kondratyev, K. Ya., Moskalenko, N. I. 'The greenhouse effect of planetary atmospheres'. *Nuovo Cimento C*, 1980, **30** (4), 436–60.

[16] Ramanathan, V. 'Climate effects of anthropogenic trace gases'. *In: Interactions of Energy and Climate*, eds. B. Bach, J. Pankrath & J. Williams, Reidel, Dordrecht, Netherlands, 1981, pp. 269–80.

[17] Gates, W. L., Cook K. H. & Schlesinger, M. E. 'Preliminary analysis of experiments on the climatic effects of increased CO_2 with the atmospheric general circulation model and a climatological ocean'. *J. Geophys. Res.*, 1981, **C86** (7), 6385–93.

[18] Morss, D. A. & Kuhn, W. R. 'Paleoatmospheric temperature structure'. *Icarus*, 1978, **33** (1), 40–9.

[19] Vupputuri, R. K.-R. The Effect of Increased Atmospheric CO_2 on Stratospheric Temperature Structure and Ozone Distribution Investigated in a 2-D Time-dependent Model. *Canadian Climate Center, Report No. 7*, 1981.

[20] Haigh, J. D. & Pyle, J. A. 'Ozone perturbation experiments in a two-dimensional circulation model'. *Q. J. Roy. Meteorol. Soc.*, 1982, **108** (457), 551–74.

[21] Kondratyev, K. Ya. The Present State, Perspectives, and the Role of Observations from Space. *Progress in Sci. and Technol., Meteorol. and Climatol.* Moscow, VINITI, 1982, Vol. 8, 274 pp.

[22] Moskalenko, N. I., Tantashev, M. V., Terzy, V. F. & Skvortsova, S. Ya. 'The optical characteristics of the atmospheric aerosols'. *In: Aerosols and Climate.* FGGE Ser., Vol. 1, Leningrad, Gidrometeoizdat, 1981, pp. 154–65 (in Russian).

[23] Ramanathan, V. 'Climatic effects of ozone change: a review'. *Low Latitude Aeron. Processes. Proc. Symp. 22nd Planetary Meet., COSPAR.*, Bangalore, 1979, Oxford, 1980, pp. 223–36.

[24] Reck, R. A. 'Carbon dioxide and climate: comparison of one- and three-dimensional models'. *Environ. Int.*, 1980, **2** (4–6), 387–91.

[25] Schneider, S. H. & Thomson, S. L. 'Cosmic conclusions from climatic models: can they be justified?' *Icarus*, 1980, **41** (3), 456–69.

[26] Hummel, J. R., Reck, R. A. 'Carbon dioxide and climate: the effects of water transport in radiative–convective models'. *J. Geophys. Res.*, 1981, **C86** (12), 12035–38.

[27] 'The problems of atmospheric carbon dioxide'. *Proc. of the Soviet-American Symp., Dushanbe*, 12–20, October 1978, Leningrad, Gidrometeoizdat, 1980, 286 pp. (in Russian).

[28] Pittock, A. B. & Salinger, M. J. 'Towards regional scenarios for a CO_2-warmed Earth'. *Clim. Change*, 1982, **4** (1), 23–40.

[29] Wang, W. S., Yung, T. L., Lacis, A. A., Mo, T. & Hansen, T. E. 'Greenhouse effects due to man-made perturbations of trace gases'. *Science*, 1976, **194** (4266), 685–91.

[30] Kukla, G. & Gavin, J. 'Summer ice and carbon dioxide'. *Science*, 1981, **214**, 497–503.

[31] Alexandrov, E. L., Karol, I. L., Rakipova, L. R., Sedunov Ya. S. & Khrgian A. Kh. *Atmospheric Ozone and Changes of Global Climate.* Leningrad, Gidrometeoizdat, 1982, 167 pp. (in Russian).

[32] Hadden, R. A. & Ramanathan, V. 'Detecting climate change due to increasing carbon dioxide'. *Science*, 1980, **209** (4458), 763–7.

[33] Kellogg, W. W. & Schware, R. *Climate Change and Society: Consequences of Increasing Atmospheric Carbon Dioxide.* Westview Press, Inc., Boulder, Colo., 1981, 178 pp.

[34] Lacis, A., Hansen, J., Lee, P., Mitchell, T. & Lebedeff, G. Greenhouse effect of trace gases, 1970–80. *Geophys. Res. Lett.*, 1981, **8** (10), 1035–8.

[35] Mitchell, C. S., Potter, G. L., Ellsaesser, H. W. & Walton, J. J. 'Case study of feedbacks and synergisms in a doubled CO_2 experiment'. *J. Atmosph. Sci.*, 1981, **38** (9), 1906–10.